QUANTUM FIELD THEORY IN CURVED SPACETIME

Quantum field theory in curved spacetime has been remarkably fruitful. It can be used to explain how the large-scale structure of the universe and the anisotropies of the cosmic background radiation that we observe today first arose. Similarly, it provides a deep connection between general relativity, thermodynamics, and quantum field theory. This book develops quantum field theory in curved spacetime in a pedagogical style, suitable for graduate students.

The authors present detailed, physically motivated derivations of cosmological and black hole processes in which curved spacetime plays a key role. They explain how such processes in the rapidly expanding early universe leave observable consequences today, and how, in the context of evaporating black holes, these processes uncover deep connections between gravitation and elementary particles. The authors also lucidly describe many other aspects of free and interacting quantized fields in curved spacetime.

LEONARD E. PARKER is Distinguished Professor Emeritus in the Physics Department at the University of Wisconsin, Milwaukee. In the 1960s, he was the first to use quantum field theory to show that the gravitational field of the expanding universe creates elementary particles from the vacuum.

DAVID J. TOMS is Reader in Mathematical Physics in the School of Mathematics and Statistics at Newcastle University. His research interests include the formalism of quantum field theory and its applications, and his most recent interests involve the use of quantum field theory methods in low energy quantum gravity.

CAMBRIDGE MONOGRAPHS ON MATHEMATICAL PHYSICS

General Editors: P. V. Landshoff, D. R. Nelson, S. Weinberg

S. J. Aarseth *Gravitational N-Body Simulations: Tools and Algorithms*
J. Ambjørn, B. Durhuus and T. Jonsson *Quantum Geometry: A Statistical Field Theory Approach*
A. M. Anile *Relativistic Fluids and Magneto-fluids: With Applications in Astrophysics and Plasma Physics*
J. A. de Azcárraga and J. M. Izquierdo *Lie Groups, Lie Algebras, Cohomology and Some Applications in Physics*[†]
O. Babelon, D. Bernard and M. Talon *Introduction to Classical Integrable Systems*[†]
F. Bastianelli and P. van Nieuwenhuizen *Path Integrals and Anomalies in Curved Space*
V. Belinski and E. Verdaguer *Gravitational Solitons*
J. Bernstein *Kinetic Theory in the Expanding Universe*
G. F. Bertsch and R. A. Broglia *Oscillations in Finite Quantum Systems*
N. D. Birrell and P. C. W. Davies *Quantum Fields in Curved Space*[†]
M. Burgess *Classical Covariant Fields*
E. A. Calzetta and B.-L. B. Hu *Nonequilibrium Quantum Field Theory*
S. Carlip *Quantum Gravity in 2+1 Dimensions*[†]
P. Cartier and C. DeWitt-Morette *Functional Integration: Action and Symmetries*
J. C. Collins *Renormalization: An Introduction to Renormalization, the Renormalization Group and the Operator-Product Expansion*[†]
M. Creutz *Quarks, Gluons and Lattices*[†]
P. D. D'Eath *Supersymmetric Quantum Cosmology*
F. de Felice and C. J. S. Clarke *Relativity on Curved Manifolds*
B. DeWitt *Supermanifolds*, 2nd edition[†]
P. G. O Freund *Introduction to Supersymmetry*[†]
J. A. Fuchs *Affine Lie Algebras and Quantum Groups: An Introduction, with Applications in Conformal Field Theory*[†]
J. Fuchs and C. Schweigert *Symmetries, Lie Algebras and Representations: A Graduate Course for Physicists*[†]
Y. Fujii and K. Maeda *The Scalar-Tensor Theory of Gravitation*
A. S. Galperin, E. A. Ivanov, V. I. Orievetsky and E. S. Sokatchev *Harmonic Superspace*
R. Gambini and J. Pullin *Loops, Knots, Gauge Theories and Quantum Gravity*[†]
T. Gannon *Moonshine beyond the Monster: The Bridge Connecting Algebra, Modular Forms and Physics*
M. Göckeler and T. Schücker *Differential Geometry, Gauge Theories and Gravity*[†]
C. Gómez, M. Ruiz-Altaba and G. Sierra *Quantum Groups in Two-Dimensional Physics*
M. B. Green, J. H. Schwarz and E. Witten *Superstring Theory Volume 1: Introduction*[†]
M. B. Green, J. H. Schwarz and E. Witten *Superstring Theory Volume 2: Loop Amplitudes, Anomalies and Phenomenology*[†]
V. N. Gribov *The Theory of Complex Angular Momenta: Gribov Lectures on Theoretical Physics*
S. W. Hawking and G. F. R. Ellis *The Large Scale Structure of Space-Time*[†]
F. Iachello and A. Arima *The Interacting Boson Model*
F. Iachello and P. van Isacker *The Interacting Boson-Fermion Model*
C. Itzykson and J. M. Drouffe *Statistical Field Theory Volume 1: From Brownian Motion to Renormalization and Lattice Gauge Theory*[†]
C. Itzykson and J. M. Drouffe *Statistical Field Theory Volume 2: Strong Coupling, Monte Carlo Methods, Conformal Field Theory and Random Systems*[†]
C. V. Johnson *D-Branes*[†]
P. S. Joshi *Gravitational Collapse and Spacetime Singularities*
J. I. Kapusta and C. Gale *Finite-Temperature Field Theory: Principles and Applications*, 2nd edition
V. E. Korepin, N. M. Bogoliubov and A. G. Izergin *Quantum Inverse Scattering Method and Correlation Functions*[†]
M. Le Bellac *Thermal Field Theory*[†]
Y. Makeenko *Methods of Contemporary Gauge Theory*
N. Manton and P. Sutcliffe *Topological Solitons*[†]
N. H. March *Liquid Metals: Concepts and Theory*
I. Montvay and G. Münster *Quantum Fields on a Lattice*[†]
L. O'Raifeartaigh *Group Structure of Gauge Theories*[†]
T. Ortín *Gravity and Strings*[†]
A. M. Ozorio de Almeida *Hamiltonian Systems: Chaos and Quantization*[†]
L. E. Parker and D. J. Toms *Quantum Field Theory in Curved Spacetime: Quantized Fields and Gravity*
R. Penrose and W. Rindler *Spinors and Space-Time Volume 1: Two-Spinor Calculus and Relativistic Fields*[†]

R. Penrose and W. Rindler *Spinors and Space-Time Volume 2: Spinor and Twistor Methods in Space-Time Geometry*[†]
S. Pokorski *Gauge Field Theories*, 2nd edition[†]
J. Polchinski *String Theory Volume 1: An Introduction to the Bosonic String*
J. Polchinski *String Theory Volume 2: Superstring Theory and Beyond*
V. N. Popov *Functional Integrals and Collective Excitations*[†]
R. J. Rivers *Path Integral Methods in Quantum Field Theory*[†]
R. G. Roberts *The Structure of the Proton: Deep Inelastic Scattering*[†]
C. Rovelli *Quantum Gravity*[†]
W. C. Saslaw *Gravitational Physics of Stellar and Galactic Systems*[†]
H. Stephani, D. Kramer, M. MacCallum, C. Hoenselaers and E. Herlt *Exact Solutions of Einstein's Field Equations*, 2nd edition
J. Stewart *Advanced General Relativity*[†]
T. Thiemann *Modern Canonical Quantum General Relativity*
D. J. Toms *The Schwinger Action Principle and Effective Action*
A. Vilenkin and E. P. S. Shellard *Cosmic Strings and Other Topological Defects*[†]
R. S. Ward and R. O. Wells, Jr *Twistor Geometry and Field Theory*[†]
J. R. Wilson and G. J. Mathews *Relativistic Numerical Hydrodynamics*

[†] Issued as a paperback

Quantum Field Theory in Curved Spacetime

Quantized Fields and Gravity

LEONARD E. PARKER
University of Wisconsin, Milwaukee

DAVID J. TOMS
University of Newcastle upon Tyne

CAMBRIDGE
UNIVERSITY PRESS

CAMBRIDGE
UNIVERSITY PRESS

University Printing House, Cambridge CB2 8BS, United Kingdom

Cambridge University Press is part of the University of Cambridge.

It furthers the University's mission by disseminating knowledge in the pursuit of education, learning and research at the highest international levels of excellence.

www.cambridge.org
Information on this title: www.cambridge.org/9780521877879

© L. E. Parker and D. J. Toms 2009

This publication is in copyright. Subject to statutory exception and to the provisions of relevant collective licensing agreements, no reproduction of any part may take place without the written permission of Cambridge University Press.

First published 2009

A catalogue record for this publication is available from the British Library

Library of Congress Cataloguing in Publication data
Parker, Leonard Emanuel, 1938–
Quantum field theory in curved spacetime : quantized fields and gravity /
Leonard E. Parker, David J. Toms.
p. cm.
Includes bibliographical references and index.
1. Quantum field theory. 2. Space and time. 3. Particles (Nuclear physics)
4. Relativity (Physics) I. Toms, David J., 1953– II. Title.
QC174.45.P367 2009
530.14′3–dc22
2008051193

ISBN 978-0-521-87787-9 Hardback

Cambridge University Press has no responsibility for the persistence or accuracy of URLs for external or third-party internet websites referred to in this publication, and does not guarantee that any content on such websites is, or will remain, accurate or appropriate.

Contents

Preface		page xi
Acknowledgments		xiii
Conventions and notation		xv
1	**Quantum fields in Minkowski spacetime**	**1**
1.1	Canonical formulation	2
1.2	Particles	13
1.3	Vacuum energy	21
1.4	Charged scalar field	24
1.5	Dirac field	27
1.6	Angular momentum and spin	32
2	**Basics of quantum fields in curved spacetimes**	**36**
2.1	Canonical quantization and conservation laws	37
2.2	Scalar field	43
2.3	Cosmological model: Arbitrary asymptotically static time dependence	47
2.4	Particle creation in a dynamic universe	51
2.5	Statistics from dynamics: Spin-0	54
2.6	Conformally invariant non-interacting field	56
2.7	Probability distribution of created particles	58
2.8	Exact solution with particle creation	61
2.9	High-frequency blackbody distribution	63
2.10	de Sitter spacetime	64
2.11	Quantum fluctuations and early inflation	73
2.12	Quantizing the inflaton field perturbations	77
2.13	A word on interacting quantized fields and on algebraic quantum field theory in curved spacetime	88
2.14	Accelerated detector in Minkowski spacetime	91
3	**Expectation values quadratic in fields**	**93**
3.1	Adiabatic subtraction and physical quantities	93
3.2	Energy-momentum tensor from trace anomaly	107
3.3	Renormalization in general spacetimes	110
3.4	Gaussian approximation to propagator	115

3.5	Approximate energy-momentum tensor in Schwarzschild, de Sitter, and other static Einstein spacetimes	118
3.6	R-summed form of propagator	129
3.7	R-summed action and cosmic acceleration	131
3.8	Normal coordinate momentum space	134
3.9	Chiral current anomaly caused by spacetime curvature	144

4 Particle creation by black holes — 152

4.1	Introduction	152
4.2	Classical considerations	153
4.3	Quantum aspects	162
4.4	Energy-momentum tensor with Hawking flux	174
4.5	Back reaction to black hole evaporation	178
4.6	Trans-Planckian physics in Hawking radiation and cosmology	180
4.7	Further topics: Closed timelike curves; closed-time-path integral	182

5 The one-loop effective action — 184

5.1	Introduction	184
5.2	Preliminary definition of the effective action	185
5.3	Regularization of the effective action	190
5.4	Effective action for scalar fields: Some examples	200
5.5	The conformal anomaly and the functional integral	216
5.6	Spinors in curved spacetime	221
5.7	The effective action for spinor fields	245
5.8	Application of the effective action for spinor fields	253
5.9	The axial, or chiral, anomaly	257

6 The effective action: Non-gauge theories — 268

6.1	Introduction	268
6.2	The Schwinger action principle	271
6.3	The Feynman path integral	277
6.4	The effective action	280
6.5	The geometrical effective action	283
6.6	Perturbative expansion of the effective action	295
6.7	Renormalization of an interacting scalar field theory	302
6.8	The renormalization group and the effective action	318
6.9	The effective potential	322
6.10	The renormalization of the non-linear sigma model	331
6.11	Formal properties of the effective action	337
	Appendix	344

7	**The effective action: Gauge theories**	**348**
7.1	Introduction	348
7.2	Gauge transformations	350
7.3	The orbit space and the gauge condition	359
7.4	Field space reparameterization and the Killing equation	368
7.5	The connection and its consequences	375
7.6	The functional measure for gauge theories	385
7.7	Gauge-invariant effective action	390
7.8	Yang–Mills theory, concluded	399
7.9	Scalar quantum electrodynamics	407
	Appendix	420

Appendix: Quantized Inflaton Perturbations   **422**

References   426
Index   445

Preface

The success of Einstein's theory of general relativity convincingly demonstrates that the classical gravitational field is a manifestation of the curvature of spacetime. Similarly, quantum field theory in Minkowski spacetime successfully describes the behavior of elementary particles over a wide range of energies. It has proved notoriously difficult to understand how gravity fits with the quantum attribute of the fields that transmit the other forces of nature. Leading attempts to combine gravitation and quantum field theory include string theory and loop quantum gravity. String theory attempts to describe elementary particles, including the graviton, as quantized excitations of systems of strings and D-branes in a higher-dimensional space. Loop quantum gravity attempts to describe the structure of spacetime itself in terms of quantized loops. At energies much below the Planck scale, these theories reduce to descriptions of quantized fields propagating in a curved spacetime having a metric described by Einstein's gravitational field equations with additional higher-order curvature corrections.

Quantum field theory in curved spacetime is the framework for describing elementary particles and gravitation at energies below the Planck scale. This theory has had striking successes. It has shown how gravitation and quantum field theory are intimately connected to give a consistent description of black holes having entropy and satisfying the second law of thermodynamics; and it has shown how the inhomogeneities and anisotropies we observe today in the cosmic microwave background and in the large-scale structure of the universe were created in a brief stage of very rapid expansion of the universe, known as inflation.

This book should give the reader a deep understanding of the principles of quantum field theory in curved spacetime and of their applications to the early universe, renormalization, black holes, and effective action methods for interacting fields in curved spacetime, including gauge fields. It is aimed at graduate students and researchers and would be appropriate as the basis for a graduate course. We have tried to be pedagogical in our presentation.

If the students have had an introduction to quantum field theory in Minkowski spacetime, then Chapter 1 could be skipped and returned to only when particular topics are unfamiliar. In that case, the instructor can expect to finish Chapters 2 through 5 in a one-semester course, depending on how much detail is covered. Particle creation by the expansion of the universe and by black holes would be covered in such a course. In a two-semester course, the instructor can expect to go through the whole book, including renormalization of interacting fields, the renormalization group, and the effective action for Yang–Mills fields in curved spacetime.

We expect the reader to have some understanding of general relativity, at least at the introductory level. The books by Hartle (2003), Misner *et al.* (1973), Wald (1984), or Weinberg (1972) provide more than sufficient background on general relativity. For additional related material on quantum field theory in curved spacetime, we recommend that the reader consult the outstandingly comprehensive treatment of the subject by Birrell and Davies (1982), and the books by Fulling (1989) and Wald (1994).

Acknowledgments

Leonard E. Parker is grateful to the late Sidney Coleman. He supervised my PhD thesis at Harvard from 1962 through 1965, in which I used quantum field theory to discover and thoroughly investigate the creation of elementary particles by the gravitational field of the expanding universe, including the role of conformal invariance and other aspects of this fundamentally important process. I am grateful as well to Steven A. Fulling, Lawrence H. Ford, Timothy S. Bunch, and my coauthor David J. Toms, who were my postdoctoral associates from 1973 through the early 1980s. Their work helped push the boundaries of quantum field theory in curved spacetime into new and fruitful territory. It is with pleasure that I thank as well my other postdoctoral associates, including the late Chaim Charach, Ian Jack, Atshushi Higuchi, Jonathan Z. Simon, Jorma Louko, Yoav Peleg, Alpan Raval, Daniel A. T. Vanzella, and Gonzalo J. Olmo. It was a privilege to work with these truly outstanding researchers. Among my other collaborators on topics involving quantum field theory in curved spacetime, I give special thanks to Paul Anderson, Jacob D. Bekenstein, Robert Caldwell, Bei-Lok Hu, and José Navarro-Salas. The PhD students who worked with me on topics related to quantum field theory in curved spacetime are Prakash Panangaden, Luis O. Pimentel, Todd K. Leen, Esteban Calzetta, Yang Zhang, Sukanta Bose, Gerald Graef, William Komp, and Laura Mersini. Matthew Glenz, one of my current PhD students, carefully read chapters 1 through 4 and pointed out many misprints and unclear passages. I am grateful to my former postdoctoral associate Alpan Raval for helping with the writing of Sections 3.5, 4.4, and 4.5. I thank the US National Science Foundation for supporting my research on quantized fields, gravitation, and cosmology for more than 35 years. This support was of great help. Above all, I thank my wife, Gloria, and children, David, Michael, and Deborah, for their support and encouragement during the writing of this book.

David Toms would like, first of all, to express his gratitude to R. C. Roeder, who first suggested to him that quantum field theory in curved spacetime would make a good topic for a PhD thesis, and to P. J. O'Donnell who supervised his PhD studies and gave him free rein to follow his interests, along with some good advice. While a postgraduate student at the University of Toronto I benefitted greatly from discussions with C. C. Dyer and E. Honig. The Natural Sciences and Engineering Research Council of Canada sponsored my first postdoctoral fellowship at Imperial College, London. While there I profited from interactions with M. J. Duff, L. H. Ford, C. J. Isham, G. Kunstatter, and M. Pilati. My first encounter with my coauthor, Leonard E. Parker, was as a postdoctoral associate at the University of Wisconsin at Milwaukee where we established a fruitful collaboration that led eventually to this book. I am also grateful to J. Friedman and D. M. Witt for valuable discussions while I was in Milwaukee. My colleagues at Newcastle deserve thanks, especially P. C. W. Davies and I. G. Moss. I would like to acknowledge my collaborators not previously mentioned, J. Balakrishnan, K. Kirsten, and H. P. Leivo, for their assistance. I have had a number of postgraduate students who worked with me on aspects of quantum field theory among whom are M. Burgess, E. J. Copeland, P. Ellicott, A. Flachi, S. R. Huggins, G. Huish, and I. Russell. P. McKay pointed out some misprints in Chapter 7. While at Newcastle I have received funding from the Sciences and Engineering Research Council and the Nuffield Foundation. My wife Linda was very understanding and supportive during the many hours taken to write this book – I express my love and gratitude for this.

Both authors are extremely grateful to Bei-Lok Hu for his advice through the years, and his comments on an early version of the manuscript.

Conventions and notation

We have tried to maintain consistency between our book and that of Birrell and Davies (1982) wherever possible. Our sign conventions are $(-\,-\,-)$ in the notation of Misner *et al.* (1973). More explicitly, an outline of our basic notation is the following:

- $\Re$ and $\Im$ denote the real and imaginary parts of any expression;
- **divp** denotes the divergent part (pole part if dimensional regularization is used);
- we use units with $c = \hbar = 1$, and often $G = 1$;
- spacetime dimension is n in general, often with $n = 4$;
- Minkowski metric: $\eta_{\mu\nu}$ is diagonal with eigenvalues $(+1, -1, \ldots, -1)$;
- ordinary partial derivative of ψ denoted by $\partial_\mu \psi$ or $\psi_{,\mu}$;
- curved spacetime metric: $g_{\mu\nu}$ with inverse metric $g^{\mu\nu}$;
- invariant volume element: $dv_x = |\det(g_{\mu\nu})|^{1/2} d^n x$;
- Christoffel connection: $\Gamma^\lambda_{\mu\nu} = \Gamma^\lambda_{\nu\mu} = \frac{1}{2} g^{\lambda\sigma}(g_{\sigma\nu,\mu} + g_{\mu\sigma,\nu} - g_{\mu\nu,\sigma})$;
- covariant derivative of ψ denoted by $\nabla_\mu \psi$ or $\psi_{;\mu}$;
- d'Alembertian, or wave, operator: $\Box = \nabla^\mu \nabla_\mu$;
- Riemann tensor: $R^\lambda{}_{\tau\mu\nu} = \Gamma^\lambda_{\tau\mu,\nu} - \Gamma^\lambda_{\tau\nu,\mu} + \Gamma^\lambda_{\nu\sigma}\Gamma^\sigma_{\tau\mu} - \Gamma^\lambda_{\mu\sigma}\Gamma^\sigma_{\tau\nu}$;
- Ricci tensor: $R_{\mu\nu} = R^\lambda{}_{\mu\lambda\nu}$;
- Dirac matrices in flat spacetime follow Bjorken and Drell (1964). (See Chapter 5 for a complete discussion.)

Other notation is introduced as needed.

1
Quantum fields in Minkowski spacetime

The theory of quantum fields in curved spacetime is a generalization of the well-established theory of quantum fields in Minkowski spacetime. To a great extent, the behavior of quantum fields in curved spacetime is a direct consequence of the corresponding flat spacetime theory. Local entities, such as the field equations and commutation relations, are to a large extent determined by the principle of general covariance and the principle of equivalence. However, global entities which are unique in Minkowski spacetime lose that uniqueness in curved spacetime. For example, the vacuum state, which in Minkowski spacetime is determined by Poincaré invariance, is not unambiguously determined in curved spacetime. This ambiguity is closely tied to the phenomenon of particle creation by certain gravitational fields, as in the expanding universe or near a black hole.

It is logical, therefore, to review the relevant aspects of flat spacetime quantum field theory. This will serve to establish the necessary background, to fix our notation, and to highlight those aspects of the theory which can be carried over to curved spacetime, as well as those which lose their meaning in curved spacetime. We will often be brief, emphasizing concepts while omitting many derivations, and only touch on particular topics. Our discussion of the curved spacetime theory in later chapters will be more detailed.

In this initial chapter, we discuss the canonical formulation, including the Schwinger action principle and the relation between symmetry transformations and conserved currents (Schwinger 1951b, 1953). We review the dynamical descriptions known as the Heisenberg picture, the Schrödinger picture, and the interaction picture. We introduce the Fock representations, in which states are described in terms of their particle content, and the Schrödinger representation, in which the states are described by field configurations. We include discussions of the Maxwell and Yang–Mills gauge fields, as well as the Dirac field, and the definitions of spin and angular momentum.

1.1 Canonical formulation

Recall that in classical mechanics the equations of motion of a particle or system of particles having independent generalized coordinates $q_i(t)$ and velocities $\dot{q}_i(t)$ are given by the principle of stationary action. This principle states that the action

$$S = \int_{t_1}^{t_2} dt\, L(q, \dot{q}) \tag{1.1}$$

is stationary under arbitrary variations of the q_i which vanish on the boundary of the region of integration. Here L is the Lagrangian of the system. The Hamiltonian is defined by

$$H(q, p) = \sum_i p_i \dot{q}_i - L,$$

where

$$p_i = \frac{\partial L}{\partial \dot{q}_i} \tag{1.2}$$

is the momentum conjugate to q_i. The system is quantized by taking the qs and ps to be Hermitian operators acting on a Hilbert space, and by imposing the canonical commutation relations

$$[q_i, q_j] = 0, \quad [p_i, p_j] = 0,$$

$$[q_i, p_j] = i\delta_{i,j}. \tag{1.3}$$

Here $\delta_{i,j}$ is the Kronecker delta. We are using units with $\hbar = c = 1$. From (1.3) it follows that if $F(q,p)$ is a function of the coordinate and momentum operators, then (assuming F can be Taylor expanded in p)

$$[q_i, F] = i\frac{\partial F}{\partial p_i}. \tag{1.4}$$

The above commutation relations imply that the q_i are a complete set of commuting observables with continuous spectra consisting (in the absence of impenetrable walls) of all real numbers. The same can be said for the p_i. (An observable is an Hermitian operator with a complete set of eigenstates.) The eigenstates[1] of the q_i are the kets $|q'>$, where q' denotes the set of eigenvalues q'_i of the operators q_i. Thus,

$$q_i|q'> = q'_i|q'>,$$

with the normalization

$$<q'|q''> = \delta(q' - q''),$$

---

[1] We use the conventions of Dirac (1958) in distinguishing the eigenvalue of an operator from the operator by using a $'$.

where $\delta(q' - q'')$ is the Dirac δ-function. It also follows that

$$<q'|p_i|q''> = -i\frac{\partial \delta(q' - q'')}{\partial q'_i},$$

and that

$$<q'|p'> = (2\pi)^{-n/2} \exp\left(i \sum_{i=1}^{n} p'_i q'_i\right),$$

where $|p'>$ is an eigenket of the p_i having delta function normalization, and we have taken the index i to run from 1 to n. (For derivations, see Messiah (1961, pp. 302–309).) If $F(q, p)$ is a function of the q_i and p_i with any well-defined ordering of factors, then

$$<q'|F(q,p)|q''> = F\left(q', -i\frac{\partial}{\partial q'}\right) \delta(q' - q''), \quad (1.5)$$

where $F(q', -i(\partial/\partial q'))$ is the same ordered function with $-i(\partial/\partial q'_i)$ replacing p_i in each position.

In the Schrödinger or configuration space representation, the abstract operators are represented by matrix elements based on the $|q'>$, such as that of p_i above, and the states are represented by functions. For example, a state $|\psi>$ is represented by the Schrödinger wave function $\psi(q') = <q'|\psi>$. An example is the wave function $<q'|p'>$ above, representing a particle of definite momentum. Similarly, in the momentum-space representation the operators are represented by matrix elements formed from the $|p'>$ and the states are represented by functions such as $<p'|\psi>$.

Up to now the description has been purely kinematical, with time playing no role. The dynamical evolution of the system is governed by the Hamiltonian $H(q, p, t)$. We have allowed for the possibility that H may have explicit time dependence, as through an interaction with an external field. The time evolution may be described in several physically equivalent ways, known as "pictures."

In the Schrödinger picture, the fundamental observables q and p do not change with time. Rather, the dynamical evolution of measurable quantities, such as expectation values of observables, is expressed through the time dependence of the ket describing the state of the system at each time. The fundamental dynamical equation is that of Schrödinger,

$$i\frac{d}{dt}|\psi(t)> = H(q, p, t)|\psi(t)>. \quad (1.6)$$

Because H may have explicitly time-dependent terms involving p and q, in general $H(t')$ and $H(t'')$ may not commute. (For brevity, we suppress the dependence of H on q and p.)

We note in passing that in the Schrödinger representation, the Schrödinger equation (1.6) becomes (using the completeness of the $|q'>$)

$$i<q'|\frac{d}{dt}|\psi(t)> = \int <q'|H(q,p,t)|q''>dq''<q''|\psi(t)>$$

or

$$i\frac{\partial}{\partial t}\psi(q',t) = H\left(q', -i\frac{\partial}{\partial q'}, t\right)\psi(q',t), \tag{1.7}$$

where $\psi(q',t) = <q'|\psi(t)>$.

The solution of (1.6) is

$$|\psi(t)> = U(t,t_0)|\psi(t_0)>,$$

with $U(t,t_0)$ satisfying

$$i\frac{d}{dt}U(t,t_0) = H(t)U(t,t_0), \tag{1.8}$$

with the boundary condition

$$U(t_0,t_0) = 1.$$

The evolution operator $U(t,t_0)$ preserves the norm of the state vector and is thus unitary, satisfying

$$UU^\dagger = U^\dagger U.$$

From $U(t_0,t)U(t,t_0) = U(t_0,t_0) = 1$, it then follows that $U(t,t_0)^\dagger = U(t_0,t)$.

In the Heisenberg picture, the ket describing the state of the system does not change with time, while the dynamical evolution of the system is expressed through the time dependence of the fundamental observables $q(t)$ and $p(t)$. By applying $U(t,t_0)^\dagger$ to the Schrödinger picture ket describing the state of the system, we obtain a time-independent ket, which can be taken as the ket describing the state of the system in the Heisenberg picture. Thus, denoting quantities in the Schrödinger picture by subscript S and those in the Heisenberg picture by subscript H, we have

$$|\psi_H> = U(t,t_0)^\dagger|\psi_S(t)> = |\psi_S(t_0)>. \tag{1.9}$$

In order that measurable expectation values remain the same as in the Schrödinger picture, the Heisenberg picture operator $F_H(t)$ corresponding to a Schrödinger picture operator F_S must be related by

$$F_H(t) = U(t,t_0)^\dagger F_S U(t,t_0). \tag{1.10}$$

Note that the Hamiltonian H in (1.8) is H_S. When H_S has no explicit time dependence, the solution of (1.8) is

$$U(t,t_0) = \exp[-i(t-t_0)H],$$

and it follows that H commutes with U so that $H_H = H_S$. When H_S does have an explicit time dependence, $H_H \neq H_S$, and we must use (1.10) to define H_H in terms of H_S. In general, F_H will depend on time through $q_H(t)$, $p_H(t)$, and through further explicit appearance of t. Denoting by $\partial F_H/\partial t$ the derivative only with respect to this further explicit appearance of t, it follows from (1.8) and (1.10) that

$$i\frac{d}{dt}F_H = [F_H, H_H] + i\frac{\partial}{\partial t}F_H. \tag{1.11}$$

This resembles the classical equation of motion with the Poisson bracket replaced by $-i$ times the commutator. Of course, we do not need to define the Schrödinger picture before the Heisenberg picture. The latter is fully defined by stating that the ket describing the state of the system is independent of time and that operators F, constructed from the q_i, p_i (dropping subscript H), and t obey the Heisenberg equation of motion (1.11).

When two systems interact through a term in the Hamiltonian, which can be regarded as a perturbative interaction, it is useful to introduce another picture of the dynamical evolution, known as the "interaction picture." In this picture, the ket describing the state of the system evolves as in the Schrödinger picture, but only under the influence of the interaction term in the Hamiltonian, while operators evolve as in the Heisenberg representation, but only under the influence of the unperturbed term in the Hamiltonian.

Let us now turn to the canonical quantization of a system of independent real fields $\phi_a(x)$, where x refers to the Minkowski space and time coordinates x^μ, and the index a includes tensor or spinor indices and internal quantum numbers of the field multiplet. We will deal here with bosons and discuss later the modification of canonical quantization required with fermions. For brevity, the index a will often be suppressed; we can usually think of ϕ as a column or row matrix, depending on where it appears in an expression. Canonical quantization proceeds as in the previously discussed quantization of a particle. One thinks of the classical field $\phi(x)$ as analogous to the classical $q_i(t)$, with the spatial coordinates $\vec{x}$ regarded as labels like i. Because we are now dealing with a continuous label, Dirac δ-functions involving $\vec{x}$ will appear where Kronecker deltas involving the label i previously appeared. As before, we assume that the system is described by an action

$$S = \int_{t_1}^{t_2} dt\, L[\phi, \partial\phi]_t, \tag{1.12}$$

where the Lagrangian L is now a functional of the field ϕ and its first derivatives $\partial\phi/\partial x^\mu \equiv \partial_\mu \phi$, which are denoted collectively by $\partial\phi$. The subscript t indicates that L is a function of t. The Lagrangian L can be expressed in terms of a Lagrangian density $\mathcal{L}$ as

$$L[\phi, \partial\phi]_t = \int dV_x\, \mathcal{L}(\phi(\vec{x}, t), \partial\phi(\vec{x}, t)),$$

where dV_x is the spatial volume element. Because L is now a functional, the momentum π conjugate to the field $\phi(\vec{x},t)$ is defined in analogy with (1.2) through the following functional derivative (regarding $\partial_0\phi$ as independent of ϕ at time t),

$$\pi(\vec{x},t) = \delta L[\phi,\partial\phi]_t/\delta\big(\partial_0\phi(\vec{x},t)\big)$$
$$= \partial\mathcal{L}\big(\phi(\vec{x},t),\partial\phi(\vec{x},t)\big)/\partial\big(\partial_0\phi(\vec{x},t)\big). \tag{1.13}$$

Here, we have used the definition of the functional derivative, which states that if $F[\chi]$ is a functional of $\chi(\vec{x})$, then under a variation $\delta\chi(\vec{x})$ of χ, which vanishes sufficiently fast at spatial infinity, we have

$$\delta F[\chi] = \int dV_x \, \frac{\delta}{\delta\chi(\vec{x})} F[\chi] \delta\chi(\vec{x}). \tag{1.14}$$

It follows that if the functional F has the form

$$F[\chi] = \int dV_x \, f\big(\chi(\vec{x}),\vec{\partial}\chi(\vec{x})\big),$$

then

$$\frac{\delta}{\delta\chi(\vec{x})} F[\chi] = \frac{\partial}{\partial\chi} f(\chi,\vec{\partial}\chi) - \partial_i\left(\frac{\partial}{\partial(\partial_i\chi)} f(\chi,\vec{\partial}\chi)\right),$$

where χ is evaluated at $\vec{x}$. The result in (1.13) follows from this with $\chi \to \partial_0\phi$. Another consequence is that

$$\frac{\delta\chi(\vec{x}')}{\delta\chi(\vec{x})} = \delta(\vec{x}' - \vec{x}).$$

One can regard the action in (1.12) as a functional depending on the space and time dependence of ϕ. Then, the Euler–Lagrange field equation below will be recognized as another application of the above result, but for a functional depending on one more dimension.

The Hamiltonian is defined by

$$H[\phi,\pi]_t = \int dV_x \, \pi_a(\vec{x},t)\partial_0\phi_a(\vec{x},t) - L[\phi,\partial\phi]_t. \tag{1.15}$$

(Although we write $H[\phi,\pi]$, dependence on spatial derivatives of ϕ (or π) is permitted.) The principle of stationary action yields upon variation of the fields in (1.12), the Euler–Lagrange field equations

$$\partial_\mu\left(\frac{\partial\mathcal{L}}{\partial(\partial_\mu\phi)}\right) - \frac{\partial\mathcal{L}}{\partial\phi} = 0, \tag{1.16}$$

where the repeated spacetime coordinate index μ is summed over its full range of values, in accordance with the Einstein summation convention.

The field is quantized in analogy with the canonical commutators of (1.3). Thus, we postulate that

1.1 Canonical formulation

$$[\phi_a(\vec{x},t), \phi_b(\vec{x}',t)] = 0, \quad [\pi_a(\vec{x},t), \pi_b(\vec{x}',t)] = 0,$$

$$[\phi_a(\vec{x},t), \pi_b(\vec{x}',t)] = i\delta_{a,b}\delta(\vec{x}-\vec{x}'), \tag{1.17}$$

where, as noted earlier, $\delta(\vec{x}-\vec{x}')$ is the Dirac δ-function. (In a theory of interacting fields, if we deal directly with the renormalized fields, then (1.17) is somewhat altered by a normalization factor. We are thus dealing here with the bare fields, which by definition satisfy the field equations with unrenormalized masses and coupling constants.) The t dependence is included in the commutation relations to emphasize that in a dynamical picture like the Heisenberg picture, in which the operators ϕ and π depend on time (as do the classical fields), the canonical commutation relations must be imposed on the fields and conjugate momenta evaluated at the same time. For it follows from (1.10) that $q_H(t)$ and $p_H(t)$ evaluated at the same time do satisfy (1.3), while that is not true in general if they are evaluated at different times. Of course, in the Schrödinger picture the fields and conjugate momenta have no time dependence, and t would not appear in (1.17).

The functional analogue of (1.4) follows from the above commutators:

$$[\phi_a(\vec{x}), F[\phi, \pi]] = i\frac{\delta}{\delta\pi_a(\vec{x})}F[\phi, \pi], \tag{1.18}$$

where we are suppressing the time dependence, taking all fields to be at a single time t. One can now set up a Schrödinger or field representation using the eigenstates of $\phi(\vec{x})$ defined by

$$\phi(\vec{x})|\phi'> = \phi'(\vec{x})|\phi'>. \tag{1.19}$$

The ket $|\phi'>$ corresponds to a state of the system in which the field has configuration $\phi'(\vec{x})$, where ϕ' is an ordinary or c-number function, unlike the field operator ϕ. Thus, we are using the analogue of the Dirac notation in which "eigenvalues" of the operator $\phi(\vec{x})$ are functions denoted by $\phi'(\vec{x})$. (Here the prime does not denote derivative, but instead distinguishes a c-number from an operator.) In ordinary quantum mechanics, it follows from (1.5) that if $|\psi>$ is an element of the Hilbert space spanned by the eigenkets $|q'>$, then

$$<q'|F(q,p)|\psi> = F\left(q', -i\frac{\partial}{\partial q'}\right)<q'|\psi>.$$

Similarly, we can show that if $|\Psi>$ is a state in the space spanned by the eigenstates $|\phi'>$, and $F[\phi, \pi]$ is a functional formed from the field operator and conjugate momentum, then

$$<\phi'|F[\phi, \pi]|\Psi> = F\left[\phi', -i\frac{\delta}{\delta\phi'}\right]<\phi'|\Psi>. \tag{1.20}$$

Here $<\phi'|\Psi> \equiv \Psi[\phi']$ is a complex number which is a functional of ϕ'. It is interpreted as the probability amplitude for finding the field observable ϕ to have the configuration or set of values given by $\phi'(\vec{x})$ when the system is in the state described by the vector $|\Psi>$.

If we work in the Schrödinger picture, then (1.20) can be used to turn the Schrödinger equation

$$i\frac{d}{dt}|\Psi(t)>= H[\phi,\pi]|\Psi(t)>$$

into the functional differential equation

$$i\frac{\partial}{\partial t}\Psi[\phi',t] = H\left[\phi',-i\frac{\delta}{\delta\phi'}\right]\Psi[\phi',t], \qquad (1.21)$$

where $\Psi[\phi',t] \equiv <\phi'|\Psi(t)>$, and ϕ, π depend on $\vec{x}$, but not on t.

On the other hand, in the Heisenberg picture the state describing the evolving system is independent of time, while a general functional F of $\phi(\vec{x},t)$ and $\pi(\vec{x},t)$ will depend on time through its dependence on ϕ and π, as well as through a possible further explicit dependence on t. Then, the Heisenberg equation of motion is

$$i\frac{d}{dt}F[\phi,\pi;t] = [F[\phi,\pi;t], H[\phi,\pi]] + i\frac{\partial}{\partial t}F[\phi,\pi;t]. \qquad (1.22)$$

The consistency of (1.22) with the Euler–Lagrange equations (1.16) can be proved in Minkowski spacetime and in the more general curved spacetime context (Parker, 1973, see Appendix B).

Finally, we expect that when there are no time- or space-dependent external parameters in the Lagrangian, then there should exist a conserved vector observable P^μ corresponding to the total energy and momentum of the system. For such a Lagrangian, by multiplying the Euler–Lagrange equations by $\partial_\mu \phi$ and using[2]

$$\partial_\mu \mathcal{L} = \frac{\partial \mathcal{L}}{\partial(\partial_\nu\phi)}\partial_\mu\partial_\nu\phi + \frac{\partial \mathcal{L}}{\partial \phi}\partial_\mu\phi,$$

we find immediately that

$$\partial_\mu T^\mu{}_\nu = 0, \qquad (1.23)$$

where

$$T^\mu{}_\nu = \frac{\partial \mathcal{L}}{\partial(\partial_\mu \phi_a)}\partial_\nu\phi_a - \delta^\mu{}_\nu \mathcal{L}. \qquad (1.24)$$

Here, summation over internal indices a of ϕ and, when the fields are treated as operators, symmetrization over fields and their conjugate momenta are understood. The tensor $T_{\mu\nu}$ is called the canonical energy-momentum or stress tensor.

We mention in passing that in order to serve as the source in the Einstein gravitational field equations, $T_{\mu\nu}$ should be symmetric under interchange of indices.

---

[2] This assumes that the Lagrangian does not have any explicit dependence on the coordinates x^μ.

(This symmetry is also required if we wish to define a conserved angular momentum in terms of the energy-momentum tensor.) However, except in the case of particular forms of $\mathcal{L}$, the expression given in (1.24) (after lowering the index ν with the Minkowski metric $\eta_{\mu\nu}$) is not a symmetric tensor. A general manifestly symmetric expression for $T_{\mu\nu}$ will be given later, when we deal with curved spacetime. (Since both of these expressions for $T^\mu{}_\nu$ will satisfy (1.23) for *any* system with no explicit time or space dependence, we expect on physical grounds that they must each yield the same conserved energy and momentum P^μ to within a constant.) Furthermore, a modification of the canonical $T_{\mu\nu}$ that makes it symmetric, and yields the same P_μ and angular momentum as the original canonical $T_{\mu\nu}$, has been given by Belinfante (1939, 1940).

From (1.23) we have

$$\int dv_x \, \partial_\mu T^\mu{}_\nu = 0,$$

where dv_x denotes the *spacetime* volume element, and the integration is over a spacetime volume bounded by spatial infinity and any two constant-time hypersurfaces. Assuming that matrix elements of physical interest will be between states in which the physical field configuration is of finite spatial extent, we obtain by the Gauss divergence theorem the conservation law

$$\frac{d}{dt}\int dV_x \, T^0{}_\nu = 0,$$

where dV_x denotes the *spatial* volume element and the integration is over any constant-time hypersurface. Hence,

$$P_\nu = \int dV_x T^0{}_\nu \tag{1.25}$$

is the conserved energy-momentum vector. The sign in this definition is chosen so that $P_0 = H$, as can be verified by comparing (1.15) with (1.24).

As a special case of the Heisenberg field equation (1.22), suppose that the functional F is an ordinary function $f(\phi(x), \partial_i\phi(x), \pi(x))$ with no explicit time dependence. Then, (1.22) can be written as

$$i\partial_0 f = [f, P_0], \tag{1.26}$$

where the partial derivative symbol is used here in the conventional manner to indicate that the x^i are held fixed. In (1.22), the partial derivative symbol denoted derivation only with respect to explicit t dependence not coming from ϕ and π. The partial derivative in (1.26) includes *all* t dependence. The result in (1.26) is the 0-component of the more general relation

$$i\partial_\mu f = [f, P_\mu], \tag{1.27}$$

The $\mu = 0$ component, as noted above, follows from (1.22). For $\mu = i$, we easily verify (1.27) for powers of ϕ and π and thus for functions which can be expanded in power series in ϕ and π. For example, suppressing the t-dependence, which is the same in all arguments, we have

$$[\pi(\vec{x})^n, P_i] = \int dV'_x\, \pi(\vec{x}')[\pi(\vec{x})^n, \partial'_i \phi(\vec{x}')]$$

$$= -i \int dV'_x\, \pi(\vec{x}') n\pi(\vec{x})^{n-1} \partial'_i \delta(\vec{x} - \vec{x}')$$

$$= i\partial_i(\pi(\vec{x})^n).$$

Equation (1.27) also follows from a powerful generalization of the action principle and of Noether's theorem, known as the Schwinger operator action principle (Schwinger 1951b, 1953). (For textbook treatments see Roman (1969) or Toms (2007).) The action of (1.12) is integrated over a spacetime volume v bounded by two constant-time hypersurfaces at t_1 and t_2. (As originally stated, the principle deals with arbitrary spacelike hypersurfaces, but we work with constant-time hypersurfaces for simplicity at this stage. See Section 6.2 for a discussion.) Consider arbitrary infinitesimal variations, δx^μ and $\delta_0 \phi(x)$, of the coordinates and field operators,

$$x^\mu \to x'^\mu = x^\mu + \delta x^\mu, \tag{1.28}$$

$$\phi(x) \to \phi'(x) = \phi(x) + \delta_0 \phi(x), \tag{1.29}$$

where $\delta_0 \phi(x)$ vanishes on the spatial boundary of integration at each time (i.e., it vanishes everywhere on the boundary of v, except on the interior of the constant-time hypersurfaces that bound v at t_1 and t_2). Then, the Schwinger action principle states that the variation of the action S of (1.12) has the form

$$\delta S = G(t_2) - G(t_1), \tag{1.30}$$

where the operator $G(t)$ is the generator of the above variation of the coordinates and fields at time t.

To say that G generates the variation means the following. For an operator functional $F[\phi, \pi]$, we have

$$i\delta_0 F = [F, G], \tag{1.31}$$

where all quantities are evaluated at the same time, and $\delta_0 F$ is the infinitesimal variation of F produced by (1.28) and (1.29). That is,

$$\delta_0 F = F[\phi + \delta_0 \phi, \partial(\phi + \delta_0 \phi)] - F[\phi, \partial \phi].$$

One can show from (1.30) that the generator has the form

$$G(t) = \int dV_x\, [\pi_a \delta\phi_a - T^0{}_\nu \delta x^\nu], \tag{1.32}$$

where $T^\mu{}_\nu$ is the energy-momentum tensor of (1.24), and

$$\delta\phi(x) \equiv \phi'(x') - \phi(x) \tag{1.33}$$

1.1 Canonical formulation

is the change of the field at a given physical point in spacetime (i.e., the local variation).

The derivation is as follows. The variation δS above is defined by

$$\delta S = S' - S,$$

where

$$S' = \int dv'_x \, \mathcal{L}(\phi'(x'), \partial'\phi'(x'))$$

and

$$S = \int dv_x \, \mathcal{L}(\phi(x), \partial\phi(x)).$$

Here, both integrals are taken over the same physical volume v, with x and x' denoting the same physical point of the system in spacetime, and thus related by (1.28). This way of viewing the transformation is the so-called passive viewpoint, in which the coordinates undergo a transformation from x to x', but the physical system does not change.

Alternatively, it is possible to adopt the active viewpoint taken by Messiah (1961), in which x and x' are regarded as describing a change of the physical system relative to a single coordinate system, such that the event at x is dragged to x'. In the active viewpoint, the volume v is infinitesimally displaced to a new physical volume v', related to v by (1.28). We can write

$$\begin{aligned} S' &= \int_{v'} dv_x \, \mathcal{L}(\phi'(x), \partial\phi'(x)) \\ &= \int_{v'} dv_x \, \mathcal{L}(\phi(x) + \delta_0\phi(x), \partial\phi(x) + \partial\delta_0\phi(x)) \\ &= \int_{v'} dv_x \mathcal{L}(\phi, \partial\phi) + \int_v dv_x \left\{ \frac{\partial\mathcal{L}}{\partial\phi}\delta_0\phi + \frac{\partial\mathcal{L}}{\partial(\partial_\mu\phi)}\partial_\mu\delta_0\phi \right\}. \end{aligned}$$

The volume of integration in the second integral has been changed to v because the integrand is already of first order.

Corresponding events on the boundaries of v and v' are related by the infinitesimal displacement δx^μ. Let $d\sigma^\mu$ denote $n^\mu d\sigma$, where n is an outward normal to the boundary of v at a point x, and $d\sigma$ is an element of surface area (i.e., hypersurface volume) on the boundary of v at x. Then, the scalar product $d\sigma_\mu \delta x^\mu$ is the volume of the cylindrical four-volume element capped at x by the surface area $d\sigma$ and extending along δx^μ from the boundary of v to that of v'. Then, denoting the boundary of v by ∂v, the surface integral,

$$\int_{\partial v} d\sigma_\mu \delta x^\mu \mathcal{L}(\phi, \partial\phi),$$

is equal to the following difference of volume integrals over v and v':

$$\int_{v'} dv_x \mathcal{L}(\phi, \partial\phi) - \int_v dv_x \mathcal{L}(\phi, \partial\phi),$$

because the difference is just the integral of $\mathcal{L}$ over the infinitesimal volume lying between the boundaries of v and v'.

Hence,
$$\delta S = S' - S$$
$$= \int_{\partial v} d\sigma_\mu \delta x^\mu \mathcal{L}(\phi, \partial\phi) + \int_v dv_x \left\{ \frac{\partial \mathcal{L}}{\partial \phi} \delta_0 \phi + \frac{\partial \mathcal{L}}{\partial(\partial_\mu \phi)} \partial_\mu \delta_0 \phi \right\}.$$

Converting the surface integral to a volume integral by the Gauss divergence theorem, and using the identity,
$$\frac{\partial \mathcal{L}}{\partial(\partial_\mu \phi)} \partial_\mu \delta_0 \phi = \partial_\mu \left\{ \frac{\partial \mathcal{L}}{\partial(\partial_\mu \phi)} \delta_0 \phi \right\} - \partial_\mu \left(\frac{\partial \mathcal{L}}{\partial(\partial_\mu \phi)} \right) \delta_0 \phi$$
in the second term, we are left with the result
$$\delta S = \int_v dv_x \, \partial_\mu \left\{ \delta x^\mu \mathcal{L} + \frac{\partial \mathcal{L}}{\partial(\partial_\mu \phi)} \delta_0 \phi \right\}.$$

Here, the Euler–Lagrange equations (1.16) have been used.[3] It is convenient to express our result in terms of the local variation $\delta\phi$ defined in (1.33). One has
$$\phi'(x) = \phi'(x' - \delta x) = \phi'(x') - \partial_\mu \phi(x) \delta x^\mu$$
to first order. Consequently, (1.29) gives
$$\delta_0 \phi = \delta \phi - (\partial_\mu \phi) \delta x^\mu.$$

Hence,
$$\delta S = \int_v dv_x \, \partial_\mu \left\{ \frac{\partial \mathcal{L}}{\partial(\partial_\mu \phi)} \delta\phi - T^\mu{}_\nu \delta x^\nu \right\}, \tag{1.34}$$
where $T^\mu{}_\nu$ is the energy-momentum tensor of (1.24). It follows that $\delta S = G(t_2) - G(t_1)$, with $G(t)$ given by (1.32), which completes the derivation.

Under an infinitesimal translation, we have $\phi'(x') = \phi(x)$, so that $\delta\phi$ vanishes. Therefore,
$$\delta_0 \phi = -(\partial_\mu \phi) \delta x^\mu.$$
Using the fact that for a translation δx^μ is constant, we now find that the generator is
$$G = -P_\mu \delta x^\mu,$$
where P_ν is the momentum operator of (1.25). One then has (1.31) for the commutator of G with an operator functional $F[\phi, \pi]$. Let us write (1.31) for the case when F is a function f formed from ϕ and the $\partial_\mu \phi$. Then,
$$\delta_0 f = \frac{\partial f}{\partial \phi} \delta_0 \phi + \frac{\partial f}{\partial(\partial_\mu \phi)} \partial_\mu \delta_0 \phi.$$

---

[3] The Euler–Lagrange equations follow, as usual, by demanding that δS is zero for the subset of variations such that $\delta_0 \phi$ vanishes on the boundary, ∂v.

Upon using the form of $\delta_0\phi$ for a translation, we find

$$\delta_0 f = -(\partial_\mu f)\delta x^\mu.$$

Using these results in (1.31), we obtain immediately (1.27).

The transformations of (1.28) and (1.29) are said to be a symmetry or invariance of the Lagrangian density $\mathcal{L}$ if

$$\mathcal{L}(\phi'(x'), \partial'\phi'(x')) = \mathcal{L}(\phi(x), \partial\phi(x)). \tag{1.35}$$

Recall that in S and S', the integrals are taken over the same *physical* volume (x and x' refer to the same physical point in spacetime). Therefore, the action is invariant when (1.35) holds, and we have $\delta S = 0$. In that case, (1.30) implies that G is a constant of the motion. Furthermore, (1.34) implies that

$$\partial_\mu J^\mu = 0, \tag{1.36}$$

with the conserved current

$$J^\mu = \frac{\partial \mathcal{L}}{\partial(\partial_\mu \phi_a)}\delta\phi_a - T^\mu{}_\nu \delta x^\nu. \tag{1.37}$$

It follows that when a translation in the x^μ direction is a symmetry of the Lagrangian density, then the operator P_μ is a constant of the motion. In Minkowski spacetime, the Lagrangian density of a free field is symmetric under space and time translations, which implies that energy and momentum are conserved. In curved spacetime, because of the presence in the Lagrangian density of the metric, we will find that for a free field (i.e., one influenced by gravitation alone), only the component of P_μ along the direction of an isometry of the spacetime is conserved.

1.2 Particles

The Schrödinger representation based on the eigenvectors of the field operator ϕ emphasizes the field aspect of the quantized field theory. Let us now turn to a representation which emphasizes the dual particle aspect of the quantum field. One expects that since the particle and wave properties of a system are known to be complementary, rather than simultaneously measurable, the observable corresponding to particle number, for example, will be constructed from both the field ϕ and its non-commuting conjugate momentum π. Furthermore, as the concept of free particle is well understood, we must first understand how to describe a system of free particles. Then, mutually interacting particles can be described through a perturbative description in which the free particles appear at early times and emerge again at late times. Therefore, we consider for now a single Hermitian free-scalar (or pseudo-scalar) field described by the Lagrangian density

$$\mathcal{L} = \frac{1}{2}(\eta^{\mu\nu}\partial_\mu\phi\partial_\nu\phi - m^2\phi^2). \tag{1.38}$$

For example, this equation describes a free neutral pion (which is a pseudo-scalar, i.e., odd under parity). The charged field is discussed below. The free spin-1 massless (Maxwell) and massive (Proca) fields and the free spin-1/2 (Dirac) field will be described in later sections. Non-Abelian massless gauge fields of spin-1 as well as spin-2 gravitational wave perturbations will be discussed in later chapters. The Euler–Lagrange field equation (1.16) gives

$$(\Box + m^2)\phi = 0, \qquad (1.39)$$

where $\Box = \eta^{\mu\nu}\partial_\mu\partial_\nu$. We shall work in the Heisenberg picture, in which the quantum field satisfies (1.39), which is known as the Klein–Gordon equation. The canonical momentum of (1.13) is

$$\pi = \partial_0\phi = \dot\phi. \qquad (1.40)$$

The field and conjugate momentum are assumed to be operators satisfying the canonical commutation relations of (1.17). From (1.24) for the stress tensor and (1.25), we find the energy and momentum observables

$$P_0 = \frac{1}{2}\int dV_x \left(\dot\phi^2 + (\vec\partial\phi)^2 + m^2\phi^2\right), \qquad (1.41)$$

$$P_i = \frac{1}{2}\int dV_x \left(\dot\phi\partial_i\phi + (\partial_i\phi)\dot\phi\right). \qquad (1.42)$$

(Recall that symmetrization over non-commuting fields was understood in (1.24), and hence in P_μ.) Because the Lagrangian density $\mathcal{L}$ is a scalar under space and time translations, the energy and momentum of the system are conserved. (One can also check, using the commutation relations, that P_μ commutes with P_0.) Because $\mathcal{L}$ is also a scalar under the group of homogeneous Lorentz transformations, there are six further conserved generators in four-dimensional spacetime, corresponding to rotations about axes perpendicular to the six planes determined by pairs taken from the four coordinate axes (i.e., three spatial rotations and three boosts or velocity transformations).

If $f_1(x)$ and $f_2(x)$ are two solutions of (1.39), then the following scalar product is conserved,

$$(f_1, f_2) = i\int dV_x \left\{f_1^*(\vec x, t)\partial_0 f_2(\vec x, t) - (\partial_0 f_1^*(\vec x, t))f_2(\vec x, t)\right\}$$

$$\equiv i\int dV_x\, f_1^* \overleftrightarrow{\partial}_0 f_2. \qquad (1.43)$$

This scalar product is linear with respect to the second argument and antilinear with respect to the first. Furthermore, we have $(f_1, f_2)^* = (f_2, f_1) = -(f_1^*, f_2^*)$.

A complete set of positive energy or positive frequency solutions of the Klein–Gordon equation (1.39) in a spacetime of dimension n is

$$g_{\vec k}(x) = (2\pi)^{-(n-1)/2}(2\omega_k)^{-1/2}\exp[i(\vec k\cdot\vec x - \omega_k t)], \qquad (1.44)$$

where
$$\omega_k = +(\vec{k}^2 + m^2)^{1/2}, \qquad (1.45)$$

and $\vec{k} \cdot \vec{x} = \sum_{i=1}^{n-1} k^i x^i$. (Note that k^i appears, and that $k^i = -k_i$ because we use the convention that $\eta_{\mu\nu} = diag(1, -1, \ldots, -1)$. In (1.44), the exponential contains the Lorentz invariant quantity, $-\eta_{\mu\nu} k^\mu x^\nu$.) These solutions are orthonormal in the sense that

$$(g_{\vec{k}}, g_{\vec{k}'}) = \delta(\vec{k} - \vec{k}'), \quad (g_{\vec{k}}, g_{\vec{k}'}^*) = 0. \qquad (1.46)$$

The general solution to (1.39) is a linear combination of positive energy solutions, $g_{\vec{k}}(x)$, and negative energy solutions, $g_{\vec{k}}^*(x)$. Expanding the Hermitian field $\phi(x)$ in these modes, we have

$$\phi(x) = \int d^{n-1}k \left\{ a(\vec{k}) g_{\vec{k}}(x) + a^\dagger(\vec{k}) g_{\vec{k}}^*(x) \right\}. \qquad (1.47)$$

This expansion resembles that of q in terms of annihilation and creation operators for a simple harmonic oscillator. Indeed, by (1.46), (1.43), and (1.40), we find

$$a(\vec{k}) = (g_{\vec{k}}, \phi) = i \int dV_x \left(g_{\vec{k}}^* \pi - i\omega_k g_{\vec{k}}^* \phi \right). \qquad (1.48)$$

Since $g_{\vec{k}}$ and ϕ are solutions of (1.39), this result is independent of time. Then, from the equal time canonical commutation relations (1.17), it follows that

$$[a(\vec{k}), a^\dagger(\vec{k}')] = \delta(\vec{k} - \vec{k}'), \quad [a(\vec{k}), a(\vec{k}')] = 0. \qquad (1.49)$$

Because a Dirac δ-function, rather than a Kronecker delta, appears in (1.49), the interpretation of the $a(\vec{k})$ operators is not as straightforward as for the simple harmonic oscillator. One can obtain a Kronecker delta, instead, by either of the two methods. First, we can form from the plane waves $g_{\vec{k}}(x)$ a complete orthonormal set of wave packets $f_j(x)$ for expanding any positive frequency solution of the Klein–Gordon equation. The index j is discrete, and the wave packets are normalized such that

$$(f_i, f_j) = \delta_{i,j}, \quad (f_i^*, f_j) = 0. \qquad (1.50)$$

Then we can write

$$\phi(x) = \sum_j \left(a_j f_j(x) + a_j^\dagger f_j^*(x) \right), \qquad (1.51)$$

and find by the same method as before that

$$[a_i, a_j^\dagger] = \delta_{i,j}, \quad [a_i, a_j] = 0. \qquad (1.52)$$

The second method is to impose spatially periodic boundary conditions, such that

$$\phi(\vec{x} + \vec{m}L, t) = \phi(\vec{x}, t), \qquad (1.53)$$

where $\vec{m} = (m_1, m_2, \cdots, m_{n-1})$ with each m_i an integer. Then, the plane wave solutions of (1.39) are

$$f_{\vec{k}}(x) = V^{-1/2}(2\omega_k)^{-1/2} \exp[i(\vec{k} \cdot \vec{x} - \omega_k t)], \qquad (1.54)$$

with $\vec{k} = 2\pi L^{-1}\vec{m}$ and $V = L^{n-1}$. Because of the periodicity, the integration in the scalar product (1.43) can be taken over each x^i from 0 to L, in which case we again have

$$(f_{\vec{k}}, f_{\vec{k}'}) = \delta_{\vec{k},\vec{k}'}, \quad (f_{\vec{k}}, f_{\vec{k}'}{}^*) = 0. \qquad (1.55)$$

Then,

$$\phi(x) = \sum_{\vec{k}} \left(a_{\vec{k}} f_{\vec{k}}(x) + a_{\vec{k}}{}^\dagger f_{\vec{k}}^*(x) \right), \qquad (1.56)$$

$$[a_{\vec{k}}, a_{\vec{k}'}{}^\dagger] = \delta_{\vec{k},\vec{k}'}, \quad [a_{\vec{k}}, a_{\vec{k}'}] = 0. \qquad (1.57)$$

After quantities of physical interest have been calculated, we can take the limit $L \to \infty$, so that the introduction of periodic boundary conditions can be regarded as a convenient mathematical device. When we write a_j or f_j in our discussion below, we can think of either positive frequency wave packets without periodic boundary conditions or of plane waves with periodic boundary conditions.

As for a simple harmonic oscillator (Messiah 1961), it follows that

$$N_j \equiv a_j{}^\dagger a_j \qquad (1.58)$$

is a number operator, with eigenvalues $0, 1, 2, \ldots$, and that $a_j{}^\dagger$ and a_j are creation and annihilation operators, increasing or decreasing by 1, respectively, the number of quanta in mode j. The representation of the Hilbert space of the system, which is based on the simultaneous eigenstates of the number operators N_j, is known as the number or Fock representation. Because

$$a_j = (f_j, \phi) = i \int dV_x \left\{ f_j^* \pi - (\partial_0 f_j^*)\phi \right\}, \qquad (1.59)$$

we find that N_j does not commute with ϕ. Thus, a state with a definite number of quanta does not have a definite field configuration, and vice versa. (This is an aspect of the wave–particle duality.)

The state containing no quanta, known as the vacuum state $|0>$, is defined by

$$a_j |0> = 0 \quad \text{for all } j. \qquad (1.60)$$

It is normalized so that $<0|0> = 1$. A complete set of orthonormal basis states is given by the set

$$|n_1(j_1), n_2(j_2), \cdots > \equiv (n_1! n_2! \cdots)^{-1/2} (a_{j_1}{}^\dagger)^{n_1} (a_{j_2}{}^\dagger)^{n_2} \cdots |0>, \qquad (1.61)$$

containing n_1 quanta in mode j_1, n_2 quanta in mode j_2, etc.

1.2 Particles

In order to ascertain the properties of the quanta created by the operators $a_j{}^\dagger$, consider (1.27) as it pertains to ϕ,

$$i\partial_\mu \phi = [\phi, P_\mu]. \tag{1.62}$$

Apply this to the vacuum state $|0>$ and assume that the Minkowski spacetime vacuum satisfies

$$P_\mu |0> = 0. \tag{1.63}$$

Using the expansion (1.51) for ϕ, we find

$$i\sum_j \partial_\mu f_j^* |1(j)> = -P_\mu \sum_j f_j^* |1(j)>.$$

Applying the bra $<1(l)|$ to this equation and taking the complex conjugate gives

$$i\partial_\mu f_l = \sum_j f_j <1(j)|P_\mu|1(l)>,$$

from which it follows that

$$(f_m, i\partial_\mu f_l) = <1(m)|P_\mu|1(l)>. \tag{1.64}$$

In the single particle quantum mechanics (Schweber 1961) with Hilbert space spanned by the set of positive energy wavefunctions f_j, and with the scalar product of (1.43), the left side of (1.64) is the matrix element of the energy-momentum vector p_μ of the particle between the single particle wave functions f_m and f_l. (Here p_μ is an operator on the single particle Hilbert space of wave functions, as distinct from P_μ.) On the right side of (1.64) is the corresponding matrix element of the total energy-momentum vector P_μ of the quantum field between single particle states $|1(m)>$ and $|1(l)>$. We conclude that the state $|1(j)>$ contains a particle having wave function f_j, and hence that the operator $a_j{}^\dagger$ creates a particle in the single particle state described by the wave function f_j.

More generally, for the generator G of the infinitesimal variations given by (1.28) and (1.29), we have

$$i\delta_0 \phi = [\phi, G]. \tag{1.65}$$

If the transformation generated by G leaves the vacuum invariant (i.e., is a symmetry of the vacuum, so that $G|0> = 0$), then the same argument as for P_μ above yields

$$(f_m, i\delta_0 f_l) = <1(m)|G|1(l)>, \tag{1.66}$$

where $\delta_0 \phi$ is defined in (1.29). It is also clear from the commutation relation, $[a_i{}^\dagger, a_j{}^\dagger] = 0$, that the many particle states of (1.61) are symmetric under interchange of any pair of particles, as is appropriate for a system of identical particles obeying Bose–Einstein statistics.

The $f_{\vec{k}}(x)$ are single particle wave functions of momentum k^i and energy ω_k. Therefore, the total momentum and energy operators P_μ should take on a simple form when expressed in terms of the $a_{\vec{k}}$ and $a_{\vec{k}}{}^\dagger$ corresponding to these wave functions. From (1.41), (1.42), (1.56), and (1.54), we find

$$P_0 = H = \frac{1}{2}\sum_{\vec{k}} \omega_k(a_{\vec{k}}{}^\dagger a_{\vec{k}} + a_{\vec{k}} a_{\vec{k}}{}^\dagger) \tag{1.67}$$

$$P_i = \sum_{\vec{k}} k_i a_{\vec{k}}{}^\dagger a_{\vec{k}} = \sum_{\vec{k}} k_i N_{\vec{k}}. \tag{1.68}$$

In obtaining (1.68), we have commuted $a_{\vec{k}}$ with $a_{\vec{k}}{}^\dagger$ (the c-number term sums to zero because positive and negative values of k^i cancel when the sum is performed). Thus, P_i is the sum of the momenta of the particles present and satisfies (1.63). On the other hand, we can write

$$P_0 = \sum_{\vec{k}} \omega_k \left(N_{\vec{k}} + \frac{1}{2}\right), \tag{1.69}$$

which shows that in addition to the sum of the single particle energies, P_0 includes the sum of the zero-point or vacuum energies of each mode of oscillation of the field. Thus, P_0 does not satisfy (1.63).

In Minkowski spacetime, this is easily dealt with by measuring the energy relative to this constant zero-point contribution (which is an infinite sum). This is equivalent to defining the energy not as P_0, but as

$$: P_0 := \sum_{\vec{k}} \omega_k N_{\vec{k}}. \tag{1.70}$$

The symbol $: P_0 :$ is the normal-ordered operator, meaning that in $: P_0 :$, products of creation and annihilation operators are rewritten with all annihilation operators to the right of all creation operators (e.g., $: a_{\vec{k}} a_{\vec{k}}{}^\dagger : = a_{\vec{k}}{}^\dagger a_{\vec{k}}$). The normal-ordered $: T^\mu{}_\nu :$ satisfies $\partial_\mu : T^\mu{}_\nu : = 0$, has no problem with infinite sums, and gives rise to a conserved energy-momentum vector which annihilates the vacuum. Normal ordering is an example of a regularization method, that is, a consistent, systematic mathematical process for redefining formally infinite expressions so that they are mathematically well defined and satisfy a set of physically motivated requirements. Normal ordering alone is not sufficient to deal with the infinities arising in interacting field theories, which require more sophisticated methods of regularization. In the case of interacting fields, regularization is a prelude to renormalization. The latter is a procedure in which parts of the regularized expressions are absorbed into the definitions of the observable coupling constants, yielding a well-defined finite theory. Renormalization with a finite number of coupling constants is only possible for a limited class of interactions (the so-called renormalizable interactions). These include the standard model of weak and electromagnetic interactions. Renormalization of interacting theories will be discussed in Chapters 6 and 7.

1.2 Particles

Even for free fields the vacuum energy will have gravitational effects, so that if the spacetime is curved it is in general not satisfactory to simply normal order, thereby dropping the vacuum energy. There is also a flat spacetime example in which the effect of the vacuum energy is measurable. If the free field in the vacuum state is considered to be between two parallel plates on which it vanishes (Dirichlet boundary conditions), then the zero-point energy will depend on the separation between the plates. The dependence of energy on separation gives rise to an attractive force exerted by one plate on the other (Casimir 1948). This Casimir effect has been observed experimentally for the electromagnetic field vacuum between conducting plates, and will be discussed in the next section.

It is not difficult to go between the discrete and continuous momentum descriptions. The expressions in (1.67)–(1.70) are of the form $\sum_{\vec{k}} B_{\vec{k}}$, where the sum is over the discrete values $k^i = 2\pi m^i / L$. Successive values of k^i have a spacing of $\Delta k^i = 2\pi/L$, which becomes small in the limit of large L. Thus,

$$\sum_{\vec{k}} B_{\vec{k}} = \left(\frac{L}{2\pi}\right)^{n-1} \sum_{\vec{k}} \Delta^{n-1} k \, B_{\vec{k}},$$

$$\to V(2\pi)^{-(n-1)} \int d^{n-1}k \, B_{\vec{k}}$$

in the limit that $L \to \infty$. (Here $\Delta^{n-1}k \equiv \prod_i \Delta k^i$ and $d^{n-1}k = \prod_i dk^i$.) Hence, in going to the continuum limit, we have the asymptotic relation

$$(2\pi)^{n-1} V^{-1} \sum_{\vec{k}} \sim \int d^{n-1}k. \tag{1.71}$$

Then, comparison of the field expansions in (1.47) and (1.56) shows that the continuum annihilation operator $a(\vec{k})$ is related to the discrete one $a_{\vec{k}}$ by

$$(2\pi)^{-(n-1)/2} V^{1/2} a_{\vec{k}} \sim a(\vec{k}). \tag{1.72}$$

The discrete and continuous commutation relations are then consistent with $(2\pi)^{-(n-1)} V \delta_{\vec{k},\vec{k}'} \sim \delta(\vec{k} - \vec{k}')$, which follows from (1.71). Defining $N(\vec{k}) \equiv a^\dagger(\vec{k}) a(\vec{k})$, and using (1.71) and (1.72), we have

$$\sum_{\vec{k}} N_{\vec{k}} \sim \int d^{n-1}k \, N(\vec{k}), \tag{1.73}$$

which makes it clear that $d^{n-1}k \, N(\vec{k})$ corresponds to the number of particles with momenta in the range dk^i near the value k^i.

Finally, let us write the vacuum state $|0>$ in the Schrödinger or field representation. The evolution operator for this free particle system follows from the equation after (1.10). In the continuum limit, we find

$$U(t,t_0) = \exp\left[-i(t-t_0)\int d^{n-1}k\, \omega_k a(\vec{k})^\dagger a(\vec{k})\right],$$

from which it follows by (1.9) that $|0>$ in the Schrödinger picture is the same as in the Heisenberg picture, and thus is independent of time. Using the plane waves $g_{\vec{k}}$ of (1.44) in (1.48) for $a(\vec{k})$ and applying the result to $|0>$, we obtain

$$\begin{aligned} 0 &= \int dV_x \left\{g_{\vec{k}}^*\pi - i\omega_k g_{\vec{k}}^*\phi\right\}|0> \\ &= \int dV_x\, g_{\vec{k}}^* \left\{\pi - i(m^2 - \vec{\partial}^2)^{1/2}\phi\right\}|0>. \end{aligned} \quad (1.74)$$

Here the operator $(m^2 - \vec{\partial}^2)^{1/2}$ is defined (Schweber 1961) through Fourier transformation (i.e., as a pseudodifferential operator) by

$$(m^2 - \vec{\partial}^2)^{1/2}\phi(\vec{x},t) = \int d^{n-1}k\, e^{i\vec{k}\cdot\vec{x}}\omega_k \tilde{\phi}(\vec{k},t),$$

where

$$\phi(\vec{x},t) = \int d^{n-1}k\, e^{i\vec{k}\cdot\vec{x}}\tilde{\phi}(\vec{k},t).$$

It is then not difficult to show that

$$\int dV_x\, g_{\vec{k}}^* \omega_k \phi = \int dV_x\, g_{\vec{k}}^* (m^2 - \vec{\partial}^2)^{1/2}\phi,$$

which we have used to obtain (1.74). Since (1.74) must hold for all $\vec{k}$, we must have

$$\left\{\pi - i(m^2 - \vec{\partial}^2)^{1/2}\phi\right\}|0> = 0.$$

As the vacuum is time-independent, let us work now in the Schrödinger picture. Applying $<\phi'|$ and using (1.20), we obtain the functional differential equation,

$$\left\{\frac{\delta}{\delta\phi'(\vec{x})} + (m^2 - \vec{\partial}^2)^{1/2}\phi'(\vec{x})\right\}<\phi'|0> = 0,$$

which has the solution (Schweber 1961)

$$<\phi'|0> = N\exp\left[-\frac{1}{2}\int dV_x\, \phi'(\vec{x})(m^2 - \vec{\partial}^2)^{1/2}\phi'(\vec{x})\right], \quad (1.75)$$

where N is a normalization factor. We can interpret (1.75) as the relative probability amplitude of finding the field configuration $\phi'(\vec{x})$ in the vacuum. Note that $\phi'(\vec{x})$ need not be zero in the vacuum state.

1.3 Vacuum energy

As noted earlier, the energy and momentum of the vacuum has gravitational effects and must be carefully taken into account in curved spacetime. In this connection, it is instructive to study a flat spacetime example in which vacuum energy is important. Consider a massless neutral scalar field in four-dimensional Minkowski spacetime satisfying

$$\Box \phi = 0. \tag{1.76}$$

Suppose there are two plates parallel to the (x, y)-plane and located, respectively, at $z = 0$ and $z = a$. Let us assume that if a is sufficiently small, the vacuum energy density outside the plates is small with respect to that between the plates, so that we can neglect the region outside the plates. We require that ϕ have the same value on each plate simultaneously in the rest frame of the plates. (This is a modified version of the Casimir problem of two neutral conducting plates with a vacuum electromagnetic field vanishing on each plate.) Let us also impose periodic boundary conditions in the x and y directions, as in (1.53), with L being large with respect to a. The field can be quantized as before, with normalized plane wave solutions having the form of the $f_{\vec{k}}(x)$ in (1.54) except that V is replaced by aL^2 and $k^z = 2\pi a^{-1}i$, while k^x and k^y have values $2\pi L^{-1}j$ as before. Let $|0>_a$ denote the vacuum or no-particle state of this system. Then, the formal vacuum expectation value of the average energy density between the plates is

$$\rho_0(a) = a^{-1}L^{-2} \,_a<0|P_0|0>_a = \frac{1}{2}a^{-1}L^{-2}\sum_{\vec{k}} \omega_k, \tag{1.77}$$

with

$$\omega_k = (\vec{k}^2)^{1/2} = \left[\left(\frac{2\pi}{L}\right)^2(i^2 + j^2) + \left(\frac{2\pi}{a}\right)^2 l^2\right]^{1/2}, \tag{1.78}$$

where i, j, and l are integers.

We can write this divergent sum as the limit of a well-defined or regularized expression by writing

$$\rho_0(a) = -\frac{1}{2}a^{-1}L^{-2} \lim_{\alpha \to 0} \frac{d}{d\alpha} \sum_{\vec{k}} e^{-\alpha \omega_k}. \tag{1.79}$$

(The quantity α has dimension of length. One could write $\alpha = \alpha'/\mu$, where μ is a constant of dimension (length)$^{-1}$ and α' is dimensionless. The arbitrary parameter μ would in the end drop out of the physical result, as may be easily verified.) We can write

$$\rho_0(a) = -\left(\frac{1}{2}\right)a^{-1}\lim_{\alpha \to 0}\frac{d}{d\alpha}S(\alpha, a),$$

where

$$S(\alpha, a) = L^{-2} \sum_{\vec{k}} e^{-\alpha \omega_k}$$

$$= (2\pi)^{-2} \sum_{l=-\infty}^{\infty} \int_{-\infty}^{\infty} dk^x dk^y \exp\left\{-\alpha \left[k^2 + \left(\frac{2\pi}{a}\right)^2 l^2\right]^{1/2}\right\}$$

$$= \pi^{-1} \sum_{l=1}^{\infty} \int_0^{\infty} dk\, k \exp\left\{-\alpha \left[k^2 + \left(\frac{2\pi}{a}\right)^2 l^2\right]^{1/2}\right\}$$

$$+ (2\pi)^{-1} \int_0^{\infty} dk\, k \exp[-\alpha k], \tag{1.80}$$

where we have let $L \to \infty$, using $\sum_i \sum_j \to L^2 (2\pi)^{-2} \int dk^x dk^y$, have separated the contribution of the mode with $l = 0$, and have introduced polar coordinates with $k = [(k^x)^2 + (k^y)^2]^{1/2}$. If we define

$$F(l) \equiv \int_0^{\infty} dk\, k \exp\left\{-\alpha \left[k^2 + \left(\frac{2\pi}{a}\right)^2 l^2\right]^{1/2}\right\}, \tag{1.81}$$

then

$$\pi S(\alpha, a) = \sum_{l=1}^{\infty} F(l) + \frac{1}{2} F(0). \tag{1.82}$$

Because we are ultimately interested in the limit as $\alpha \to 0$ and derivatives of $F(l)$ with respect to l introduce factors of α, it is advantageous here to use the Euler–Maclaurin formula, which states that if $F(l)$ is analytic for $b \le l < \infty$ and if the infinite sum converges, then

$$\frac{1}{2} F(b) + \sum_{l=1}^{\infty} F(b+l) = \int_b^{\infty} F(l) dl - \sum_{m=1}^{\infty} \frac{B_{2m}}{(2m)!} F^{(2m-1)}(b), \tag{1.83}$$

where $F^{(j)}(b)$ is the jth derivative with respect to l evaluated at $l = b$, and B_j is the jth Bernoulli number. (See, for example, Gradshteyn and Ryzhik 1965.) One has $B_2 = 1/6$, $B_4 = -1/30$, $B_6 = 1/42$, $B_8 = -1/30$, and $B_{10} = 5/66$. In the present case, the Euler–Maclaurin formula with $b = 0$ gives

$$\pi S(\alpha, a) = \int_0^{\infty} F(l) dl - \sum_{m=1}^{\infty} \frac{B_{2m}}{(2m)!} F^{(2m-1)}(0). \tag{1.84}$$

The integral in (1.81) yields

$$F(l) = \left(\frac{1}{\alpha} \frac{2\pi}{a} l + \frac{1}{\alpha^2}\right) \exp\left(-\alpha \frac{2\pi}{a} l\right), \tag{1.85}$$

1.3 Vacuum energy

from which we find $F^{(1)}(0) = 0$, $F^{(3)}(0) = 2\alpha(2\pi/a)^3$, and $F^{(j)}(0) = O(\alpha^2)$ for $j \geq 5$. Furthermore, since $F(l)$ depends only on l/a, we have

$$\int_0^\infty F(l)dl = aG(\alpha), \tag{1.86}$$

where $G(\alpha)$ does not depend on a. Therefore,

$$\pi S(\alpha, a) = aG(\alpha) + \frac{\pi^3}{45a^3}\alpha + O(\alpha^2) \tag{1.87}$$

and

$$\rho_0(a) = -\frac{1}{2}a^{-1} \lim_{\alpha \to 0} \frac{d}{d\alpha} S(\alpha, a)$$

$$= \lim_{\alpha \to 0} \frac{d}{d\alpha}\left[-\frac{1}{2\pi}G(\alpha)\right] - \frac{\pi^2}{90a^4}. \tag{1.88}$$

The term of order α^2 in $S(\alpha, a)$ did not contribute in the limit $\alpha \to 0$. The term involving $G(\alpha)$ diverges in the limit $\alpha \to 0$, but is independent of a for each value of α. Therefore, the same procedure with a replaced by L and then with L taken to infinity would yield a vacuum energy density for Minkowski spacetime of

$$\rho_0(\infty) = \lim_{\alpha \to 0} \frac{d}{d\alpha}\left[-\frac{1}{2\pi}G(\alpha)\right].$$

If we assume that in Minkowski spacetime the physical vacuum energy density is zero, then it is natural to define the vacuum energy density $\rho(a)$ between the plates by

$$\rho(a) = \lim_{\alpha \to 0}[\rho_0(\alpha, a) - \rho_0(\alpha, \infty)], \tag{1.89}$$

where

$$\rho_0(\alpha, a) = -\frac{1}{2}a^{-1}\frac{d}{d\alpha}S(\alpha, a),$$

from which it follows that $\rho_0(\alpha, \infty) = -(2\pi)^{-1}\frac{d}{d\alpha}G(\alpha)$. The average vacuum energy density of a massless scalar field with periodic boundary conditions may be found from (1.89) to be

$$\rho(a) = -\frac{\pi^2}{90a^4}. \tag{1.90}$$

We will derive this result in a different way in Section 5.4.2.

Essentially, the same calculation can be done with vanishing boundary conditions on the plates at $z = 0$ and a. The only differences in the calculation are (1) that $k^z = \pi l/a$ instead of $2\pi l/a$, so that π^3 replaces $(2\pi)^3$ in $F^{(3)}(0)$, and (2) that l does not run over the negative integers because $+l$ and $-l$ no longer correspond to linearly independent mode functions. This introduces a factor of

(1/8)(1/2) into the previous result. One then obtains for a massless scalar field with vanishing boundary conditions the average density,

$$\rho(a) = -\frac{\pi^2}{1440a^4}. \tag{1.91}$$

(If the energy-momentum tensor of the massless scalar field is taken to have zero trace, as is the case for the electromagnetic field, then the energy density is constant between the plates (Birrell and Davies (1982).) For the electromagnetic field between conducting plates, a similar calculation, taking into account the two independent polarization states, yields the energy density,

$$\rho_{em}(a) = -\frac{\pi^2}{720a^4}. \tag{1.92}$$

(A factor of 2 appears as a result of the polarization states; the zero mode does not depend on a and therefore does not influence this result.) The vacuum energy of the electromagnetic field contained in a cylindrical volume between two opposite regions of area A on each plate is $aA\rho_{em}(a)$. Because this energy increases as the plates are separated, an attractive force is exerted between the plates. The force per unit area in the z direction on the plate at $z = a$ is (with factors of $\hbar$ and c restored by dimensional analysis)

$$\frac{F}{A} = -\frac{d}{da}(a\rho_{em}(a)) = -\frac{\hbar c \pi^2}{240a^4}. \tag{1.93}$$

This force was first predicted by Casimir (1948). Laboratory experiments have confirmed the existence of Casimir forces (Sparnaay 1958; Lamoreaux 1997; Mohideen and Roy 1998; Lamoreaux 1999; Mohideen and Roy 1999).

1.4 Charged scalar field

In Section 1.2, we considered an Hermitian scalar field, which describes a neutral spin-0 particle. A charged spin-0 particle and its oppositely charged antiparticle can be described by a pair of independent Hermitian scalar fields ϕ_1 and ϕ_2, with the total Lagrangian being a sum of Lagrangians of the form of (1.38), one for each field. This Lagrangian has a symmetry which is most clearly brought out by defining the complex field

$$\phi(x) = \frac{1}{\sqrt{2}}[\phi_1(x) - i\phi_2(x)]. \tag{1.94}$$

Then,

$$\mathcal{L} = \eta^{\mu\nu}\partial_\mu \phi^\dagger \partial_\nu \phi - m^2 \phi^\dagger \phi, \tag{1.95}$$

which is invariant under the transformation

$$\phi(x) \to \phi'(x) = \exp(i\alpha)\phi(x), \tag{1.96}$$

together with its Hermitian conjugate. (Here α is a real constant.)

1.4 Charged scalar field

This theory may be canonically quantized by means of (1.17). The result is the same if we consider (ϕ_1, ϕ_2) or $(\phi, \phi^\dagger)$ as the independent fields. Using the latter pair, we find from (1.13) that $\pi = \partial_0 \phi^\dagger$ and $\pi^\dagger = \partial_0 \phi$ are the momenta conjugate to ϕ and $\phi^\dagger$, respectively. Expanding in terms of periodic plane waves $f_{\vec{k}}$ of (1.54),

$$\phi(x) = \sum_{\vec{k}} \{a_{\vec{k}} f_{\vec{k}}(x) + b_{\vec{k}}^\dagger f_{\vec{k}}^*(x)\} \tag{1.97}$$

$$\phi^\dagger(x) = \sum_{\vec{k}} \{b_{\vec{k}} f_{\vec{k}}(x) + a_{\vec{k}}^\dagger f_{\vec{k}}^*(x)\}, \tag{1.98}$$

we find as in Section 1.2 that as a consequence of the canonical commutation relations of (1.17),

$$[a_{\vec{k}}, a_{\vec{k}'}^\dagger] = \delta_{\vec{k},\vec{k}'}, \quad [b_{\vec{k}}, b_{\vec{k}'}^\dagger] = \delta_{\vec{k},\vec{k}'} \tag{1.99}$$

and all other independent commutators involving $a_{\vec{k}}$ and $b_{\vec{k}}$ are zero. Therefore,

$$N_{\vec{k}}^+ = a_{\vec{k}}^\dagger a_{\vec{k}}, \quad N_{\vec{k}}^- = b_{\vec{k}}^\dagger b_{\vec{k}} \tag{1.100}$$

are number operators, representing two types of quanta associated with the complex field.

Following the general discussion after (1.35), a conserved current J^μ of (1.37) and generator G of (1.32) are associated with the transformation (1.96). The infinitesimal form of the latter corresponds to $\delta x^\mu = 0$ and $\delta \phi = \delta_0 \phi = i\alpha\phi$, so that the conserved current is (after symmetrization over non-commuting quantities)

$$J^\mu = i\alpha \frac{1}{2} \eta^{\mu\lambda} [(\partial_\lambda \phi^\dagger)\phi + \phi(\partial_\lambda \phi^\dagger)] - i\alpha \frac{1}{2} \eta^{\mu\lambda} [(\partial_\lambda \phi)\phi^\dagger + \phi^\dagger(\partial_\lambda \phi)]$$

and the conserved generator is

$$G = i\alpha \frac{1}{2} \int dV_x [(\partial_0 \phi^\dagger)\phi + \phi(\partial_0 \phi^\dagger) - (\partial_0 \phi)\phi^\dagger - \phi^\dagger(\partial_0 \phi)].$$

Note that indices indicating a sum over the independent field components (in this case ϕ and $\phi^\dagger$) are present in (1.37) and (1.32). Substituting (1.97) and (1.98) into this expression for G, and using (1.54) and (1.99), we find that

$$G = \alpha \sum_{\vec{k}} (N_{\vec{k}}^- - N_{\vec{k}}^+).$$

(The contributions from commuting creation and annihilation operators cancel.) From (1.31), with $F = F[\phi, \phi^\dagger, \pi, \pi^\dagger]$, we have

$$i\delta_0 F = [F, G],$$

where $\delta_0 F$ is the change induced in F by the infinitesimal changes $\delta_0 \phi = i\alpha\phi$ and the corresponding changes in $\phi^\dagger$, π, and $\pi^\dagger$.

In view of these conserved quantities, the operators $a_{\vec{k}}^\dagger$ and $b_{\vec{k}}^\dagger$ can be regarded as creating particles of opposite charges $+e$ and $-e$, respectively. The conserved total charge Q and current j^μ are then obtained from G and J^μ by replacing α by $-e$:

$$Q = \frac{1}{2}ie \int dV_x \left(\{\partial_0\phi, \phi^\dagger\} - \{\partial_0\phi^\dagger, \phi\} \right) = e \sum_{\vec{k}} \left(N_{\vec{k}}^+ - N_{\vec{k}}^- \right) \tag{1.101}$$

$$j^\mu = \frac{1}{2}ie(\{\partial^\mu\phi, \phi^\dagger\} - \{\partial^\mu\phi^\dagger, \phi\}), \tag{1.102}$$

where $\{A, B\} \equiv AB + BA$ is the anticommutator.

The interaction of the charged field with an external electromagnetic field, $F_{\mu\nu} = \partial_\mu A_\nu - \partial_\nu A_\mu$, is introduced through the minimal coupling prescription, in which $\partial_\mu \phi$ is replaced in $\mathcal{L}$ by $(\partial_\mu + ieA_\mu)\phi$, so that

$$\mathcal{L} = \eta^{\mu\nu}(\partial_\mu \phi^\dagger - ieA_\mu \phi^\dagger)(\partial_\nu \phi + ieA_\nu \phi) - m^2 \phi^\dagger \phi. \tag{1.103}$$

This Lagrangian is invariant under the transformation

$$A_\mu(x) \to A'_\mu(x) = A_\mu(x) - e^{-1}\partial_\mu \theta(x) \tag{1.104}$$

$$\phi(x) \to \phi'(x) = \exp[i\theta(x)]\phi(x), \tag{1.105}$$

where $\theta(x)$ is an arbitrary scalar function. In particular, the quantity

$$(\partial_\mu + ieA_\mu)\phi \equiv D_\mu \phi \tag{1.106}$$

transforms under the gauge transformations of (1.104) and (1.105) in the same way as does ϕ itself. Therefore, D_μ is known as the gauge-covariant derivative. The transformation given by (1.104) and (1.105) is known as a local $U(1)$ gauge transformation because θ depends on x, and because successive transformations generated by θ_1 and θ_2 yield the transformation generated by $\theta_1+\theta_2$, thus forming a representation of the group $U(1)$. Similarly, (1.96) is known as a global or rigid $U(1)$ gauge transformation because α is the same at all points.

If the field A_μ is to be a dynamical field, we must add to $\mathcal{L}$ the Lagrangian of the Maxwell field, so that

$$\mathcal{L} = -\frac{1}{4}F^{\mu\nu}F_{\mu\nu} + D^\mu \phi^\dagger D_\mu \phi - m^2 \phi^\dagger \phi. \tag{1.107}$$

Note that $[D_\mu, D_\nu] = ieF_{\mu\nu} \equiv ie(\partial_\mu A_\nu - \partial_\nu A_\mu)$.

If we start with a Lagrangian such as (1.95), which is only invariant under the global $U(1)$ gauge group, the vector potential $A_\mu(x)$ and gauge-covariant derivative must be introduced if we wish to make the Lagrangian invariant under the local $U(1)$ gauge group. Therefore, A_μ is called a $U(1)$ gauge field. The generalization of this procedure to non-Abelian Lie groups, such as $SU(n)$, is well known. Here, we give a brief summary to fix notation.

If the field multiplet ϕ transforms under a Lie group G of dimension N, then

$$\phi'(x) = \exp[i\theta^a(x)T^a]\phi(x), \tag{1.108}$$

where a is summed from 1 to N, and the T^a are the N generators of G, satisfying

$$[T^a, T^b] = if^{abc}T^c, \tag{1.109}$$

with the f^{abc} (real constants antisymmetric in indices a, b) being the structure constants of the group. If the representation G_R of the group has dimension d_R, then $\phi(x)$ can be represented by a column matrix of d_R elements, and T^a is a $d_R \times d_R$ matrix. The covariant derivative under the local gauge transformation (1.108) is

$$D_\mu \phi = (\partial_\mu + igA_\mu)\phi, \tag{1.110}$$

where the Yang–Mills field A_μ is a $d_R \times d_R$ matrix given by

$$A_\mu(x) = A_\mu{}^a(x)T^a, \tag{1.111}$$

which transforms under the gauge transformation (1.108) as

$$A_\mu \to A'_\mu = UA_\mu U^{-1} - (ig)^{-1}(\partial_\mu U)U^{-1}, \tag{1.112}$$

with $U = \exp[i\theta^a(x)T^a]$. The quantity g is the Yang–Mills coupling constant, which is the generalization of the electric charge of the $U(1)$ theory. The analogue of the electromagnetic field is the $d_R \times d_R$ matrix

$$\begin{aligned} F_{\mu\nu} &= (ig)^{-1}[D_\mu, D_\nu] \\ &= \partial_\mu A_\nu - \partial_\nu A_\mu + ig[A_\mu, A_\nu]. \end{aligned} \tag{1.113}$$

In view of (1.109) and (1.111), we have $[A_\mu, A_\nu] = if^{abc}A_\mu{}^a A_\nu{}^b T^c$. A Lagrangian invariant under the local gauge group G is

$$\mathcal{L} = -\frac{1}{4}F^{a\mu\nu}F^a{}_{\mu\nu} + D^\mu \phi^\dagger D_\mu \phi - V(\phi^\dagger, \phi), \tag{1.114}$$

where $F^a{}_{\mu\nu}$ is defined by $F_{\mu\nu} = F^a{}_{\mu\nu}T^a$, and $V(\phi^\dagger, \phi)$ is a gauge-invariant expression such as $m^2\phi^\dagger\phi$. We will return to the study of Yang–Mills theory in Chapter 7.

1.5 Dirac field

In dealing with the canonical quantization of the free Dirac electron, a suitable action is $S = \int dv_x \, \mathcal{L}$, with Lagrangian density

$$\mathcal{L} = \bar{\psi}(i\gamma^\mu \partial_\mu - m)\psi. \tag{1.115}$$

(Spinor indices have been suppressed; our conventions agree with Bjorken and Drell (1964).) The matrices γ^μ satisfy the anticommutation rules

$$\{\gamma^\mu, \gamma^\nu\} = 2\eta^{\mu\nu}, \tag{1.116}$$

and the relations $\gamma^{0\dagger} = \gamma^0$, $\gamma^{i\dagger} = -\gamma^i$. The Pauli adjoint $\bar\psi$ is defined as $\bar\psi = \psi^\dagger \gamma^0$, and transforms under the Lorentz group in such a way that $\mathcal{L}$ is a scalar. The Lagrangian in (1.116) differs only by a perfect divergence from the Hermitian Lagrangian $\frac{1}{2}(\mathcal{L} + \mathcal{L}^\dagger)$. Varying the action with respect to $\bar\psi$, treated as independent of ψ, we obtain the Dirac equation

$$(i\gamma^\mu \partial_\mu - m)\psi = 0. \tag{1.117}$$

The momentum conjugate to ψ is given by (1.13) as

$$\pi = i\psi^\dagger. \tag{1.118}$$

Thus, ψ and π exhaust the independent degrees of freedom of this first-order theory. (Indeed, the momentum conjugate to $\bar\psi$ is zero.)

It was shown by Schwinger that the operator action principle of (1.30), with G acting as generator in (1.31) of the infinitesimal variations of (1.28)–(1.29), will hold valid not only if π and ϕ satisfy the commutation relations of (1.17), but also if they satisfy instead the following anticommutation relations:

$$\{\phi_a(\vec{x},t), \phi_b(\vec{x}',t)\} = 0, \{\pi_a(\vec{x},t), \pi_b(\vec{x}',t)\} = 0,$$

$$\{\phi_a(\vec{x},t), \pi_b(\vec{x}',t)\} = i\delta_{a,b}\delta(\vec{x}-\vec{x}'). \tag{1.119}$$

In the case of the spinor field, quantization should proceed by imposing these canonical anticommutation relations. The reason is that otherwise the energy of the free field is not bounded from below and we also have difficulty with causality requirements. Similar problems arise if we attempt to quantize the scalar field using the above anticommutators instead of the commutators of (1.17). In the present case, (1.119) becomes

$$\{\psi_\alpha(\vec{x},t), \psi_\beta^\dagger(\vec{x}',t)\} = \delta_{\alpha\beta}\delta(\vec{x}-\vec{x}'), \tag{1.120}$$

with the other anticommutators vanishing. Here α,β refer to spinor indices.

If ψ_1 and ψ_2 are two solutions of the Dirac equation, the following scalar product is conserved:

$$(\psi_1, \psi_2) = \int dV_x\, \psi_1^\dagger(\vec{x},t)\psi_2(\vec{x},t). \tag{1.121}$$

We will expand the field ψ in terms of a complete set of periodic plane wave solutions analogous to those in the scalar case. These solutions will be orthonormal with respect to the above scalar product integrated over the volume $V = L^{n-1}$, where L is the periodicity interval. (The continuum limit can be taken in the same way as in the scalar case described earlier.) In the case when $m \neq 0$, we take the complete set of solutions (Bjorken and Drell 1964, 1965)

$$u_{\vec{k},s}(x) = \left(\frac{m}{\omega_k V}\right)^{\frac{1}{2}} u(\vec{k},s)\, \exp(-ik_\mu x^\mu), \tag{1.122}$$

1.5 Dirac field

$$v_{\vec{k},s}(x) = \left(\frac{m}{\omega_k V}\right)^{\frac{1}{2}} v(\vec{k},s) \exp(ik_\mu x^\mu), \qquad (1.123)$$

where the spinors $u(\vec{k},s)$ satisfy as a consequence of (1.115)

$$(\gamma^\mu k_\mu - m)u(\vec{k},s) = 0,$$

$$(\gamma^\mu k_\mu + m)v(\vec{k},s) = 0,$$

and $k_\mu = (\omega_k, -\vec{k})$ with $\omega_k = (\vec{k}^2 + m^2)^{\frac{1}{2}}$. In our notation, the argument s takes the values $\pm 1/2$.

In discussing the spin, we will work with three spatial dimensions. Let s^μ represent a unit vector (with $s^\mu s_\mu = -1$), which in the rest frame of the particle described by wave function $u_{\vec{k},s}$ or $v_{\vec{k},s}$ reduces to a given spatial unit vector $(0, \vec{s})$ describing the direction of the spin of the particle. More precisely, in the single particle Dirac theory the operator $\Sigma^i \equiv \frac{1}{2}i\epsilon^{ijk}\Sigma^{jk}$ (j,k summed from 1 to 3), where

$$\Sigma^{\mu\nu} \equiv \frac{1}{4}[\gamma^\mu, \gamma^\nu] \qquad (1.124)$$

represents the intrinsic angular momentum or spin of the particle. It is easily verified that for $j \neq k$, we have $(\Sigma^{jk})^2 = -1/4$, so that the Hermitian operator Σ^i has eigenvalues $\pm 1/2$. We require that in the rest frame of the particle, where $\vec{k} = 0$ and $s^\mu = (0, \vec{s})$, the following relations hold

$$\vec{\Sigma} \cdot \vec{s}\, u\left(\vec{0}, \pm\frac{1}{2}\right) = \pm\frac{1}{2} u\left(\vec{0}, \pm\frac{1}{2}\right), \qquad (1.125)$$

$$\vec{\Sigma} \cdot \vec{s}\, v\left(\vec{0}, \pm\frac{1}{2}\right) = \mp\frac{1}{2} v\left(\vec{0}, \pm\frac{1}{2}\right). \qquad (1.126)$$

These relations can be written in an arbitrary inertial frame as

$$\tau_\mu s^\mu u_{\vec{k},\pm\frac{1}{2}}(x) = \pm\frac{1}{2} u_{\vec{k},\pm\frac{1}{2}}(x), \qquad (1.127)$$

$$\tau_\mu s^\mu v_{\vec{k},\pm\frac{1}{2}}(x) = \pm\frac{1}{2} v_{\vec{k},\pm\frac{1}{2}}(x), \qquad (1.128)$$

where $\tau_\mu s^\mu \equiv -s^\mu (2m)^{-1} \epsilon_{\sigma\mu\nu\lambda} \Sigma^{\nu\lambda} \partial^\sigma$ reduces to $\vec{\Sigma} \cdot \vec{s}$ or $-\vec{\Sigma} \cdot \vec{s}$ when operating in the rest frame on u or v, respectively. It is clear that $u_{\vec{k},s}(x)$ is the spinor wave function of a particle of momentum $\vec{k}$, energy ω_k, and spin s along the $\vec{s}$ direction in the rest frame, while $v_{\vec{k},s}(x)$ describes a particle of momentum $-\vec{k}$, spin $-s$ along the $\vec{s}$ direction in the rest frame, and negative energy $-\omega_k$. (Here s takes the values $\pm 1/2$ and should not be confused with s^μ or $\vec{s}$.) In the Dirac single particle theory, the negative energy states are all filled in the vacuum, and the absence of a particle in a negative energy state (i.e., a hole) is observed as an antiparticle having properties opposite to those of a particle in the negative energy state. In the quantum field theory, the properties of the antiparticle emerge naturally.

The interpretation is clarified by writing the field expansion

$$\psi(x) = \sum_s \sum_{\vec{k}} \{c_{\vec{k},s} u_{\vec{k},s}(x) + d_{\vec{k},s}{}^\dagger v_{\vec{k},s}(x)\}, \qquad (1.129)$$

where the sum on s is over the values $\pm 1/2$. Thus, $v_{\vec{k},s}$ is analogous to $f_{\vec{k}}^*$ in the expansion (1.97) of the charged scalar fields, and $d_{\vec{k},s}$ is interpreted as annihilating an antiparticle having momentum $\vec{k}$, spin s along the $\vec{s}$ direction in the rest frame, and positive energy ω_k. The normalization of $u_{\vec{k},s}$ and $v_{\vec{k},s}$ is such that with the conserved scalar product of (1.121)

$$(u_{\vec{k},s}, u_{\vec{k}',s'}) = \delta_{\vec{k},\vec{k}'}\delta_{s,s'}, \quad (v_{\vec{k},s}, v_{\vec{k}',s'}) = \delta_{\vec{k},\vec{k}'}\delta_{s,s'}, \qquad (1.130)$$

and scalar products between u and v are zero. Then, we can use the scalar product to express the operators $c_{\vec{k},s}$ and $d_{\vec{k},s}$ in terms of ψ and $\bar\psi$.

The anticommutation relations of (1.120) then give

$$\{c_{\vec{k},s}, c_{\vec{k}',s'}{}^\dagger\} = \delta_{\vec{k},\vec{k}'}\delta_{s,s'}, \qquad (1.131)$$

$$\{d_{\vec{k},s}, d_{\vec{k}',s'}{}^\dagger\} = \delta_{\vec{k},\vec{k}'}\delta_{s,s'}, \qquad (1.132)$$

with the other independent anticommutators, such as $\{c_{\vec{k},s}, c_{\vec{k}',s'}\}$, all vanishing. Therefore, these particles obey Fermi–Dirac statistics. It was shown by Pauli (1940) that in a local, Lorentz-invariant quantum field theory particles of odd half-integer spin must obey Fermi–Dirac statistics and particles of integer spin must obey Bose–Einstein statistics. In the next chapter, we will show that curved spacetime quantum field theory provides a fundamentally different proof of this spin-statistics theorem.

The Lagrangian of (1.115) is symmetric under the global gauge transformation

$$\psi(x) \to \psi'(x) = \exp(i\alpha)\psi(x) \qquad (1.133)$$

and its conjugate. Then (1.37) and (1.32) with $\delta x^\mu = 0$ and $\delta\psi = i\alpha\psi$ yield

$$J^\mu = -\alpha\bar\psi\gamma^\mu\psi$$

and

$$G = -\alpha \int dV_x\, \psi^\dagger \psi.$$

Since the quantum fields $\psi_\alpha{}^\dagger$ and ψ_β do not commute, some kind of symmetrization of the classical expression is indicated. (In practice, this is only carried out when it makes an essential difference.) Here, antisymmetrization over anticommuting fermion fields is appropriate (for example, antisymmetrization leaves $\psi_\alpha\psi_\beta$ unchanged, whereas symmetrization would replace it by $1/2\{\psi_\alpha,\psi_\beta\}=0$).

Thus, an expression of the form $\psi_\alpha{}^\dagger A_{\alpha\beta}\psi_\beta$, where $A_{\alpha\beta}$ is a matrix of complex numbers, not an operator in the Hilbert space, should be replaced by $1/2 A_{\alpha\beta}(\psi_\alpha{}^\dagger \psi_\beta - \psi_\beta \psi_\alpha{}^\dagger) \equiv 1/2[\psi^\dagger, A\psi] \equiv 1/2[\psi^\dagger A, \psi]$. If we replace α by the magnitude of the electron charge, we obtain the appropriate current and charge operators for a system of electrons and positrons,

$$j^\mu = -\frac{1}{2}e[\bar\psi, \gamma^\mu \psi], \tag{1.134}$$

$$Q = -\frac{1}{2}e \int dV_x\, [\psi^\dagger, \psi]. \tag{1.135}$$

Using (1.129) and (1.130), we find that

$$Q = -e \sum_s \sum_{\vec k} (c^\dagger{}_{\vec k,s} c_{\vec k,s} - d^\dagger{}_{\vec k,s} d_{\vec k,s}), \tag{1.136}$$

which shows that $c^\dagger{}_{\vec k,s} c_{\vec k,s}$ is the number operator for electrons of charge $-e$ and $d^\dagger{}_{\vec k,s} d_{\vec k,s}$ for positrons of charge e. Introduction of an electromagnetic interaction through minimal coupling preserves the $U(1)$ gauge symmetry, so that Q remains conserved, and production or annihilation processes must occur through electron–positron pairs. Because of the antisymmetrization over pairs of fermion fields in (1.134) and (1.135), the expressions contain no divergent terms.

The same result is obtained by normal ordering, but with the additional condition for fermion creation and annihilation operators that a change of sign occurs with each interchange of operators; for example, $: d_{\vec k,s} d^\dagger{}_{\vec k,s} := -d^\dagger{}_{\vec k,s} d_{\vec k,s}$.

From (1.23) and (1.24), we obtain the conserved energy-momentum vector

$$P_\mu = \int dV_x\, (i\psi^\dagger \partial_\mu \psi - \delta^0{}_\mu \mathcal{L}).$$

With (1.115) and (1.117) this becomes

$$P_\mu = i \int dV_x\, \psi^\dagger \partial_\mu \psi. \tag{1.137}$$

This expression is Hermitian. (For $\mu = i$, we can integrate by parts, and for $\mu = 0$, the Dirac equation replaces $i\partial_0$ by an Hermitian–Hamiltonian operator.) In (1.137), changing $\psi^\dagger \partial_\mu \psi$ into $1/2[\psi^\dagger, \partial_\mu \psi]$ would not remove the divergent zero-point energy in this case, nor alter the result given below. Equations (1.118) and (1.119) give

$$P_\mu = \sum_s \sum_{\vec k} k_\mu \{c^\dagger{}_{\vec k,s} c_{\vec k,s} - d_{\vec k,s} d^\dagger{}_{\vec k,s}\},$$

resulting in

$$P_i = \sum_s \sum_{\vec k} k_i \{c^\dagger{}_{\vec k,s} c_{\vec k,s} + d^\dagger{}_{\vec k,s} d_{\vec k,s}\}, \tag{1.138}$$

because positive and negative contributions of k_i cancel each other in the zero-point term. Also

$$P_0 = \sum_s \sum_{\vec{k}} \omega_k \{c^\dagger_{\vec{k},s} c_{\vec{k},s} + d^\dagger_{\vec{k},s} d_{\vec{k},s} - 1\}.$$

Note that the zero-point term is of the opposite sign from that of the scalar field. In general, fermions and bosons have zero-point energies of opposite sign (which cancel each other in supersymmetric theories containing bosons and fermion multiplets of the same mass). The zero-point contribution is removed in Minkowski spacetime by normal ordering of P_μ in (1.137),

$$: P_0 : = \sum_s \sum_{\vec{k}} \omega_k \{c^\dagger_{\vec{k},s} c_{\vec{k},s} + d^\dagger_{\vec{k},s} d_{\vec{k},s}\}. \tag{1.139}$$

These expressions show that $c_{\vec{k},s}$ and $d_{\vec{k},s}$ annihilate particles of momentum $\vec{k}$ and energy ω_k. Let us next consider the spin, first in general and then for the Dirac field.

1.6 Angular momentum and spin

The total angular momentum of a system is, by definition, proportional to the generator of spatial rotations and is conserved if the Lagrangian is rotationally invariant, or, more generally, Lorentz invariant. A Lorentz transformation is a change of spacetime coordinates $x^\mu \to x'^\mu$ that satisfies the equation $\eta_{\mu\nu} x^\mu x^\nu = \eta_{\mu\nu} x'^\mu x'^\nu$. An infinitesimal homogeneous Lorentz transformation of the coordinates has the form

$$x^\mu \to x'^\mu = x^\mu + \epsilon^\mu{}_\nu x^\nu, \quad \epsilon_{\mu\nu} = -\epsilon_{\nu\mu} \tag{1.140}$$

to first order in $\epsilon_{\mu\nu}$. Similarly, if the set of fields $\phi_a(x)$ (e.g., the components of the Dirac spinor $\psi(x)$) form the basis of a representation of the Lorentz group, then under the above infinitesimal Lorentz transformation, the fields undergo the infinitesimal transformation

$$\phi_a(x) \to \phi'_a(x') = \phi_a(x) + \frac{1}{2} \epsilon_{\mu\nu} \Sigma^{\mu\nu}{}_{ab} \phi_b(x), \tag{1.141}$$

where the $\Sigma^{\mu\nu}$ are constant matrices, antisymmetric under interchange of μ and ν, serving to generate the Lorentz transformation in the representation of the fields ϕ_a. (Here, summation over b from 1 to the dimension of the representation is understood, where the $\Sigma^{\mu\nu}{}_{ab}$ denote the matrix elements of $\Sigma^{\mu\nu}$.) In any representation of the proper Lorentz group, the matrices $\Sigma^{\mu\nu}$ satisfy the commutation relations,

$$[\Sigma^{\mu\nu}, \Sigma^{\lambda\sigma}] = -\eta^{\nu\lambda} \Sigma^{\nu\sigma} - \eta^{\nu\sigma} \Sigma^{\mu\lambda} + \eta^{\nu\lambda} \Sigma^{\mu\sigma} + \eta^{\mu\sigma} \Sigma^{\nu\lambda}, \tag{1.142}$$

as may be checked by writing (1.140) in the form

1.6 Angular momentum and spin

$$x'^\omega = x^\omega + \frac{1}{2}\epsilon_{\mu\nu}(\Sigma^{\mu\nu})^\omega{}_\tau x^\tau,$$

with the matrices $(\Sigma^{\mu\nu})^\omega{}_\tau = \eta^{\mu\omega}\delta^\nu{}_\tau - \eta^{\nu\omega}\delta^\mu{}_\tau$ satisfying (1.142). (In this case, we write the ω index on $(\Sigma^{\mu\nu})^\omega{}_\tau$ as up, but in general we write $(\Sigma^{\mu\nu})_{ab}$ with index a down, since there is no raising or lowering of group indices. In the vector case, we can simply define $(\Sigma^{\mu\nu})_{\omega\tau} \equiv (\Sigma^{\mu\nu})^\omega{}_\tau$.)

In the notation of (1.28) and (1.33), we have

$$\delta x^\mu = \epsilon^\mu{}_\nu x^\nu, \quad \delta\phi_a = \frac{1}{2}\epsilon^{\mu\nu}\Sigma^{\mu\nu}{}_{ab}\phi_b(x).$$

From (1.32) the generator of the above infinitesimal Lorentz transformation is

$$G = \frac{1}{2}\epsilon_{\mu\nu}M^{\mu\nu}, \tag{1.143}$$

with

$$M^{\mu\nu} \equiv \int dV_x \left(\pi_a \Sigma^{\mu\nu}{}_{ab}\phi_b + x^\mu T^{0\nu} - x^\nu T^{0\mu}\right), \tag{1.144}$$

and from (1.24),

$$T^{0\mu} = \pi_a \partial^\mu \phi_a - \eta^{0\mu}\mathcal{L}.$$

Because $M^{\mu\nu}$ is antisymmetric, the conservation of G implies that

$$\frac{d}{dt}M^{\mu\nu} = 0. \tag{1.145}$$

The spatial components M^{ij} are identified with the components of the conserved total angular momentum generating rotations in the (i,j) plane. In three spatial dimensions, we have a total angular momentum

$$\vec{J}_{ang} = (M^{23}, M^{31}, M^{12}).$$

The first term in (1.144) is independent of the choice of origin and is therefore the intrinsic angular momentum or spin of the particles,

$$S^{\mu\nu} = \int dV_x \, \pi_a(x)\Sigma^{\mu\nu}{}_{ab}\phi_b(x). \tag{1.146}$$

The second term of (1.144) is the orbital angular momentum.

Using the identity $[AB, C] = A[B, C] + [A, C]B$ followed, in the case of fermion fields satisfying (1.119), by $[A, BC] = \{A, B\}C - B\{A, C\}$, we obtain for bosons or fermions

$$[S^{\mu\nu}, S^{\lambda\sigma}] = i \int dV_x \, \pi(x)[\Sigma^{\mu\nu}, \Sigma^{\lambda\sigma}]\phi(x),$$

where matrix notation, suppressing the Lorentz group indices, is used. Then from the Lorentz group algebra of (1.142) we find that

$$[S^{\mu\nu}, S^{\lambda\sigma}] = i(-\eta^{\nu\lambda}S^{\nu\sigma} - \eta^{\nu\sigma}S^{\mu\lambda} + \eta^{\nu\lambda}S^{\mu\sigma} + \eta^{\mu\sigma}S^{\nu\lambda}). \tag{1.147}$$

It follows that
$$[S^{23}, S^{31}] = iS^{12}. \tag{1.148}$$

This, and the relations obtained by cyclic interchange of indices, show that the spin operators of the field satisfy the standard angular momentum commutation relations.

According to (1.31), we have $i\delta_0\phi_a = [\phi_a, G]$, where $\delta_0\phi_a(x) = \phi'_a(x) - \phi_a(x)$. Then, (1.143) gives

$$\phi'_a(x) = \phi_a(x) - i\left[\phi_a, \frac{1}{2}\epsilon_{\mu\nu}M^{\mu\nu}\right], \tag{1.149}$$

which is the infinitesimal form of

$$\phi'_a(x) = \exp\left(i\frac{1}{2}a_{\mu\nu}M^{\mu\nu}\right)\phi_a(x)\exp\left(-i\frac{1}{2}a_{\mu\nu}M^{\mu\nu}\right), \tag{1.150}$$

where $a_{\mu\nu}$ is the matrix for a finite Lorentz transformation,

$$x'^{\mu} = a^{\mu}{}_{\nu}x^{\nu}.$$

For an infinitesimal Lorentz transformation, from (1.140), we have $a_{\mu\nu} = \eta_{\mu\nu} + \epsilon_{\mu\nu}$. Since $\eta_{\mu\nu}M^{\mu\nu}$ vanishes by symmetry, we recover (1.149) in the infinitesimal case. Relations like (1.149) and (1.150) also hold between $\pi'_a(x)$ and $\pi_a(x)$. The unitary relation between the fields and momenta in different inertial frames assures that the canonical commutation or anticommutation relations have the same form in each inertial frame, as is necessary for the Lorentz invariance of the theory. The requirement that expectation values transform covariantly then implies that Hilbert space state vectors, $|a>$ and $|a>'$, representing the same physical state in two different inertial frames must be related by the unitary transformation $|a>' = \exp(i\frac{1}{2}a_{\mu\nu}M^{\mu\nu})|a>$ corresponding to the Lorentz transformation from the unprimed to the primed frame. So far our considerations have been general.

For scalar fields, we have $\phi'(x') = \phi(x)$, so that $\Sigma^{\mu\nu} = 0$ and the spin is 0. For the Dirac field, (1.141) takes the form

$$\psi'(x') = \psi(x) + \frac{1}{2}\epsilon_{\mu\nu}\Sigma^{\mu\nu}\psi(x),$$

with

$$\Sigma^{\mu\nu} = \frac{1}{4}[\gamma^{\mu}, \gamma^{\nu}]. \tag{1.151}$$

This follows from the requirement that the Dirac equation transform covariantly under Lorentz transformations (see, for example, Bjorken and Drell 1964). That is, we require that as a consequence of the Dirac equation in one inertial frame,

$$(i\gamma^{\mu}\partial_{\mu} - m)\psi(x) = 0,$$

1.6 Angular momentum and spin

the corresponding Dirac equation in another inertial frame should follow,

$$(i\gamma^\mu \partial'_\mu - m)\psi'(x') = 0,$$

where $\psi'(x')$ is given by (1.141) and ∂'_μ follows from (1.140). (The gamma matrices can be defined as being unaltered in passing from one frame to the other, since if they were altered by other than a similarity transformation preserving (1.116), we would obtain different physical results in different inertial frames.) $\Sigma^{\mu\nu}$ is the operator introduced in (1.124). Writing $\vec{S} = (S^{23}, S^{31}, S^{12})$ and $\vec{\Sigma} = i(\Sigma^{23}, \Sigma^{31}, \Sigma^{12})$, we have from (1.146)

$$\vec{S} = i \int dV_x : \psi^\dagger(x) \vec{\Sigma} \psi(x) : . \qquad (1.152)$$

One can show (Bjorken and Drell (1965), p. 61) that

$$(\vec{s} \cdot \vec{S}) d^\dagger_{\vec{0}, \pm 1/2} |0> = \pm \frac{1}{2} d^\dagger_{\vec{0}, \pm 1/2} |0>,$$

so that $d^\dagger_{\vec{k}, \pm 1/2}$ creates a positron, which in its proper frame has its spin aligned along the $\pm \vec{s}$ director (the same holds for the electron created by $c^\dagger_{\vec{k}, \pm 1/2}$). By applying $(\vec{s} \cdot \vec{S})$ to $d^\dagger_{\vec{0}, \pm 1/2} |0>$, we effectively project out the $\vec{k} = 0$ part of the field expansion in $\vec{S}$, so (1.125), (1.126), and then (1.130) can be used (terms involving three creation operators require special consideration). It is clear from this last equation that the particles associated with the field ψ have spin 1/2.

2
Basics of quantum fields in curved spacetimes

The successful predictions of general relativity are convincing evidence that gravitational phenomena are most clearly understood by regarding spacetime as curved. In general relativity matter exerts its gravitational influence by curving spacetime, and we study the propagation of particles and waves on this curved background. It is then natural to study the propagation of quantum fields in curved spacetimes in order to search for new effects of gravitation. At this level, the gravitational field itself is not quantized, and the methods of Minkowski spacetime quantum field theory are carried over as much as possible. As we shall see, this modest extension of quantum field theory has turned out to be richer in consequences than we could have anticipated.

Among other things, it gives rise to the physically important processes of particle creation in cosmological and black hole spacetimes. The same amplification process that creates particles in an expanding universe is responsible for creating, in the context of an early inflationary expansion, the primordial fluctuations that are now observed with astonishing accuracy in the cosmic microwave background (CMB) radiation. These same primordial fluctuations also appear responsible for the large-scale structure of the universe. The creation of particles by black holes is necessary for maintaining the second law of thermodynamics in their presence. This process of radiation and evaporation of black holes is an important facet in the fundamental search for a microscopic explanation of the entropy of black holes; a search which appears to be leading to new and exciting physics connecting gravitation and quantum theory. Quantum field theory in curved spacetime also provides a new dynamical explanation of the connection between spin and statistics, and brings out some new features of Minkowski spacetime physics, such as the excitation of accelerated detectors. This partial listing is sufficient motivation to turn now to the basis of the theory of quantized fields in curved spacetime.

2.1 Canonical quantization and conservation laws

Consider a set of fields $\phi_a(x)$ propagating in a curved spacetime with invariant line element

$$ds^2 = g_{\mu\nu}(x)dx^\mu dx^\nu. \tag{2.1}$$

The metric $g_{\mu\nu}(x)$ will be treated as a given unquantized external field. We will assume that the spacetime has a well-defined causal structure and set of Cauchy hypersurfaces. The set of fields $\phi_a(x)$ to be quantized may include linearized gravitational wave perturbations propagating on the background $g_{\mu\nu}(x)$. Let n denote the dimension of spacetime, with x^0 being the time coordinate and $x^1, \ldots, x^{n-1}$ being the spatial coordinates.

The action S is constructed from the field ϕ_a, so that it is invariant under general coordinate transformations (diffeomorphisms):

$$S[\phi'(x'), \nabla'\phi'(x'), g'_{\mu\nu}(x')] = S[\phi(x), \nabla\phi(x), g_{\mu\nu}(x)]. \tag{2.2}$$

The simplest way to construct such an action is to start with the Minkowski spacetime action and replace ordinary derivatives ∂_μ by covariant derivatives ∇_μ, $\eta^{\mu\nu}$ by $g^{\mu\nu}$, and $d^n x$ by the invariant volume element $d^n x |g|^{1/2}$, where $g = \det(g_{\mu\nu})$. This is called the minimal coupling prescription, and is consistent with the Einstein principle of equivalence, according to which local gravitational effects are not present in a neighborhood of the spacetime origin of a locally inertial frame of reference. Occasionally, we can further increase the symmetry by the addition to the Lagrangian of a term which does not vanish at the origin of a locally inertial frame, and such a possibility will also be considered. (The action already involves $\partial_\lambda g_{\mu\nu}$ through $\nabla_\mu \phi_a$; when additional terms are included, it may also involve higher derivatives of $g_{\mu\nu}$.)

The requirement that variations of the action

$$S = \int d^n x \mathcal{L}(\phi, \nabla\phi, g_{\mu\nu}) \tag{2.3}$$

vanish with respect to variations of the fields ϕ_a, which are zero on the boundary of integration, then yields the Euler–Lagrange equations

$$\partial_\mu \left(\frac{\partial \mathcal{L}}{\partial(\partial_\mu \phi_a)} \right) - \frac{\partial \mathcal{L}}{\partial \phi_a} = 0. \tag{2.4}$$

The general covariance of (2.4) is insured by the invariance of the action S. Note that $\mathcal{L}$ is a scalar density, transforming like $|g|^{1/2}$.

Variation of the action with respect to the external field $g_{\mu\nu}$ does not in general vanish because we have not included an additional term proportional to the scalar curvature R in (2.3). (Such a term gives rise to the geometric part of the Einstein gravitational field equations. See Misner et al. (1973) or Weinberg (1972) for example.) However, because of the invariance of S under general coordinate transformation, δS will be zero under the change in $g_{\mu\nu}$ induced by an infinitesimal coordinate transformation

$$x^\mu \to x'^\mu = x^\mu - \epsilon^\mu(x), \tag{2.5}$$

where x and x' refer to the same event in spacetime. Under this transformation we have

$$g_{\mu\nu}(x) \to g'_{\mu\nu}(x') = \frac{\partial x^\lambda}{\partial x'^\mu}\frac{\partial x^\sigma}{\partial x'^\nu} g_{\lambda\sigma}(x),$$

which yields the variation

$$\delta_0 g_{\mu\nu}(x) \equiv g'_{\mu\nu}(x) - g_{\mu\nu}(x) = \mathcal{L}_\epsilon g_{\mu\nu}, \tag{2.6}$$

where $\mathcal{L}_\epsilon$ denotes the Lie derivative,

$$\mathcal{L}_\epsilon g_{\mu\nu} = \nabla_\mu \epsilon_\nu + \nabla_\nu \epsilon_\mu. \tag{2.7}$$

Let us assume that $\epsilon^\mu(x)$ and $\partial_\lambda \epsilon^\mu(x)$ are zero on the boundary of the region of integration defining the action S of (2.3). Then, under the above infinitesimal coordinate transformation, we have

$$\delta S = \int d^n x \frac{\delta S}{\delta g_{\mu\nu}(x)} \delta_0 g_{\mu\nu}(x),$$

because variations in S produced by the changes in the dynamical fields ϕ_a vanish as a consequence of (2.4) and the boundary conditions on ϵ_μ. With $\delta_0 g_{\mu\nu}$ given by (2.6) and (2.7), the invariance of S under coordinate transformations requires that $\delta S = 0$. Hence, with $dv_x \equiv d^n x |g|^{1/2}$,

$$\delta S = -\int dv_x T^{\mu\nu} \nabla_\mu \epsilon_\nu = 0, \tag{2.8}$$

where we have defined the tensor

$$T^{\mu\nu} \equiv -2|g|^{-1/2} \frac{\delta S}{\delta g_{\mu\nu}(x)}, \tag{2.9}$$

and have used its symmetry under interchange of indices. Then from (2.8) and

$$\nabla_\mu(T^{\mu\nu}\epsilon_\nu) = |g|^{-1/2}\partial_\mu(|g|^{1/2}T^{\mu\nu}\epsilon_\nu)$$
$$= (\nabla_\mu T^{\mu\nu})\epsilon_\nu + T^{\mu\nu}\nabla_\mu \epsilon_\nu,$$

it follows that

$$\int dv_x (\nabla_\mu T^{\mu\nu})\epsilon_\nu = 0,$$

and because ϵ_ν is arbitrary, we must have

$$\nabla_\mu T^{\mu\nu} = 0. \tag{2.10}$$

This is the generally covariant generalization of (1.23). Furthermore, $T^{\mu\nu}$ as defined by (2.9) is symmetric, which is not true in general for the canonical

2.1 Canonical quantization and conservation laws

energy-momentum tensor $\Theta^{\mu\nu}$ of (1.24). (In Chapter 1, the symbol $T^{\mu\nu}$ was used for the canonical energy-momentum tensor, $\Theta^{\mu\nu}$, but from now on $T^{\mu\nu}$ will refer to the symmetric energy-momentum tensor.) From $\delta(g^{\mu\nu}g_{\mu\lambda}) = 0$, it follows that $g_{\mu\lambda}g_{\nu\sigma}\frac{\delta}{\delta g} = -\frac{\delta}{\delta g^{\mu}}$, so

$$T_{\mu\nu} = +\frac{2}{|g|^{1/2}}\frac{\delta S}{\delta g^{\mu\nu}}. \tag{2.11}$$

The sign convention in our definition of $T^{\mu\nu}$ in (2.9) is chosen so that T_{00} will be positive for the classical electromagnetic field. If we were to use the opposite metric signature, then the signs on the right-hand sides of (2.9) and (2.11) would also be opposite.

We can calculate the symmetric energy-momentum tensor $T^{\mu\nu}$ in curved spacetime and then go to the flat spacetime limit, thereby obtaining a symmetric energy-momentum tensor satisfying $\partial_\mu T^{\mu\nu} = 0$ in Minkowski spacetime. For any isolated system in Minkowski spacetime, both $\Theta^{\mu\nu}$ and $T^{\mu\nu}$ yield a conserved energy-momentum vector p_μ (which, as noted earlier, is unique). In curved spacetime it is the symmetric energy-momentum tensor $T^{\mu\nu}$ defined in (2.9) which describes the matter and radiation and couples to the gravitational field through the Einstein field equations. (In a general curved spacetime, the tensor density $\Theta^{\mu\nu}$ of weight $1/2$ defined by (1.24) does not satisfy $\partial_\mu \Theta^\mu{}_\nu = 0$ as it does in Minkowski spacetime, nor any simple generalization of that equation.)

The Schwinger operator action principle continues to hold in curved spacetime for an arbitrary infinitesimal transformation of the form given in (1.28) and (1.29), provided that under the transformation $\delta_0 g_{\mu\nu}(x) = 0$, or

$$g'_{\mu\nu}(x) = g_{\mu\nu}(x). \tag{2.12}$$

In that case, the derivation leading to (1.34) goes through as before (with the canonical stress tensor now called $\Theta^\mu{}_\nu$); because we have $\delta_0 g_{\mu\nu} = 0$, it is not necessary for the Euler–Lagrange equation to hold for the external field $g_{\mu\nu}$. (Note that the metric $g_{\mu\nu}$ is not included here among the fields ϕ_a of (1.34).) Thus, under a transformation of the coordinates and fields given in (1.28) and (1.29), and for which (2.12) holds, we have as in Section 1.1

$$\delta S = G(t_2) - G(t_1), \tag{2.13}$$

with

$$G(t) = \int d^{n-1}x [\pi_a \delta \phi_a - \Theta^o{}_\nu \delta x^\nu], \tag{2.14}$$

where, as noted above, $\Theta^o{}_\nu$ is defined by the right-hand side of (1.24). Here the integration is on a constant time hypersurface, and

$$\pi_a = \frac{\partial \mathcal{L}}{\partial(\partial_0 \phi_a)}. \tag{2.15}$$

As before, with suitable operator ordering and regularization, G acts as the generator of the transformation, satisfying

$$i\delta_0 F = [F, G], \qquad (2.16)$$

where F is a functional of the ϕ_a and π_a.

As in Minkowski spacetime, (2.16) will hold with

$$[\phi_a(\vec{x},t), \phi_b(\vec{x}',t)] = 0, \quad [\pi_a(\vec{x},t), \pi_b(\vec{x}',t)] = 0,$$

$$[\phi_a(\vec{x},t), \pi_b(\vec{x}',t)] = i\delta_{a,b}\delta(\vec{x}-\vec{x}'), \qquad (2.17)$$

which is appropriate for bosons, or

$$\{\phi_a(\vec{x},t), \phi_b(\vec{x}',t)\} = 0, \quad \{\pi_a(\vec{x},t)\pi_b(\vec{x}',t)\} = 0,$$

$$\{\phi_a(\vec{x},t), \pi_b(\vec{x}',t)\} = i\delta_{a,b}\delta(\vec{x}-\vec{x}'), \qquad (2.18)$$

which is appropriate for fermions. Here $\delta(\vec{x}-\vec{x}')$ is the Dirac δ-function satisfying $\int d^{n-1}x\, \delta(\vec{x}-\vec{x}')f(\vec{x}) = f(\vec{x}')$ with the integral being over the spacelike hypersurface $t =$ constant. The above commutation relations are imposed on independent field components. (They may be modified when gauge conditions are present.) One can show that $\pi(\vec{x}',t)$ and $\delta(\vec{x}-\vec{x}')$ each transform as spatial scalar densities under transformations of the spatial coordinates on the constant-t hypersurface. Hence, (2.17) and (2.18) are covariant under transformations of the spatial coordinates on the hypersurface, and are therefore the spatially covariant generalization of the corresponding relations that hold in flat spacetime. One can also show that they are consistent with the equations of motion of the fields, in the sense that if they hold on one constant-t spatial hypersurface, then they also will hold on the other constant-t spatial hypersurfaces.

Furthermore, for bosons we can define a complete set of commuting Hermitian fields, and define a basis $|\phi'>$ for the Hilbert space of state vectors, as in Minkowski spacetime, through (1.19). This gives a field or Schrödinger representation. For a system of fermions, we can define a set of Hermitian commuting quantities which are bilinear in the fields and build a Hilbert space of state vectors spanned by the simultaneous eigenvectors of these bilinear operators.

There are several cases of interest when G is conserved. The simplest is when $\delta x^\mu = 0$ and $\delta_0 \phi_a$ is a symmetry of $\mathcal{L}$. Then (2.12) is trivially satisfied, and the symmetry of $\mathcal{L}$ implies that $\delta S = 0$ so that G is independent of time. One also has, in that case, $\partial_\mu J^\mu = 0$, with

$$J^\mu = \frac{\partial \mathcal{L}}{\partial(\partial_\mu \phi_a)} \delta\phi_a. \qquad (2.19)$$

(Note that J^μ is a vector density of weight $1/2$, so that we have $\nabla_\mu(|g|^{-1/2} J^\mu) = 0$.) Thus, in curved spacetime the electric charge and the generators of internal symmetries continue to be conserved.

2.1 Canonical quantization and conservation laws

The generator G is also conserved when $\delta x^\mu \neq 0$, but is such that (2.12) holds, and the fields ϕ_a are components of a spacetime tensor. Then, invariance under coordinate transformations implies that $\delta S = 0$, and (2.13) implies that G is constant. For example, if it is possible to choose a coordinate system in which a particular coordinate, say x^λ, does not appear in $g_{\mu\nu}(x)$, then under a translation in the x^λ direction (2.12) holds, and furthermore we have $\delta\phi(x) = 0$. Then, as in Section 1.1, it follows that

$$p_\lambda = \int d^{n-1}x \Theta^0{}_\lambda \qquad (2.20)$$

is constant. (Since $g^{\lambda\mu}$ and the other components of p_μ may not be constants, it does not follow that $p^\lambda = g^{\lambda\mu}p_\mu$ is constant.) A similar expression involving the symmetric energy-momentum tensor $T_{\mu\nu}$ also holds. In deriving that result, we will also work in a more covariant language.

A coordinate transformation for which (2.12) holds is called an isometry of the spacetime. For an infinitesimal isometry of the form

$$x^\mu \to x'^\mu = x^\mu - \epsilon \xi^\mu(x), \qquad (2.21)$$

with $\epsilon \ll 1$, it follows from (2.6) and (2.12) that

$$\mathcal{L}_\xi g_{\mu\nu} \equiv \nabla_\mu \xi_\nu + \nabla_\nu \xi_\mu = 0. \qquad (2.22)$$

A vector field satisfying (2.22) is called a Killing vector of the spacetime. As a consequence of the symmetry of $T^{\mu\nu}$ and (2.10) we have

$$\nabla_\mu(T^{\mu\nu}\xi_\nu) = 0. \qquad (2.23)$$

But since $T^{\mu\nu}\xi_\nu$ is a vector, we have $\nabla_\mu(T^{\mu\nu}\xi_\nu) = |g|^{-1/2}\partial_\mu(|g|^{1/2}T^{\mu\nu}\xi_\nu)$, and it follows that

$$\partial_\mu(|g|^{1/2}T^{\mu\nu}\xi_\nu) = 0. \qquad (2.24)$$

Hence

$$P_\xi \equiv \int dV_x T^0{}_\nu(x)\xi^\nu(x) \qquad (2.25)$$

is constant, where $dV_x = d^{n-1}x|g|^{1/2}$. In the case when the coordinates are such that $g_{\mu\nu}(x)$ is independent of a particular coordinate, say x^λ, then $\xi^\nu = \delta^\nu{}_\lambda$ is a Killing vector field, and (2.25) reduces to $P_\lambda = \int dV_x T^0{}_\lambda$ being constant. In such a case, (2.20) should yield the same p_λ to within possibly an additive constant independent of the field configuration.

Thus, in a general curved spacetime, we have shown that $\nabla_\mu T^{\mu\nu} = 0$, and in a spacetime having special symmetries or isometries as implied by the existence of one or more Killing vector fields ξ^μ, we have $\partial_\mu(|g|^{1/2}T^\mu{}_\nu \xi^\nu) = 0$ and $P_\xi \equiv \int dV_x T^0{}_\nu \xi^\nu$ is constant.

Let us now consider again a general curved spacetime without special isometries. An invariance of the action which is of interest is curved spacetime conformal invariance. Certain fields, such as the electromagnetic and massless Dirac fields in curved spacetime, exhibit an invariance of the action under conformal transformations of the metric and field, as defined below. An action $S[\phi, g_{\mu\nu}]$ (suppressing derivatives for brevity) is conformally invariant if

$$S[\phi, g_{\mu\nu}] = S[\tilde{\phi}, \tilde{g}_{\mu\nu}] + \text{surface integral}, \qquad (2.26)$$

where

$$\tilde{g}_{\mu\nu}(x) = \Omega^2(x) g_{\mu\nu}(x), \qquad (2.27)$$

$$\tilde{\phi}(x) = \Omega^{2p}(x) \phi(x). \qquad (2.28)$$

Here $\Omega(x)$ is an arbitrary function and p is a dimensionless constant. Consider an infinitesimal conformal transformation with

$$\Omega^2(x) = 1 + \lambda(x), \quad |\lambda(x)| << 1. \qquad (2.29)$$

Then $\delta x^\mu = 0$, and from (2.27) and (2.28),

$$\delta_0 g_{\mu\nu}(x) = \tilde{g}_{\mu\nu}(x) - g_{\mu\nu}(x) = \lambda(x) g_{\mu\nu}(x), \qquad (2.30)$$

$$\delta_0 \phi(x) = \tilde{\phi}(x) - \phi(x) = p\lambda(x) \phi(x). \qquad (2.31)$$

From (2.26), if $\lambda(x)$ vanishes sufficiently rapidly on the boundary of the region of integration,

$$0 = \delta S = \int d^n x \left\{ \frac{\delta S}{\delta \phi} \delta_0 \phi + \frac{\delta S}{\delta g_{\mu\nu}} \delta_0 g_{\mu\nu} \right\}.$$

If we assume that ϕ satisfies the Euler–Lagrange equation, $\delta S/\delta \phi = 0$, then it follows that

$$0 = \int d^n x \frac{\delta S}{\delta g_{\mu\nu}(x)} g_{\mu\nu}(x) \lambda(x),$$

and because $\lambda(x)$ is arbitrary,

$$\frac{\delta S}{\delta g_{\mu\nu}(x)} g_{\mu\nu}(x) = 0,$$

from which we have

$$g_{\mu\nu} T^{\mu\nu} = 0. \qquad (2.32)$$

Thus, conformal invariance of the action implies that the trace of the energy-momentum tensor is zero. We will find later that a theory based on a classical or "bare" action which is conformally invariant will in general lose its conformal invariance in the quantum theory as a result of renormalization. The energy-momentum tensor thus acquires a non-vanishing trace, known as the trace or conformal anomaly.

In this section, we have discussed the action, field equations, symmetric energy-momentum tensor, generators of field transformations, commutation or anticommutation relations, Hilbert space of state vectors in the field representation, isometries and conservation laws, and conformal invariance. Having built a foundation for our discussion, let us consider next the curved spacetime generalization of the free neutral scalar field.

2.2 Scalar field

Following the minimal coupling prescription based on the principle of equivalence and described in Section 2.1, the action of the free neutral scalar field based on the Lagrangian density of (1.38) becomes in curved spacetime

$$S = \int d^n x \mathcal{L} \tag{2.33}$$

with

$$\mathcal{L} = \frac{1}{2}|g|^{1/2}(g^{\mu\nu}\partial_\mu\phi\partial_\nu\phi - m^2\phi^2). \tag{2.34}$$

(For the scalar field we have $\nabla_\mu \phi = \partial_\mu \phi$.) This is a special case of the more general Lagrangian density

$$\mathcal{L} = \frac{1}{2}|g|^{1/2}(g^{\mu\nu}\partial_\mu\phi\partial_\nu\phi - m^2\phi^2 - \xi R \phi^2), \tag{2.35}$$

where ξ is a dimensionless constant and R is the scalar curvature of the spacetime, $R \equiv g^{\mu\nu} R_{\mu\nu}$. (As noted in the section on notation, we use the following conventions: metric signature -2, $R^\mu{}_{\nu\lambda\sigma} = \partial_\sigma \Gamma^\mu{}_{\nu\lambda} - \partial_\lambda \Gamma^\mu{}_{\nu\sigma} + \Gamma^\tau{}_{\nu\lambda}\Gamma^\mu{}_{\sigma\tau} - \Gamma^\tau{}_{\nu\sigma}\Gamma^\mu{}_{\lambda\tau}$, and $R_{\mu\nu} = R^\lambda{}_{\mu\lambda\nu}$, in agreement with Birrell and Davies (1982), or $(-,-,-)$ in the notation of Misner et al. (1973).) We will carry out our discussion for this more general Lagrangian. One reason is that when we include an interaction term such as $|g|^{1/2}\lambda\phi^4$, with λ a constant (dimensionless in four-dimensional spacetime), then a term of the form $|g|^{1/2}\xi R\phi^2$ is needed for renormalization. (This will be shown in Section 6.7.) Another reason is that when $m = 0$ and $\xi = 1/6$ (in four dimensions), the action is invariant under the curved spacetime conformal transformations. The case $\xi = 0$ is referred to as minimal coupling. The magnitude of ξ cannot be very large, because if ξ has a non-zero value, then the term in the Lagrangian that is proportional to $R\phi^2$ can cause the effective gravitational constant to vary with time and position as a result of such variations in ϕ.

Consider an infinitesimal conformal transformation of the form of (2.29)–(2.31) with $p = -1/2$:

$$g_{\mu\nu}(x) \to \tilde{g}_{\mu\nu}(x) = (1 + \lambda(x))g_{\mu\nu}(x), \tag{2.36}$$

$$\phi(x) \to \tilde{\phi}(x) = \left(1 - \frac{1}{2}\lambda(x)\right)\phi(x). \tag{2.37}$$

Under this transformation, we have in four dimensions to first order in λ

$$|g|^{1/2} \to |\tilde{g}|^{1/2} = (1 + 2\lambda)|g|^{1/2},$$
$$g^{\mu\nu} \to \tilde{g}^{\mu\nu} = (1 - \lambda)g^{\mu\nu},$$
$$\Gamma^{\mu}{}_{\nu\tau} \to \tilde{\Gamma}^{\mu}{}_{\nu\tau} = \Gamma^{\mu}{}_{\nu\tau} + \frac{1}{2}(\delta^{\mu}{}_{\tau}\partial_{\nu}\lambda + \delta^{\mu}{}_{\nu}\partial_{\tau}\lambda - g^{\mu\sigma}g_{\nu\tau}\partial_{\sigma}\lambda),$$
$$R \to \tilde{R} = (1 - \lambda)R + 3\Box\lambda. \tag{2.38}$$

Then, working to first order in λ, the transformed Lagrangian density with $\xi = 1/6$ and $m = 0$ becomes

$$\begin{aligned}\tilde{\mathcal{L}} &= \frac{1}{2}|\tilde{g}|^{1/2}\left(\tilde{g}^{\mu\nu}\partial_{\mu}\tilde{\phi}\partial_{\nu}\tilde{\phi} - \frac{1}{6}\tilde{R}\tilde{\phi}^2\right) \\ &= \frac{1}{2}(1+2\lambda)|g|^{1/2}\left\{(1-\lambda)g^{\mu\nu}\partial_{\mu}\left[\left(1-\frac{1}{2}\lambda\right)\phi\right]\partial_{\nu}\left[\left(1-\frac{1}{2}\lambda\right)\phi\right]\right. \\ &\quad \left. - \frac{1}{6}[(1-\lambda)R + 3\Box\lambda](1-\lambda)\phi^2\right\} \\ &= \frac{1}{2}|g|^{1/2}\left\{g^{\mu\nu}\partial_{\mu}\phi\partial_{\nu}\phi - g^{\mu\nu}\phi\partial_{\mu}\lambda\partial_{\nu}\phi - \frac{1}{6}R\phi^2 - \frac{1}{2}(\Box\lambda)\phi^2\right\} \\ &= \mathcal{L} - \frac{1}{2}|g|^{1/2}g^{\mu\nu}\phi\partial_{\mu}\lambda\partial_{\nu}\phi - \frac{1}{4}|g|^{1/2}(\Box\lambda)\phi^2 \\ &= \mathcal{L} - \partial_{\mu}\left(\frac{1}{4}|g|^{1/2}g^{\mu\nu}\phi^2\partial_{\nu}\lambda\right),\end{aligned} \tag{2.39}$$

since

$$|g|^{1/2}\Box\lambda = \partial_{\mu}\left(|g|^{1/2}g^{\mu\nu}\partial_{\nu}\lambda\right).$$

Hence, the conformally related actions differ only by a surface term and the field equation for ϕ is form invariant under conformal transformations. Since a finite conformal transformation with $\Omega^2(x) = \exp[\lambda(x)]$ can be built from an infinite product of infinitesimal transformations (with λ being the sum of the infinitesimal λ's), we must have for a finite conformal transformation with

$$g_{\mu\nu} \to \tilde{g}_{\mu\nu} = \Omega^2 g_{\mu\nu}$$
$$\phi \to \tilde{\phi} = \Omega^{-1}\phi$$

that

$$\tilde{\mathcal{L}} = \mathcal{L} - \partial_{\mu}\left(\frac{1}{2}|g|^{1/2}g^{\mu\nu}\phi^2\partial_{\nu}\ln\Omega\right). \tag{2.40}$$

This has the same form in terms of λ as in the infinitesimal case; and again the field equation is form invariant.

In n dimensions, the value of ξ in (2.35) which makes the classical theory conformally invariant under the transformation of (2.27) and (2.28), with $p = (2-n)/2$, is $\xi = [4(n-1)]^{-1}(n-2)$. (See, for example, Birrell and Davies 1982.)

Working with ξ and m arbitrary, the Euler–Lagrange equation for the Lagrangian density (2.35) reads

$$(\Box + m^2 + \xi R)\phi = 0, \qquad (2.41)$$

and (2.9) gives the symmetric energy-momentum tensor,

$$\begin{aligned} T^{\mu\nu} &= \nabla^\mu \phi \nabla^\nu \phi - \frac{1}{2} g^{\mu\nu} \nabla^\rho \phi \nabla_\rho \phi + \frac{1}{2} g^{\mu\nu} m^2 \phi^2 - \xi\left(R^{\mu\nu} - \frac{1}{2} g^{\mu\nu} R\right)\phi^2 \\ &\quad + \xi[g^{\mu\nu}\Box(\phi^2) - \nabla^\mu \nabla^\nu (\phi^2)]. \end{aligned} \qquad (2.42)$$

This satisfies, as a consequence of (2.41),

$$\nabla_\mu T^{\mu\nu} = 0, \qquad (2.43)$$

and in four dimensions

$$T^\mu{}_\mu = 0, \quad \text{when} \quad \xi = \frac{1}{6} \quad \text{and} \quad m = 0. \qquad (2.44)$$

As we know from the previous section, (2.43) is a consequence of coordinate invariance and (2.44) of conformal invariance for the unquantized field. In the quantized theory, (2.44) may not hold as we will see.

The calculation of (2.42) proceeds briefly as follows. Vary $g_{\mu\nu}$ in the action (2.33) with $\mathcal{L}$ given by (2.35), using the identities

$$\delta g^{\mu\nu} = -g^{\mu\rho} g^{\nu\sigma} \delta g_{\rho\sigma}, \qquad (2.45)$$

$$\delta |g|^{1/2} = \frac{1}{2} |g|^{1/2} g^{\mu\nu} \delta g_{\mu\nu}, \qquad (2.46)$$

$$\delta R = -R^{\mu\nu} \delta g_{\mu\nu} + g^{\rho\sigma} g^{\mu\nu} (\delta g_{\rho\sigma;\mu\nu} - \delta g_{\rho\mu;\sigma\nu}). \qquad (2.47)$$

This gives

$$\begin{aligned} \delta S &= \frac{1}{2} \int d^n x |g|^{1/2} \Big\{ \frac{1}{2} g^{\mu\nu} \delta g_{\mu\nu} (g^{\rho\sigma} \partial_\rho \phi \partial_\sigma \phi - m^2 \phi^2 - \xi R \phi^2) \\ &\quad - \delta g_{\mu\nu} \nabla^\mu \phi \nabla^\nu \phi \\ &\quad - \xi [-R^{\mu\nu} \delta g_{\mu\nu} + g^{\rho\sigma} g^{\mu\nu} (\delta g_{\rho\sigma;\mu\nu} - \delta g_{\rho\mu;\sigma\nu})] \phi^2 \Big\}. \end{aligned}$$

Using the identities (taking $\delta g_{\mu\nu}$ and its first derivative to be zero on the boundary of the region of integration)

$$\int d^n x |g|^{1/2} g^{\rho\sigma} g^{\mu\nu} \delta g_{\rho\sigma;\mu\nu} \phi^2 = \int d^n x |g|^{1/2} g^{\rho\sigma} \delta g_{\rho\sigma} \Box(\phi^2)$$

and

$$\int d^n x |g|^{1/2} g^{\rho\sigma} g^{\mu\nu} \delta g_{\rho\mu;\sigma\nu} \phi^2 = \int d^n x |g|^{1/2} g^{\sigma\mu} g^{\lambda\nu} \delta g_{\mu\nu} \nabla_\sigma \nabla_\lambda (\phi^2),$$

we then obtain

$$\delta S = -\frac{1}{2} \int d^n x |g|^{1/2} T^{\mu\nu} \delta g_{\mu\nu},$$

with $T^{\mu\nu}$ given by (2.42), as was to be shown.

The scalar field is quantized by imposing the canonical commutation relation of (2.17). The appropriate generalization of the scalar product (1.43) is

$$(f_1, f_2) \equiv i \int d^{n-1} x |g|^{1/2} g^{0\nu} f_1^*(\vec{x}, t) \overleftrightarrow{\partial_\nu} f_2(\vec{x}, t), \qquad (2.48)$$

where the integral is taken over a constant t hypersurface. If f_1 and f_2 are solutions of the field equation (2.41) which vanish at spatial infinity, then (f_1, f_2) is conserved, since

$$\frac{d}{dt}(f_1, f_2) = i \int d^{n-1} x \partial_0 (|g|^{1/2} g^{0\nu} f_1^* \overleftrightarrow{\partial_\nu} f_2)$$

$$= i \int d^{n-1} x |g|^{1/2} \nabla_\mu (g^{\mu\nu} f_1^* \overleftrightarrow{\partial_\nu} f_2)$$

$$- i \int d^{n-1} x \partial_i (|g|^{1/2} g^{i\nu} f_1^* \overleftrightarrow{\partial_\nu} f_2)$$

$$= 0.$$

The first term of the second equality above is zero by virtue of the field equation, and the second term gives a contribution at spatial infinity which vanishes (this also vanishes if we use normalization in a cube with periodic boundary conditions on f_1 and f_2). We have used the basic identity $\nabla_\mu V^\mu = |g|^{-1/2} \partial_\mu (|g|^{1/2} V^\mu)$, valid for any vector field V^μ, in the derivation.

In terms of a general spacelike hypersurface σ with future-directed unit normal n^μ and hypersurface element $d\sigma$, the scalar product (2.48) is

$$(f_1, f_2) = i \int_\sigma d\sigma |g|^{1/2} n^\nu f_1^* \overleftrightarrow{\partial_\nu} f_2. \qquad (2.49)$$

We can show that this scalar product is conserved under deformations of σ. Suppose $\sigma \to \sigma'$ such that σ and σ' form the spacelike boundaries of a volume v (there may also be timelike boundaries of v at spatial infinity). Then by the Gauss divergence theorem

$$(f_1, f_2)_{\sigma'} - (f_1, f_2)_\sigma = i \int_v d^n x \, \partial^\mu (|g|^{1/2} f_1^* \overleftrightarrow{\partial_\mu} f_2)$$

$$= i \int_v d^n x |g|^{1/2} \nabla^\mu (f_1^* \overleftrightarrow{\nabla}_\mu f_2)$$

$$= 0,$$

as a consequence of the field equation.

Before continuing our discussion of the scalar field in a general curved spacetime, it will be instructive to look at a specific model. This will bring out new features which arise in curved spacetime.

2.3 Cosmological model: Arbitrary asymptotically static time dependence

An understanding of the quantum effects that can be expected in curved spacetime was first arrived at in Parker (1965) by studying the spatially flat isotropically changing metric

$$ds^2 = dt^2 - a^2(t)(dx^2 + dy^2 + dz^2), \tag{2.50}$$

and we will find it instructive to investigate this model before continuing with the discussion of more general spacetimes. We will suppose that the cosmological scale factor has an arbitrary time dependence that asymptotically approaches constant values at early and late values of the cosmic time t. This cosmic time t is the proper time of a set of clocks on geodesic worldlines that remain at constant values of the spatial coordinates (x, y, z). We take

$$a(t) \sim \begin{cases} a_1 & \text{as } t \to -\infty, \\ a_2 & \text{as } t \to +\infty. \end{cases} \tag{2.51}$$

We assume that $a(t)$ is sufficiently smooth and approaches constant values sufficiently fast that the statements we make below are well defined. (An exactly soluble example will be given in the next section.) We will suppose that $a(t)$ approaches the constant values a_1 and a_2 sufficiently rapidly that the asymptotic forms we write below are actually approached. (We can, in fact, suppose that $a(t) = a_1$ and $a(t) = a_2$ for arbitrarily long initial and final time intervals, respectively; but such a strong condition is not necessary.)

Within the context of this model, we will illustrate the following points (Parker 1965, 1968, 1969):

(1) the creation of particles by the changing metric;
(2) the role of Bogolubov transformations (linear transformation of creation and annihilation operators) in describing this particle creation;
(3) the ambiguity of "the vacuum state" in curved spacetime;
(4) that the connection between spin and statistics follows from the requirement of consistent dynamical evolution of the field in curved spacetime;
(5) the absence of particle creation for the conformally invariant non-interacting field; and
(6) the vacuum state for the conformally invariant field.

Here is a brief summary of our route through the rest of this chapter. In this section and the next, for simplicity, we will take $m = 0$ and $\xi = 0$ – the massless, minimally coupled case. The methods that we introduce, including the use of Bogolubov transformations to describe the creation of particles by the gravitational field, apply equally well to the case when m and ξ are non-zero. The derivation of the canonical commutation relations of (2.62) that is given in the paragraph following that equation is sufficiently general that it applies to the case of arbitrary mass m and coupling constant ξ, and can even be extended to an inhomogeneous universe. In Section 2.5 on statistics from dynamics, we return explicitly to the case with arbitrary mass and coupling constant. After Section 2.6 on conformal invariance, we obtain the probability distribution of created particles in terms of the Bogolubov coefficients when m and ξ are arbitrary. Then we give a class of dynamical universes, characterized by four parameters, in which the Bogolubov coefficients of the massless, minimally coupled field can be found exactly as explicit functions of the parameters. We show that for an expansion with a sufficiently large redshift, the spectrum of created particles that would be observed at late times is thermal. Finally, we turn to a discussion of de Sitter spacetime and the inflationary universe.

Let us now consider the massless, minimally coupled field, which is governed in the Heisenberg picture by the field equation

$$\Box \phi = 0. \tag{2.52}$$

(The components of gravitational wave perturbations in a Robertson–Walker universe do, in fact, satisfy this equation in an appropriate gauge.) With the metric of (2.50), we have

$$a^{-3}\partial_t(a^3 \partial_t \phi) - a^{-2} \sum_i \partial_i^2 \phi = 0. \tag{2.53}$$

It is convenient to impose periodic boundary conditions in a cube having sides of coordinate length L and coordinate volume $V = L^3$. As in Minkowski spacetime, this is a mathematical device, with L taken to infinity after physical quantities are calculated. Then we can expand the field operator ϕ in the form

$$\phi = \sum_{\vec{k}} \left\{ A_{\vec{k}} f_{\vec{k}}(x) + A_{\vec{k}}^\dagger f_{\vec{k}}^*(x) \right\}, \tag{2.54}$$

where

$$f_{\vec{k}} = V^{-1/2} e^{i\vec{k}\cdot\vec{x}} \psi_k(\tau). \tag{2.55}$$

Here $k^i = 2\pi n^i/L$ with n^i an integer, $k = |\vec{k}|$, and we have defined

$$\tau = \int^t a^{-3}(t') dt'. \tag{2.56}$$

It follows from (2.53) that

$$\frac{d^2 \psi_k}{d\tau^2} + k^2 a^4 \psi_k = 0. \tag{2.57}$$

We impose the initial condition that in the initial flat spacetime with $a = a_1$, which is approached as $t \to -\infty$, we have the Minkowski spacetime field expansion given in (1.51), but with the constant scale factor a_1 taken into account. This gives from (1.54) the condition that as $t \to -\infty$

$$f_{\vec{k}} \sim (V a_1^3)^{-1/2} (2\omega_{1k})^{-1/2} \exp[i(\vec{k} \cdot \vec{x} - \omega_{1k} t)] \tag{2.58}$$

2.3 Cosmological model: Arbitrary asymptotically static time dependence

with $\omega_{1k} = k/a_1$. In the initial Minkowski spacetime the metric is that of (2.50) with $a = a_1$. The coordinates can be rescaled with $x^i \to x'^i = a_1 x^i$, so that we have the usual Minkowski metric, and x'^i is the physical or measured distance. The appropriate rescaled physical momentum is then $k'^i = k^i/a_1$, and the physical energy of a particle is $|\vec{k}'| = k/a_1 = \omega_{1k}$. It is the physical volume $V a_1^3$ and energy ω_{1k} which must appear in (2.58). Comparison of (2.55) with (2.58) gives an asymptotic condition on ψ_k as $t \to -\infty$,

$$\psi_k(\tau) \sim (2a_1{}^3 \omega_{1k})^{-1/2} \exp(-i\omega_{1k} a_1{}^3 \tau), \tag{2.59}$$

where we have used (2.56) to replace t by $a_1^3 \tau$ + constant, and have absorbed the constant phase factor into the definition of $A_{\vec{k}}$ (or into the choice of the time origin).

From (2.58), it follows that with the scalar product of (2.48) we have, as in (1.55),

$$(f_{\vec{k}}, f_{\vec{k}'}) = \delta_{\vec{k}, \vec{k}'}, \quad (f_{\vec{k}}, f_{\vec{k}'}^*) = 0. \tag{2.60}$$

Because the scalar product is conserved, these relations must be satisfied at all times by the $f_{\vec{k}}$ of (2.55).

Similarly, the Minkowski spacetime quantization in the initial flat spacetime implies that the operator $A_{\vec{k}}$ of (2.54) satisfy

$$\left[A_{\vec{k}}, A_{\vec{k}'}^\dagger\right] = \delta_{\vec{k}, \vec{k}'}, \quad [A_{\vec{k}}, A_{\vec{k}'}] = 0, \tag{2.61}$$

and that $A_{\vec{k}}$ annihilates particles of momentum $\vec{k}/a_1$ and energy ω_{1k} in the initial Minkowski spacetime. The operators $A_{\vec{k}}$ are time-independent, and as a consequence (2.61) is valid at all times. From (2.60) and (2.61), it follows that the canonical commutation relations hold,

$$[\phi(\vec{x},t), \phi(\vec{x}',t)] = 0, \quad [\pi(\vec{x},t), \pi(\vec{x}',t)] = 0,$$

$$[\phi(\vec{x},t), \pi(\vec{x}',t)] = i\delta(\vec{x} - \vec{x}'), \tag{2.62}$$

with π given from (2.15) and (2.50) as

$$\pi = a^3 \partial_t \phi = \partial_\tau \phi. \tag{2.63}$$

The proof that (2.62) hold is as follows. From (2.54) and (2.61) we find

$$[\phi(\vec{x},t), \pi(\vec{x}',t)] = a^3(t) \sum_{\vec{k}} \left\{ f_{\vec{k}}(\vec{x},t) \partial_t f_{\vec{k}}^*(\vec{x}',t) - f_{\vec{k}}^*(\vec{x},t) \partial_t f_{\vec{k}}(\vec{x}',t) \right\}.$$

An arbitrary solution of $\Box h = 0$ can be expanded in the form

$$h(\vec{x},t) = \sum_{\vec{k}} \left\{ f_{\vec{k}}(\vec{k},t)(f_{\vec{k}},h) - f^*_{\vec{k}}(\vec{x},t)(f^*_{\vec{k}},h) \right\}$$

$$= -i \int d^3x' \, a^3(t) \sum_{\vec{k}} \left\{ f_{\vec{k}}(\vec{x},t)\partial_t f^*_{\vec{k}}(\vec{x}',t) - f^*_{\vec{k}}(\vec{x},t)\partial_t f_{\vec{k}}(\vec{x}',t) \right\} h(\vec{x}',t)$$

$$+ i \int d^3x' a^3(t) \sum_{\vec{k}} \{f_{\vec{k}}(\vec{x},t) f^*_{\vec{k}}(\vec{x}',t) - f^*_{\vec{k}}(\vec{x},t) f_{\vec{k}}(\vec{x}',t)\} \partial_t h(\vec{x}',t),$$

where the $a^3(t)$ comes from $|g|^{1/2}$ in the scalar product of (2.48). Since h is an arbitrary solution of this second-order equation, it follows that

$$a^3(t) \sum_{\vec{k}} \left\{ f_{\vec{k}}(\vec{x},t)\partial_t f^*_{\vec{k}}(\vec{x}',t) - f^*_{\vec{k}}(\vec{x},t)\partial_t f_{\vec{k}}(\vec{x}',t) \right\} = i\delta(\vec{x} - \vec{x}'), \qquad (2.64)$$

where $\int d^3x' \delta(\vec{x}-\vec{x}') h(\vec{x}',t) = h(\vec{x},t)$, and that

$$\sum_{\vec{k}} \left\{ f_{\vec{k}}(\vec{x},t) f^*_{\vec{k}}(\vec{x}',t) - f^*_{\vec{k}}(\vec{x},t) f_{\vec{k}}(\vec{x}',t) \right\} = 0. \qquad (2.65)$$

The canonical commutation relation of ϕ and π follows from (2.64), and that of ϕ with ϕ from (2.65). To obtain the equal time commutator of $\pi(\vec{x},t)$ with $\pi(\vec{x}',t)$, we take the time derivative of the previous expansion of $h(\vec{x},t)$, recalling that the scalar products are conserved. Then

$$\partial_t h(\vec{x},t) = \sum_{\vec{k}} \left\{ \partial_t f_{\vec{k}}(\vec{x},t)(f_{\vec{k}},h) - \partial_t f^*_{\vec{k}}(\vec{x},t)(f^*_{\vec{k}},h) \right\}.$$

Expanding the scalar products as before, and using (2.64), we obtain

$$\sum_{\vec{k}} \left\{ \left(\partial_t f^*_{\vec{k}}(\vec{x},t)\right) \partial_t f_{\vec{k}}(\vec{x}',t) - \left(\partial_t f^*_{\vec{k}}(\vec{x}',t)\right) \partial_t f_{\vec{k}}(\vec{x},t) \right\} = 0, \qquad (2.66)$$

from which we find $[\pi(\vec{x},t), \pi(\vec{x}',t)] = 0$.

Alternatively, we could have used the explicit form of $f_{\vec{k}}$ in (2.55) and the conserved Wronskian of (2.57) to show that the canonical commutators hold, but we used the present approach because it is easily extended to more general spacetimes.

Note that for the results of the present section, both asymptotically flat regions were not required. It would be sufficient, for example, to suppose that at some time in the distant future the universe becomes asymptotically flat, although it may have never been flat at earlier times. Then the canonical commutation relations would have to hold when $a(t)$ is changing rapidly if they are to hold in the far future. *Causality then implies that the canonical commutation relations must hold even if the universe never becomes asymptotically flat.*

Thus, we see that the canonical commutation relations hold in this curved spacetime with any $a(t)$ as a consequence of their holding in Minkowski spacetime.

2.4 Particle creation in a dynamic universe

We have seen that for an asymptotically static $a(t)$ satisfying (2.51), the function $\psi_k(\tau)$ defined in (2.55) satisfies (2.57) with the asymptotic condition of (2.59). This means that in the initial Minkowski spacetime, $f_{\vec{k}}$ is a positive frequency solution of the field equation (2.52), and that $A_{\vec{k}}$ of (2.54) annihilates a particle at early times. Suppose now that the state vector describing the system in the Heisenberg picture is such that no particles are present at early times. Denoting this state vector by $|0\rangle$, we have

$$A_{\vec{k}}|0\rangle = 0, \quad \text{for all } \vec{k}. \tag{2.67}$$

The time development of $\psi_{\vec{k}}(\tau)$ is governed by the ordinary second-order differential equation (2.57). This has two linearly independent solutions $\psi_k^{(\pm)}(\tau)$ such that as $t \to +\infty$

$$\psi_k^{(\pm)} \sim (2a_2{}^3 \omega_{2k})^{-1/2} \exp(\mp i \omega_{2k} a_2{}^3 \tau),$$

where $\omega_{2k} \equiv k/a_2$. Therefore, the particular solution $\psi_k(\tau)$ of (2.57) with early time behavior given by (2.59) must have late time behavior such that as $t \to +\infty$

$$\psi_k(\tau) = \alpha_k \psi_k^{(+)}(\tau) + \beta_k \psi_k^{(-)}(\tau)$$
$$\psi_k(\tau) \sim (2a_2{}^3 \omega_{2k})^{-1/2} \left[\alpha_k e^{-i a_2{}^3 \omega_{2k} \tau} + \beta_k e^{i a_2{}^3 \omega_{2k} \tau} \right], \tag{2.68}$$

where α_k and β_k are complex constants, with values depending on the detailed form of $a(t)$. The Wronskian of (2.57) gives the conserved quantity

$$\psi_k \partial_\tau \psi_k^* - \psi_k^* \partial_\tau \psi_k = i, \tag{2.69}$$

where the right-hand side is determined by the early time asymptotic form of (2.59). Then the asymptotic form at late times of (2.69) requires that

$$|\alpha_k|^2 - |\beta_k|^2 = 1. \tag{2.70}$$

From (2.55) and (2.68), we find that at late times

$$f_{\vec{k}} \sim (V a_2{}^3)^{-1/2} (2\omega_{2k})^{-1/2} e^{i \vec{k} \cdot \vec{x}} \left[\alpha_k e^{-i \omega_{2k} t} + \beta_k e^{i \omega_{2k} t} \right], \tag{2.71}$$

where we have used $a_2{}^3 \tau \sim t +$ constant at late t and have absorbed the constant phase factors into the constants α_k and β_k. Then, the asymptotic form at late times of ϕ, after regrouping (2.54) according to late time positive and negative frequency parts, shows that $\phi(x)$ can be written as

$$\phi(x) = \sum_{\vec{k}} \left\{ a_{\vec{k}} g_{\vec{k}}(x) + a_{\vec{k}}^\dagger g_{\vec{k}}^*(x) \right\}, \qquad (2.72)$$

with $g_{\vec{k}}(x)$ being a solution of the field equation which is positive frequency at late times,

$$g_{\vec{k}}(x) \sim (V a_2{}^3)^{-1/2} (2\omega_{2k})^{-1/2} \exp[i(\vec{k} \cdot \vec{x} - \omega_{2k} t)], \qquad (2.73)$$

and with

$$a_{\vec{k}} = \alpha_k A_{\vec{k}} + \beta_k^* A^\dagger_{-\vec{k}}. \qquad (2.74)$$

Clearly, the $a_{\vec{k}}$ can be interpreted as annihilation operators for particles of momentum $\vec{k}/a_2$ and energy $\omega_{2k} = k/a_2$ at late times. This interpretation is consistent, since (2.74) yields

$$[a_{\vec{k}}, a_{\vec{k}'}^\dagger] = \delta_{\vec{k},\vec{k}'}(|\alpha_k|^2 - |\beta_k|^2) = \delta_{\vec{k},\vec{k}'} \qquad (2.75)$$

by virtue of (2.70) and the canonical commutation relations (2.61) satisfied by the $A_{\vec{k}}$. Transformation of annihilation and creation operators such as that in (2.74) appears also in condensed matter physics and are known as Bogolubov transformations.

Using the $a_{\vec{k}}$ and the state vector $|0\rangle$ we can calculate the expectation value of number of particles present at late times in mode $\vec{k}$:

$$\langle N_{\vec{k}} \rangle_{t \to \infty} = \langle 0 | a_{\vec{k}}^\dagger a_{\vec{k}} | 0 \rangle = |\beta_k|^2. \qquad (2.76)$$

On the other hand, at early times

$$\langle N_{\vec{k}} \rangle_{t \to -\infty} = \langle 0 | A_{\vec{k}}^\dagger A_{\vec{k}} | 0 \rangle = 0. \qquad (2.77)$$

Thus, if $a(t)$ is such that $|\beta_k|^2$ is non-zero, as is generally the case, particles are created by the changing scale factor of the universe. A particular example, in which an exact solution is possible, is given in Section 2.8.

The above discussion is readily extended to the case when the mass m and coupling constant ξ are non-zero (Parker 1965, 1968, 1969, 1971). The main difference is that in the asymptotic expressions for $\psi_{\vec{k}}$, $f_{\vec{k}}$, and $g_{\vec{k}}$ at early and late times, $\omega_{1k} = \sqrt{(k/a_1)^2 + m^2}$ and $\omega_{2k} = \sqrt{(k/a_2)^2 + m^2}$. With that change, the equations obtained in this section remain valid.

Instead of introducing the coordinate τ in (2.56), when the mass or coupling constant is non-zero it is physically more natural to work with the proper time t that appears in the original metric and with the original form of the field equation. Another convenient coordinate that is often used is the conformal time η, defined by requiring that $dt^2/a(t)^2 = d\eta^2$. (The cosmological metric under consideration is manifestly conformally flat in terms of η time.) We have introduced the coordinate τ in the case of the minimally coupled massless field because of its convenience later when we consider that field in (2.57) with a particular class of function $a(t)$ for which we obtain an exact analytic solution of the field equation. This explicit solution clearly demonstrates that the particle number associated with such a field changes as a result of the time dependence of $a(t)$. Furthermore, we use this exact solution to demonstrate, in Section 2.9, some interesting properties of the probability distribution and spectrum of created particles.

The following considerations follow from the expressions we have obtained and also apply to the case with non-zero values of m and ξ. Particles are created, rather than annihilated, regardless of the relation between a_1 and a_2. This occurs, despite the time reversal invariance of the field equation, because we have chosen the state vector such that no particles are present at early times. In the time-reversed situation, in which particles are annihilated so that none are present at late times, we would have to take the state vector to be one in which initially there are correlated pairs of particles present. Such an initial state appears unnatural in a physical context because of the correlations required.

During a rapid change of $a(t)$ in which particle creation is occurring, the particle number of the system is not operationally well defined (Parker 1965, 1969). Suppose that we try to measure the particle number in a comoving volume (one bounded by geodesics of the spacetime), and that the measurement process takes a time interval Δt. If Δt is very small, a significant number of particles will be created by the measurement process because of the time-energy uncertainty relation. But if Δt is large, then a significant number of particles will be created by the change of $a(t)$ during the time interval of the measurement. There is no value of Δt for which the minimum uncertainty in the measured particle number is 0. The irreducible imprecision in the measured particle number will become large during a process of rapid particle creation. This uncertainty is reflected in the theory by the absence of an unambiguous or unique definition of a positive frequency solution corresponding to physical particles during a period when $a(t)$ is changing. This ambiguity of the particle interpretation of quantum field theory naturally carries over to more general non-static curved spacetimes as well as to spacetimes with event horizons. The lack of a unique particle interpretation means that in a general curved spacetime, in contrast to Minkowski spacetime, there is no physically unambiguous unique Heisenberg state vector which can be identified as "the vacuum state." A particular case was our cosmological example; although there were no non-gravitational interactions present, the state vector containing no particles at early times was different from the state vector containing no particles at late times. This was the original context in which the irreducible ambiguity of the vacuum state of a quantized field in curved spacetime was demonstrated (Parker 1965, 1969). It was also shown in this context that the early time and late time vacuum states are orthogonal to one another, thus giving unitary inequivalent representations of the commutation relations in curved spacetime (Parker 1969). In addition, it was shown that there exists a well-defined vacuum state (i.e., the conformal vacuum state) for the free conformally coupled massless scalar field (see Section 2.6) and for the free massless Dirac and two-component neutrino fields (Parker 1965, 1969, 1971).

2.5 Statistics from dynamics: Spin-0

In curved spacetime, a connection between statistics and dynamics appears that is not present in Minkowski spacetime. The statistics is determined by the algebra of the creation operators: commuting operators give Bose–Einstein statistics and anticommuting operators give Fermi–Dirac statistics. We will show that for the spin-0 field only Bose–Einstein statistics is consistent with the curved spacetime dynamics.

Consider a spin-0 field with $\mathcal{L}$ given by (2.35). The field equation is

$$\Box\phi + (m^2 + \xi R)\phi = 0. \tag{2.78}$$

In the cosmological spacetime of (2.50), the scalar curvature is

$$R = 6[(\dot{a}/a)^2 + (\ddot{a}/a)] \tag{2.79}$$

and the field equation is

$$a^{-3}\partial_t(a^3\partial_t\phi) - a^{-2}\sum_i \partial_i^2\phi + (m^2 + \xi R)\phi = 0. \tag{2.80}$$

Suppose that $a(t)$ satisfies (2.51); for simplicity suppose that $a(t)$ is actually constant at early times and again at late times. (There is also no need for a_1 to differ from a_2.)

As in Section 2.3, we can write

$$\phi = \sum_{\vec{k}} \left\{ A_{\vec{k}} f_{\vec{k}}(x) + A_{\vec{k}}^\dagger f_{\vec{k}}^*(x) \right\}$$

where $f_{\vec{k}}(x)$ reduces to a positive frequency Minkowski spacetime solution at early times,

$$f_{\vec{k}} \sim (Va_1^3)^{-1/2}(2\omega_{1k})^{-1/2} \exp[i(\vec{k}\cdot\vec{x} - \omega_{1k}t)]$$

with $\omega_{1k} = [(k/a_1)^2 + m^2]^{1/2}$. One can also write, as in Section 2.4,

$$\phi = \sum_{\vec{k}} \left\{ a_{\vec{k}} g_{\vec{k}}(x) + a_{\vec{k}}^\dagger g_{\vec{k}}^*(x) \right\},$$

where $g_{\vec{k}}(x)$ reduces to a positive frequency Minkowski solution at late times,

$$g_{\vec{k}} \sim (Va_2^3)^{-1/2}(2\omega_{2k})^{-1/2} \exp[i(\vec{k}\cdot\vec{x} - \omega_{2k}t)],$$

with

$$\omega_{2k} = [(k/a_2)^2 + m^2]^{1/2}.$$

2.5 Statistics from dynamics: Spin-0

The field equation implies that at late times the asymptotic form of $f_{\vec{k}}$ is given by (2.71), so that the late time annihilation operator $a_{\vec{k}}$ is related to the early time annihilation operator $A_{\vec{k}}$ by a Bogolubov transformation,

$$a_{\vec{k}} = \alpha_k A_{\vec{k}} + \beta_k^* A^\dagger_{-\vec{k}}. \qquad (2.81)$$

Because only k^2 enters into the field equation, α_k and β_k depend only on $k = |\vec{k}|$. Furthermore, the conserved scalar product $(f_{\vec{k}}, f_{\vec{k}}) = 1$, calculated at late times using (2.71), implies as before that

$$|\alpha_k|^2 - |\beta_k|^2 = 1. \qquad (2.82)$$

Note that this relation is purely a consequence of the field equation.

Now consider the possible commutation rules

$$[A_{\vec{k}}, A_{\vec{k}'}]_\pm = 0, \quad [A^\dagger_{\vec{k}}, A^\dagger_{\vec{k}'}]_\pm = 0, \qquad (2.83)$$

$$[A_{\vec{k}}, A^\dagger_{\vec{k}'}]_\pm = \delta_{\vec{k}, \vec{k}'}, \qquad (2.84)$$

where the $+$ sign denotes the anticommutator corresponding to Fermi–Dirac statistics and the $-$ sign denotes the commutator corresponding to Bose–Einstein statistics. We assume that, if possible, the particles at late times should obey the same statistics as at early times. As a consequence of (2.81), (2.83), and (2.84),

$$[a_{\vec{k}}, a_{\vec{k}'}]_\pm = \alpha_k \beta^*_{k'} [A_{\vec{k}}, A^\dagger_{-\vec{k}'}]_\pm + \alpha_{k'} \beta^*_k [A^\dagger_{-\vec{k}}, A_{\vec{k}'}]_\pm$$
$$= (\alpha_k \beta^*_k \pm \alpha_k \beta^*_k) \delta_{\vec{k}, -\vec{k}'}, \qquad (2.85)$$

and similarly

$$[a_{\vec{k}}, a^\dagger_{\vec{k}'}]_\pm = (|\alpha_k|^2 \pm |\beta_k|^2) \delta_{\vec{k}, \vec{k}'}. \qquad (2.86)$$

If β_k is not zero, then the particles at late times will obey the same statistics as those at early times only in the case of Bose–Einstein statistics. In that case, (2.82) can be used on the right side of (2.86), and the right side of (2.85) vanishes identically. Since $a(t)$ can vary arbitrarily between early and late times, and in general β_k is not zero, only Bose–Einstein statistics is consistent with the dynamics of the spin-0 field.

This is a truly curved-spacetime derivation of the relation between spin and statistics, because if $a(t)$ remains constant so that we do not leave Minkowski spacetime then $\beta_k = 0$, and this connection between dynamics and statistics is absent. In the same way, we find that for a spin-1/2 field only Fermi–Dirac statistics is consistent with curved spacetime dynamics (Parker 1965, 1971). The same method of proving the spin-statistics theorem has been extended to higher spin fields (Parker and Wang 1989) and to ghost fields (Higuchi et al., 1990), and has been used to rule out the altered commutation relations associated with certain types of generalized statistics (Goodison and Toms 1993).

2.6 Conformally invariant non-interacting field

In Section 2.2, we showed that the massless scalar field with

$$\mathcal{L} = \frac{1}{2}|g|^{1/2}\left(g^{\mu\nu}\partial_\mu\phi\partial_\nu\phi - \frac{1}{6}R\phi^2\right) \tag{2.87}$$

is invariant (to within a perfect divergence) under a curved spacetime conformal transformation $g_{\mu\nu} \to \tilde{g}_{\mu\nu}, \phi \to \tilde{\phi}$ with

$$\tilde{g}_{\mu\nu}(x) = \Omega^2(x)g_{\mu\nu}(x), \tag{2.88}$$
$$\tilde{\phi}(x) = \Omega^{-1}(x)\phi(x). \tag{2.89}$$

(We are working in four-dimensional spacetime.) In the Lagrangian of (2.87) it turns out, as we will see in detail in a Chapter 3, that no counterterms involving ϕ are required for renormalization, although counterterms involving the metric will be necessary. Therefore, the field equations governing ϕ and $\tilde{\phi}$ have the same form in the classical and quantized theory. (The same is true for the non-interacting field with arbitrary m and ξ.) The conformal invariance of the action $S = \int d^4x \mathcal{L}$ implies that

$$\frac{\delta S}{\delta \varphi} = \frac{\delta \tilde{S}}{\delta \varphi} = \frac{\delta \tilde{S}}{\delta \tilde{\varphi}} \Omega^{-1},$$

where $\tilde{S} = \int d^4x \tilde{\mathcal{L}}$ with $\tilde{\mathcal{L}}$ obtained from $\mathcal{L}$ by the above conformal transformation. Using $|\tilde{g}|^{1/2} = \Omega^4 |g|^{1/2}$, this gives

$$\left(\Box + \frac{1}{6}R\right)\phi = \Omega^3 \left(\tilde{\Box} + \frac{1}{6}\tilde{R}\right)\tilde{\phi}.$$

Hence, if $\tilde{\phi}$ is a solution of the conformally transformed field equation

$$\left(\tilde{\Box} + \frac{1}{6}\tilde{R}\right)\tilde{\phi} = 0 \tag{2.90}$$

then $\phi = \Omega\tilde{\phi}$ is a solution of the original field equation

$$\left(\Box + \frac{1}{6}R\right)\phi = 0. \tag{2.91}$$

Similar considerations show that the Einstein gravitational action that is proportional to the Ricci scalar curvature R is not conformally invariant. On the other hand, the contracted square of the Weyl tensor is invariant under conformal transformation. In Adler (1982), it was suggested that at very high energies the gravitational action involves the conformally invariant contracted square of the Weyl tensor, and that at lower energies a mechanism such as symmetry breaking induces an Einstein term in the action proportional to R which dominates at low energies, giving rise to Einstein–Newton gravitation. This idea of induced gravity is attractive, as it holds out the hope of relating the Newtonian gravitational constant to more fundamental physics (see also Sakharov 1967).

2.6 Conformally invariant non-interacting field

Consider again the metric of (2.50) with $a(t)$ satisfying the asymptotic conditions of (2.51). Define a new time coordinate

$$\eta = \int^t a^{-1}(t')dt', \tag{2.92}$$

in terms of which the line element is

$$ds^2 = a^2(t)(d\eta^2 - dx^2 - dy^2 - dz^2). \tag{2.93}$$

The conformal transformation of (2.88) and (2.89) with $\Omega = a^{-1}(t)$ takes this $g_{\mu\nu}$ into the Minkowski metric $\tilde{g}_{\mu\nu} = \eta_{\mu\nu}$. Then $\tilde{R} = 0$, and (2.90) is the wave equation in Minkowski spacetime, with independent solutions $\tilde{f}_{\vec{k}}$ and $\tilde{f}_{\vec{k}}^*$, where

$$\tilde{f}_{\vec{k}}(x) = V^{-1/2}(2k)^{-1/2} \exp[i(\vec{k} \cdot \vec{x} - k\eta)].$$

Independent solutions of the original field equation (2.91) are given by $f_{\vec{k}}$ and $f_{\vec{k}}^*$, where

$$f_{\vec{k}}(x) = a^{-1}(t)\tilde{f}_{\vec{k}}(x) = (2Va^3(t)\omega_k(t))^{-1/2} \exp\left[i\left(\vec{k} \cdot \vec{x} - \int^t \omega_k(t')dt'\right)\right]. \tag{2.94}$$

We have defined $\omega_k(t) \equiv k/a(t)$ and have written $k\eta = \int^t \omega_k(t')dt'$.

Since the $f_{\vec{k}}(x)$ are clearly positive frequency solutions at early times, in the expansion

$$\phi = \sum_{\vec{k}} \left\{ A_{\vec{k}} f_{\vec{k}}(x) + A_{\vec{k}}^\dagger f_{\vec{k}}^*(x) \right\}$$

the $A_{\vec{k}}$ annihilate particles at early times. However, the solution (2.94) is also positive frequency at late times, so that the $A_{\vec{k}}$ annihilate particles at late times. Therefore, we have $a_{\vec{k}} = A_{\vec{k}}$ and $\beta_k = 0$, so the early and late time vacua are the same, and no particles of the massless, conformally invariant free field are created by the change in $a(t)$ (Parker 1965, 1968, 1969). This is true regardless of the manner in which $a(t)$ changes smoothly from a_1 at early times to a_2 at late times. Therefore, in this case $A_{\vec{k}}$ can be interpreted at any time as annihilating a particle having positive energy wave function $f_{\vec{k}}(x)$ of (2.94). For this field, the particle concept is well defined at all times, and there is a unique vacuum state $|0\rangle$ defined by

$$A_{\vec{k}}|0\rangle = 0 \quad \text{for all } \vec{k}.$$

This state is called the conformal vacuum state.

This result also holds for Robertson–Walker (isotropic, homogeneous) universes which are not spatially flat (since all Robertson–Walker universes are conformally flat), and for massless fields of higher spin which obey conformally invariant free field equations. These include massless neutrinos and photons. However, gravitons regarded as linearized gravitational wave perturbations of a background Robertson–Walker metric in the Einstein equations do not satisfy a conformally

2.7 Probability distribution of created particles

Consider any case in which particles are created such that the late time creation and annihilation operators $a_{\vec{k}}$ are related to the early time operators $A_{\vec{k}}$ by a relation of the form

$$a_{\vec{k}} = \alpha_k A_{\vec{k}} + \beta_k^* A^\dagger_{-\vec{k}}. \qquad (2.95)$$

We will suppose that the boson commutation relations hold, and that

$$|\alpha_k|^2 - |\beta_k|^2 = 1. \qquad (2.96)$$

Let us calculate the probability distribution of the particles created, assuming that no particles are present at early times. Thus, the state vector is $|0\rangle$, defined by

$$A_{\vec{k}}|0\rangle = 0 \quad \text{for all } \vec{k}.$$

Let $|0)$ denote the late time vacuum, defined by

$$a_{\vec{k}}|0) = 0 \quad \text{for all } \vec{k}. \qquad (2.97)$$

The late time Fock space is constructed by acting on $|0)$ with the $a^\dagger_{\vec{k}}$ operators, as in (1.61).

The probability amplitude for finding the state $|n(\vec{k}), n(-\vec{k}))$, with n particles in mode $\vec{k}$ and n in mode $-\vec{k}$, is

$$(n(\vec{k}), n(-\vec{k})|0\rangle = (n!)^{-1}(0|\,(a_{-\vec{k}})^n (a_{\vec{k}}^n)|0\rangle.$$

From (2.95) we have

$$a_{\vec{k}}|0\rangle = \beta_k^* A^\dagger_{-\vec{k}}|0\rangle = \beta_k^*(\alpha_k^*)^{-1} a^\dagger_{-\vec{k}}|0\rangle,$$

(By (2.96), α_k is non-zero.)

$$(n(\vec{k}), n(-\vec{k})|0\rangle = (n!)^{-1}(\beta_k^*/\alpha_k^*)^n (0|(a_{-\vec{k}})^n (a^\dagger_{-\vec{k}})^n|0\rangle$$
$$= (n!)^{-1/2}(\beta_k^*/\alpha_k^*)^n (n(-\vec{k})|(a^\dagger_{-\vec{k}})^n|0\rangle,$$

---

[1] In a R–W universe, as shown by Lifshitz (1946), in the Lifshitz gauge, each of the two components of the linearized gravitational wave satisfies exactly the same equation as the minimally coupled scalar field. As a result, gravitons of each polarization are created in the same way as the minimally coupled scalar particles we already discussed (Grishchuk 1974; Ford and Parker 1977b). We note that in Parker (1968, 1969), the "gravitons" referred to were clearly spin-2 massless particles satisfying the conformally invariant free spin-2 equation, not the physically relevant Einstein gravitons that are created exactly like the minimally coupled massless scalar particles.

2.7 Probability distribution of created particles

where we have used the adjoint of the relation

$$|n(-\vec{k})\rangle = (n!)^{-1/2}(a^\dagger_{-\vec{k}})^n|0\rangle.$$

Now repeated use of the adjoint of $a_{-\vec{k}}|n(-\vec{k})\rangle = n^{1/2}|(n-1)(-\vec{k})\rangle$ yields

$$(n(\vec{k}), n(-\vec{k})|0) = (\beta^*_k/\alpha^*_k)^n (0|0). \tag{2.98}$$

By the same method, we find that amplitudes of the form

$$(m(\vec{k}), n(-\vec{k})|0\rangle$$

with $m \neq n$ are zero because the resulting expression has a factor $\langle 0|a^\dagger_{\pm\vec{k}}$ which vanishes.

Thus, the particles are created in pairs with equal and opposite momenta. Because creation and annihilation operators in different modes commute, it follows that the most general late time state having a non-zero matrix element with $|0\rangle$ is a tensor product of states $|n(\vec{k}), n(-\vec{k})\rangle$ over different modes $\vec{k}$. For brevity, let $|\{n_j(\vec{k}_j)\}\rangle$ denote such a state, containing n_1 pairs, each with one particle in mode $\vec{k}_1$ and the other particle in mode $-\vec{k}_1$, n_2 pairs in modes $(\vec{k}_2, -\vec{k}_2)$, etc. Then, as for (2.98) we find

$$(\{n_j(\vec{k}_j)\}|0\rangle = \prod_j \left(\beta^*_{k_j}/\alpha^*_{k_j}\right)^{n_j} (0|0). \tag{2.99}$$

These give all the non-vanishing matrix elements between the late time Fock basis and the state $|0\rangle$, so

$$|0\rangle = \sum_{\{n_j(\vec{k}_j)\}} |\{n_j(\vec{k}_j)\}\rangle(\{n_j(\vec{k}_j)\}|0\rangle,$$

where the sum is over all possible sets $\{n_j(\vec{k}_j)\}$. Then the normalization condition $\langle 0|0\rangle = 1$ gives $|\langle 0|0\rangle|$:

$$1 = \sum_{\{n_j(\vec{k}_j)\}} \left|(\{n_j(\vec{k}_j)\}|0\rangle\right|^2$$

$$= \left(\sum_{\{n_j(\vec{k}_j)\}} \prod_j |\beta_{k_j}/\alpha_{k_j}|^{2n_j}\right) |\langle 0|0\rangle|^2.$$

Now $\sum_{\{n_j\}} \prod_j x^{n_j}$, with the sum being over all sets of integers $\{n_j\}$, is equal to $\prod_j \left(\sum_{n_j=0}^{\infty} x^{n_j}\right)$, assuming convergence of sums and products. (This is readily checked in the case when j and n_j have finite ranges.) Hence

$$1 = |\langle 0|0\rangle|^2 \prod_j \sum_{n_j=0}^{\infty} |\beta_{k_j}/\alpha_{k_j}|^{2n_j}$$

$$= |\langle 0|0\rangle|^2 \prod_j \left(1 - |\beta_{k_j}/\alpha_{k_j}|^2\right)^{-1}$$

$$= |\langle 0|0\rangle|^2 \prod_j |\alpha_{k_j}|^2,$$

where we have used (2.96). Finally, we have

$$|\langle 0|0\rangle|^2 = \prod_j |\alpha_{k_j}|^{-2}, \qquad (2.100)$$

and from (2.99) we obtain the probability

$$\left|\langle \{n_j(\vec{k}_j)\}|0\rangle\right|^2 = \prod_j \left(|\beta_{k_j}/\alpha_{k_j}|^{2n_j} |\alpha_{k_j}|^{-2}\right). \qquad (2.101)$$

Thus, the production of pairs in different modes are independent events. The probability of observing at late times n of these bosons in mode $\vec{k}$ is

$$P_n(\vec{k}) = |\beta_k/\alpha_k|^{2n} |\alpha_k|^{-2}. \qquad (2.102)$$

(Since the particles occur in pairs there will also be n particles in mode $-\vec{k}$.) The production of particles in a given mode are not independent events. For example, we find that $P_2(\vec{k}) = P_1(\vec{k})^2/P_0(\vec{k}) > P_1(\vec{k})^2$. The probability of observing two particles in mode $\vec{k}$ (i.e., two pairs of particles occupying modes $\vec{k}$ and $-\vec{k}$) is greater than the square of the probability of observing one particle in mode $\vec{k}$ (i.e., one pair of particles occupying modes $\vec{k}$ and $-\vec{k}$).

Using (2.96), we obtain easily

$$\sum_{n=0}^{\infty} P_n(\vec{k}) = 1$$

and

$$\langle N_{\vec{k}}\rangle_{t\to\infty} = \sum_{n=0}^{\infty} n P_n(\vec{k}) = |\beta_k|^2,$$

in agreement with (2.76). This is the average number in mode $\vec{k}$ in the volume $(La_2)^3$. The average particle density summed over all modes in the continuum limit is

$$\langle N\rangle_{t\to\infty} = \lim_{L\to\infty} (La_2)^{-3} \sum_{\vec{k}} |\beta_k|^2$$

$$= (2\pi^2 a_2^3)^{-1} \int_0^{\infty} dk\, k^2 |\beta_k|^2. \qquad (2.103)$$

2.8 Exact solution with particle creation

We now return to the minimally coupled massless spin-0 field discussed in Sections 2.3 and 2.4, in order to obtain an exact solution exhibiting particle production.

The form of the scale factor a will include several free parameters, so that a wide class of changing scale factors may be approximated by the one considered here with appropriately chosen parameters. We will use the notation of Sections 2.3 and 2.4. We seek a solution of (2.57) subject to the asymptotic condition of (2.59).

We take $a(\tau)$ such that

$$a(\tau) = \{a_1^4 + e^\zeta[(a_2^4 - a_1^4)(e^\zeta + 1) + b](e^\zeta + 1)^{-2}\}^{1/4}, \quad (2.104)$$

with

$$\zeta = \tau s^{-1}, \quad (2.105)$$

and a_1, a_2, b, and s being positive parameters. This function approaches a_1 as $\tau \to -\infty$ and a_2 as $\tau \to \infty$, and passes through a maximum if $b > |a_2^4 - a_1^4|$. With $a(\tau)$ given by (2.104) the equation for $\psi_k(\tau)$ is equivalent to the one considered by Epstein (1930), in connection with reflection of radio waves by the ionosphere, and by Eckart (1930), in connection with reflection of electrons by a potential barrier. The spatial coordinate in those works corresponds to time in our discussion. For $b = 0$, $a(\tau)$ reduces to the form

$$a(\tau) = \left\{\frac{a_2^4 + a_1^4}{2} + \frac{a_2^4 - a_1^4}{2}\tanh(\zeta/2)\right\}^{1/4},$$

which changes monotonically from a_1 to a_2 in a τ-time interval of order s. For $a_1 = a_2$, $a(\tau)$ reduces to

$$a(\tau) = \left\{a_1^4 + \frac{b}{4\cosh^2(\zeta/2)}\right\}^{1/4},$$

which approaches a_1 at $\tau \to \pm\infty$ and rises to a maximum in between. We will follow Epstein's discussion, as applied to $a(\tau)$ of (2.104).

With

$$u = \exp(\zeta), \quad (2.106)$$

(2.57) takes the form (with prime denoting d/du)

$$0 = \psi_k'' + u^{-1}\psi_k'$$
$$+ k^2 s^2 \{u^{-2}a_1^4 + u^{-1}(1+u)^{-2}[(a_2^4 - a_1^4)(1+u) + b]\}\psi_k. \quad (2.107)$$

The coefficients in this equation have singularities at $u = 0$ and $u = -1$. The behavior of ψ_k for u near 0 and near -1 is readily found to be $\psi_k \sim u^{\pm iksa_1^2}$ and $\psi_k \sim (1+u)^d$, respectively, with

$$d = \frac{1}{2}[1 - (1 + 4k^2s^2b)^{1/2}]. \quad (2.108)$$

Therefore, we define $f(u)$ by

$$\psi_k = (1+u)^d u^{-c_1} f(u), \qquad (2.109)$$

with

$$c_1 = iksa_1^2. \qquad (2.110)$$

It is also convenient to define

$$c_2 = iksa_2^2. \qquad (2.111)$$

The reason $-c_1$ rather than c_1 or $\pm c_2$ appears in (2.109) is that we seek a solution satisfying the asymptotic condition of (2.59).

Substituting (2.109) into (2.107) gives the hypergeometric differential equation,

$$u(u+1)f'' + [(2d - 2c_1 + 1)u + (1 - 2c_1)]f'$$
$$+ (d - c_1 + c_2)(d - c_1 - c_2)f = 0. \qquad (2.112)$$

The solution is the hypergeometric function $F(d-c_1+c_2, d-c_1-c_2; 1-2c_1; -u)$ in the notation of Abramowitz and Stegun (1972, see #15.5.3 and #15.5.1). Hence, the solution of (2.57) with the desired asymptotic form at early times is

$$\psi_k = N_1 u^{-c_1}(1+u)^d F(d - c_1 + c_2, d - c_1 - c_2; 1 - 2c_1; -u), \qquad (2.113)$$

where $N_1 = (2a_1^3 \omega_{1k})^{-1/2} = a_1^{-1}(2k)^{-1/2}$.

In order to identify α_k and β_k, we must compare the late time asymptotic form of ψ_k with (2.68). For that purpose we use the identity (Abramowitz and Stegun (1972), #15.3.7, p. 559)

$$F(d - c_1 + c_2, d - c_1 - c_2; 1 - 2c_1; -u)$$
$$= (u)^{-d+c_1-c_2} A F(d - c_1 + c_2, d + c_1 + c_2; 1 + 2c_1; -u^{-1})$$
$$+ (u)^{-d+c_1+c_2} B F(d - c_1 - c_2, d + c_1 - c_2; 1 - 2c_1; -u^{-1}),$$

with

$$A = \frac{\Gamma(1-2c_1)\Gamma(-2c_2)}{\Gamma(d-c_1-c_2)\Gamma(1-c_1-c_2-d)},$$
$$B = \frac{\Gamma(1-2c_1)\Gamma(2c_2)}{\Gamma(d-c_1+c_2)\Gamma(1-c_1+c_2-d)}. \qquad (2.114)$$

Since for large u we have $F(k_1, k_2; k_3; -u^{-1}) \sim 1$ and $(1+u)^d \sim u^d$, we then find from (2.113) that as $u \to \infty$,

$$\psi_k \sim N_1 A u^{-c_2} + N_1 B u^{c_2}.$$

To facilitate comparison with the asymptotic form in (2.68), we note that

$$u^{-c_2} = \left(e^{\tau s^{-1}}\right)^{-iksa_2^2} = e^{-ia_2^2 k\tau} = e^{-ia_2^3 \omega_{2k}\tau}$$

and that $(2a_2{}^3\omega_{2k})^{-1/2} = a_2{}^{-1}(2k)^{-1/2}$. Comparison then yields

$$\alpha_k = (a_2/a_1)A,$$

$$\beta_k = (a_2/a_1)B.$$

Because d is real and c_1, c_2 are imaginary, we find that

$$\left|\frac{\beta_k}{\alpha_k}\right|^2 = \frac{\Gamma(d-c_1-c_2)\Gamma(1+c_1+c_2-d)\Gamma(d+c_1+c_2)\Gamma(1-c_1-c_2-d)}{\Gamma(d-c_1+c_2)\Gamma(1+c_1-c_2-d)\Gamma(d+c_1-c_2)\Gamma(1-c_1+c_2-d)}$$

$$= \frac{\sin\pi(d-c_1+c_2)\sin\pi(d+c_1-c_2)}{\sin\pi(d-c_1-c_2)\sin\pi(d+c_1+c_2)},$$

where we used the identity $\Gamma(z)\Gamma(1-z) = \pi/\sin\pi z$. The identities $\sin(\theta_1 - \theta_2)\sin(\theta_1 + \theta_2) = \sin^2\theta_1 - \sin^2\theta_2$, and $\sin(i\theta) = -i\sinh\theta$ now yield

$$\left|\frac{\beta_k}{\alpha_k}\right|^2 = \frac{\sin^2\pi d + \sinh^2[\pi ks(a_1{}^2 - a_2{}^2)]}{\sin^2\pi d + \sinh^2[\pi ks(a_1{}^2 + a_2{}^2)]}, \quad (2.115)$$

where d is defined in (2.108) and we have used (2.110) and (2.111) for c_1 and c_2. Thus, in general β_k is non-zero. Next we will examine the spectrum and probability distribution of the created particles.

2.9 High-frequency blackbody distribution

According to (2.102), the probability of observing at late times n particles in mode $\vec{k}$ is

$$P_n(\vec{k}) = |\beta_k/\alpha_k|^{2n}(1 - |\beta_k/\alpha_k|^2), \quad (2.116)$$

where we have used $|\alpha_k|^{-2} = 1 - |\beta_k/\alpha_k|^2$. For the function a considered in the previous section, $|\beta_k/\alpha_k|^2$ is given by (2.115). Let us examine the high-frequency behavior of this probability distribution when $a_1 \neq a_2$.

Let $a_<$ denote the smaller and $a_>$ the greater of a_1 and a_2, and suppose that k is sufficiently large so that

$$\exp[\pi ks(a_>{}^2 - a_<{}^2)] \gg 1. \quad (2.117)$$

Then the $\sin^2\pi d$ terms in (2.115) can be neglected relative to the sinh terms, which can be approximated by exponentials. Thus,

$$|\beta_k/\alpha_k|^2 \approx \exp[2\pi ks(a_>{}^2 - a_<{}^2)]/\exp[2\pi ks(a_>{}^2 + a_<{}^2)]$$
$$= \exp[-\mu k] \quad (2.118)$$

with

$$\mu \equiv 4\pi s a_<{}^2. \quad (2.119)$$

(Note that $\exp[\mu k]$ itself need not be large; for example $a_>{}^2$ could be much larger than $a_<{}^2$.) This exponential behavior of $|\beta_k/\alpha_k|^2$ for large k appears to be generic to a much wider class of functions for a than we have considered here (see Parker (1977) for more discussion). The rate of change of a is proportional to s^{-1}. From (2.118), it is evident that $|\beta_k/\alpha_k|^2$ and hence $|\beta_k|^2$ approaches zero faster than any power of s^{-1} as s^{-1} approaches zero, that is, it has an essential singularity at $s^{-1}=0$ and cannot be expanded in powers of s^{-1}.

In an expanding universe, for which $a_2 = a_>$, $a_1 = a_<$, with $a_2 \gg a_1$, the low-frequency end of the spectrum is effectively redshifted away so that (2.118) would describe the spectrum at all but extremely low frequencies. The probability distribution given by (2.116) and (2.118) is

$$P_n(\vec{k}) = \exp(-n\mu k)[1 - \exp(-\mu k)]. \qquad (2.120)$$

The average number in mode $\vec{k}$, $\langle N_{\vec{k}} \rangle$, is given by

$$|\beta_k|^2 = \frac{|\beta_k/\alpha_k|^2}{1 - |\beta_k/\alpha_k|^2} = \frac{1}{\exp(\mu k) - 1}, \qquad (2.121)$$

and the average particle density in the limit of infinite volume by

$$\langle \mathcal{N} \rangle = (2\pi^2 a_2^3)^{-1} \int_0^\infty dk\, k^2 [\exp(\mu k) - 1]^{-1}. \qquad (2.122)$$

The energy of one of these massless particles in mode $\vec{k}$ at late times is k/a_2. Therefore, it will be recognized that the probability of observing n particles in mode $\vec{k}$ is the same as for blackbody radiation of temperature

$$T = (k_B\, \mu a_2)^{-1}, \qquad (2.123)$$

where k_B is the Boltzmann constant. The average number in mode k and average number density are one-half that for electromagnetic blackbody radiation because this is scalar radiation with fewer degrees of freedom.

As we have seen in Section 2.7, the state $|0\rangle$ is actually a coherent superposition of states containing pairs of particles at late times, in contrast to blackbody radiation which is an incoherent mixture. However, local observations would be unable to distinguish between this created radiation and strict blackbody radiation because the correlated pairs of particles would have separations of a cosmological scale. In addition, interactions with other systems would tend to destroy the correlations between members of a pair. This thermal creation of particles in an expanding universe is discussed in Parker (1975b, 1976, 1977).

2.10 de Sitter spacetime

The maximally symmetric four-dimensional solution to the vacuum Einstein field equations with a positive cosmological constant is known as de Sitter spacetime. It is well known that the de Sitter metric can be represented in suitable coordinates as an exponentially expanding Robertson–Walker universe; and it is of interest to understand how our previous considerations relate to results based on the group of symmetries of de Sitter spacetime. (See Schrödinger (1956) or Hawking and Ellis (1973) for discussions of de Sitter spacetime.)

We also discuss the inflationary universe and how quantum field theory in curved spacetime acts as the original source of perturbations that are sufficiently large and have the correct spectrum to account for detailed characteristics of the observed CMB radiation. Without these very early quantum fluctuations, the early universe would also be too smooth to have given rise to the large-scale structures such as galaxies that are observed.

Observations of the fluctuations in the nearly isotropic 2.73 K cosmic microwave background radiation are consistent with an early inflationary stage of the expanding universe (Peiris et al., 2003). The early stage of inflation can be approximated by the exponentially expanding form of the de Sitter metric. The early inflation amplifies vacuum fluctuations of a scalar field through the same process as is responsible for particle production by the expanding universe. These quantum fluctuations induce curvature perturbations that eventually set the initial conditions for acoustic oscillations in the plasma present during the hot radiation-dominated stage of the expanding universe that follows the inflationary stage. The acoustic fluctuations leave their mark on the surface of last scattering that corresponds to the time at which the temperature has cooled sufficiently for neutral atoms to form. The radiation that finally propagates to us as the CMB has small variations of about 1 part in 10^5 in its average temperature of about 2.725 K (after the effect of the earth's motion is subtracted). These accurately measured temperature variations carry information about the temperature variations on the surface of last scattering, and provide evidence of the quantum field theory process of particle creation by the expanding universe that we have been discussing in this chapter. The origin of the amplified vacuum fluctuations can be traced to this process, and their spectrum and magnitude can be estimated during the inflationary era by calculating the expectation value of ϕ^2, where ϕ is the quantized part of a scalar field (called the inflaton field) thought to be responsible for causing the early inflationary expansion.[2] In what quantum state should this expectation value be calculated? Our discussion below will first show how the group of isometry transformations in de Sitter spacetime helps to determine a suitable state vector. The relevant state vector is known as the Bunch–Davies vacuum state (Bunch and Davies 1979). As we shall see, in the case when $m = 0$ and $\xi = 1/6$, it reduces to the conformal vacuum state already discussed.

de Sitter spacetime can be represented as a four-dimensional hypersphere embedded in a five-dimensional flat spacetime. It has been extensively investigated both classically and in quantum field theory. (A selection of references

---

[2] As explained in the next section, the inflaton field is written as $\phi_0 + \delta\phi$, where ϕ_0 is the non-zero vacuum expectation value of the field and $\delta\phi$ is the quantized part of the field. $\delta\phi$ has no vacuum expectation value and is treated as a perturbation to linear order. The field $\delta\phi$ is analogous to the field ϕ of the present section.

is de Sitter (1917); Schrödinger (1956); Chernikov and Tagirov (1968); Fulling (1972); Adler (1973); Hawking and Ellis (1973); Candelas and Raine (1975); Dowker and Critchley (1976); Gibbons and Hawking (1977); Bunch and Davies (1979); Birrell and Davies (1982); Vilenkin and Ford (1982); Ford (1985); Allen (1985); Allen and Folacci (1987); Higuchi (1987a,b); and Habib et al. (1999).) In terms of the flat spacetime coordinates, (W, X, Y, Z, T), the invariant line element is

$$ds^2 = dT^2 - dW^2 - dX^2 - dY^2 - dZ^2. \tag{2.124}$$

de Sitter spacetime is the four-dimensional pseudospherical hypersurface satisfying

$$W^2 + X^2 + Y^2 + Z^2 - T^2 = H^{-2}, \tag{2.125}$$

where H^{-1} is the radius of the pseudosphere. The symmetries of this pseudosphere are the rotations and Lorentz transformations in the ten planes formed by pairs of the five coordinates of the flat embedding spacetime. This ten parameter de Sitter symmetry group, $SO(1,4)$, is instrumental in defining a preferred vacuum state, just as the ten parameter Poincaré group is in Minkowski spacetime.

It is well known that half of the de Sitter pseudosphere can be covered by coordinates (x, y, z, t), each running from $-\infty$ to ∞, in which the line element takes the form (see, for example, Hawking and Ellis (1973, pp. 124–131) or Birrell and Davies (1982, p. 130)):

$$ds^2 = dt^2 - e^{2Ht}(dx^2 + dy^2 + dz^2). \tag{2.126}$$

This is the line element of a spatially flat Robertson–Walker universe with scale factor $a(t) = \exp(Ht)$. The constant H is the Hubble constant, $\dot{a}/a$, in this universe, and the scalar curvature is $R = 12H^2$. It is thought that the very early universe may have experienced a stage of exponentially rapid expansion (i.e., inflation) similar to that described by this line element.

With the transformation $\eta = -H^{-1}\exp(-Ht)$, this line element takes the manifestly conformally flat form,

$$ds^2 = (H\eta)^{-2}(d\eta^2 - dx^2 - dy^2 - dz^2), \tag{2.127}$$

with η between $-\infty$ and 0. There is a coordinate singularity at $\eta = 0$. The other half of de Sitter spacetime can be covered by extending η from 0 to ∞. Then $\eta \to -\infty$ and $\eta \to +\infty$ approach the same null hypersurface from opposite sides while $\eta \to 0^-$ and $\eta \to 0^+$ approach hypersurfaces of infinite three-volume at opposite ends of the de Sitter pseudosphere. Because $\eta \to \pm\infty$ approach the same null hypersurface, we will not regard these as asymptotic regions in discussing the vacuum state in de Sitter spacetime.

Instead, we will make use of the de Sitter group invariance and the known high-frequency behavior of the field to determine the preferred vacuum state. We

2.10 de Sitter spacetime

will work in the coordinate system of (2.126) in which t is the proper time along a set of geodesics. Thus, high-frequency behavior is defined invariantly as that measured by observers on these geodesics.

Consider a scalar field ϕ satisfying (2.41) for arbitrary ξ and m. With the metric of (2.126), the field ϕ satisfies

$$\partial_t^2 \phi + 3H \partial_t \phi - e^{-2Ht} \sum_i \partial_i^2 \phi + M^2 \phi = 0, \tag{2.128}$$

where $M^2 \equiv m^2 + 12\xi H^2$. For simplicity, let M^2 be positive. Expand the field as in (2.54) and let

$$f_{\vec{k}} = (2V\, e^{3Ht})^{-1/2} e^{i \vec{k} \cdot \vec{x}} h_k, \tag{2.129}$$

where h_k depends on t alone. Because $a(t) = \exp(Ht)$, we can identify the quantity $V \exp(3Ht)$ as the physical volume of the cube on which periodic boundary conditions were imposed. With the definition

$$v \equiv kH^{-1} \exp(-Ht), \tag{2.130}$$

we find that h_k satisfies Bessel's equation,

$$v^2 \frac{d^2}{dv^2} h_k + v \frac{d}{dv} h_k + (v^2 - \nu^2) h_k = 0. \tag{2.131}$$

Here

$$\nu \equiv (9/4 - M^2/H^2)^{1/2}. \tag{2.132}$$

The sign in (2.130) is such that v increases as t decreases. (One could equally well use as the independent variable $u \equiv -v$, which is negative and increases as t increases. In that case, in (2.135) below, we would replace $H_\nu^{(1)}(v)$ by $H_\nu^{(2)}(u)$ and $H_\nu^{(2)}(v)$ by $H_\nu^{(1)}(u)$. The constants $E(k)$ and $F(k)$ would then differ only by a constant phase from those of (2.135), and the argument leading to (2.136) would be essentially unchanged.)

The physical momentum of a particle with wave number k is $k/a(t) = k \exp(-Ht)$. Therefore v of (2.130) is the ratio of the momentum to the expansion rate H.

2.10.1 Meaning of the Hubble horizon

A worldline along which the coordinates (x, y, z) are constant is a geodesic in de Sitter spacetime. Consider a set of such geodesics. At a particular cosmic time t_1, the proper distance between the geodesic at $\vec{x_1} \equiv (x_1, y_1, z_1)$ and another geodesic at $\vec{x_2} \equiv (x_2, y_2, z_2)$ is $\exp(Ht_1)|\vec{x_1} - \vec{x_2}|$, where $|\vec{x_1} - \vec{x_2}|$ is the coordinate separation between the geodesics. Suppose a spherical outgoing electromagnetic wave originates at event $(\vec{x_1}, t_1)$. It is easy to show that this electromagnetic wave will never reach any geodesic that at time t_1 is at a proper distance greater than

H^{-1} from the geodesic on which the signal originated. Thus, observers on the geodesics at $\vec{x}_1$ and $\vec{x}_2$ cannot communicate via a signal sent at cosmic time t_1 if their proper separation distance at t_1 is greater than H^{-1}. In this sense, the distance H^{-1} is referred to as the (future) Hubble horizon of an observer on such a geodesic in de Sitter spacetime.[3]

It is important to realize that this horizon is a limit on communication with other geodesics in the future, but not in the past. The observer at event $(\vec{x}_1, t_1)$ can receive signals that were sent from any of the other geodesics under consideration at a sufficiently early cosmic time. It is this fact that makes an exponentially expanding inflationary stage of the universe a possible explanation of why distant regions of our universe were able to reach nearly the same temperature by the time of recombination, when the cosmic background radiation originated.

In a spatially flat universe that is radiation- or matter-dominated after reheating, the scale factor $a(t)$ is proportional to $t^{1/2}$ or $t^{2/3}$, respectively. In either of these cases, H^{-1} does not act as a horizon limiting future communication. Therefore, we will refer to H^{-1} in general as the "Hubble radius" or the "Hubble scale," and in the de Sitter case as the "Hubble horizon." In the literature on inflation, "Hubble horizon" is often used synonymously with "Hubble radius" even in the universe after reheating has occurred.

2.10.2 Reasons to study an early inflationary stage of the universe

If the universe were radiation- or matter-dominated without an earlier inflationary stage of its expansion, then an observer at an event $(\vec{x}_1, t_1)$ on a geodesic of constant $\vec{x}_1$ would not be able to receive signals that were sent from arbitrarily distant geodesics because the big-bang singularity would limit how far back in time the past light cone of event $(\vec{x}_1, t_1)$ extended. This limit can be shown to be so stringent that any two regions that we observe in the sky with angular separation of a few degrees of arc would not have been able to communicate by the time of recombination. Consequently, the fact that the CMB radiation that we observe coming from any two such regions is very nearly at the same temperature could not be explained as a consequence of the tendency of interacting thermodynamic systems to approach the same temperature. A sufficient long early period of nearly exponential inflation allows time for thermodynamic equilibrium to have been reached.

However, the statistical mechanics of such regions in thermodynamic equilibrium would not give fluctuations (in the absence of inflation) that were large enough to explain the features that we observe today in the universe. But as we

---

[3] Gibbons and Hawking (1977) showed that a detector on such a geodesic would be excited as though it were bathed by isotropic thermal radiation coming from the geodesic's Hubble horizon. The de Sitter Gibbons–Hawking temperature of this thermal excitation of the detector is $T_{GH} = H/(2\pi k_B)$.

shall see, the quantum amplification of zero-point fluctuations that took place during the early period of inflation explains the statistical properties of the anisotropies observed today in the CMB radiation and in the large-scale structure of the universe.

Early cosmological inflation explains

(1) how separated regions reached thermodynamic equilibrium, and
(2) how quantum field theory in curved spacetime could give perturbations of sufficient magnitude and the right spectrum to lead to the observed fluctuations of the CMB radiation and the observed large-scale structure of the universe.

Taken together, these are strong reasons for us to study, in this section, a quantized field in an exponentially expanding universe, and in the following sections, the inflationary model of the universe and the quantum field theory explanation of perturbations produced by inflation.

Returning to our discussion of a quantized scalar field in the spacetime of (2.126), we can also think of v as the ratio of the proper radius H^{-1} of the de Sitter Hubble horizon to $\lambda/(2\pi)$, where λ is the wavelength measured by these observers. When v is large, the wavelength is small with respect to the radius of the event horizon, or to the radius of curvature of the de Sitter spacetime; or equivalently, the frequency is large with respect to the rate of expansion of the universe. We therefore assume that the positive frequency solutions $f_{\vec{k}}$ have the positive frequency adiabatic form for large v, as in a universe with a slowly changing $a(t)$.

The adiabatic form of $f_{\vec{k}}$ for slowly varying $a(t)$ is the (approximate) WKB solution of the scalar field equation. This WKB solution has the same form as (2.94) for general $a(t)$, and requires that h_k have the asymptotic form:

$$h_k \sim (\omega_k(t))^{-1/2} \exp\left[-i \int^t \omega_k(t')dt'\right], \qquad (2.133)$$

where $\omega_k(t) \sim k/a(t)$ in the limit of large k. (The WKB solution would involve a mass term in $\omega_k(t)$, but here we are considering k sufficiently large that the mass term can be neglected.) In the present case, with $a(t) = \exp(Ht)$, this gives to within a constant phase,

$$h_k \sim (Hv)^{-1/2} \exp(iv) \qquad (2.134)$$

as $k \to \infty$. The most general solution of (2.131) with this asymptotic form is

$$h_k(t) = \sqrt{\pi/2H} \left\{ E(k) H_\nu^{(2)}(v) + F(k) H_\nu^{(1)}(v) \right\}. \qquad (2.135)$$

with

$$E(k) \sim 0, \quad \text{and} \quad F(k) \sim 1, \qquad (2.136)$$

both taken as $k \to \infty$.

We invoke de Sitter invariance to determine $E(k)$ and $F(k)$ for all k (Higuchi 1987b). The ten de Sitter group symmetries (or isometries) of the metric of (2.126) include the obvious three rotations about the spatial axes and three translations along the spatial directions. These symmetries are already satisfied to within a constant phase by the solution given in (2.129) and (2.135). However, the metric is also symmetric under the transformation:

$$t \to t' = t + t_0, \tag{2.137}$$

$$\vec{x} \to \vec{x}' = \exp(-Ht_0)\vec{x}, \tag{2.138}$$

where t_0 is a constant. Define

$$\vec{k}' \equiv \vec{k}\exp(Ht_0), \tag{2.139}$$

so $\vec{k}'/a(t') = \vec{k}/a(t)$ and $\vec{k} \cdot \vec{x} = \vec{k}' \cdot \vec{x}'$. Then we find from (2.130) and (2.135) that

$$h_{k'}(t') = \sqrt{\pi/2H}\left\{E(k')H_\nu^{(2)}(v) + F(k')H_\nu^{(1)}(v)\right\}, \tag{2.140}$$

since v is unchanged when $t \to t'$ and $k \to k'$. Because the physical volume, $V\exp(3Ht)$, that appears in (2.129) is unchanged by this coordinate transformation, form invariance of the single particle wave function $f_{\vec{k}}$ under the transformation of (2.137) and (2.138) requires (to within a phase) that

$$h_{k'}(t') = h_k(t). \tag{2.141}$$

It follows that

$$E(k') = E(k) \quad \text{and} \quad F(k') = F(k). \tag{2.142}$$

Because t_0 is arbitrary, in view of the asymptotic forms in (2.136), we find to within a phase that

$$E(k) = 0 \quad \text{and} \quad F(k) = 1, \tag{2.143}$$

so

$$h_k(t) = \sqrt{\pi/2H}\,H_\nu^{(1)}(v), \tag{2.144}$$

and

$$f_{\vec{k}}(\vec{x},t) = \frac{1}{2}\sqrt{\frac{\pi}{H}}\,V^{-\frac{1}{2}}e^{-\frac{3}{2}Ht}\,H_\nu^{(1)}(kH^{-1}e^{-Ht})\,e^{i\vec{k}\cdot\vec{x}}. \tag{2.145}$$

These mode functions determine the creation and annihilation operators, and hence the vacuum state, of the quantum field theory. This vacuum has been determined by specifying the high frequency asymptotic form of the mode functions and requiring de Sitter invariance. Only seven of the ten local isometry transformations of the de Sitter group have been used to determine this state. It is sometimes referred to as the de Sitter invariant vacuum, and also as the

Bunch–Davies vacuum (Bunch and Davies 1979). It is not de Sitter invariant in the sense defined in Allen (1985) because invariance is required there under a class of space- and time-reflections that take the full de Sitter spacetime into itself. The Bunch–Davies vacuum is probably the most natural state in which to calculate the expectation value of the square of the inflaton field that determines the growth of curvature perturbations during the inflationary stage of the expansion. As alluded to earlier, the result is consistent with observations of the power spectrum of the temperature perturbations of the CMB radiation. The same quantum growth of perturbations during early inflation is also thought to be the original source of perturbations sufficiently large to have ultimately given rise, through gravitational clumping, to the large-scale structure that we observe in the distribution of clusters of galaxies. Observational cosmology is lending credence to the validity of quantum field theory in curved spacetime and to its predictions concerning the very early universe.

Because of the importance of this topic, we will devote the next section to a discussion of inflation, explaining how the dispersion obtained from the two-point function of the scalar inflaton field is related to the spectrum of perturbations that serve as initial conditions for the plasma- and radiation-dominated stage of the expanding universe. The two-point function can be regularized using the methods developed in the next chapter. Among the examples done there is the two-point function of a scalar field in a universe undergoing exponential inflation. As discussed in the next section, there is a question as to whether the dispersion of the inflaton field that determines the spectrum of inflaton perturbations should be taken to be the regularized or unregularized two-point function.

Before closing the present section, we discuss two interesting topics that concern the massless scalar field in de Sitter spacetime. The first concerns the massless, conformally coupled ($\xi = 1/6$) scalar field, and the second concerns the massless, minimally coupled ($\xi = 0$) scalar field.

We first demonstrate that the conformal vacuum state, which was shown to be a preferred state for the conformally coupled scalar field in any FLRW universe (Parker 1965, 1969), coincides with the Bunch–Davies vacuum state in de Sitter spacetime. In the case when $\xi = 1/6$ and $m = 0$, the scalar field equation is conformally invariant, as discussed in Section 2.6. Based on conformal invariance, a preferred vacuum state, known as the conformal vacuum, was determined by the exact mode functions given in (2.94). For the metric of (2.126) we have $a(t) = \exp(Ht)$, and the conformally preferred mode function of (2.94) is, to within a phase,

$$f_{\vec{k}} = (V\, e^{3Ht})^{-\frac{1}{2}} (2k e^{-Ht})^{-\frac{1}{2}} \exp(ikH^{-1}e^{-Ht})\, e^{i\vec{k}\cdot\vec{x}}. \tag{2.146}$$

When $\xi = 1/6$ and $m = 0$, we find that $M^2 = 2H^2$ and $\nu = 1/2$ (see (2.132)). Using

$$H^{(1)}_{\frac{1}{2}}(v) = -i\sqrt{\frac{2}{\pi v}}\, e^{iv}, \tag{2.147}$$

we immediately find that the de Sitter mode functions of (2.145) agree, to within a phase, with the conformal mode functions of (2.146). Thus, the de Sitter spacetime Bunch–Davies vacuum that is defined for general m and ξ satisfying $M^2 > 0$ agrees with the conformal vacuum that is defined for $m = 0$, $\xi = 1/6$, and general scale factor $a(t)$.

In fact, a quick way to derive the Bunch–Davies de Sitter vacuum state is to assume that to within constant phase factors, $E(k)$ and $F(k)$ in (2.135) do not depend on the value of ν, at least for $0 < \nu < 3/2$. This is reasonable because the symmetries of the mode functions under transformations of the de Sitter group should not be affected by small changes in the value of ν. Then appeal to our earlier arguments that the correct vacuum for the massless, conformally coupled field, for which $\nu = 1/2$, is the conformal vacuum. It follows immediately from these two assumptions that $E(k) = 0$ and $F(k) = 1$ for other values of ν, and hence for other values of m and ξ.

Finally, we note that for the massless, minimally coupled scalar field in de Sitter spacetime, it has been argued that the Bunch–Davies vacuum is not physically acceptable because it leads to a two-point function that is infinite as a result of an infrared divergence. This was first argued (before the Bunch–Davies vacuum was defined) by Ford and Parker (1977a) (see discussion after (3.26) of that paper). Then Allen (1985) (see also Allen and Folacci 1987) showed that the same was true for any vacuum having the full set of de Sitter invariances, including time and space reflection invariance on the full de Sitter hypersphere. Looking at (2.132), with $\xi = 0$ and $m = 0$, we see that the Hankel function $H_\nu^{(1)}(v)$ appears in the mode function of (2.144) with $\nu = 3/2$. For minimal coupling with non-zero positive m^2, the value of ν is smaller than $3/2$. As m decreases, it is at exactly the value $\nu = 3/2$ that the Hankel function acquires a higher power of v in its denominator, thus resulting in the infrared singularity in the two-point function $\langle \phi(\vec{x}, t)^2 \rangle$.

However, for the minimally coupled massless scalar field, we shall find that the adiabatic subtractions that make $\langle \phi(\vec{x}, t)^2 \rangle$ converge at ultraviolet frequencies have the remarkable effect of also canceling the divergences that occur at infrared frequencies (Parker 2007). In fact, in the case that $m = 0$ and $\xi = 0$, instead of being infinite, the two-point function $\langle \phi(\vec{x}, t)^2 \rangle$ is *exactly* zero! We shall show this in detail when we consider the quantized perturbations of the inflaton field in the next two sections of this chapter. There is no freedom to change the adiabatic subtraction terms because they are determined by the ultraviolet behavior. Hence, this seemingly fortuitous cancelation of the infrared divergences is noteworthy. (This agrees with the results of Habib *et al.* (1999).) If the regularized two-point function is taken to define the dispersion of the inflaton field, then the vanishing of the two-point function of ϕ implies that in the de Sitter vacuum state ϕ has no dispersion. The uncertainty principle would then imply that the dispersion in the conjugate momentum, $\dot\phi$, should become infinite, corresponding to the extreme limit of a "squeezed state," that is, a state in which the uncertainty in one of a pair of conjugate observables is smaller than the value it would have in the minimal uncertainty state of the system.

2.11 Quantum fluctuations and early inflation

In this section, we give a brief overview of the mechanism of inflation, in which the quantized fluctuations of a scalar field, known as the inflaton field, are amplified during a period of exponentially rapid expansion of the universe (or at least of the part of the universe that we can observe today). In the next section, we explain the inflationary mechanism in more detail and show how the fluctuations in the inflaton field may be calculated. Quantum field theory in curved spacetime is the foundation on which the amplification of these quantum fluctuations by the inflating universe is built, so it is fitting that inflation be discussed in the present context.

The spatially isotropic, exponentially expanding metric of (2.126) is a solution of the Einstein gravitational field equations for a universe dominated by a positive cosmological constant Λ. As we discussed in the previous section, the coordinates of (2.126), in which the expansion is exponentially fast, cover part of the maximally symmetric four-dimensional de Sitter hypersphere. In the previous section, we considered a quantized scalar field in the exponentially expanding four-dimensional de Sitter spacetime. We used the de Sitter symmetries analogous to space- and time-translations to arrive at the de Sitter vacuum state known as the Bunch–Davies vacuum.

Beginning in the late 1970s, interest increased in the possibility of an early exponential expansion of the universe, sparked by developments in quantum field theory in curved spacetime and in high-energy physics. An exponentially inflating FLRW solution in the early universe was found in the context of quantum field theory in curved spacetime by Starobinsky (1979, 1980). This Starobinsky inflation is driven by the trace anomaly of a conformally invariant free scalar field. (The trace anomaly of such a field is derived in Chapters 3 and 5.) A model for the big bang that involved particle creation and an early de Sitter stage of expansion was proposed by Brout et al. (1978, 1980). The vacuum states of a massless minimally coupled scalar field in isotropically expanding universes undergoing power-law inflation and exponential inflation, with emphasis on the infrared behavior of the two-point functions of the scalar field, were studied by Ford and Parker (1977a).

In 1981, Alan H. Guth wrote an influential paper (Guth (1981) (see also Sato 1981) discussing the virtues of an early exponential expansion of the universe that inflated it by a huge factor. First, it solves the horizon problem; a period of exponential inflation permits regions to be causally connected that otherwise would never have been in causal contact since the big bang. Regions of the sky today that have angular separations of more than about 2° of arc would never have been in causal contact were it not for early inflation. Second, it solves the flatness problem; during the period of rapid exponential inflation the dominant source in the Einstein equations must have the characteristics of a large positive cosmological constant, namely, a large positive energy density and a negative pressure of equal magnitude. In grand unified elementary particle theories with massive scalar fields, the huge inflation could be caused by the large energy density and negative pressure of the scalar field poised near the top of its potential. As the expectation value of the scalar field moved or tunneled toward the minimum of its potential (taken to be at zero energy), energy would be released causing a reheating of the universe. The scalar field near the top of its potential would act like a cosmological constant, while after inflation the dominant energy density and pressure would come from more conventional matter and radiation produced by the reheating. After sufficient inflation, the spatial curvature terms in Einstein's equations would be very small, so that the energy density produced by reheating would have to be very near the critical density required for a spatially flat solution of Einstein's equations. Guth also pointed out that such a period of inflation would suppress the density of magnetic monopoles that can arise as a consequence of symmetry breaking in grand unified theories.

Guth's original proposal for the scalar-field-potential causing inflation did not work because the separate bubble-like regions in which the scalar field was at the bottom of its potential would not be able to merge to form the observed universe (Guth and Weinberg 1983). This problem was overcome by introducing new inflationary models with different potentials and initial conditions for the scalar field (Steinhardt 1982; Albrecht and Steinhardt 1982; Linde 1982a,b; Hawking and Moss 1982). Soon after, it was discovered that inflation also gave rise to the types of perturbations that eventually could grow into the large-scale structure of galaxies in our universe (Mukhanov and Chibisov 1981; Guth and Pi 1982; Hawking 1982; Starobinsky 1982; Bardeen et al., 1983). These same perturbations arising from inflation also serve to explain the power spectrum of the small anisotropies of the CMB temperature that we observe (Peiris et al., 2003; Spergel et al., 2007). Although inflation gives a good explanation for the power spectrum of primordial perturbations, a detailed explanation of the observed magnitude of these perturbations is elusive and will be one of the tests that must be met as we try to extend our understanding of the gravitational and high-energy physics of the earliest universe.

In addition to the potentials proposed for the inflaton field in the references given above, we mention that fine-tuning constraints that apply to reasonable inflaton potentials have been found and studied by Adams et al. (1991). The implications of these constraints have been evaluated for various inflationary potentials, including that of "natural inflation" produced by a very light pseudo-Nambu–Goldstone boson (Adams et al., 1993; Freese and Kinney 2004). Finally, we mention that various mechanisms for ending inflation have been proposed, including one involving quantum gravitational back reaction (Tsamis and Woodard 1998).

There are a number of books that cover inflation in detail (Liddle and Lyth 2000; Dodelson 2003; Mukhanov 2005); see also (Mukhanov et al., 1992; Kolb and Turner 1994; Carroll 2004). Here, we will consider only inflation driven by a single scalar field. Our main emphasis will be on the generation of quantized perturbations of the scalar field driving inflation. The calculation of the two-point function giving the spectrum of perturbations of the inflaton field will also be discussed.

As we explain in mathematical detail in the next section, the homogeneous part of the inflaton field acts through a large inflaton potential energy as the source in the Einstein gravitational field equations, producing a period of nearly exponential expansion, known as inflation. This period of inflation is thought to have lasted for at least 50 e-foldings, or in cosmic time, t, for a period of at least $50H^{-1}$, where H is the Hubble constant during inflation.[4] It is during this period of inflation that the inhomogeneous quantized fluctuations of the inflaton field were built up by the same processes that we have studied in connection with particle creation. After 50 or more e-foldings of inflation, a process is thought to have occurred that caused the universe (or our part of it) to become radiation-dominated with a plasma of relativistic particles. This process is called "reheating."

Prior to reheating, the quantum fluctuations of the inflaton field lead to curvature perturbations. After reheating, these curvature perturbations produce perturbations in the density of the hot plasma of relativistic particles present in the radiation-dominated stage of the expansion. These density perturbations serve as the initial conditions for acoustic plasma oscillations that continue until the universe has cooled sufficiently for neutral atoms to form. The three-space at this time of "recombination" is known as the three-dimensional hypersurface of last scattering. The photons present at recombination have a temperature of about 4000 K and propagate with almost no further scattering. They form the cosmic background radiation.

By the present time, the cosmic background radiation has cooled to an average temperature of about 2.73 K with frequencies in the microwave range. The small anisotropies of about one part in 10^5 observed in the CMB indicate that the initial spectrum of perturbations at the beginning of the radiation-dominated stage of the expansion was nearly scale-invariant. An early exponentially expanding stage of the universe prior to radiation domination would explain why the perturbations are nearly scale-independent. The reason is that, as we shall see, the mode function of each quantized perturbation of the scalar inflaton field depends on the ratio of the wavelength of that perturbation to the inflationary Hubble horizon length.

The physical radius of the Hubble horizon is nearly constant during inflation, while the physical wavelength of each mode of the perturbation grows

---

[4] For example, if H^{-1} is, say, $10^5 t_{\text{Planck}}$, then inflation would last for a time of at least 10^{-38} seconds before reheating.

exponentially in time with the inflating universe. After the perturbation's wavelength has become large with respect to the Hubble horizon, the perturbation's amplitude is imprinted[5] as a corresponding perturbation of a mode of the dimensionless spatial curvature (Lyth 1985). This dimensionless spatial curvature δK is analogous in the Einstein equations to the constant that appears in the spatial curvature term in a Friedmann, Lemaitre, Robertson, Walker (FLRW) universe. The amplitude of each mode of this dimensionless spatial curvature remains nearly constant as inflation continues (Lyth 1985; Liddle and Lyth 2000; Dodelson 2003). After reheating occurs, the expansion rate goes as $t^{1/2}$, as does the wavelength of each mode of the perturbation, while the Hubble scale H^{-1} grows more rapidly, as t. The Hubble scale continues to expand until its radius catches up with the more slowly expanding physical wavelengths of the modes of the metric or curvature perturbations. A mode is said to "reenter" the Hubble scale at the time when the Hubble scale catches up with the mode's wavelength. Reentry occurs at later times for modes of longer wavelengths. It is the spectrum of these dimensionless curvature perturbations at reentry that sets the initial conditions for the density perturbations of the plasma present after reheating.

In turn, the spectrum of these curvature perturbations has been set by the spectrum of the quantized perturbations of the scalar inflaton field. This inflaton perturbation spectrum results from the quantum amplification process that occurs during the period when the inflaton perturbation wavelengths expand from a value much smaller than the Hubble horizon, H^{-1}, to a value large with respect to H^{-1}. When the wavelengths reach a value sufficiently large with respect to H^{-1}, the basically classical curvature perturbations have developed and remain constant at least up to and beyond the time of reheating, after which they induce density perturbations in the plasma of matter and radiation that was produced by the reheating process.

The modes k that are of interest to us are those that today correspond to length scales, $k/a(t_0)$, ranging from a small fraction of the present Hubble scale, $H(t_0)^{-1}$, up to about $H(t_0)^{-1}$. These length scales include structures ranging from the CMB anisotropies to the large-scale structure of the matter in the observable universe. During the entire early time period during which these modes underwent quantum amplification as quantized inflaton perturbations and then induced a spectrum of classical curvature perturbations, δK, the value of the Hubble parameter $H(t)$ is thought to have been very nearly constant (and many orders-of-magnitude larger than $H(t_0)$). Therefore, during this early period we expect that each mode underwent very nearly the same evolution when expressed as a function of the ratio of its wavelength to the radius H^{-1} of the event horizon. The process in which an inflaton perturbation generates the corresponding curvature perturbation should be completed for each mode by the time its wavelength has

---

[5] In a comoving coordinate system, with respect to which the perturbations carry no energy-momentum flux along the comoving spatial hypersurfaces.

expanded to a sufficiently large given number of Hubble horizon lengths, *the same number for each mode*. The evolution of each mode function, to good approximation, is the same during this period and should therefore result in the same size curvature perturbation when the wavelength of the mode has reached the given number of Hubble horizon lengths. Therefore, the spectrum of curvature perturbations produced should be independent of k to good approximation. Because the amplitudes of those curvature perturbations then remain constant up to the times when they induce density perturbations in the plasma after reheating, it follows that the amplitudes of those initial density perturbations are essentially the same for each mode k, and hence nearly scale-independent.

In summary, the scale independence of the relevant ratio of wavelength to Hubble horizon is what implies that the spectrum of quantized perturbations is scale-independent at the times when they get frozen into the spectrum of curvature perturbations. The scale-independent spectrum of these time-independent dimensionless curvature perturbations present after inflation serves as the initial conditions for the acoustic oscillations of the plasma present after reheating. The small anisotropies observed in the CMB carry information about the amplitudes of these acoustic oscillations of different wavelengths at the time of last scattering. The observational data (Spergel et al., 2007) show that the spectrum of the initial perturbations of the radiation-dominated stage of the expanding universe were indeed nearly (but not quite) scale-invariant.

2.12 Quantizing the inflaton field perturbations

2.12.1 Perturbative expansions and zero-order equations

First we explain the notation used in the literature on inflation. We limit our discussion to inflation that is driven by a single field. The inflaton field is a scalar field present in the inflationary stage of the expanding universe. It has a Lagrangian like that of (2.35), but with $\frac{1}{2}m^2\phi^2$ replaced by a more general potential $V(\phi)$:

$$\mathcal{L} = \frac{1}{2}|g|^{1/2}(g^{\mu\nu}\partial_\mu\phi\partial_\nu\phi - 2V(\phi)). \tag{2.148}$$

For simplicity, and to avoid a time-dependent Newtonian gravitational constant in Einstein's equations, we have taken $\xi = 0$ in the term $\xi R\phi^2$ appearing in the Lagrangian of (2.35). The scalar field ϕ is the inflaton field and $V(\phi)$ is called the inflaton potential. The metric and field are expanded perturbatively to first order:

$$g_{\mu\nu} = g_{0\mu\nu} + \delta g_{\mu\nu}. \tag{2.149}$$

$$\phi = \phi_0 + \delta\phi. \tag{2.150}$$

The zeroth order metric $g_{0\mu\nu}$ is taken to have the spatially flat FLRW form of (2.50)

$$ds^2 = dt^2 - a^2(t)(dx^2 + dy^2 + dz^2). \tag{2.151}$$

The zeroth order scalar field ϕ_0 is taken to be homogeneous, depending only on t. Therefore, it satisfies (see (2.41)):

$$a^{-3}\partial_t(a^3 \partial_t \phi_0) + dV/d\phi_0 = 0, \tag{2.152}$$

which can be written as

$$\ddot{\phi}_0 + 3H(t)\dot{\phi}_0 + V'(\phi_0) = 0, \tag{2.153}$$

with

$$H(t) \equiv \dot{a}/a \quad \text{and} \quad V' \equiv dV/d\phi. \tag{2.154}$$

The homogeneous field $\phi_0(t)$ is regarded as classical. The inflaton potential $V(\phi_0)$ that appears in (2.153) appears in the zero-order part of the energy-momentum tensor of the inflaton field and thus finds its way into the zero-order Einstein equations. The potential can be chosen so that $V(\phi_0)$ is large, positive, and nearly constant at very early times when inflation occurred. At those times it acts like a positive cosmological constant, $\Lambda = 8\pi G V/3$, sufficiently large to cause an exponential expansion with $a(t) = \exp[Ht]$, where $H = \sqrt{8\pi G V/3}$ is nearly constant for a sufficient time to cause a period of inflation that lasts for a cosmic time interval of the order of at least $50H^{-1}$. The time interval H^{-1} is the time for $a(t)$ to increase by one factor of e.

The Einstein equations for the zero-order metric of (2.151) are

$$H^2 = (8\pi G/3)\rho \tag{2.155}$$

and

$$\ddot{a}/a = \dot{H} + H^2 = -(4\pi G/3)(\rho + 3p). \tag{2.156}$$

Here, the energy density ρ and pressure p of the inflaton field drive the inflationary expansion. From $\mathcal{L}$, these are

$$\rho = \frac{1}{2}(\dot{\phi}_0)^2 + V(\phi_0), \quad p = \frac{1}{2}(\dot{\phi}_0)^2 - V(\phi_0). \tag{2.157}$$

The Einstein equations then imply that

$$H^2 = (8\pi G/3)\left(\frac{1}{2}(\dot{\phi}_0)^2 + V(\phi_0)\right) \tag{2.158}$$

and

$$\dot{H} = -4\pi G \dot{\phi}_0^2. \tag{2.159}$$

2.12.2 Conditions for nearly exponential inflation

During the period of inflation, it is expected that conditions such as the following hold. First, let

$$\frac{1}{2}(\dot{\phi}_0)^2 \ll V(\phi_0) \tag{2.160}$$

so that $V(\phi_0)$ can act like a slowly changing positive cosmological constant in (2.158), implying that H is nearly constant, with

$$H^2 \approx (8\pi G/3)V(\phi_0). \tag{2.161}$$

Second, during a period H^{-1} of one e-folding of inflation, the change in $V(\phi_0)$ is of order $\Delta V \approx V'(\phi_0)\dot{\phi}_0 H^{-1}$. For $V(\phi_0)$ to act like a cosmological constant during the period of inflation, require that $|\Delta V| \ll V(\phi_0)$, or

$$|V'(\phi_0)\dot{\phi}_0 H^{-1}| \ll V(\phi_0). \tag{2.162}$$

Third, assume that $\ddot{\phi}_0$ can be neglected in (2.153). This is motivated by the analogy between (2.153) and the equation for a "particle" of unit mass and "position coordinate" $\phi_0(t)$ moving under the influence of a resistance force $-3H\dot{\phi}_0(t)$ acting opposite to its direction of "velocity" and a force $-V'(\phi_0)$ that is assumed to be such as to allow the particle to reach a slowly changing "terminal velocity" with

$$\dot{\phi}_0 \approx -V'(\phi_0)/(3H) \tag{2.163}$$

as the particle slowly rolls down the potential $V(\phi_0)$. (If the slope $V'(\phi_0)$ is negative then the velocity $\dot{\phi}_0$ is positive, as we would have for a potential that decreases as ϕ_0 increases. Conversely, a positive slope of the potential would imply a negative velocity as the particle slowly rolls down the potential.)

Then substituting (2.163) in (2.160), and using (2.161), we find that

$$\frac{1}{6}\frac{1}{8\pi G}\left(\frac{V'}{V}\right)^2 \ll 1. \tag{2.164}$$

To within a factor of order 1, this can be written as

$$\epsilon \ll 1, \tag{2.165}$$

where the first slow-roll parameter ϵ is defined by

$$\epsilon \equiv \frac{1}{2}\frac{1}{8\pi G}\left(\frac{V'}{V}\right)^2. \tag{2.166}$$

For the slowly rolling particle to reach the terminal velocity of (2.163), consistency demands that $\ddot{\phi}_0 \approx 0$, which may be written as

$$-\frac{1}{3H}V''(\phi_0)\dot{\phi}_0 + \frac{\dot{H}}{3H^2}V'(\phi_0) \approx 0. \tag{2.167}$$

Multiplying this by $\dot\phi_0$ and imposing the condition of (2.162), we obtain

$$|V''(\phi_0)(\dot\phi_0)^2| \ll |\dot H V(\phi_0)|. \tag{2.168}$$

Finally, make use of (2.159) to write this as

$$\left|\frac{1}{4\pi G}\frac{V''(\phi_0)}{V(\phi_0)}\right| \ll 1. \tag{2.169}$$

To within a factor of order 1, this can be written as

$$|\eta| \ll 1, \tag{2.170}$$

where the second slow-roll parameter η is defined by

$$\eta \equiv \frac{1}{8\pi G}\frac{V''(\phi_0)}{V(\phi_0)}. \tag{2.171}$$

In appropriate units, H is thought to have a value within a few orders of magnitude of 10^{15} GeV. After the period of inflation, it is often assumed that ϕ_0 has fallen to a minimum of the potential $V(\phi_0)$, and that "reheating" has caused the universe to become radiation-dominated.

2.12.3 Quantization of $\delta\phi(\vec x, t)$

Our main interest here is to discuss the quantized part of the inflaton field during the period of exponential inflation. This quantized field is the first-order term $\delta\phi(\vec x, t)$ in the perturbative expression for the inflaton field.[6] As we shall see, the expectation value of $\delta\phi(\vec x, t)^2$ is sufficiently large and has the type of spectral distribution necessary to account for the original density perturbations that gave rise to the large-scale structure of the universe and to the anisotropies observed in the CMB radiation. Substituting (2.149)–(2.151) into the Lagrangian of (2.148) and expanding to first order will yield the equation of motion satisfied by $\delta\phi$. During the inflationary period prior to reheating, the inflaton field and its perturbations are well described by neglecting the first-order perturbations of the metric. (This is no longer true at the end of the inflationary period.) Therefore, during inflation we will take $\delta\phi(\vec x, t)$ to satisfy

$$\partial_t^2\delta\phi + 3H\partial_t\delta\phi - e^{-2Ht}\sum_i \partial_i^2\delta\phi + V''(\phi_0)\delta\phi = 0, \tag{2.172}$$

where we are neglecting a term of first order in the perturbation of the metric, which would give a small correction to the result. This equation is just (2.128), with the scalar field $\phi(\vec x, t)$ of that equation replaced by the inflaton perturbation field $\delta\phi(\vec x, t)$ and m^2 replaced by $V''(\phi_0)$. The slow-roll assumptions of (2.165)

---

[6] For a vacuum state in which the expectation value of $\delta\phi(\vec x, t)$ is 0, the homogenous classical field $\phi_0(t)$ is the vacuum expectation value of the inflaton field ϕ of (2.150), and $\delta\phi(\vec x, t)$ describes the inhomogeneous perturbations of the inflaton field about $\phi_0(t)$.

2.12 Quantizing the inflaton field perturbations

and (2.170) are consistent with a period of exponential inflation, $a(t) = \exp(Ht)$, with H and $m^2 \equiv V''(\phi_0)$ each nearly constant. Depending on the potential $V(\phi_0)$, the quantity m^2 can have either sign. The slow-roll condition of (2.169) requires, in conjunction with the Einstein equation (2.161), that $|m^2|$ is small with respect to H^2.

The quantization of $\delta\phi(\vec{x}, t)$ proceeds in the standard way. Expand $\delta\phi(\vec{x}, t)$ as in (2.54):

$$\delta\phi = \sum_{\vec{k}} \left\{ A_{\vec{k}} f_{\vec{k}}(x) + A_{\vec{k}}^\dagger f_{\vec{k}}^*(x) \right\}, \tag{2.173}$$

where, for consistency with (2.129), we let

$$f_{\vec{k}} = (2Va^3(t))^{-1/2} e^{i\vec{k}\cdot\vec{x}} h_k(t). \tag{2.174}$$

Here $k^i = 2\pi n^i/L$ with n^i an integer, $k = |\vec{k}|$, and we have imposed periodic boundary conditions on a cube of coordinate volume $V = L^3$ and physical volume $Va^3(t)$. (The limit as $L \to \infty$ is to be taken after physical quantities are calculated.) If $(k/a(t))^2$ is sufficiently large that the physical wavelength is much smaller than the curvature scales of the spacetime (in this case measured by quantities like $(\dot{a}/a)^2$ and $(\ddot{a}/a)$), then we require that

$$f_{\vec{k}} \sim (Va^3(t))^{-1/2} \left(\frac{2k}{a(t)}\right)^{-1/2} \exp\left[i\left(\vec{k}\cdot\vec{x} - \int^t \frac{k}{a(t')} dt'\right)\right], \tag{2.175}$$

where we have also assumed that $(k/a(t))^2$ is large with respect to $m^2 = V''(\phi_0)$. This condition is the same as that of (2.133); its relation to the WKB approximation was discussed in connection with that equation. We present a similar adiabatic condition in connection with regularization and renormalization in Chapter 3.

It is important to realize that the condition in (2.175) does *not* uniquely determine the mode function solutions $f_{\vec{k}}$. For example, the exact solution that we gave in Section 2.8 satisfies this condition at early times *and* at late times, despite the fact that there is particle creation that causes a mixing of positive and negative frequencies. The negative frequency term that results from particle creation vanishes sufficiently rapidly as k increases that it is not determined by the condition of (2.175). We explain this in more detail in Section 3.1.

Although (2.175) by itself is not sufficient to uniquely determine the solution $f_{\vec{k}}(\vec{x}, t)$, we showed that in the de Sitter exponentially expanding spacetime considered earlier, if we also impose the six spatial symmetries and the de Sitter analogue of time-translation symmetry, then we do arrive at the unique solution given in (2.145) for $f_{\vec{k}}(\vec{x}, t)$.

We are considering an inflaton potential $V(\phi_0)$ such that $a(t) = \exp(Ht)$, with H and $m^2 \equiv V''(\phi_0)$ each nearly constant, and with m^2 small with respect to H. Therefore, it is reasonable to consider first the approximation in which H and m^2 are constant. We will deal with this approximation here, without going into further analysis of the effect of the slow variation of those quantities as the inflaton field ϕ_0 changes with time. Then it is reasonable to take as our solution the one already found in (2.145) corresponding to the Bunch–Davies de Sitter vacuum. We take

$$f_{\vec{k}}(\vec{x},t) = \frac{1}{2}\sqrt{\frac{\pi}{H}}\, V^{-\frac{1}{2}} e^{-\frac{3}{2}Ht}\, H_\nu^{(1)}(kH^{-1}e^{-Ht})\, e^{i\vec{k}\cdot\vec{x}}, \qquad (2.176)$$

where $\nu = \sqrt{\frac{9}{4} - \frac{m^2}{H^2}}$. When H and m^2 change sufficiently slowly with time, this solution should remain a good approximation to the mode function, with $H = H(t)$ and $m^2 = m(t)^2$ determined by the inflaton potential $V(\phi_0)$. In a more general treatment, the state vector could be based on mode functions $f_{\vec{k}}(\vec{x},t)$ that involve a suitable linear combination of $H_\nu^{(1)}(kH^{-1}e^{-Ht})$ and $H_\nu^{(2)}(kH^{-1}e^{-Ht})$, but we will only consider the Bunch–Davies vacuum state here.

We note in passing that, to within a phase factor that involves ν, we can replace $H_\nu^{(1)}(kH^{-1}e^{-Ht})$ in (2.176) with $H_\nu^{(2)}(-kH^{-1}e^{-Ht})$, although we must be careful about the cut on the negative real axis in the complex plane that is used in defining the Hankel functions. Because ν is so slowly changing in time, this replacement would not affect quantities of physical interest.

2.12.4 Inflaton perturbation spectrum and two-point function

The inflaton perturbation spectrum can be found from the coincident two-point function, $\langle \delta\phi(\vec{x},t)^2 \rangle$, of the inflaton field in the chosen state. This quantity is a measure of the dispersion of the values of $\delta\phi(\vec{x},t)$ that would be expected if the field were measured in the vicinity of the spacetime point $(\vec{x},t)$ in the given state (in the present discussion, the de Sitter Bunch–Davies vacuum). Over a range of positions at a given time t it indicates the dispersion in the values of the field that would be measured. Because of the homogeneity of the background spacetime, this dispersion will be the same for all values of $\vec{x}$ at cosmic time t. (Recall that we are neglecting the inhomogeneous perturbation of the metric in our field equation during inflation; this metric perturbation would introduce a small inhomogeneity into the dispersion of the inflaton perturbation field.)

By definition, the coincident two-point correlation function is related to the *spectrum of inflaton perturbations*, denoted as $\mathcal{P}_\phi(k,t)$ (and also as $\Delta_\phi^2(k,t)$) by Liddle and Lyth (2000) and Peiris et al., (2003):

$$\langle 0| |\delta\phi(\vec{x},t)|^2 |0\rangle = \int_0^\infty \mathcal{P}_\phi(k,t) k^{-1} dk. \qquad (2.177)$$

This definition of the power spectrum $\mathcal{P}_\phi(k,t)$ is independent of whether we use the discrete or continuous momentum representation to define the mode functions and field expansion.

In the Appendix (p. 446), we discuss the field expansion and power spectrum using both the discrete and continuous momentum representations. There, we also define operators $\delta\phi_k(t)$ that correspond to momentum-space perturbation components for the discrete and continuous cases. Here, we work in the discrete momentum representation, taking the continuum limit $L \to \infty$ at the end.

The momentum-space spectrum of the inflaton perturbations at time t will be found below in terms of the momentum-space mode functions of the inflaton perturbation field. As described in the previous section, this inflaton perturbation spectrum sets up the initial conditions that eventually give rise to the power spectrum of the CMB radiation that we observe today.

To obtain the spectrum of inflaton perturbations, evaluate the expectation value of $\langle \delta\phi(\vec{x},t)\delta\phi(\vec{x}',t)\rangle$ using the field expansion of (2.173). The state under consideration is annihilated by the operators $A_{\vec{k}}$, so

$$\langle 0|\,\delta\phi(\vec{x},t)\delta\phi(\vec{x}',t')\,|0\rangle = \sum_{\vec{k}} f_{\vec{k}}(\vec{x},t) f_{\vec{k}}^*(\vec{x}',t')$$

$$= \frac{1}{2(2\pi)^3}[a(t)a(t')]^{-\frac{3}{2}} \int d^3k\, e^{i\vec{k}\cdot(\vec{x}-\vec{x}')} h_k(t) h_k^*(t'), \quad (2.178)$$

where in the last step we used (2.174) and took the continuum limit, replacing $V^{-1}\sum_{\vec{k}}$ by $(2\pi)^{-3}\int d^3k$. When $\vec{x} = \vec{x}'$ this becomes

$$\langle 0|\,\delta\phi(\vec{x},t)^2\,|0\rangle = \left(4\pi^2 a^3(t)\right)^{-1} \int_0^\infty dk\, k^2\, |h_k(t)|^2. \quad (2.179)$$

As expected for this homogeneous background spacetime, this coincident two-point function depends on t alone. The function $|h_k(t)|^2$ is proportional to k^{-1} in the limit of large k, so this integral has a quadratic ultraviolet divergence. Such divergences are the subject of the next chapter. However, because they could be important to theories of inflation, we will discuss them later in the present section.

By comparing (2.177) with (2.179), we find the spectrum:

$$\mathcal{P}_\phi(k,t) = (4\pi^2 a^3(t))^{-1} k^3 |h_k(t)|^2. \quad (2.180)$$

In the absence of some form of regularization and renormalization, this spectrum diverges as k^2 for large k. In the case when $t = t'$, (2.178) and (2.180) show that the non-coincident equal-time correlation function can be written in terms of this same spectrum as

$$\langle 0|\,\delta\phi(\vec{x}_1,t)\delta\phi(\vec{x}_2,t)\,|0\rangle = \int_0^\infty \mathcal{P}_\phi(k,t) \frac{\sin(k|\vec{x}_1 - \vec{x}_2|)}{k|\vec{x}_1 - \vec{x}_2|} \frac{dk}{k} \quad (2.181)$$

Here we have taken the continuum limit and integrated over the angle between $\vec{k}$ and $(\vec{x}_1 - \vec{x}_2')$. In the coincidence limit, $\vec{x}_1 \to \vec{x}_2$, this reduces to (2.177), provided that the expression for the spectrum has been regularized so that infinitely large values of k do not contribute significantly to the integral.

There are various approaches that can be used to regularize and renormalize the two-point correlation functions in (2.179) and (2.181). For example, we can use point-splitting regularization and renormalization (Bunch and Davies 1979; Vilenkin and Ford 1982) to define the coincidence limit of $\langle 0| \, \delta\phi(\vec{x}_1,t)\delta\phi(\vec{x}_2,t) \, |0\rangle$. Other methods, such as adiabatic regularization (Parker 2007), work directly with $\langle 0| \, \delta\phi(\vec{x},t)^2 \, |0\rangle$. We expect that they all lead to expressions for $\langle 0| \, \delta\phi(\vec{x},t)^2 \, |0\rangle$ that are essentially the same. (See the first three paragraphs of Section 2.13.)

The expression for the spectrum $\mathcal{P}_\phi(k,t)$ that appears implicitly in the regularized expression for $\langle 0| \, \delta\phi(\vec{x},t)^2 \, |0\rangle$ is different from the unregularized spectrum that appears in (2.180). Which expression for the spectrum is the one of physical significance?

In the past literature on inflation, the value of the spectrum has generally been taken to be that in (2.180) with no regularization, except perhaps to assume that there is a "smoothing scale" or cut-off in the inflaton spectrum at a value of k that is sufficiently large that the spectrum is not significantly altered at values of k that would be of observational significance in the present universe.

The possibility that regularization and renormalization will affect the spectrum at values of k that are of current observational significance has recently been considered by Parker (2007) who finds that for the same inflaton potential, the predicted amplitude of the inflaton perturbations taking into account regularization may be much smaller than the predicted amplitude obtained directly from (2.180) without regularization. This difference in amplitude would be of importance when testing fundamental physical theories that give rise to definite inflationary potentials. The amplitude of the anisotropies in the CMB temperature is accurately measured, and this amplitude depends on the amplitude of the early inflaton perturbations. Thus, we may be able to deduce from the observed CMB anisotropies, the strength of the inflaton self-interaction potential. Before this can be done with confidence, the question of whether or not to use the regularized expression for $\mathcal{P}_\phi(k,t)$ must be settled, and it must be determined how important the difference would be. (For example, some supersymmetric theories of particles or strings may require no regularization of the two-point function.)

One could argue that the unregularized spectrum should be used because it appears in the point-separated two-point function of (2.181), which has a well-defined value in a distributional sense with the unregularized spectrum. However, as the spatial points get close together, the two-point function distribution will grow large and diverge as the points merge. For consistency between the regularized coincident two-point function and the point-separated one, it seems necessary to use the regularized spectrum in both expressions.

Having already presented the unregularized expression for the spectrum in (2.180), we will proceed to present the regularized expression, and to compare the two following Parker (2007).

The method of adiabatic regularization is the most direct way of obtaining the coincident two-point function in a form in which the regularized spectrum of perturbations is immediately evident. It is particularly suited for homogeneous background spacetimes such as that of (2.151). The method is presented in detail in the first part of the next chapter and is applied to quantities such as the coincident two-point function and the energy-momentum tensor. The calculation is the same as that given for the two-point function of a scalar field (denoted there by ϕ) in (3.48). The method was developed for cases in which $m^2 \geq 0$, and we will assume that is the case in the following. (It may also apply for a range of negative values of m^2, but we will not investigate that question here.) The result is given for general ξ in (3.53). By comparing the definitions of $h_k(t)$ in (3.4) and (2.174), we see that the definitions agree. Therefore (3.53) can be used. In the present context, (3.53) becomes

$$\langle 0| \delta\phi(\vec{x},t)^2 |0\rangle_{phys} = \left(4\pi^2 a^3(t)\right)^{-1} \int_0^\infty dk\, k^2 \Big\{ |h_k(t)|^2$$
$$- \omega_k(t)^{-1} - \left(W_k(t)^{-1}\right)^{(2)} \Big\} \qquad (2.182)$$

with $\omega_k(t) = \sqrt{\frac{k^2}{a(t)^2} + m^2}$ and $\left(W^{-1}(t)\right)^{(2)}$ given by (3.51). The integrand with the adiabatic subtraction terms converges.[7] For the present exponentially expanding universe and minimally coupled ($\xi = 0$) field of (2.172), the second subtraction term is

$$\left(W_k(t)^{-1}\right)^{(2)} = -\frac{5H^2 m^4}{8\omega_k(t)^7} + \frac{3H^2 m^2}{4\omega_k(t)^5} + \frac{H^2}{\omega_k(t)^3}, \qquad (2.183)$$

and $h_k(t)$ is given by (2.144) as

$$h_k(t) = \sqrt{\frac{\pi}{2H}} H_\nu^{(1)}(v). \qquad (2.184)$$

The coincident two-point function is related to $\mathcal{P}_\phi(k,t)$ by (2.177). Comparing with (2.182), we find

$$\mathcal{P}_\phi(k,t) = (4\pi^2 a^3(t))^{-1} k^3 \left(|h_k(t)|^2 - \omega_k(t)^{-1} - (W_k(t)^{-1})^{(2)} \right). \qquad (2.185)$$

Thus, if the physically meaningful coincident two-point function is used to define the k-space spectrum of perturbations of the inflaton field, it follows that the spectrum is given by (2.185).

Replacing $k/a(t)$ by Hv in the last equation, the spectrum takes the form

---

[7] As explained in the next chapter, minimal subtraction is used to define the physical quantity. One further subtraction is required when dealing with the energy-momentum tensor.

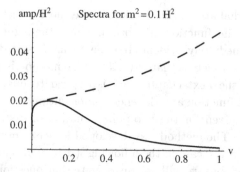

Fig 2.1. Plot of the inflaton spectrum $\mathcal{P}_\phi$ for $m^2 = 0.1H^2$. The amplitude is in units of H^2. The quantity $v = kH^{-1}\exp(-Ht)$ is 1 when $\lambda(k,t) = 2\pi H^{-1}$. The dashed curve is the spectrum without adiabatic subtractions.

$$\mathcal{P}_\phi(k,t) = \frac{H^2 v^3}{32\pi^2}\left(4\pi\left|H_\nu^{(1)}(v)\right|^2 \right. $$
$$\left. - \frac{8m_H^6 + 3m_H^4(3+8v^2) + 2m_H^2 v^2(11+12v^2) + 8(v^4+v^6)}{(m_H^2+v^2)^{7/2}}\right) $$
(2.186)

where $\nu = \sqrt{(9/4) - m_H^2}$ and $m_H \equiv m/H$.

A graph showing the spectral amplitude $\mathcal{P}_\phi$ as a function of v appears in Fig. 2.1 as the solid curve. The dashed curve is a plot of the term in (2.186) that involves $4\pi\left|H_\nu^{(1)}(v)\right|^2$ without any of the adiabatic subtractions. It is easy to see that the integral over k in (2.177) will have an ultraviolet divergence unless the adiabatic subtractions are included in the integrand.

Notice that the spectrum only depends on the ratio

$$v = kH^{-1}\exp(-Ht) = 2\pi H^{-1}/\lambda(k,t).$$

This is proportional to the ratio of the radius of the de Sitter event horizon, H^{-1}, to the wavelength, $\lambda(k,t)$, of the perturbation. But as we showed in Section 2.10 on de Sitter spacetime, the value of v is not changed under the subgroup of de Sitter transformations that include spatial rotations and translations, as well as the de Sitter generalization of time translation. Under the latter transformation, taking k to k' and t to t', this means that the wavelength is invariant, that is, $\lambda(k,t) = \lambda(k',t')$.

The wavelength λ of any given mode k expands with the universe. Because the horizon radius H^{-1} is fixed during exponential de Sitter inflation, the value of v corresponding to a given k decreases as the universe expands.

It can be shown (Liddle and Lyth 2000; Dodelson 2003) that there is a quantity that remains constant after v has fallen below a particular value, which we denote by v_1. The value, C, of this constant is determined for any given k by the value of the inflaton perturbation spectral amplitude evaluated at v_1. But as we have seen, the spectral amplitude depends on k and t only through the de Sitter subgroup invariant quantity v. Hence the value of the constant C will be the same for any k for which exponential de Sitter inflation is still going on at the time when v falls to the value v_1. (For a sufficiently long period of inflation, this includes all k corresponding to length scales in our observable universe.)

After reheating has occurred, the value of the matter density perturbation of mode k in the initial radiation- or matter-dominated universe[8] is determined by the value of the constant C for that mode. But for all modes k relevant to our observable universe, the values of C are the same. Hence, the spectral amplitudes of the initial density perturbations are also the same. This means that the initial spectrum of density perturbations is independent of k, or scale invariant. In slow-roll inflation, as described earlier in this section, the value of H and of m^2 may slowly change, so the scale invariance of the initial spectrum of inflaton and density perturbations would not be expected to be exact.

The initial matter density perturbations set up acoustic oscillations under the influence of pressure and gravity. These result in time-dependent temperature variations in the plasma. The frequencies of these temperature variations depend on the mode k. We can "see" the spectrum that these temperature variations have at the time when the temperature of the plasma falls sufficiently for neutral atoms to form. At this time of "recombination," the opacity drops, allowing electromagnetic radiation to propagate freely as the CMB radiation. Because the different modes k of the acoustic waves have different wavelengths and frequencies, the spectrum of temperature variations encoded in the CMB radiation at the time of recombination is not scale invariant, but can be calculated from the spectrum of initial matter density perturbations that we have just discussed.

Observations of the CMB radiation show that its present-day temperature anisotropies are consistent with a nearly scale-invariant spectrum of initial density fluctuations. This is consistent with the consequences of early cosmological inflation that we have just described. For a thorough analysis of the observations, including their implications for a stage of early cosmological inflation, see Spergel et al., (2007).

The value of v_1 is not precisely known. If $v_1 \approx 1$, then from Fig. 2.1 we can see that the spectral amplitude is small as compared with the result without the minimal adiabatic subtractions. But if $v \approx m^2/H^2$, the approximate value at which the maximum of the adiabatically subtracted spectrum occurs is about the same as for the unsubtracted spectrum. The latter has been used in much of the past literature to determine the amplitude of the spectrum, but we see that the subtractions could conceivably make the perturbations several orders of magnitude smaller than without the subtractions. In analyzing the CMB data (Spergel et al.,

---

[8] For larger k the density perturbations are set up at earlier times than for small k; and these times span both the radiation-dominated and the early matter-dominated stage of the expansion of our universe.

2007), the amplitude of the initial perturbations is taken as a parameter set by the observations. Thus, this amplitude is precisely known, but understandably difficult to predict from our current limited knowledge of fundamental physics at the very high energies relevant to cosmological inflation.

As in any theory where infinities occur, we could regard the infinities as a manifestation of the incompleteness of the theory. This may be the case here. However, once the amplitude of the initial nearly scale-invariant perturbations can be predicted from fundamental physics, the correctness of including the adiabatic subtractions will become a question that can be tested by observation.

As noted earlier, we have only considered here inflationary potentials having non-negative effective mass, $m^2 \geq 0$. If $V''(\phi_0(t)) < 0$ were negative, as occurs for some inflationary potentials, the effective value of m^2 evidently would be negative. A separate analysis would be necessary in such a case. We do not discuss that further here.

2.13 A word on interacting quantized fields and on algebraic quantum field theory in curved spacetime

Canonical quantization of quantum field theory in curved spacetime has opened our eyes to new phenomena in which general relativity and quantum field theory are intertwined in a convincing and sometimes remarkable manner. As we have explained above, particle creation in the early inflationary universe is the most likely *source* of the perturbations that led to the observed anisotropies of the CMB radiation and of the observed statistical features of the large-scale structure of the universe. As discussed in the previous sections, accurate measurements of these features continue to support this explanation. A second phenomenon that probes the deepest reaches of current physics is the remarkable temperature of a black hole. This is a consequence of quantum field theory in curved spacetime and is necessary for the second law of thermodynamics to encompass systems in general relativity in which black holes exist. Such systems currently serve as a testing ground for developing new fundamental physical theories that combine gravitation and quantum theory.

The canonical approach also showed us that the vacuum state is not unique in general curved spacetimes, and that there are many possible states that can serve as "the vacuum state." This was already shown in Parker (1965, 1968, 1969), where the use of Bogolubov transformations to describe particle creation in curved spacetimes was first introduced. It was proved in Parker (1969) that in curved spacetime there are unitarily inequivalent Fock spaces based on orthogonal vacuum states that give rise to inequivalent representations of the canonical commutation rules. The fact that the particle concept itself becomes ambiguous in a general curved spacetime was also discovered in the above cited works of Parker and is related to the ambiguity of the vacuum and the irreducible

2.13 Interacting quantized fields and algebraic quantum field theory

uncertainty in the measurement of particle number. These results were obtained for the spacetimes of isotropically expanding universes; and therefore must be a feature of general curved spacetimes in the absence of special symmetries. The adiabatic condition (discussed earlier in this chapter and in more detail in the next chapter) was introduced in Parker (1965) and further developed in Parker and Fulling (1974). The adiabatic condition serves as a requirement on physically acceptable states, and determines the singularity structure of expectation values of products of fields. The Hadamard condition[9] also specifies the singularity structure of the symmetrized two-point function of the quantized field. Both conditions give natural generalizations of the singularity structure that is found in Minkowski spacetime. Are they equivalent? It has been shown, by comparing the expansion of the Hadamard form of such expectation values and the corresponding expectation values formed from states satisfying the adiabatic condition (i.e., "adiabatic vacua"), in spacetimes where they are both defined, that the expansions are the same to all orders (Pirk 1993). As we will see in the next chapter, these series expansions are in general asymptotic and not convergent, so an "adiabatic vacuum" or a "Hadamard state" is not unique, but corresponds to a large class of acceptable states. The Hadamard condition appears to be a natural generalization of the adiabatic condition to arbitrary curved spacetimes. When sufficient symmetries are present, as in the Robertson–Walker spacetimes, the adiabatic condition gives a relatively simple and direct way to deal with the infinities that appear in the expectation values of products of fields in the limit that the fields are evaluated (with a suitable measure) at the same point.

Lüders and Roberts (1990) showed that there is a unique local quasi-equivalence class of physically relevant states, and that this class can be specified in a Robertson–Walker spacetime by using the concept of an adiabatic vacuum state. It was shown in Junker and Schrohe (2002) that the definition of adiabatic vacuum states can be generalized to a general curved spacetime manifold by using the Sobolev wavefront set, and that this definition is also applicable to interacting field theories. Interestingly, Hadamard states form a special subclass of the adiabatic vacuum states defined by this method.

The fact that there is no unique vacuum state in a curved spacetime having insufficient symmetries motivates the search for a way to formulate quantum field theory in curved spacetime in a way that does not single out any particular state to serve as the basis of a Fock or Hilbert space formulation. In Streater and Wightman (1964), it was rigorously shown that a field theory in Minkowski spacetime is defined by the *vacuum expectation values* of products of field operators. The algebraic approach to quantum field theory in Minkowski spacetime makes use of such expectation values to define the theory starting from the algebra of products of field operators and from the symmetries of the vacuum state

---

[9] See Chapter 3.

in Minkowski spacetime.[10] Arthur Wightman, around 1971, suggested that the methods of algebraic quantum field theory that had been developed in flat spacetime, when suitably generalized to curved spacetime, would give a rigorous way to formulate quantum field theory in curved spacetime without reference to a particular vacuum state in the absence of sufficient symmetries of the curved spacetime. This suggestion of Wightman was the motivation for the discussion following Eq. (18) in Parker and Fulling (1973), where the algebraic approach to quantum field theory in curved spacetime was described. Aspects of the algebraic approach to quantum field theory in curved spacetime are also discussed by Fulling (1989). A satisfactory understanding of how to formulate the theory of a *free* quantized field in curved spacetime by means of the algebraic approach was obtained by the mid-1980s, as developed in the works of Fulling (1972), Ashtekar and Magnon (1975), Kay (1978), Sewell (1982), Kay (1985), and others. The algebraic formulation does not supplant the canonical formulation of quantum field theory in curved spacetime, but serves to frame the theory in a way that does not single out any particular state vector.

Starting in the early 1970s, the canonical and path integral approaches were used to define quantities, such as expectation values of energy-momentum tensors of free fields, that involve formal products of fields evaluated at the same spacetime point. We will take this up in the next chapter in Section 3.1 through Section 3.3. Analogous formal products of field operators also appear in interacting quantum field theory. They present many more problems than do free fields when renormalization is considered. At about the same time as the energy-momentum tensor of free fields were being studied, investigations of interacting quantized fields in various curved spacetimes were undertaken. (See, for example, Drummond 1975; Birrell and Davies 1978; Drummond and Shore 1979a,b; and Birrell and Ford 1979.)

In order to use the momentum-space methods of quantum field theory that had been developed in flat spacetime, Bunch and Parker (1979) introduced a *local* momentum-space representation using Riemann normal coordinates in a general curved spacetime. This local momentum-space representation is explained in Section 3.8. Using this method, they carried out the renormalization of a scalar field with a quartic self-interaction to one-loop order in a general curved spacetime. This is non-trivial because there are curvature terms that are not present in flat spacetime and cannot be canceled by means of counterterms in the Lagrangian. Nevertheless the curvature terms do cancel one another. This raised the conjecture that physically viable interacting theories that are renormalizable in Minkowski spacetime are also renormalizable in curved spacetime. For the quartically self-interacting scalar field, this was proved to all orders in a general curved spacetime by Bunch (1981b). We will discuss the use of the local momentum-space method to analyze $\lambda \phi^4$ theory to two-loop order in Section 6.7.

---

[10] See, for example, Emch (1972) or Haag (1992).

Using the local momentum-space representation of Bunch and Parker (1979), quantum electrodynamics (Panangaden 1980, 1981) as well as non-Abelian gauge theories (Leen 1982, 1983) were shown to be renormalizable to one-loop order in a general curved spacetime. Other aspects of interacting quantized fields and effective actions in a general curved spacetime, including scaling and renormalization group properties, were also studied (Toms 1982; Parker and Toms 1984, 1985b,c; Calzetta 1985, 1986; Calzetta et al., 1985). We further discuss interacting quantized fields in curved spacetime in Section 3.9, and in Chapters 5, 6, and 7.

In addition to not singling out a particular "vacuum" state in a general curved spacetime, the algebraic approach emphasizes mathematical rigor. In the context of the algebraic formulation of quantum field theory in curved spacetime rigorous methods of micro-local analysis (Duistermaat and Hörmander 1972) have been applied to defining expectation values of energy-momentum tensors and of observables in interacting field theories in curved spacetime (Brunetti et al., 1996; Radzikowski 1996; Hollands and Wald 2005). An historical review of the development of quantum field theory in curved spacetime and of the algebraic approach to it is given in Wald (2006). The use of operator product expansions to do perturbative interacting quantum field theory in curved spacetime within the rigorous algebraic viewpoint is discussed in Hollands and Wald (2008), where further references can be found.

2.14 Accelerated detector in Minkowski spacetime

A fundamental process related to quantum field theory in curved spacetime is the radiation detected by an accelerating observer in Minkowski spacetime. Quantum field theory in the coordinate system (known as Rindler coordinates) appropriate to a set of uniformly accelerated observers in Minkowski spacetime was first studied by Fulling (1972, 1973). The Minkowski metric when expressed in Rindler coordinates remains static, permitting the definition of creation and annihilation operators appropriate to the spacetime of the accelerated observers. Fulling defined these operators and found the Bogolubov transformation relating them to the usual Minkowski creation and annihilation operators of a set of inertial observers. He discovered that the Minkowski spacetime vacuum, having no particles with respect to the inertial creation and annihilation operators, appeared to have particles with respect to the creation and annihilation operators of the accelerated observers. It remained unclear how to interpret those "particles." In Davies (1975), it was pointed out that the spectrum of the latter particles (obtained from the Bogolubov transformation found by Fulling) was a blackbody spectrum having a temperature given by $a/(2\pi)$, where a is the constant acceleration of the observers. The correct interpretation of these "particles" was discovered by Unruh (1976). By considering an accelerated detector coupled to a quantized scalar field Unruh showed that the detector would be excited with the same probability distribution as a similar detector bathed in blackbody radiation. Therefore, we will refer to the thermal response of an accelerated detector as the Fulling–Davies–Unruh effect.

There are excellent treatments of these accelerated detectors in the literature, beginning with Unruh's own exposition (Unruh (1976). Other notable treatments are in DeWitt (1975), Gibbons and Hawking (1977), Birrell and Davies (1982), and Higuchi *et al.* (1993). The instability of accelerated particles in Minkowski spacetime, such as the acceleration-induced weak-interaction decay of a proton into a neutron, positron, and neutrino are studied in Vanzella and Matsas (2001).

In Section 3.5, we derive the Fulling–Davies–Unruh effect by applying the Page approximation to a zero-temperature field in Minkowski spacetime. The Page approximation is *exact* in Minkowski spacetime and shows that a uniformly accelerated observer detects a local temperature given by $a/(2\pi)$. See the derivation leading to (3.141).

3
Expectation values quadratic in fields

3.1 Adiabatic subtraction and physical quantities

A number of quantities of physical interest, such as the action S and the energy-momentum tensor $T_{\mu\nu}$, are quadratic in the fields and their derivatives evaluated at a single point. As discussed in Chapter 1, the expectation values of such quantities diverge and can be regularized in the case of free fields in Minkowski spacetime by normal ordering. In curved spacetime, even for free fields, the implicit gravitational interaction introduces additional divergences. Furthermore, vacuum energy must be treated more carefully because it can give rise to gravitational effects through the gravitational field equations.

Various methods have been developed to regularize and renormalize quantities that involve squares or higher powers of fields or their derivatives evaluated at a single point of spacetime. Among them are proper-time regularization, dimensional regularization, zeta-function regularization, point-splitting regularization, particularly by the Hadamard method, and adiabatic regularization in homogeneous spacetimes. In this chapter we employ several of the above methods of regularization, including adiabatic, Hadamard, point-splitting, proper-time, and dimensional regularization as applied to curved spacetime. The trace anomaly of the energy-momentum tensor of the conformally coupled free field in a Robertson–Walker spacetime is derived using adiabatic regularization and compared with the equivalent result obtained from the proper-time series. It is shown that the expectation values of all the components of the energy-momentum tensor then follow from the trace anomaly by using the conformal symmetry of the spacetime. The Gaussian approximation to the propagator in curved spacetime is discussed and used to approximate the energy-momentum tensor in several spacetimes. It also motivates the study of possible non-perturbative effects that come from summing the scalar curvature terms in the proper-time series to all orders in curved spacetime. Such effects may be related to cosmological acceleration observed today. Next, Hadamard regularization is presented and used to derive the trace anomaly while preserving the conservation of the energy-momentum tensor. The Gaussian approximation is related to the innovative and powerful Page approximation, which we take up next and present in some detail. The normal coordinate local momentum-space expansion is introduced, as it allows the method of dimensional regularization to be applied in curved spacetime to interacting as well as free fields in much the same way as in flat spacetime. This will be studied in detail in Chapter 6. Finally, we present the theory of a Dirac field in curved spacetime and use it to obtain the chiral current anomaly in curved spacetime. In the derivation we use the normal coordinate local momentum-space method with the proper-time series.

We start by describing the method called adiabatic subtraction, or regularization, as applied to the free scalar field in a Robertson–Walker universe. This follows naturally from the basis we have already built, provides a clear understanding of some of the physical issues, and allows us to lead easily into more general approaches. Furthermore, the method of adiabatic regularization, when applicable, is perhaps the most efficient method for calculating the finite expectation values of quadratic field quantities, rather than just the form of the divergences. It involves a mode-by-mode subtraction process, and is particularly suitable for cases in which numerical methods must be used. It was first introduced in connection with expectation values of the particle number in an expanding universe by Parker (1965, 1968) and later generalized and applied to the energy-momentum tensor (Fulling and Parker 1974; Fulling et al., 1974; Parker and Fulling 1974). It has been further developed and applied in numerous papers (Hu and Parker 1977, 1978; Birrell 1978; Bunch 1978; Hu 1978; Bunch 1980; Berger 1982, 1984; Birrell and Davies 1982; Anderson and Parker 1987; Suen and Anderson 1987; Anderson and Eaker 1999). In Zeldovich and Starobinsky (1972), a regularization method (called n-wave regularization) that involves a mode sum and subtractions of terms corresponding to large mass particles (reminiscent of Pauli–Villars regularization) was introduced. As pointed out in Parker and Fulling (1974), this method can be shown to give the same result as adiabatic regularization. In Birrell (1978), it was shown that adiabatic regularization in homogeneous universes will give the same result as point-splitting regularization.

The metric under consideration is that of (1.130),

$$ds^2 = dt^2 - a^2(t)d\vec{x}^2.$$

The scalar field has the Lagrangian density of (2.35),

$$\mathcal{L} = \frac{1}{2}|g|^{1/2}(g^{\mu\nu}\partial_\mu\phi\partial_\nu\phi - m^2\phi^2 - \xi R\phi^2),$$

the energy-momentum tensor of (2.42), and satisfies the field equation (2.41),

$$(\Box + m^2 + \xi R)\phi = 0.$$

The first question we must ask is how to construct the Hilbert space of state vectors for this system in the absence of an isometry or Killing vector in the timelike direction. In such cases, we cannot uniquely specify a preferred set of positive frequency solutions of the field equation. Nevertheless, we can construct a suitable space of state vectors. A physical state of the system can be specified, following Dirac (1958), by a set of eigenvalues of a complete set of commuting observables. A fundamental requirement of a space of state vectors is that there exist states which can reproduce any physically allowed set of expectation values. In the absence of special symmetries or other conditions that serve to specify a unique vacuum state, any space of state vectors satisfying the above requirement may be used in setting up the quantum field theory (just as any coordinate system may be used to label events).

Nevertheless, one condition that must be met is that particles should not be created in the limit when the single particle energy is large with respect to the energy scale determined by the spacetime curvature. For the present cosmological model this means that in modes such that $k^2/a^2(t) + m^2$ is large with respect to $(\dot{a}/a)^2$ and $\ddot{a}/a$, the particle number should remain nearly constant as $a(t)$ changes, even if $a(t)$ changes by a large fraction. Equivalently, we may say that the particle number in a given mode should not change if the rate of change of $a(t)$ is sufficiently slow, or adiabatic, as measured by the above condition.

Thus, we can set up our theory as follows. Choose any complete set of solutions $f_{\vec{k}}(x)$ of the field equation that are orthonormal with respect to the conserved scalar product and satisfy an asymptotic condition, which we shall call the adiabatic condition. This condition will insure that observables such as the energy-momentum tensor and particle number behave as expected in the limit of infinitely slow expansion, or in more general terms, in the limit of infinitely small curvature of the spacetime. In order to express the adiabatic condition in a clear way, it is useful to temporarily introduce a dimensionless slowness parameter T by replacing the function $a(t)$ by a one parameter family of functions $a_T(t) \equiv a(t/T)$. Then in the limit of large T, the derivatives of $a_T(t)$ with respect to t all approach zero. We shall call the power of T^{-1} in an expression the adiabatic order. In the present case, this is equal to the number of time derivatives of $a(t)$ appearing in an expression. Therefore, we will not introduce T explicitly, but merely count time derivatives of $a(t)$ to determine adiabatic order. Nevertheless, it is convenient for us to write $O(T^{-n})$ to indicate terms of adiabatic order n or higher. For more general metrics, such as the Robertson–Walker universes with spatial curvature, we imagine the metric parameterized so that $g_{\mu\nu}(\vec{x},t) \to g_{\mu\nu}(\vec{x},t)_T \equiv g_{\mu\nu}\left(\frac{\vec{x}}{T}, \frac{t}{T}\right)$. Then spatial derivatives of the metric also increase the adiabatic order by 1, and we find, for example, that in the scalar curvature of the spatially curved Robertson–Walker universe, the term proportional to a^{-2} is of second adiabatic order because it comes from two spatial derivatives of the metric. The adiabatic expansion is most generally an expansion in powers of the curvature tensor.

In the present context, we state the *adiabatic condition* as follows. Orthonormal solutions $f_{\vec{k}}(x)$ of the field equation, which will be used to expand the quantum field and to define the basis of the space of state vectors, should in lowest adiabatic order have the form[1]

$$f_{\vec{k}}(x) \sim (Va(t)^3)^{-1/2}(2\omega_k(t))^{-1/2} \exp\left[i(\vec{k}\cdot\vec{x} - \int^t \omega_k(t')dt')\right] \quad (3.1)$$

with $\omega_k(t) = [k^2/a(t)^2 + m^2]^{1/2}$. As we will show, this condition determines the adiabatic expansion of $f_{\vec{k}}(x)$ uniquely to as many adiabatic orders as there are well-defined derivatives of $a(t)$. However, in general, the adiabatic expansion yields asymptotic, not convergent, series. Therefore $f_{\vec{k}}(x)$ is not uniquely determined by (3.1).

A motivation for the adiabatic condition is that it is satisfied at all times if $a(t)$ approaches a constant at early (or late) times and $f_{\vec{k}}(x)$ is taken to approach the Minkowski spacetime positive frequency solution at early (or late) times. This can be verified, for example, for the exact solution obtained in Section 2.8. The positive frequency solutions found for de Sitter space also satisfy the adiabatic condition, as is evident from the way they were determined.

We require that the adiabatic condition hold for the $f_{\vec{k}}(x)$ on which the Fock space is constructed, even when there are no asymptotically static regions. The adiabatic condition is clearly also consistent with the exact positive frequency solutions found in Section 2.6 for the conformally invariant field equation.

Using any complete orthonormal set $f_{\vec{k}}$ of solutions of the field equation (2.41) which satisfy the adiabatic condition, we continue our construction of the space of state vectors by writing

$$\phi(x) = \sum_{\vec{k}}\left\{A_{\vec{k}}f_{\vec{k}}(x) + A_{\vec{k}}^\dagger f_{\vec{k}}^*(x)\right\}. \quad (3.2)$$

As a consequence of the canonical commutation relations of the field and conjugate momentum, $A_{\vec{k}}^\dagger$ and $A_{\vec{k}}$ are creation and annihilation operators. Then the "vacuum" state, on which the Fock basis of state vectors is constructed by repeated application of $A_{\vec{k}}^\dagger$, is defined by

$$A_{\vec{k}}|0\rangle = 0 \quad \text{for all } \vec{k}. \quad (3.3)$$

A vacuum state defined in this way is called an adiabatic vacuum state. Because the adiabatic condition does not uniquely determine the solution $f_{\vec{k}}(x)$, as we shall see later, the "adiabatic" Fock basis of state vectors constructed here is not

---

[1] V is the coordinate volume of a cube with periodic boundary conditions. As $V \to \infty$, the continuum limit is approached.

unique. It is analogous to a coordinate system in which the metric satisfies certain asymptotic conditions. Such a coordinate system is not unique, but nevertheless can be used to label events. Analogous to the physically significant events are the sets of expectation values and matrix elements of observables. Any adiabatic basis of state vectors may be used to calculate physically significant expectation values and their time evolution.

The adiabatic condition can be shown to be satisfied for *any* initially static expansion $a(t)$, in which the $f_{\vec{k}}(x)$ are taken to be positive frequency in the initially flat spacetime. Therefore, it is natural to impose it in general, regardless of the form of $a(t)$.

In Section 2.8, we considered a class of asymptotically static expansions given by (2.104). We took the $f_{\vec{k}}(x)$ to approach positive frequency solutions at early times. The parameter s in (2.105) clearly plays the role of the slowness parameter T. The Bogolubov coefficients α_k and β_k of (2.74) satisfy (2.115), which in the limit of large s or k has the same asymptotic form as $\exp[-4\pi s k a_<^2]$, where $a_<$ is the smaller of a_1 and a_2. As s^{-1} or k^{-1} approaches zero, it is clear that $|\beta_k/\alpha_k|^2$ approaches zero faster than any power of s^{-1} or k^{-1}. From the equation, $|\alpha_k|^2 - |\beta_k|^2 = 1$, it follows that $|\beta_k|^2$ also approaches zero faster than any power of s^{-1} or k^{-1}, that is, faster than any adiabatic order. This is a special case of a theorem of Kulsrud (1957) from which it follows that $|\beta_k/\alpha_k|^2$ is zero to as many orders in the rate of change of a as a has continuous derivatives. In this case, it follows that $f_{\vec{k}}(x)$ differs from the solution $g_{\vec{k}}(x)$, which is positive frequency at late times, by terms of infinite adiabatic order. One can show that both $g_{\vec{k}}(x)$ and $f_{\vec{k}}(x)$ satisfy the adiabatic condition at all times.

The vacuum states $|0\rangle_f$ and $|0\rangle_g$ defined by

$$A_{\vec{k}}|0\rangle_f = 0, \quad a_{\vec{k}}|0\rangle_g = 0 \quad \text{for all } \vec{k}$$

are different states (the in and out vacua, respectively), but each is an adiabatic vacuum state.

Although the adiabatic condition does not yield a unique basis for the space of state vectors, it does uniquely determine the large k behavior of the $f_{\vec{k}}(x)$, or more precisely, the asymptotic expansion of $f_{\vec{k}}(x)$ in powers of T^{-1} or k^{-1} (to as many orders as $a(t)$ has continuous derivatives).

This follows from the fact (as we show later) that the higher adiabatic orders of the expansion of $f_{\vec{k}}(x)$ follow by iteration from the leading term in the expansion, which is given by (3.1). Furthermore, if $a(t)$ has at least two continuous derivatives, then the Fock space of state vectors determined by different adiabatic vacua are all unitarily equivalent (Fulling and Parker 1974; Parker and Fulling 1974) for finite L, so that they are equally good in the absence of a distinguishing symmetry for calculating expectation values of physical quantities. (One can then obtain densities in the limit $L \to \infty$. The unitary inequivalence of the Fock bases for infinite volume merely reflects the fact that different densities correspond to infinitely differing particle numbers in a homogeneous infinite universe.)

We now show that if $f_{\vec{k}}(t)$ satisfies the adiabatic condition of (3.1), then its adiabatic expansion is uniquely determined to as many orders as there are derivatives of $a(t)$. We can see this, and at the same time obtain a convenient closed "summed" form (in (3.26)–(3.30)) of the adiabatic expansion to any given finite adiabatic order, as follows. Let

$$f_{\vec{k}}(x) = (2V)^{-1/2} a(t)^{-3/2} h_k(t) e^{i\vec{k}\cdot\vec{x}}, \qquad (3.4)$$

where $f_{\vec{k}}$ is a solution of the field equation (2.41). Then $h_k(t)$ satisfies (Parker (1965), Eq. (16) with $m^2 \to m^2 + \xi R$):

$$\frac{d^2}{dt^2} h_k + \Omega_k^2 h_k = 0, \qquad (3.5)$$

with

$$\Omega_k^2 = \omega_k^2 + \sigma, \qquad (3.6)$$

$$\omega_k(t) = (k^2/a(t)^2 + m^2)^{1/2}, \qquad (3.7)$$

and

$$\sigma(t) = (6\xi - 3/4)(\dot{a}/a)^2 + (6\xi - 3/2)\ddot{a}/a. \qquad (3.8)$$

We have used $R = 6[(\dot{a}/a)^2 + \ddot{a}/a]$. Note that σ is of second adiabatic order. (For the Robertson–Walker universes of positive and negative spatial curvature, we have an additional term in R of $\pm 6/a^2$, which is of second adiabatic order because it arises by taking two spatial derivatives of the metric in calculating R.)

Dropping subscripts k, let us define new variables (Chakraborty 1973)

$$t_1 = \int^t \Omega dt', \qquad (3.9)$$

$$h_1 = \Omega^{1/2} h. \qquad (3.10)$$

One finds, after some calculation, that

$$\frac{d^2}{dt_1^2} h_1 + \Omega_1^2 h_1 = 0, \qquad (3.11)$$

with

$$\Omega_1^2 = 1 + \epsilon_2, \qquad (3.12)$$

and

$$\epsilon_2 = -\Omega^{-1/2} \frac{d^2}{dt_1^2} \Omega^{1/2}. \qquad (3.13)$$

Since ϵ_2 contains terms of second and higher adiabatic order, Ω_1^2 is unity to first adiabatic order, and we have independent solutions of (3.11)

$$h_1 = \exp(\pm i t_1) + O(T^{-2}). \qquad (3.14)$$

3.1 Adiabatic subtraction and physical quantities

In view of (3.9) and (3.10), the corresponding solutions of (3.5) are

$$h = \Omega^{-1/2} \exp(\pm i \int^t \Omega dt') + O(T^{-2}), \tag{3.15}$$

Here $O(T^{-2})$ means "terms of second or higher adiabatic order."

But (3.11) is of the same form as (3.5), so we can repeat the steps between them. Thus, let

$$t_2 = \int^{t_1} \Omega_1 dt'_1, \tag{3.16}$$

$$h_2 = \Omega_1^{1/2} h_1. \tag{3.17}$$

Then

$$\frac{d^2}{dt_2^2} h_2 + \Omega_2^2 h_2 = 0, \tag{3.18}$$

with

$$\Omega_2^2 = 1 + \epsilon_4, \tag{3.19}$$

and

$$\epsilon_4 = -\Omega_1^{-1/2} \frac{d^2}{dt_2^2} \Omega_1^{1/2}. \tag{3.20}$$

The quantity ϵ_4 contains terms of fourth and higher adiabatic order, so that two independent solutions of (3.18) are

$$h_2 = \exp(\pm i t_2) + O(T^{-4}), \tag{3.21}$$

and the corresponding solutions of (3.5) are

$$h = W_2^{-1/2} \exp\left(\pm i \int^t W_2 dt'\right) + O(T^{-4}), \tag{3.22}$$

with

$$W_2 = \Omega(1 + \epsilon_2)^{1/2}. \tag{3.23}$$

The next iteration is straightforward and gives

$$h = W_4^{-1/2} \exp\left(\pm i \int^t W_4 dt'\right) + O(T^{-6}), \tag{3.24}$$

with

$$W_4 = \Omega(1 + \epsilon_2)^{1/2}(1 + \epsilon_4)^{1/2}. \tag{3.25}$$

For adiabatic regularization in four dimensions this result is sufficient.

Repeated iteration will clearly yield

$$h = W_{2n}^{-1/2} \exp\left(\pm i \int^t W_{2n} dt'\right) + O(T^{-2n-2}), \tag{3.26}$$

with
$$W_{2n} = \Omega(1+\epsilon_2)^{1/2}(1+\epsilon_4)^{1/2}\cdots(1+\epsilon_{2n})^{1/2} \tag{3.27}$$

and
$$\epsilon_{2j} = -(1+\epsilon_{2j-2})^{-1/4}\frac{d^2}{dt_j^2}(1+\epsilon_{2j-2})^{+1/4}. \tag{3.28}$$

Here
$$t_j = \int^{t_{j-1}} \Omega_{j-1}dt'_{j-1} = \int^{t} W_{2j-2}dt', \tag{3.29}$$

so
$$\epsilon_{2j} = -(1+\epsilon_{2j-2})^{-1/4}(W_{2j-2})^{-1}\frac{d}{dt}\left[(W_{2j-2})^{-1}\frac{d}{dt}(1+\epsilon_{2j-2})^{+1/4}\right]. \tag{3.30}$$

This iteration process can be continued as long as the derivatives involved in forming the ϵ_{2j} exist. Even when $a(t)$ has continuous derivatives to all orders, the series obtained from (3.26) is in general asymptotic, but not convergent.

The adiabatic condition, (3.1), requires that the adiabatic expansion of h is given to order $2n$ by expanding the solution for h in (3.26) with the *minus* sign in the exponential. Thus, we have shown that the lowest order term in the adiabatic expansion, as specified by the adiabatic condition, uniquely determines the adiabatic expansion to all orders for which derivatives of $a(t)$ are defined. In addition, we have given an explicit expression from which the adiabatic expansion of h_k and hence $f_{\vec{k}}$ follows immediately.

A second, equivalent, method of obtaining the adiabatic expansion will be given before we finally apply the result. Consider the function h_0 defined by
$$h_0(t) = W(t)^{-1/2}\exp\left(-i\int^t W(t')dt'\right), \tag{3.31}$$

where W is twice differentiable, but otherwise arbitrary. (The function $h_0(t)$ depends on k, and should not be confused with $h_k(t)$ evaluated at $k=0$.) One finds that h_0 satisfies the equation (Parker (1969), Eqs. (19)–(22))
$$\frac{d^2h_0}{dt^2} + \left[W^2 - W^{1/2}\left(\frac{d^2}{dt^2}W^{-1/2}\right)\right]h_0 = 0. \tag{3.32}$$

Now (3.5) can be written as
$$\frac{d^2h}{dt^2} + \left[W^2 - W^{1/2}\left(\frac{d^2}{dt^2}W^{-1/2}\right)\right]h = 2WSh, \tag{3.33}$$

where we have defined a new function S by
$$2WS \equiv W^2 - W^{1/2}\frac{d^2}{dt^2}W^{-1/2} - \Omega^2. \tag{3.34}$$

From (3.26) we know that it is possible to choose W so that h_0 is a solution of (3.5) or (3.33) to adiabatic order $2n$ if $2n$ derivatives of the metric exist. But

3.1 Adiabatic subtraction and physical quantities

this implies that if W is chosen to be of adiabatic order $2n$, then S is of higher adiabatic order (as can be shown by subtracting (3.32) from (3.33), and noting that $h - h_0$ is of adiabatic order larger than $2n$). Then we can find the adiabatic series for W to as many orders as there exist derivatives of the metric by iterating (3.34) with S replaced by zero. This justifies the procedure used in Bunch (1980); an equivalent procedure was used in Parker (1965) to obtain the first three terms in the adiabatic expansion of W.

Thus, using (3.6) and assuming that W is to be chosen so that S is zero to the desired adiabatic order, we obtain from (3.34):

$$W^2 = \omega^2 + \sigma + W^{1/2} \frac{d^2}{dt^2} W^{-1/2}. \tag{3.35}$$

Writing

$$W = \omega^{(0)} + \omega^{(1)} + \omega^{(2)} + \omega^{(3)} + \omega^{(4)} + \cdots, \tag{3.36}$$

and recalling that σ of (3.8) is of second adiabatic order, we find readily that to fourth order

$$\omega^{(0)} = \omega, \tag{3.37}$$

$$\omega^{(1)} = \omega^{(3)} = 0, \tag{3.38}$$

$$\omega^{(2)} = \frac{1}{2} \omega^{-1/2} \frac{d^2}{dt^2} \omega^{-1/2} + \frac{1}{2} \omega^{-1} \sigma, \tag{3.39}$$

$$\omega^{(4)} = \frac{1}{4} \omega^{(2)} \omega^{-3/2} \frac{d^2}{dt^2} \omega^{-1/2} - \frac{1}{2} \omega^{-1} (\omega^{(2)})^2$$
$$- \frac{1}{4} \omega^{-1/2} \frac{d^2}{dt^2} \left[\omega^{-3/2} \omega^{(2)} \right]. \tag{3.40}$$

Substituting the expressions for ω and σ into (3.39), we obtain after some calculation,

$$\omega^{(2)} = -\frac{3\dot{a}^2}{8a^2\omega} - \frac{3\ddot{a}}{4a\omega} - \frac{3k^2\dot{a}^2}{4a^4\omega^3} + \frac{k^2\ddot{a}}{4a^3\omega^3} + \frac{5k^4\dot{a}^2}{8a^6\omega^5} + \frac{3\xi\dot{a}^2}{a^2\omega} + \frac{3\xi\ddot{a}}{a\omega}. \tag{3.41}$$

One can check whether the dimensions are consistent, taking dimension$(a) = 1$, dimension$(\dot{a}) = $ length^{-1}, dimension$(\omega) = $ length^{-1}, and dimension$(k) = $ length^{-1}. The corresponding expression for $\omega^{(4)}$ has 30 terms and will not be written explicitly here. (We only use $\omega^{(4)}$ in calculating the fourth-order part of the adiabatic expansion of W^{-1} in (3.46) below, for the case when $\xi = 1/6$. In that case, the expression for $\omega^{(4)}$ simplifies considerably, and we leave it to the reader to verify the result in (3.46).)

We will need also the adiabatic expansion of W^{-1}, which can be found from

$$W^{-1} = \omega^{-1} + \left(W^{-1}\right)^{(2)} + \left(W^{-1}\right)^{(4)} + O(T^{-6}), \tag{3.42}$$

with

$$(W^{-1})^{(2)} = -\omega^{-2}\omega^{(2)}, \tag{3.43}$$

$$(W^{-1})^{(4)} = \omega^{-3}\left(\omega^{(2)}\right)^2 - \omega^{-2}\omega^{(4)}. \tag{3.44}$$

An important case to consider is when $\xi = 1/6$. We know that the *exact* solution in the case when $\xi = 1/6$ and $m = 0$ is obtained by letting $W = \omega = k/a$. Therefore, $\omega^{(2)}$ and all higher-order $\omega^{(j)}$ vanish in that case. We expect simplification to occur, even when the mass m is *non-zero*, if we set $\xi = 1/6$ in the above expressions and also express k^2, in the numerators of expressions like (3.41), as $a^2(\omega^2 - m^2)$, in order to make it manifest that the higher-order terms will vanish if m vanishes. Then we find, for the case $\xi = 1/6$,

$$(W^{-1})^{(2)}\left(\xi = \frac{1}{6}\right) = \frac{m^2\dot{a}^2}{2a^2\omega^5} + \frac{m^2\ddot{a}}{4a\omega^5} - \frac{5m^4\dot{a}^2}{8a^2\omega^7} \tag{3.45}$$

and

$$(W^{-1})^{(4)}\left(\xi = \frac{1}{6}\right) =$$
$$-\frac{m^2\dot{a}^4}{2a^4\omega^7} - \frac{7m^2\ddot{a}^2}{16a^2\omega^7} - \frac{m^2\dddot{a}}{16a\omega^7} - \frac{33m^2\dot{a}^2\ddot{a}}{16a^3\omega^7}$$
$$-\frac{11m^2\dot{a}\,\dddot{a}}{16a^2\omega^7} + \frac{49m^4\dot{a}^4}{8a^4\omega^9} + \frac{21m^4\ddot{a}^2}{32a^2\omega^9} + \frac{35m^4\dot{a}^2\ddot{a}}{4a^2\omega^9}$$
$$+\frac{7m^4\dot{a}\,\dddot{a}}{8a^2\omega^9} - \frac{231m^6\dot{a}^4}{16a^4\omega^{11}} - \frac{231m^6\dot{a}^2\ddot{a}}{32a^3\omega^{11}} + \frac{1155m^8\dot{a}^4}{128a^4\omega^{13}}. \tag{3.46}$$

Now consider the quantity $\langle 0|\,\phi(x)\phi(x')\,|0\rangle$. From (3.2), (3.3), and (3.4) and the commutation rules, we obtain

$$\langle 0|\,\phi(x)\phi(x')\,|0\rangle = \sum_{\vec{k}} f_{\vec{k}}(x) f_{\vec{k}}^*(x')$$

$$= \frac{1}{2(2\pi)^2}[a(t)a(t')]^{-\frac{3}{2}} \int d^3k\, e^{i\vec{k}\cdot(\vec{x}-\vec{x}')} h_k(t) h_k^*(t'), \tag{3.47}$$

where in the last step we took the continuum limit, replacing $V^{-1}\Sigma_{\vec{k}}$ by $(2\pi)^{-3}\int d^3k$. When $x = x'$ this becomes

$$\langle 0|\,\phi(x)^2\,|0\rangle = \left(4\pi^2 a^3(t)\right)^{-1}\int_0^\infty dk\, k^2\,|h_k(t)|^2. \tag{3.48}$$

The ultraviolet divergences in this quantity can be ascertained from the adiabatic expansion of h_k. To see this, let $Q^{(n)}$ denote the term of adiabatic order n in the adiabatic expansion of a quantity Q. Let $Q_1^{(n)}$ be the term in $Q^{(n)}$ which in the limit of large k has the smallest power of k^{-1}. Then $Q_1^{(n+1)}$ will have at least one more power of k^{-1} than does $Q_1^{(n)}$, as can be seen from the adiabatic expansions we have already given.

3.1 Adiabatic subtraction and physical quantities

In lowest adiabatic order and for large k, we have

$$\langle 0| \phi(x)\phi(x') |0\rangle \propto \int d^3k \, e^{i\vec{k}\cdot(\vec{x}-\vec{x}')} k^{-1} e^{-ik\int_{t'}^{t} a^{-1} dt''}.$$

When $\vec{x} \neq \vec{x}'$ or $t \neq t'$, the oscillating exponentials make this well defined as a distribution (e.g., introduce $e^{-\alpha k}$ with $\alpha \to 0^+$). However, when $x = x'$ this diverges quadratically (if $\int_0^\infty dk$ is replaced by $\lim_{K\to\infty} \int_0^K dk$), and is not defined even in the sense of a distribution. Divergences occurring in the limit of large K are called ultraviolet (UV) divergences.

In adiabatic regularization, or subtraction, the physically relevant finite expression is obtained from the formal one containing UV divergences by subtracting mode by mode (i.e., under the integral sign) each term in the adiabatic expansion of the integrand that contains at least one UV divergent part for arbitrary values of the parameters of the theory (in this case ξ and m). Thus, if $Q^{(n)}$ denotes the term of adiabatic order n in the expansion of the integrand, we must subtract all of $Q^{(n)}$ – including parts which are not UV divergent, if any term in $Q^{(n)}$ gives rise to a UV divergence. One must subtract all of $Q^{(n)}$ because otherwise the definition of the finite part to retain would be ambiguous. For example, a logarithmically divergent term such as $\int_0^\infty dk \, k^2 \omega_k^{-3}$ can be written as $\int_0^\infty dk \, k^2 \omega_k^{-5} \omega_k^2 = a^{-2} \int_0^\infty dk \, k^4 \omega_k^{-5}$
$+ m^2 \int_0^\infty dk \, k^2 \omega_k^{-5}$, which consists of another logarithmically divergent term plus a finite term. By subtracting all of $Q^{(n)}$, we give a unique characterization of the terms to be subtracted. We subtract only the minimum number of terms required (for general values of the parameters) to obtain a finite result because that procedure retains as much as possible of the original formal expression. Also, in cases when the procedure can be justified in terms of renormalization of coupling constants, subtracting the minimum number of terms requires the fewest number of coupling constants in the theory. In a case when $\int_0^\infty d\mu(k) Q^{(n)}(\xi_0, m_0)$ converges for some particular values ξ_0, m_0 of the parameters, but diverges for general values of ξ, m, we subtract $Q^{(n)}$ even at (ξ_0, m_0). (Here $d\mu(k)$ is the relevant measure; in the present example $d\mu(k) = dk \, k^2$.) This procedure assures that the physical expectation value of the quantity quadratic in the fields will be continuous as a function of the parameters (except perhaps in exceptional circumstances).

Let us now apply that procedure to (3.48). We know from (3.26) that we can express the adiabatic expansion of $h_{\vec{k}}(t)$, to any finite order (up to the number of metric derivatives which exist), in the form

$$h_k(t) \sim W_k^{-1/2}(t) \exp\left(-i \int^t W_k(t') dt'\right),$$

where the adiabatic expansion of W_k to fourth order was given in (3.36)–(3.40). This gives for the adiabatic expansion of (3.48)

$$\langle 0| \phi(x)^2 |0\rangle \sim (4\pi^2 a(t)^3)^{-1} \int_0^\infty dk\, k^2\, W_k(t)^{-1}. \tag{3.49}$$

Then if we use (3.42)

$$\langle 0| \phi(x)^2 |0\rangle \sim (4\pi^2 a(t)^3)^{-1} \int_0^\infty dk\, k^2 \Big\{ (W_k^{-1})^{(0)}$$
$$+ (W_k^{-1})^{(2)} + (W_k^{-1})^{(4)} + O(T^{-6}) \Big\}. \tag{3.50}$$

Here $(W_k^{-1})^{(0)} = \omega_k^{-1} \underset{k\to\infty}{\sim} k^{-1}$. From our previous discussion we must have $(W_k^{-1})^{(2)} \underset{k\to\infty}{\sim} k^{-p_2}$ with $p_2 \geq 3$ and $(W_k^{-1})^{(4)} \underset{k\to\infty}{\sim} k^{-p_4}$ with $p_4 \geq 5$. Therefore the zeroth-order term diverges quadratically, the second-order term diverges at worst logarithmically, and the fourth-order term converges.

In fact, from (3.45), we see that when $\xi = 1/6$ we have

$$(W^{-1})^{(2)} \underset{k\to\infty}{\sim} k^{-5},$$

so the second-order term converges in that case. However, before deciding whether to subtract this term, we must consider arbitrary values of ξ. When non-gravitational interactions are present, it may be *necessary* to leave ξ arbitrary as it may get renormalized. In addition, the case of minimal coupling ($\xi = 0$) is often also of interest. Therefore, we next consider the case when $\xi \neq 1/6$. From (3.41) and (3.43), we find that

$$(W^{-1})^{(2)} = \frac{m^2 \dot{a}^2}{2a^2\omega^5} + \frac{m^2 \ddot{a}}{4a\omega^5} - \frac{5m^4 \dot{a}^2}{8a^2\omega^7} - \frac{3\xi' \dot{a}^2}{a^2\omega^3} - \frac{3\xi' \ddot{a}}{a\omega^3}, \tag{3.51}$$

where

$$\xi' \equiv \xi - \frac{1}{6}. \tag{3.52}$$

Hence, for general values of ξ (including the value 1/6) we have

$$\langle 0| \phi(x)^2 |0\rangle_{\text{phys}} = (4\pi^2 a(t)^3)^{-1} \int_0^\infty dk\, k^2 \Big\{ |h_k(t)|^2$$
$$- \omega_k(t)^{-1} - (W_k(t)^{-1})^{(2)} \Big\} \tag{3.53}$$

with $(W^{-1})^{(2)}$ given by (3.51). Because $h_k(t)$ satisfies the adiabatic condition, by subtracting mode by mode (i.e., under the integral sign), we have obtained a result which is well defined. Thus, the subtraction procedure carried out in this way has acted also as a regularization procedure; hence the name adiabatic regularization. We could have instead used another regularization procedure (e.g., a cut-off, point-splitting, or dimensional regularization) to first make all divergent quantities well defined, and then subtracted, finally taking the appropriate limit (e.g., $K \to \infty$, or $x \to x'$, in dimension 4). The result is the same. Thus, as defined here, adiabatic regularization serves as both a subtraction and a regularization method, yielding the physically relevant finite result. Usually, the regularization is separate from the subtraction or renormalization, and is used to produce well-defined quantities upon which the subtraction or renormalization can be carried out. For example, if we wish to study the form of the infinite quantities which are subtracted we would have to use some other regularization procedure, such as dimensional regularization, to replace them by well-defined quantities. This would, for example, be necessary if we want to justify the subtractions through infinite renormalization of coupling constants.

In the flat space limit $h_k(t)$ goes over to the Minkowski spacetime positive frequency solution and (3.53) gives zero, in agreement with normal ordering. In the case of de Sitter spacetime, the de Sitter invariant vacuum, $h_k(t)$ is given by (2.144), and (3.53) will give a non-zero result.

As a second application of adiabatic regularization, we calculate

$$\langle 0 | T^\mu{}_\mu | 0 \rangle$$

for a scalar field with $\xi = 1/6$ in the limit that $m = 0$. From (2.42), we obtain

$$T^\mu{}_\mu = -\nabla^\mu \phi \nabla_\mu \phi + 2m^2 \phi^2 + \frac{1}{6} R \phi^2 + \frac{1}{2} \Box (\phi^2),$$

which, with the help of (2.41) satisfied by the field, becomes

$$T^\mu{}_\mu = m^2 \phi^2. \tag{3.54}$$

Thus, formally we have

$$\langle 0 | T^\mu{}_\mu | 0 \rangle = m^2 \langle 0 | \phi^2 | 0 \rangle. \tag{3.55}$$

However, this does *not* imply that $\langle 0 | T^\mu{}_\mu | 0 \rangle_{\text{phys}} = m^2 \langle 0 | \phi^2 | 0 \rangle_{\text{phys}}$! The reason is that the divergences in the individual components $\langle 0 | T_\mu{}^\nu | 0 \rangle$ are worse than those in $\langle 0 | \phi^2 | 0 \rangle$. For example, $\langle 0 | T^0{}_0 | 0 \rangle$ will have a term proportional to $\int_0^\infty dk\, k^2 |\dot{h}_k(t)|^2$, which has an adiabatic expansion given by

$$\lim_{K \to \infty} \int_0^K dk\, k^2 W_k \sim \lim_{K \to \infty} \left\{ K^4 O(T^0) + K^2 O(T^{-2}) + \ln K\, O(T^{-4}) + \cdots \right\},$$

where the additional terms are finite.

Thus, the physical expectation values of $T_\mu{}^\nu$ are obtained by subtracting up to adiabatic order T^{-4}, inclusive. When $\langle 0 | T^\mu{}_\mu | 0 \rangle_{\text{phys}}$ is formed, we will of course include the sum of the subtractions up to fourth adiabatic order. This will yield

$$\langle 0 | T^\mu{}_\mu | 0 \rangle_{\text{phys}} = m^2 \left(4\pi^2 a(t)^3 \right)^{-1} \int_0^\infty dk\, k^2 \Big\{ |h_k(t)|^2 - \omega_k(t)^{-1}$$
$$- \left(W_k(t)^{-1} \right)^{(2)} - \left(W_k(t)^{-1} \right)^{(4)} \Big\}, \tag{3.56}$$

which is what we would obtain from (3.48) and (3.55) by subtracting terms up to fourth adiabatic order (compare with (3.53)). In (3.56) the quantity $\int_0^\infty dk\, k^2 \left(W_k(t)^{-1}\right)^{(4)}$ is convergent, but remains as a remnant from the subtractions which were necessary in the individual components.

Thus,

$$\langle 0|T^\mu{}_\mu|0\rangle_{\text{phys}} = m^2 \langle 0|\phi^2|0\rangle_{\text{phys}} - m^2(4\pi^2 a(t)^3)^{-1} \int_0^\infty dk\, k^2 \left(W_k(t)^{-1}\right)^{(4)}. \tag{3.57}$$

Now let us take the limit of this as $m \to 0$. One expects that

$$m^2 \langle 0|\phi^2|0\rangle_{\text{phys}} \to 0$$

as $m \to 0$. It is easily seen, by changing the variable of integration to $k/(ma)$ in (3.53), that the subtracted terms in the definition of $\langle 0|\phi^2|0\rangle_{\text{phys}}$ do not have divergences caused by m, in the limit that $m \to 0$. Furthermore, there is no reason to expect the term involving $h_k(t)$ in (3.53) to give rise to such divergences in general. Thus,

$$\langle 0|T^\mu{}_\mu|0\rangle_{\text{phys}} = -(4\pi^2 a(t)^3)^{-1} \lim_{m\to 0} m^2 \int_0^\infty dk\, k^2 \left(W_k(t)^{-1}\right)^{(4)}, \tag{3.58}$$

where $\left(W_k(t)^{-1}\right)^{(4)}$ with $\xi = 1/6$ is given in (3.46). Thus, we obtain

$$\langle 0|T^\mu{}_\mu|0\rangle_{\text{phys}} = (4\pi^2 a^3)^{-1} \lim_{m\to 0} \int_0^\infty dk\, k^2 \Bigg\{ \frac{7m^4 \ddot{a}^2}{16a^2\omega^7} + \frac{m^4 \dddot{a}}{16a\omega^7}$$
$$+ \frac{m^4 \dot{a}^4}{2a^4\omega^7} + \frac{11m^4 \dot{a}\,\dddot{a}}{16a^2\omega^7} + \frac{33m^4 \dot{a}^2 \ddot{a}}{16a^3\omega^7} - \frac{21m^6 \ddot{a}^2}{32a^2\omega^9} - \frac{49m^6 \dot{a}^4}{8a^4\omega^9}$$
$$- \frac{7m^6 \dot{a}\,\dddot{a}}{8a^2\omega^9} - \frac{35m^6 \dot{a}^2 \ddot{a}}{4a^3\omega^9} + \frac{231m^8 \dot{a}^4}{16a^4\omega^{11}} + \frac{231m^8 \dot{a}^2 \ddot{a}}{32a^3\omega^{11}} - \frac{1155m^{10}\dot{a}^4}{128a^4\omega^{13}} \Bigg\}. \tag{3.59}$$

By changing the variable of integration to $k/(ma)$ it is evident that these terms are independent of the value of m. The integration is straightforward and yields the result

$$\langle 0|T^\mu{}_\mu|0\rangle_{\text{phys}} = \frac{1}{480\pi^2}\left[\frac{\ddot{a}^2}{a^2} + \frac{\dddot{a}}{a} - 3\frac{\dot{a}^2\ddot{a}}{a^3} + 3\frac{\dot{a}\,\dddot{a}}{a^2}\right]. \tag{3.60}$$

This is the well-known trace or conformal anomaly for a free scalar field with $m = 0$ and $\xi = 1/6$. It is called an anomaly because for this field we have classically that $T^\mu{}_\mu = 0$, as mentioned in Section 2.2. It is clear that the anomaly arises from the subtractions which were necessary to arrive at the finite physical expectation values of the components of the stress tensor.

As $T^\mu{}_\mu$ is a scalar, we should be able to express the right side of (3.60) as a linear combination of the scalars

$$R^2 = \left[6\left(\frac{\dot{a}^2}{a^2} + \frac{\ddot{a}}{a}\right)\right]^2,$$

$$\Box R = 6\left(\frac{\ddot{a}^2}{a^2} + \frac{\dddot{a}}{a} - \frac{5\dot{a}^2\ddot{a}}{a^3} + \frac{3\dot{a}\,\dddot{a}}{a^2}\right),$$

and

$$R_{\mu\nu}R^{\mu\nu} = 12\left(\frac{\dot{a}^4}{a^4} + \frac{\ddot{a}^2}{a^2} + \frac{\dot{a}^2\ddot{a}}{a^3}\right).$$

We find

$$\langle 0|T^\mu{}_\mu|0\rangle_{\text{phys}} = \frac{1}{2880\pi^2}\left\{\Box R - \left(R^{\mu\nu}R_{\mu\nu} - \frac{1}{3}R^2\right)\right\}, \tag{3.61}$$

which is the generally covariant form of the trace anomaly of the massless conformal scalar field. This result can be shown to hold for the massless conformal scalar field in any conformally flat spacetime with certain restrictions on the state vector analogous to the adiabatic condition. We show this in Section 3.3, where we derive the trace anomaly of the conformal scalar field in a general curved spacetime.[2] The conformal anomaly will also be analyzed in Chapter 5 using functional integral methods and a different regularization procedure.

We will summarize the expressions for the trace anomaly of the energy-momentum tensor of conformally invariant spin-0, spin-1/2, and spin-1 fields in equations (3.93)–(3.98). Using those expressions, the derivation that we give in the next section, of the complete energy-momentum tensor from the spin-0 trace anomaly in spatially flat Friedmann–Robertson–Walker (FRW) universes, is easily generalized to higher spin. The same method also works for positively and negatively curved FRW universes.

3.2 Energy-momentum tensor from trace anomaly

The expression for $\langle T_\mu{}^\mu\rangle$ of the massless conformal scalar field in (3.60), together with

$$\nabla_\mu\langle 0|T^{\mu\nu}|0\rangle_{\text{phys}} = 0, \tag{3.62}$$

is sufficient to determine $\langle 0|T_{\mu\nu}|0\rangle_{\text{phys}}$ in the FRW spacetime under consideration. It can be shown that when adiabatic regularization is carried out, the expression for $\langle 0|T^{\mu\nu}|0\rangle_{\text{phys}}$ satisfies (3.62). This is because the terms subtracted at each adiabatic order, as well as the formal expression, $\langle 0|T^{\mu\nu}|0\rangle$, independently have vanishing covariant four-divergence. In the adiabatic expansion

$$\langle 0|T^{\mu\nu}|0\rangle \sim \langle 0|T^{\mu\nu}|0\rangle^{(0)} + \langle 0|T^{\mu\nu}|0\rangle^{(2)}T^{-2} + \langle 0|T^{\mu\nu}|0\rangle^{(4)}T^{-4} + O(T^{-6}),$$

---

[2] See (3.91) and (3.92).

application of ∇_μ to the left side (with any value of the mass m) formally gives zero. This result can hold for arbitrary slowness parameter T only if application of ∇_μ causes each adiabatic order to vanish separately on the right side.

The following argument (Parker 1979b) now determines $\langle T^{\mu\nu} \rangle$ from the trace anomaly, which is independent of state for the massless conformally invariant field. Thus $\langle T^{\mu\nu} \rangle$ below refers to the expectation value in any physical state. For the metric of (1.130) there exists a vector field

$$\xi^\mu = a(t)\delta^\mu{}_0 \tag{3.63}$$

which satisfies

$$\pounds_\xi g_{\mu\nu}(x) = \lambda(x) g_{\mu\nu}(x), \tag{3.64}$$

with

$$\lambda(x) = 2\dot{a}(t). \tag{3.65}$$

Here $\pounds_\xi$ is the Lie derivative in the ξ direction, defined by[3]

$$\pounds_\xi g_{\mu\nu} = \xi^\alpha \partial_\alpha g_{\mu\nu} + g_{\mu\alpha} \partial_\nu \xi^\alpha + g_{\alpha\nu} \partial_\mu \xi^\alpha = 2\nabla_{(\mu} \xi_{\nu)}. \tag{3.66}$$

We use the notation $\nabla_{(\mu} \xi_{\nu)} \equiv (1/2)(\nabla_\mu \xi_\nu + \nabla_\nu \xi_\mu)$. A vector field satisfying (3.64) is known as a conformal Killing vector field. From (3.62) and (3.64) it follows that

$$\nabla_\mu (\langle T^{\mu\nu} \rangle \xi_\nu) = \langle T^{\mu\nu} \rangle \nabla_{(\mu} \xi_{\nu)} = \frac{1}{2} \lambda \langle T^{\mu\nu} \rangle g_{\mu\nu}. \tag{3.67}$$

Integrating this over the four-volume bounded by constant time hypersurfaces at t_1 and t_2 gives

$$\frac{1}{2} \int d^4x \sqrt{-g} \lambda \langle T^\mu{}_\mu \rangle = \int_{t_2} d^3x \sqrt{-g} \langle T^{0\nu} \rangle \xi_\nu - \int_{t_1} d^3x \sqrt{-g} \langle T^{0\nu} \rangle \xi_\nu, \tag{3.68}$$

where the divergence theorem has been used, and the surface terms at spatial infinity, or on the cube of side L, have been dropped (in the case of periodic boundary conditions on the cube of side L they cancel one another). For the conformal Killing vector field (3.63), we have

$$\int d^4x\, a^3(t)\dot{a}(t) \langle T^\mu{}_\mu \rangle = \int_{t_2} d^3x\, a^4(t_2) \langle T_{00} \rangle - \int_{t_1} d^3x\, a^4(t_1) \langle T_{00} \rangle.$$

We now restrict the state to be the vacuum or to contain a spatially homogeneous distribution of particles. Since $\langle T_{\mu\nu} \rangle$ is a function only of t as a result of the spatial homogeneity, the spatial integrations cancel on each side (e.g., we can keep L finite before cancelation), and we have

$$\int_{t_1}^{t_2} dt\, a^3(t)\dot{a}(t) \langle T^\mu{}_\mu(t) \rangle = a^4(t_2) \langle T_{00}(t_2) \rangle - a^4(t_1) \langle T_{00}(t_1) \rangle. \tag{3.69}$$

---

[3] See (2.6) and (2.7).

With $\langle T^\mu{}_\mu \rangle$ given by (3.60), we find

$$\int_{t_1}^{t_2} dt\, a^3 \dot{a} \langle T^\mu{}_\mu \rangle = \frac{1}{480\pi^2} \int_{t_1}^{t_2} dt\, \dot{a}(a\ddot{a}^2 + a^2\dddot{a} - 3\dot{a}^2\ddot{a} + 3a\dot{a}\ddot{a})$$

$$= \frac{1}{480\pi^2} \int_{t_1}^{t_2} dt\, \frac{d}{dt}\left(-\dot{a}^4 + a^2\dot{a}\dddot{a} + a\dot{a}^2\ddot{a} - \frac{1}{2}a^2\ddot{a}^2\right)$$

$$= g(t_2) - g(t_1),$$

where

$$g(t) = \frac{1}{480\pi^2}\left(-\dot{a}^4 + a^2\dot{a}\dddot{a} + a\dot{a}^2\ddot{a} - \frac{1}{2}a^2\ddot{a}^2\right). \tag{3.70}$$

It follows from (3.69) and (3.70) that

$$a^4(t)\langle T_{00}(t)\rangle = g(t) + E, \tag{3.71}$$

where E is a constant. Thus, the expectation value of the energy density is (for $\xi = 1/6$, $m = 0$)

$$\langle T_{00}(t)\rangle = \frac{1}{480\pi^2}\left(-\frac{\dot{a}^4}{a^4} + \frac{\dot{a}\dddot{a}}{a^2} + \frac{\dot{a}^2\ddot{a}}{a^3} - \frac{1}{2}\frac{\ddot{a}^2}{a^2}\right) + \frac{E}{a^4}. \tag{3.72}$$

By symmetry, we have $\langle T^1{}_1 \rangle = \langle T^2{}_2 \rangle = \langle T^3{}_3 \rangle$, so

$$\langle T^1{}_1 \rangle = \frac{1}{3}\left(\langle T^\mu{}_\mu \rangle - \langle T^0{}_0 \rangle\right)$$

$$= \frac{1}{1440\pi^2}\left\{\frac{\dot{a}^4}{a^4} + 2\frac{\dot{a}\dddot{a}}{a^2} - 4\frac{\dot{a}^2\ddot{a}}{a^3} + \frac{3}{2}\frac{\ddot{a}^2}{a^2} + \frac{\dddot{a}}{a}\right\} - \frac{E}{3a^4}. \tag{3.73}$$

Since the state is taken to be invariant under the isometries of the Robertson–Walker line element, the off-diagonal components of $\langle T_{\mu\nu} \rangle$ are zero.

The constant E included above allows for the possibility that the state contains an isotropic distribution of particles of the massless field under consideration. We see that there is no particle creation because in a statically bounded expansion of the universe from scale factor a_1 to a_2, the initial massless fluid of energy density $(480\pi^2)^{-1}Ea_1^{-4}$ simply expands into a final massless fluid of energy density $(480\pi^2)^{-1}Ea_2^{-4}$, as would be expected for a gas of massless particles in the absence of particle creation. The terms in $\langle T_{00}(t)\rangle$ and $\langle T^1{}_1(t)\rangle$ which vanish when $a(t)$ is constant do not correspond to particle creation because they depend only on the value of a and its first four derivatives at time t and not on the past history of the universe. Therefore, such terms must be viewed as a vacuum contribution to $\langle T^\mu{}_\nu \rangle$ and not as the result of an accumulation of created particles. This is consistent with our earlier proof that there is no particle creation for the free scalar field with $\xi = 1/6$ and $m = 0$ in Robertson–Walker universes.

The above derivation of $\langle T_{\mu\nu}\rangle$ from $\langle T^{\mu}{}_{\mu}\rangle$ based on the conformal Killing vector is carried out for general FRW universes (including curved three-spaces) and fields up to spin-1 in Parker (1979b). If the Lagrangian includes a self-interaction term $\lambda\phi^4$, then, although it is classically conformally invariant, renormalization will affect the value of ξ, breaking conformal invariance and causing particle creation to occur (Birrell and Davies (1982), pp. 301–317, and references cited there).

3.3 Renormalization in general spacetimes

Let us consider the free neutral scalar field in an n-dimensional spacetime with line element

$$ds^2 = g_{\mu\nu}(x)dx^\mu dx^\nu. \tag{3.74}$$

The field satisfies

$$(\Box + m^2 + \xi R)\phi = 0 \tag{3.75}$$

and is expanded as before in terms of mode functions

$$\phi(x) = \sum_j \left(A_j f_j(x) + A_j^\dagger f_j^*(x)\right), \tag{3.76}$$

where, as in Minkowski spacetime, the j can run over a discrete or continuous set. If the Cauchy surface consists of several separately labeled surfaces, such as future null infinity and the future event horizon of a black hole, then j includes an index specifying the Cauchy surface for which f_j and f_j^* form a complete orthonormal set.

The analogue of the adiabatic condition will be specified as a condition on the Feynman Green function or propagator (or any of the related Green functions satisfying the same equation). Let $|0\rangle$ be defined as

$$A_j|0\rangle = 0, \quad \text{for all } j. \tag{3.77}$$

The Feynman propagator is defined as

$$G_F(x, x') = -i\langle 0|T(\phi(x)\phi(x'))|0\rangle, \tag{3.78}$$

where T is the time-ordering operator for the coordinate x^0. As a consequence of (3.75) and (3.78), G_F satisfies

$$(\Box_x + m^2 + \xi R(x))G_F(x, x') = -\delta(x, x'), \tag{3.79}$$

where

$$\delta(x, x') \equiv |g(x)|^{-1/2}\delta(x - x'). \tag{3.80}$$

3.3 Renormalization in general spacetimes

The minus sign in (3.79) is a convention, and $\delta(x, x')$ and G_F are biscalars, meaning that they transform in the same way as the product of a scalar at x with another scalar at x'.

The analogue of the adiabatic postulate can now be stated by requiring that the $f_j(x)$ are such that $G_F(x, x')$ has an adiabatic expansion (i.e., an expansion in numbers of derivatives of the metric) with leading term of the form (in n dimensions)

$$G_F(x, x') \sim \Delta^{1/2}(x, x')(4\pi)^{-n/2} \int_0^\infty ds(is)^{-n/2} \exp\left[-im^2 s + \frac{\sigma(x, x')}{2is}\right], \tag{3.81}$$

where

$$\Delta(x, x') = -|g(x)|^{-1/2} \det[-\partial_\mu \partial_{\nu'} \sigma(x, x')]|g(x')|^{-1/2} \tag{3.82}$$

is the biscalar Van Vleck–Morette determinant (Van Vleck 1928; Morette 1951), and

$$\sigma(x, x') = \frac{1}{2}\tau(x, x')^2, \tag{3.83}$$

where $\tau(x, x')$ is the proper distance along the geodesic between x and x'. We assume that x' is in a normal neighborhood of x, meaning that geodesics emanating from x do not intersect, so only one geodesic in the normal neighborhood goes from x to x'. In (3.81), s is a scalar parameter.

Given the first term, (3.81) in the adiabatic expansion of G_F, we can obtain the other terms by iteration. The general form of the expansion is (Minakshishundaram and Pleijel 1949; DeWitt 1964a)

$$G_F(x, x') \sim -i\Delta^{1/2}(x, x')(4\pi)^{-n/2} \int_0^\infty ids(is)^{-n/2}$$
$$\times \exp\left[-im^2 s + \frac{\sigma(x, x')}{2is}\right] F(x, x'; is), \tag{3.84}$$

with

$$F(x, x'; is) \sim a_0(x, x') + a_1(x, x')(is) + a_2(x, x')(is)^2 + \cdots. \tag{3.85}$$

We will only need the coincidence limits of a_0 through a_2 here:

$$a_0(x) = 1, \tag{3.86}$$

$$a_1(x) = \left(\frac{1}{6} - \xi\right) R, \tag{3.87}$$

$$a_2(x) = \frac{1}{180} R_{\alpha\beta\gamma\delta} R^{\alpha\beta\gamma\delta} - \frac{1}{180} R^{\alpha\beta} R_{\alpha\beta}$$
$$- \frac{1}{6}\left(\frac{1}{5} - \xi\right) \Box R + \frac{1}{2}\left(\frac{1}{6} - \xi\right)^2 R^2, \tag{3.88}$$

where $a_i(x) \equiv \lim_{x' \to x} a_i(x, x')$. It is clear from (3.81) that $a_0(x, x') = 1$. The expressions for a_1 and a_2 are derived by the normal coordinate momentum-space method in Section 3.8. A derivation using Riemann normal coordinates in configuration space is given in Parker (1979b).

The adiabatic expansion is the same for all the different Green functions, just as all acceptable mode functions considered earlier have the same adiabatic expansion. Also, m^2 is understood to be $m^2 - i\epsilon$ with ϵ a positive infinitesimal, so there is no divergence in (3.84) as $s \to \infty$. Although this choice picks out a particular propagator (e.g., the Feynman propagator in flat spacetime), it does not affect the adiabatic expansion. Therefore, we can use this convenient choice to determine the subtractions that must be made in doing renormalization for any Green function. Our focus in this section is on these subtractions. (To find the specific form of the Green function after subtraction, we must of course choose boundary conditions appropriate to the particular problem under consideration.)

The divergences at $s = 0$ are the short distance UV divergences because they occur for $\sigma(x, x') \to 0$. From (3.84) and (3.85), we see that on a general spacetime, $G(x, x')$ will have a divergence in the s integration as $s \to 0$ when $\sigma(x, x') \to 0$ for the jth term in the expansion if $j - \frac{n}{2} \leq -1$. Thus, in four dimensions ($n = 4$), we have divergences for $j = 0$ and $j = 1$ (i.e., up to adiabatic order T^{-2}, as we found earlier using adiabatic regularization). The adiabatic regularization or subtraction method then tells us that in four dimensions

$$\langle 0|\phi(x)\phi(x')|0\rangle_{\text{phys}} = \lim_{n \to 4} \Big\{ \langle 0|\phi(x)\phi(x')|0\rangle + i \times i\Delta^{1/2}(x, x')(4\pi)^{-\frac{n}{2}}$$
$$\times \int_0^\infty ids(is)^{-\frac{n}{2}} \exp\left[-im^2 s + \frac{\sigma(x, x')}{2is}\right]$$
$$\times (1 + a_1(x, x')(is)) \Big\}. \qquad (3.89)$$

This has a smooth well-defined limit as $x' \to x$ in any spacetime. The limit as $n \to 4$ is necessary here because the individual terms on the right-hand side have poles at $n = 4$, but are well defined (i.e., regularized) when we take $n \neq 4$. The pole terms (which diverge as $n \to 4$) should cancel, so the limit exists and is finite. This manifestly covariant procedure of taking $n \neq 4$ to make individual terms well defined is known as dimensional regularization. In adiabatic regularization earlier, subtraction mode by mode made such a procedure unnecessary.

Next consider $\langle 0|T^\mu{}_\mu|0\rangle$ for $\xi = 1/6$, $m = 0$. As before, we have for $m \neq 0$,

$$\langle 0|T_\mu{}^\mu(x)|0\rangle = m^2 \langle 0|\phi^2(x)|0\rangle, \qquad (3.90)$$

and as explained in connection with adiabatic regularization, making the minimum subtractions necessary to cancel (for general ξ) divergences in the individual components of $\langle 0|T_\mu{}^\nu(x)|0\rangle$, we have

3.3 Renormalization in general spacetimes

$$\langle 0|T_\mu{}^\mu(x)|0\rangle_{\text{phys}} = \lim_{\substack{n\to 4\\ m\to 0}} m^2 \Big\{ \langle 0|\phi^2(x)|0\rangle$$
$$+ i\times i(4\pi)^{-\frac{n}{2}} \int_0^\infty ids(is)^{-\frac{n}{2}} \exp[-im^2 s] \times$$
$$(1 + a_1(x)(is) + a_2(x)(is)^2)\Big\}$$
$$= \lim_{m\to 0} m^2 \langle 0|\phi^2(x)|0\rangle_{\text{phys}}$$
$$- \lim_{\substack{n\to 4\\ m\to 0}} m^2 (4\pi)^{-\frac{n}{2}} \int_0^\infty ids(is)^{-\frac{n}{2}} \exp(-im^2 s) a_2(x)(is)^2$$
$$= \lim_{m\to 0} \Big\{ m^2 (4\pi)^{-2} \int_0^\infty ids\, \exp(-im^2 s) \Big\} a_2(x)$$
$$= -(4\pi)^{-2} a_2(x)\big|_{\xi=1/6}, \qquad (3.91)$$

where we used $\Delta(x,x) = 1$, $\sigma(x,x) = 0$, and

$$\int_0^\infty ds\, \exp[-i(m^2 - i\epsilon)s] = \frac{1}{im^2}.$$

Here a_2 is (3.88) evaluated for $\xi = 1/6$:

$$a_2(x)\big|_{\xi=1/6} = \frac{1}{180} R_{\alpha\beta\gamma\delta} R^{\alpha\beta\gamma\delta} - \frac{1}{180} R_{\alpha\beta} R^{\alpha\beta} - \frac{1}{180} \Box R. \qquad (3.92)$$

This expression is valid for the conformally invariant scalar field. The result does not require the spacetime to be conformally flat.

Finally, let us compare this with the result obtained by adiabatic regularization in the spatially flat Robertson–Walker universe. The FRW universes are conformally flat, so that the Weyl tensor $C^\alpha{}_{\beta\gamma\delta}$ is zero. In four dimensions

$$C_{\alpha\beta\gamma\delta} C^{\alpha\beta\gamma\delta} = R_{\alpha\beta\gamma\delta} R^{\alpha\beta\gamma\delta} - 2 R_{\alpha\beta} R^{\alpha\beta} + \frac{1}{3} R^2, \qquad (3.93)$$

so in an FRW universe

$$R_{\alpha\beta\gamma\delta} R^{\alpha\beta\gamma\delta} = 2 R_{\alpha\beta} R^{\alpha\beta} - \frac{1}{3} R^2, \qquad (3.94)$$

and

$$a_2(x)\big|_{\xi=1/6} = \frac{1}{180}\Big(R_{\alpha\beta} R^{\alpha\beta} - \frac{1}{3} R^2\Big) - \frac{1}{180} \Box R.$$

Then

$$\langle 0|T^\mu{}_\mu|0\rangle_{\text{phys}} = \frac{1}{2880\pi^2} \Big\{\Box R - \Big(R_{\alpha\beta} R^{\alpha\beta} - \frac{1}{3} R^2\Big)\Big\}, \qquad (3.95)$$

in agreement with (3.61).

3.3.1 Trace anomaly for fields of spins 1/2 and 1

A massless free spin-1/2 field satisfies a conformally invariant field equation. Therefore, the trace of its classical energy-momentum tensor vanishes. Just as for the conformally invariant free massless scalar field, the free massless spin-1/2 field when quantized has an analogous trace anomaly. The same is true of the massless field of spin-1 (the photon) in four spacetime dimensions. Classically, the Maxwell field is conformally invariant. When quantized in curved spacetime, its energy-momentum tensor also has a trace anomaly. Here we will summarize the values that have been obtained for those anomalies in a general curved spacetime. (We do not consider non-trivial topologies or boundaries.)

The trace anomaly of the free conformally invariant spin-0 field in a general four-dimensional curved spacetime is given by (3.91) and (3.92). Let us rewrite that result in terms of the contracted square of the Weyl tensor given in (3.93), and the integrand of the Gauss–Bonnet topological invariant:

$$G \equiv R_{\alpha\beta\gamma\delta} R^{\alpha\beta\gamma\delta} - 4 R_{\alpha\beta} R^{\alpha\beta} + R^2. \tag{3.96}$$

Eliminating, $R_{\alpha\beta\gamma\delta} R^{\alpha\beta\gamma\delta}$ and $R_{\alpha\beta} R^{\alpha\beta}$ in favor of G and $C_{\alpha\beta\gamma\delta} C^{\alpha\beta\gamma\delta}$, we can write the spin-0 trace anomaly as

$$\langle 0 | T_\mu{}^\mu(x) | 0 \rangle_{\text{phys}} = \frac{1}{4\pi^2} \left\{ \frac{1}{120} C_{\alpha\beta\gamma\delta} C^{\alpha\beta\gamma\delta} - \frac{1}{360} G + \frac{1}{180} \Box R \right\}. \tag{3.97}$$

Various people have calculated the trace anomalies for the spin-1/2 and spin-1 conformally invariant fields.[4] The results have the same general form as for the spin-0 case, but with different coefficients. Thus, for a massless Dirac spin-1/2 particle and a massless neutral spin-1 particle the results can be expressed as

$$\langle 0 | T_\mu{}^\mu(x) | 0 \rangle_{\text{phys}} = \frac{1}{4\pi^2} \left\{ A_s C_{\alpha\beta\gamma\delta} C^{\alpha\beta\gamma\delta} + B_s G + C_s \Box R \right\}, \tag{3.98}$$

with $A_{1/2} = 1/20$, $B_{1/2} = -11/360$, and $A_1 = 1/10$, $B_1 = -31/180$ respectively. The values of the C coefficients can be changed by adding a term of the form R^2 to the action. The variation of such a term with respect to the metric would give terms having a trace proportional to $\Box R$ which could change the coefficient of the $\Box R$ term in the trace anomaly. The other terms in the anomaly cannot be altered by adding analogous local terms formed from contractions of two Riemann tensors. The value of the C coefficient obtained from various regularization methods yield $C_{1/2} = 1/30$, but give values for C_1 that seem to depend on the method of regularization. Point-splitting and zeta-function regularization give $C_1 = -1/10$, and dimensional regularization gives $C_1 = 1/15$ (Duff 1994; Hawking et al., 2001). For a two-component massless Pauli neutrino, the result is $A_{1/2} = 1/40$, $B_{1/2} = -11/720$, $C_{1/2} = 1/60$ which is half that given above

---

[4] See Birrell and Davies (1982) and references therein.

for a massless Dirac fermion because the Pauli neutrino has half the number of helicity states (Parker 1979b).

It was proved in Parker (1979b), using the conservation law based on the conformal Killing vector field in Robertson–Walker universes, that the above trace anomalies imply that there are no free massless conformally invariant scalar, Dirac, or spin-1 (photon) particles created. In addition to the three terms that appear in the trace anomaly of (3.97) and (3.98), there could be an independent additional term proportional to R^2. As shown in Parker (1979b), the absence of such a term in the trace anomaly is necessary and sufficient for there to be no creation of such particles in Robertson–Walker universes. (The proof given in Section 3.2 is easily extended to the spin-1/2 and spin-1 cases, or more generally to any values of the coefficients A, B, and C. But if there was a further R^2 term in the anomaly, then the proof would not go through.) The absence of an additional R^2 term in the trace anomaly is necessary for there to be consistency with the earlier proof based on a conformal transformation that showed there are no such particles created in a Robertson–Walker universe, and that the corresponding (conformal) vacuum state is uniquely determined for those fields.[5]

3.4 Gaussian approximation to propagator

An interesting approximation to the Feynman propagator, $G_F(x, x')$, can be obtained by defining the kernel, $\langle x, s | x', 0 \rangle$, through the equation (the Dirac bra-ket notation will become clear shortly):

$$G_F(x, x') = -i \int_0^\infty ds \exp\left(-im^2 s\right) \langle x, s | x', 0 \rangle, \tag{3.99}$$

where $\mathrm{Im}(m^2) < 0$ is understood. As a consequence of (3.79), the kernel $\langle x, s | x', 0 \rangle$ satisfies a Schrödinger equation:

$$i \frac{\partial}{\partial s} \langle x, s | x', 0 \rangle = (\Box_x + \xi R(x)) \langle x, s | x', 0 \rangle \tag{3.100}$$

with the boundary condition that $\lim_{s \to 0} \langle x, s | x', 0 \rangle = \delta(x, x')$. As noted by Schwinger (1951a), the kernel $\langle x, s | x', 0 \rangle$ represents the probability amplitude for a fictitious particle to propagate, in a "proper-time" interval s, from point x to point x' on a hypersurface having the number of dimensions of the original spacetime. Equations (3.84)–(3.88) come from the asymptotic expansion of the kernel, $\langle x, s | x', 0 \rangle$, in powers of s. This asymptotic expansion follows by iteration (DeWitt 1964a) from (3.100). As already discussed, the leading terms in this asymptotic expansion give a generally covariant way of expressing the *divergent* terms that appear in the expectation values of products of quantum fields at a single spacetime point.

---

[5] See Chapter 2 and Parker (1965, 1968, 1969, 1971, 1972).

A different approach, aimed at finding a good approximation to the propagator at widely separated points, was initiated by Bekenstein and Parker (1981); see also Parker (1979a). The basic idea is to express the kernel $\langle x, s | x', 0 \rangle$ in the form of a Feynman path integral solution of Schrödinger's equation, and to use Fermi coordinates with respect to the path between x and x' that extremizes the action that appears in the path integral. (If there are a number of distinct paths between x and x' that extremize the action, then we can add their contributions.) The main contribution to the path integral comes from paths that lie close to the path that extremizes the action, since the amplitudes of such nearby paths have nearly the same phase. The key point is that in Fermi coordinates, the metric at a point in a neighborhood of the extremal path can be expressed as a series in powers of the distance from the extremal path and of the Riemann tensor evaluated at the point *on* the extremal path from which that distance is measured. A good approximation to the value of the path integral is found by keeping terms up to second order in the distance from the extremal path. The path integral can then be evaluated by doing a set of Gaussian integrations (Bekenstein and Parker 1981). When the scalar curvature R is constant along the extremal path, this Gaussian approximation has a particularly simple form, which in a four-dimensional spacetime is

$$\langle x, s | x', 0 \rangle_{\text{Gauss}} = -i \Delta^{1/2}(x, x') (4\pi s)^{-2} \exp\left[\frac{\sigma(x, x')}{2is} - i\left(\xi - \frac{1}{6} \right) Rs \right], \tag{3.101}$$

where $\Delta(x, x')$ is the Van Vleck–Morette determinant of (3.82). Examples of spacetimes with constant scalar curvature, R, are the static Einstein universe (a static universe with spherical three-dimensional spatial cross sections) and the de Sitter universe (a four-dimensional spacetime of maximal symmetry). The Gaussian approximation can be shown to give an *exact* solution for any pair of separated points in the Einstein static universe and a very good approximation for widely separated points in the de Sitter universe (Bekenstein and Parker 1981).

When $\xi = 1/6$ in an arbitrary spacetime, or when ξ is arbitrary in any spacetime having $R = 0$, (3.101) reduces to

$$\langle x, s | x', 0 \rangle_{\text{Gauss}} = -i \Delta^{1/2}(x, x') (4\pi s)^{-2} \exp\left[\frac{\sigma(x, x')}{2is} \right]. \tag{3.102}$$

This last result has the same form as the first term in the proper-time Schwinger–DeWitt series for the kernel of the propagator. The latter series can be read off from the definition of the kernel (3.99), and (3.84)–(3.88). Although the series is not in general convergent, its first few terms give the form of the ultraviolet-divergent terms in quantities such as the energy-momentum tensor. (This fact was used to renormalize the trace of the energy-momentum tensor in the previous section.) For the purpose of renormalization, only the form of the series for nearby points and small s is required. There is no reason to expect a truncated form of the

3.4 Gaussian approximation to propagator

Schwinger–DeWitt series to give a good approximation to the actual expression for the kernel of the propagator for widely separated points or for large values of s.

However, the Gaussian approximation to the kernel of the Feynman propagator was derived by a method (as described above) that is valid for widely separated points and does not require s to be small. This fact means that the Gaussian approximation should give a good approximation to the actual kernel for arbitrary values of s, x, and x' (Bekenstein and Parker 1981). Hence, integration over s in (3.99) should yield a good approximation to the Feynman propagator, $G_F(x, x')$. However, as we know from the previous section, the trace anomaly comes from the term of order s^2 in the Schwinger–DeWitt series, so we do not expect the Gaussian approximation to give an energy-momentum tensor that has a trace anomaly. On the other hand, as noted by Page (1982), we can expect the Gaussian approximation to be particularly good in spacetimes in which the known expression for the trace anomaly vanishes (i.e., in which the term of order s^2 in the series is zero).

This observation motivated Page (1982) to use the Gaussian approximation as the basis for an ingenious way of approximating the energy-momentum tensor of a massless conformally coupled quantized field in a class of spacetimes that are conformally related to an ultrastatic spacetime (defined in the next section). Because this class of spacetimes includes the Schwarzschild black hole, he was able to find a good analytic approximation to the energy-momentum tensor of a Schwarzschild black hole, a problem that had proved too difficult for an exact analytic solution. Page starts from an ultrastatic spacetime that is conformally related to the Schwarzschild spacetime. He shows that the trace anomaly is zero in ultrastatic spacetimes, so the Gaussian approximation should be particularly accurate. He uses the Gaussian approximation to find the regularized energy-momentum tensor in the conformally ultrastatic spacetime, and then conformally transforms it to the Schwarzschild spacetime, taking into account the breaking of conformal invariance that yields the trace anomaly. Page works in a state of thermal equilibrium at a given temperature. (Page also considers vacuum spacetimes with a non-zero cosmological constant, including de Sitter spacetime.) The Page approximation and related work will be discussed in the next section.

The Gaussian approximation is based on an expansion for the metric in powers of the distance from the extremal path. However, its derivation does *not* involve an expansion in powers of s. The fact that it is non-perturbative in s is evident in the term involving the exponential of Rs in (3.101). As proposed by Parker and Toms (1985a) and proved by Jack and Parker (1985, 1987), this exponential is in fact related to the sum of *all* terms in the Schwinger–DeWitt proper-time series for the kernel, $\langle x, s | x, 0 \rangle$, that involve at least one factor of the scalar curvature $R(x)$ (in the form of the series that is independent of the dimension of spacetime). By factoring out this exponential involving $R(x)s$, the terms of the proper-time series are considerably simplified. In addition, the exponential term involving $R(x)s$ incorporates non-perturbative effects of quantum fields in curved spacetimes. This form of the kernel is called the "partially-summed" or "R-summed" form of the kernel. It will be presented in Section 3.6, followed by a discussion of the propagator and effective action that it implies.

The R-summed form of the propagator gives rise to non-perturbative vacuum effects which may have important cosmological implications (Parker and Toms 1985b,c; Parker and Raval 1999a,b, 2000, 2001). It has been argued (Parker and Raval 1999a,b, 2000, 2001; Parker et al., 2003; Parker and Vanzella 2004) that a free quantized field of very small mass will give rise, through this non-perturbative effect, to an acceleration of the universe that fits the current cosmological data, including the observations of high redshift type-Ia supernovae, and of fluctuations in the cosmic background radiation that imply that the universe is (or nearly is) spatially flat. The effective action of such a field appears to cause a rapid transition to an accelerating expansion in which R becomes a constant value related to the mass of the particle. Furthermore, Caldwell et al. (2006) consider a more general class of cosmological transitions in which a linear combination of curvature invariants become constant and find that, of the class of transitions under consideration, the transition to a constant R gives the best fit to the observational data; a fit that is acceptable without a cosmological constant. They also show how the presence of a sufficiently small non-zero vacuum expectation value of the low-mass scalar field (which acts like a cosmological constant) would still allow the transition to constant R and could agree better with the data than a cosmological constant alone.

3.5 Approximate energy-momentum tensor in Schwarzschild, de Sitter, and other static Einstein spacetimes

We consider here the subject of analytic approximations to the expectation value of the energy-momentum tensor. A well known such approximation for conformally invariant fields in static spacetimes that satisfy the vacuum Einstein equations (with a possible cosmological constant term) is due to Page (1982), and is partly based upon the Gaussian approximation.

We first discuss the topic of Hadamard regularization (Section 3.5.1) before developing the Page approximation (Section 3.5.2) in detail. We then briefly mention other approximation methods and applications. We consider for simplicity a scalar field ϕ satisfying the conformally invariant field equation

$$\left(\Box + \frac{1}{6}R\right)\phi = 0. \qquad (3.103)$$

3.5.1 Hadamard regularization

The method of Hadamard regularization is a general method for regularizing both $\langle \phi^2 \rangle$ and $\langle T_{\mu\nu} \rangle$. It is implemented at the level of the Hadamard Green

3.5 Approximate energy-momentum tensor in ...

function $G^{(1)}(x, x') \equiv (1/2)\langle \phi(x)\phi(x') + \phi(x')\phi(x)\rangle$, which satisfies (3.103) above. The method consists of subtracting from $G^{(1)}$ a locally determined elementary solution that has the same singularity structure (in the coincidence limit $x' \to x$) as $G^{(1)}$ itself, as outlined below. The resulting subtracted Green function is then finite in the coincidence limit.

In four spacetime dimensions, the Green function $G^{(1)}$ has the formal Hadamard series solution (Hadamard 1923; DeWitt 1972)

$$G^{(1)}(x,x') = \frac{\Delta^{1/2}(x,x')}{(4\pi)^2}\left(\frac{2}{\sigma(x,x')} + v(x,x')\ln\sigma(x,x') + w(x,x')\right), \quad (3.104)$$

where $\sigma(x, x')$ is the geodetic biscalar of (3.83), and v and w are analytic functions of σ, admitting the power series expansions[6]

$$v(x,x') = \sum_{n=0}^{\infty} v_n(x,x')\sigma^n, \quad (3.105)$$

$$w(x,x') = \sum_{n=0}^{\infty} w_n(x,x')\sigma^n. \quad (3.106)$$

The coefficients v_n and w_n satisfy the following recursion relations (Adler et al., 1977), as can be obtained by substituting (3.104) into the field equation:

$$v_0 + v_{0,\mu}\sigma^{,\mu} = (1/6)R - \Delta^{-1/2}(\Delta^{1/2})_{,\mu}{}^{\mu},$$

$$v_n + \frac{v_{n,\mu}\sigma^{,\mu}}{n+1} = -\frac{1}{2n(n+1)}\left(\Delta^{-1/2}(\Delta^{1/2}v_{n-1})_{,\mu}{}^{\mu} - (1/6)Rv_{n-1}\right),$$

$$w_n + \frac{w_{n,\mu}\sigma^{,\mu}}{n+1} = -\frac{1}{2n(n+1)}\Big(\Delta^{-1/2}(\Delta^{1/2}w_{n-1})_{,\mu}{}^{\mu}$$
$$- (1/6)Rw_{n-1}\Big) - \frac{v_n}{n+1}$$
$$- \frac{1}{2n^2(n+1)}\Big(\Delta^{-1/2}(\Delta^{1/2}v_{n-1})_{,\mu}{}^{\mu}$$
$$- (1/6)Rv_{n-1}\Big), \quad (3.107)$$

where $n \geq 1$. The above recursion relations uniquely determine all the coefficients v_n but leave the coefficient w_0 completely arbitrary. This coefficient is determined by the global boundary conditions on the Green function. Once it is known, the subsequent coefficients w_n for $n \geq 1$ are uniquely determined by the recursion relations. Thus, following Adler et al., (1977), it is natural to split the Hadamard series for $G^{(1)}$ into a *locally determined* part G^L, which is given by (3.104) *with*

---

[6] The singularity structure of $G^{(1)}$ as given in (3.104) can be rigorously proved in curved spacetime only for massive fields (DeWitt 1964a), and for massive and massless fields in flat spacetime. We assume here that the same singularity structure holds in the curved spacetime massless limit, as is assumed in Wald (1978) and Adler et al., (1977).

$w_0 = 0$, and a boundary-condition-dependent remainder. Since G^L is independent of boundary conditions, and contains the full singularity structure of $G^{(1)}$, we can define a Hadamard-regularized Green function by

$$G^{(1)}_{\text{reg}}(x,x') \equiv G^{(1)}(x,x') - G^L(x,x'). \tag{3.108}$$

The regularized expectation value of ϕ^2 is then given by the coincidence limit of the above expression:

$$\langle \phi^2(x) \rangle_{\text{reg}} \equiv \lim_{x' \to x} G^{(1)}_{\text{reg}}(x,x'). \tag{3.109}$$

Since the regularization procedure is carried out before taking the coincidence limit, the method outlined above is an example of regularization by "point-splitting."

To obtain the regularized expectation value of the energy-momentum tensor, we first note that the classical expression for the energy-momentum tensor of a conformally invariant scalar field is

$$T_{\mu\nu} = \frac{2}{3} \nabla_\mu \phi \nabla_\nu \phi - \frac{1}{6} g_{\mu\nu} \nabla_\alpha \phi \nabla^\alpha \phi - \frac{1}{3} \phi \nabla_\mu \nabla_\nu \phi$$
$$+ \frac{1}{3} g_{\mu\nu} \phi \nabla_\alpha \nabla^\alpha \phi - \frac{1}{6} \left(R_{\mu\nu} - \frac{1}{2} g_{\mu\nu} R \right) \phi^2. \tag{3.110}$$

The formal expectation value of the quantum energy-momentum tensor is given by the coincidence limit of the corresponding "point-split" expression, that is,

$$\langle T_{\mu\nu}(x) \rangle = \lim_{x' \to x} \mathcal{T}_{\mu\nu}(x,x'), \tag{3.111}$$

where

$$\mathcal{T}_{\mu\nu}(x,x') = \frac{1}{3} \left(\nabla_\mu \nabla_{\nu'} G^{(1)} + \nabla_{\mu'} \nabla_\nu G^{(1)} \right) - \frac{1}{6} g_{\mu\nu} \nabla_\alpha \phi \nabla^{\alpha'} G^{(1)}$$
$$- \frac{1}{6} \left(\nabla_\mu \nabla_\nu G^{(1)} + \nabla_{\mu'} \nabla_{\nu'} G^{(1)} \right)$$
$$+ \frac{1}{6} g_{\mu\nu} \left(\nabla_\alpha \nabla^\alpha G^{(1)} + \nabla_{\alpha'} \nabla^{\alpha'} G^{(1)} \right)$$
$$- \frac{1}{6} (R_{\mu\nu} - \frac{1}{2} g_{\mu\nu} R) G^{(1)}, \tag{3.112}$$

and the unprimed indices refer to covariant derivatives with respect to x and primed indices refer to covariant derivatives with respect to x'. The limit in (3.111), of course, does not exist because $G^{(1)}$ and its derivatives are singular in the coincidence limit.

Following the Hadamard regularization of $\langle \phi^2 \rangle$ that we outlined earlier, we would expect to get a finite result for $\langle T_{\mu\nu} \rangle$ if we were to replace $G^{(1)}$ in the above expression by $G^{(1)}_{\text{reg}}$. This was indeed the procedure carried out in Wald (1977) and Adler et al., (1977). Let us call the resulting finite expression for expectation value of the energy-momentum tensor $\langle T_{\mu\nu} \rangle^{(0)}_{\text{reg}}$.

Unfortunately, as pointed out by Wald (1978), $\langle T_{\mu\nu}\rangle_{\text{reg}}^{(0)}$ is not generally a conserved quantity[7], that is, $\nabla^\mu \langle T_{\mu\nu}\rangle_{\text{reg}}^{(0)} \neq 0$. Wald shows that the tensor $\langle T_{\mu\nu}\rangle_{\text{reg}}$, defined by

$$\langle T_{\mu\nu}\rangle_{\text{reg}} \equiv \langle T_{\mu\nu}\rangle_{\text{reg}}^{(0)} - \frac{1}{2(4\pi)^2} g_{\mu\nu} v_1(x,x) \qquad (3.113)$$

is a conserved quantity. Also, as shown earlier in Wald (1977) and Adler *et al.*, (1977), $\langle T^\mu_\mu\rangle_{\text{reg}}^{(0)} = 0$ for conformally invariant fields. Therefore,

$$\langle T^\mu_\mu\rangle_{\text{reg}} = -\frac{2}{(4\pi)^2} v_1(x,x). \qquad (3.114)$$

It is straightforward to derive from the recursion relations for the v_n's that $v_1(x,x) = -(1/2)a_2(x,x)$. Thus, $\langle T^\mu_\mu\rangle_{\text{reg}} = a_2/(4\pi)^2$, which agrees with the trace anomaly derived earlier.

The Hadamard regularization procedure for the expectation value of the energy-momentum tensor therefore consists of replacing $G^{(1)}$ by $G^{(1)}_{\text{reg}}$ in (3.111) and (3.112) and then adding the extra term of (3.113) to the resulting expression. This yields a conserved energy-momentum tensor and the correct trace anomaly.

We now turn to Page's approximation for the stress tensor, which builds upon the results of the previous section and the Hadamard regularization method developed here.

3.5.2 Page approximation

An important analytic approximation for $\langle\phi^2\rangle_{\text{reg}}$ and $\langle T_{\mu\nu}\rangle_{\text{reg}}$ in static spacetimes that satisfy the Einstein equations has been developed by Page (1982). This approximation actually holds for expectation values in a thermal state. The vacuum expectation values are the zero-temperature limits of the corresponding finite-temperature quantities. Physically, a thermal state is the equilibrium state of the field when it interacts with a finite-temperature bath at a definite temperature, T. Although we will deal with thermal states in greater detail in a subsequent chapter, we will here assume one important property of a thermal state: namely, that the Green functions in a thermal state of temperature T are periodic in imaginary time with period $i\beta$, where $\beta = (k_B T)^{-1}$ (k_B is the Boltzmann constant). That is,

$$G_\beta(t, \mathbf{x}; t', \mathbf{x}') = G_\beta(t + i\beta, \mathbf{x}; t', \mathbf{x}'). \qquad (3.115)$$

A special type of thermal state is the so-called Hartle–Hawking state (Hartle and Hawking 1976), defined in static black hole spacetimes as a thermal state with temperature

---

[7] The lack of conservation of $\langle T_{\mu\nu}\rangle_{\text{reg}}^{(0)}$ is due to the fact that G^L is not a symmetric function of x and x', and hence does not satisfy the field equation in x'. See Wald (1978) for details.

$$T = \frac{\hbar \kappa}{2\pi k_B}, \qquad (3.116)$$

where κ is the surface gravity of the black hole. For Schwarzschild black holes, $\kappa = (4M)^{-1}$, where M is the mass of the black hole. This state of a quantized field describes a black hole that is in thermal equilibrium with the field. The Hartle–Hawking state has the important property that all expectation values of observables (including the expectation value of the energy-momentum tensor) are finite on the past and future event horizons of the black hole. As we shall show below, the Page approximation is an excellent one for the expectation value of the stress tensor in the Hartle–Hawking state of a Schwarzschild black hole.

To develop the general form of the Page approximation, we consider a conformally invariant scalar field $\overline{\phi}$ in an arbitrary *static* background spacetime whose metric, $\overline{g}_{\mu\nu}$, satisfies the Einstein equations

$$\overline{R}_{\mu\nu} = \Lambda \overline{g}_{\mu\nu}. \qquad (3.117)$$

The approximation scheme is implemented in a conformally related metric $g_{\mu\nu}$, such that

$$g_{\mu\nu} = \Omega^{-2} \overline{g}_{\mu\nu}, \qquad (3.118)$$

and the conformal factor Ω^{-2} is chosen so that $g_{tt} = 1$. (This means that Ω has no dependence on t.) The metric $g_{\mu\nu}$ is therefore an *ultrastatic metric*, that is, a static metric in which the timelike Killing vector field has constant norm. It is called the *optical metric* of the physical metric $\overline{g}_{\mu\nu}$. Because the theory is conformally invariant, all unregularized Green functions in the physical metric may be obtained from those in the optical metric via the transformation rule

$$\overline{G}(x,x') = \Omega^{-1}(x) G(x,x') \Omega^{-1}(x'). \qquad (3.119)$$

We assume that we can analytically continue the imaginary time coordinate $\tau = -it$ to real values. Correspondingly, we also define an imaginary proper time coordinate $u = is$ and analytically continue it to real values. The Euclidean Green function $G_E(\tau, \mathbf{x}; \tau', \mathbf{x}') \equiv iG_F(-it, \mathbf{x}; -it', \mathbf{x}')$ may then be expressed as

$$G_E(x,x') = \int_0^\infty du\, K(x, x' u), \qquad (3.120)$$

with the kernel function K obeying the Schrodinger-like equation[8]

$$\left(\frac{\partial}{\partial u} + \Box + \frac{1}{6} R \right) K(x, x', u) = 0, \qquad (3.121)$$

with $K(x, x', 0) = \delta(x, x')$. This kernel function obeys the same equation as the quantity $\langle x, -iu \mid x', 0 \rangle$ that appears in (3.100).

---

[8] Note that because our convention for the signature of the metric is opposite to that of Page (1982), our $\Box$ operator appears with the opposite sign to his.

3.5 Approximate energy-momentum tensor in ...

Page's approximation consists of the following steps. We first apply the Gaussian approximation to the kernel function in the optical metric, taking into account the periodic boundary conditions that define a thermal state. We then find the Gaussian-approximated Green function in the optical metric using (3.120). Finally we transform back to the physical metric as in (3.119). This procedure gives an unregularized approximate Green function in the physical metric. Hadamard regularization is used to give the finite expectation values of ϕ^2 and $T_{\mu\nu}$ in the physical metric.

Consider the kernel function K in the optical metric. Because the optical metric is ultrastatic, the Schrodinger-like equation for K is separable and K may be factorized as

$$K(x, x', u) = K_1(\tau, \tau', u) K_3(\mathbf{x}, \mathbf{x}', u), \qquad (3.122)$$

where

$$\left(\frac{\partial}{\partial u} - \frac{\partial^2}{\partial \tau^2}\right) K_1(t, t', u) = 0, \qquad (3.123)$$

and

$$\left(\frac{\partial}{\partial u} - \nabla^i \nabla_i + \frac{1}{6} R\right) K_3(\mathbf{x}, \mathbf{x}', u) = 0. \qquad (3.124)$$

Latin indices run over the spatial variables in (3.124). With the initial condition $K_1(\tau, \tau', 0) = \delta(\tau - \tau')$ and periodic boundary conditions on K_1 (with period β), (3.123) has the solution

$$K_1(\tau, \tau', u) = \beta^{-1} \sum_{n=-\infty}^{+\infty} \exp\left(-\frac{4u\pi^2 n^2}{\beta^2} - \frac{2\pi i n}{\beta}(\tau - \tau')\right). \qquad (3.125)$$

The Gaussian approximation is now used to give an approximate solution to (3.124). Taking account of the fact that K_3 is the kernel function for a three-dimensional Green function, and that $\xi = 1/6$, we obtain

$$K_{3\text{Gauss}}(\mathbf{x}, \mathbf{x}', u) = (4\pi u)^{-3/2} \Delta^{1/2}(\mathbf{x}, \mathbf{x}') \exp\left(-\frac{{}^{(3)}\sigma(\mathbf{x}, \mathbf{x}')}{2u}\right), \qquad (3.126)$$

where ${}^{(3)}\sigma$ is the three-dimensional geodetic biscalar, and $\Delta(\mathbf{x}, \mathbf{x}')$ is the corresponding three-dimensional Van Vleck–Morette determinant. Note that, because the four-dimensional geodetic biscalar $\sigma = {}^{(3)}\sigma + (1/2)(\tau - \tau')^2$, $\Delta(\mathbf{x}, \mathbf{x}')$ has the same value if ${}^{(3)}\sigma$ were replaced by σ in its definition, that is, Δ can be taken to be the full four-dimensional Van Vleck–Morette determinant.

Equation (3.126) above may be compared with the Schwinger–DeWitt expansion for K_3, which gives

$$K_3(\mathbf{x}, \mathbf{x}', u) = K_{3\text{Gauss}}(\mathbf{x}, \mathbf{x}', u) \sum_{n=0}^{\infty} a_n(\mathbf{x}, \mathbf{x}') u^n. \qquad (3.127)$$

The Gaussian approximation for K_3 is therefore equivalent to retaining only the first term, a_0, in the Schwinger–DeWitt expansion. The reason that this approximation is a good one can be briefly stated as follows.

Let us introduce the notation $[f]$ to denote the coincidence limit $x' \to x$ of a function $f(x, x')$. Since we ultimately wish to compute $\langle \phi^2(x) \rangle$ and $\langle T_{\mu\nu}(x) \rangle$, we only require $K_3(\mathbf{x}, \mathbf{x}', u)$ when $\mathbf{x}'$ is close to $\mathbf{x}$. Therefore, we need only the first few terms in a power series expansion of $a_n(\mathbf{x}, \mathbf{x}')$ in the separation of $\mathbf{x}$ and $\mathbf{x}'$. Because $\xi = 1/6$, we automatically have $[a_1] = [a_{1;\mu}] = 0$. Also, as pointed out by Page (1982), $[a_{1;\nu}^{;\mu}] = [a_2] = 0$ *for the optical metric of any static Einstein space*. The Gaussian approximation is therefore a much better approximation in such metrics than others.

Equations (3.125) and (3.126) can be substituted into the expression for the Euclidean Green function (3.120) to yield the Gaussian approximation for the Euclidean Green function, which we denote by G_{Gauss}:

$$G_{\text{Gauss}}(\tau, \mathbf{x}; 0, \mathbf{x}') = \frac{\Delta^{1/2}}{4\pi\beta r} \frac{\sinh(2\pi r/\beta)}{\cosh(2\pi r/\beta) - \cos(2\pi\tau/\beta)}, \quad (3.128)$$

where $r = (2\,^{(3)}\sigma)^{1/2}$.

To obtain the renormalized value of $\langle \bar\phi^2 \rangle_{\text{Gauss}}$ in the physical metric, we can now apply the conformal transformation of (3.119) to (3.128), subtract the locally determined Hadamard Green function G^L in the physical metric $\bar g_{\mu\nu}$, and finally take the coincidence limit. The details of this procedure are worked out in Page (1982). Here, we give the result (using our sign conventions, which differ from Page (1982)):

$$\langle \bar\phi^2 \rangle_{\text{Gauss}} = (48\pi^2)^{-1} \left(\frac{4\pi^2}{\beta^2 \Omega^2} - \frac{\Box \Omega}{\Omega^3} \right) \quad (3.129)$$

$$= (288\pi^2 \Omega^2)^{-1} \left(\frac{24\pi^2}{\beta^2} + R - \Omega^2 \bar R \right), \quad (3.130)$$

where R is the scalar curvature in the optical metric $g_{\mu\nu}$ and $\bar R$ is the scalar curvature in the physical metric $\bar g_{\mu\nu}$. The above expression for $\langle \bar\phi^2 \rangle_{\text{Gauss}}$ can be put in a more suggestive form. Note that the optical metric has $R_{00} = 0$, so if the physical metric obeys the Einstein equations, we have the relation

$$\bar R_{00} = \Lambda \bar g_{00} = \Lambda \Omega^2$$

$$= R_{00} - 2\Omega(\Omega^{-1})_{;00} + \frac{1}{2}\Omega^{-2}\Box(\Omega^2)g_{00}$$

$$= \Omega^{-1}\Box\Omega + \Omega^{-2}\Omega^{;\alpha}\Omega_{;\alpha}, \quad (3.131)$$

where the time independence of Ω has been used. In addition, raised indices and $\Box$ in these expressions are based on the optical metric $g^{\alpha\beta}$. Solving these equations for $\Omega^{-3}\Box\Omega$, and substituting the result into the previous expression for $\langle \bar\phi^2 \rangle$, we find that

$$\langle \bar{\phi}^2 \rangle_{\text{Gauss}} = (48\pi^2)^{-1} \left(\frac{4\pi^2}{\beta^2 \Omega^2} + \Omega^{-4} \Omega^{;\alpha} \Omega_{;\alpha} - \Lambda \right). \qquad (3.132)$$

Now the local inverse temperature in the physical metric is given by taking account of the redshift factor $\bar{g}_{00}^{1/2}$, as

$$\beta_{\text{loc}} = \beta \, \bar{g}_{00}^{1/2} = \beta \Omega. \qquad (3.133)$$

Therefore, $\langle \bar{\phi}^2 \rangle$ can be expressed as

$$\langle \bar{\phi}^2 \rangle_{\text{Gauss}} = \frac{1}{12} \left(\beta_{\text{loc}}^{-2} + \frac{\Omega^{-4} \Omega^{;\alpha} \Omega_{;\alpha}}{4\pi^2} - \frac{\Lambda}{4\pi^2} \right). \qquad (3.134)$$

The second term in the above expression can be written in terms of the acceleration of the timelike Killing vector orbit in the physical metric. This acceleration a is given by

$$a = \left(-\bar{g}^{\alpha\beta} \left(\ln \bar{g}_{00}^{1/2} \right)_{,\alpha} \left(\ln \bar{g}_{00}^{1/2} \right)_{,\beta} \right)^{1/2} = \left(-\Omega^{-4} g^{\alpha\beta} \Omega_{,\alpha} \Omega_{,\beta} \right)^{1/2}. \qquad (3.135)$$

Thus the value of $\langle \bar{\phi}^2 \rangle$ in the Gaussian approximation can be expressed directly in terms of physical quantities in a static Einstein metric as

$$\langle \bar{\phi}^2 \rangle_{\text{Gauss}} = \frac{1}{12} \left(\beta_{\text{loc}}^{-2} - \left(\frac{a}{2\pi} \right)^2 \right) - \frac{\Lambda}{48\pi^2}. \qquad (3.136)$$

Consider now the application of the above expression to a zero-temperature field in Minkowski spacetime (which has $\Lambda = 0$) for which we must have $\langle \bar{\phi}^2 \rangle = 0$. In Minkowski spacetime, the Page approximation is exact because there are no curvature-dependent corrections in the kernel function. In usual Minkowski coordinates, $\Omega = 1$ and $\beta_{\text{loc}}^{-1} = 0$, so each of the first two terms in (3.136) vanish individually and we obtain the correct result. However, we could choose a different set of coordinates for Minkowski spacetime that are static and have the property that the orbits of the locally timelike Killing vector field are worldlines of constant acceleration. Such coordinates are the Rindler (Rindler 1966) coordinates (η, ξ, y, z), in which the metric of Minkowski spacetime has the form

$$d\bar{s}^2 = e^{2\alpha\xi} \left(d\eta^2 - d\xi^2 - e^{-2\alpha\xi} dy^2 - e^{-2\alpha\xi} dz^2 \right). \qquad (3.137)$$

The Rindler coordinates can be obtained from the usual Minkowski coordinates (t, x, y, z) via the transformations (Birrell and Davies 1982)

$$t = \alpha^{-1} e^{\alpha\xi} \sinh \alpha\eta, \qquad (3.138)$$
$$x = \alpha^{-1} e^{\alpha\xi} \cosh \alpha\eta. \qquad (3.139)$$

The acceleration of an observer moving along the worldline having as its tangent the timelike Killing vector $\partial/\partial\eta$ is

$$a = \alpha e^{\alpha\xi}. \qquad (3.140)$$

In these coordinates, the second term of (3.136) will not vanish but it must be still true that $\langle\bar{\phi}^2\rangle = 0$. This leads us to the remarkable result that, in these coordinates, an observer moving along the worldline $\partial/\partial\eta$ detects a local temperature given by

$$\beta_{\text{loc}}^{-1} = \frac{a}{2\pi}. \tag{3.141}$$

This result is the so-called Fulling–Davies–Unruh effect (Fulling 1973; Davies 1975; Unruh 1976), although the derivation of this effect presented above is different from the way it was originally discovered. The effect shows that the Minkowski vacuum state appears as a thermal state to an observer undergoing constant acceleration. We will generally refer to the temperature of this thermal state as the Unruh acceleration temperature, defined by $\beta_{\text{acc}} \equiv 2\pi/a$.

For a general static Einstein spacetime, we may then express $\langle\bar{\phi}^2\rangle_{\text{Gauss}}$ in the form

$$\langle\bar{\phi}^2\rangle_{\text{Gauss}} = \frac{1}{12}\left(\beta_{\text{loc}}^{-2} - \beta_{\text{acc}}^{-2}\right) - \frac{\Lambda}{48\pi^2}. \tag{3.142}$$

To obtain the renormalized value of $\langle T_\nu^\mu\rangle$ in the physical metric, we can proceed in a manner similar to that for finding $\langle\bar{\phi}^2\rangle$, that is, we can apply the conformal transformation of (3.119) to (3.128), subtract G^L (i.e., (3.104) with $w_0 = 0$) from G_{Gauss} to regularize it, calculate the derivatives in (3.112) (after replacing $G^{(1)}$ by $G_{\text{Gauss}} - G^L$), take the coincidence limit of (3.111), and finally apply Wald's correction of (3.113) to obtain the expectation value of the energy-momentum tensor in the Gaussian approximation.

A calculationally simpler route, as suggested by Page (1982), is to first compute the Hadamard-regularized value of $\langle T_\nu^\mu\rangle$ in the optical metric, and to transform this directly to the physical metric. Page finds that the Gaussian-approximated Hadamard-regularized stress tensor in the optical metric has the particularly simple form

$$T_\nu^\mu = \frac{\pi^2}{90}\beta^{-4}\left(\delta_\nu^\mu - 4\delta_0^\mu\delta_\nu^0\right). \tag{3.143}$$

To transform the stress tensor in the optical metric to the one in the physical metric, Page suggests making use of the functional-differential equation for the change in the energy-momentum tensor under a conformal change in the metric described by Brown and Cassidy (1977):

$$\tilde{g}_{\alpha\beta}(x')\frac{\delta}{\delta\tilde{g}_{\alpha\beta}(x')}\left(|\tilde{g}|^{1/2}T_\nu^\mu(x)\right) = \tilde{g}_{\nu\gamma}(x)\frac{\delta}{\delta\tilde{g}_{\mu\gamma}(x)}\left(|\tilde{g}|^{1/2}T_\lambda^\lambda(x')\right), \tag{3.144}$$

where $\tilde{g}_{\mu\nu}$ is an arbitrary metric. Let $\tilde{g}_{\mu\nu} = \omega^2 g_{\mu\nu}$ (recall that $g_{\mu\nu}$ is the optical metric). Then the left-hand side of (3.144) is proportional to the functional derivative of $|\tilde{g}|^{1/2}T_\nu^\mu(x)$ with respect to $\omega(x)$. For conformally invariant fields, such as the one considered here, the right-hand side of (3.144) is given entirely by the trace anomaly in the metric $\tilde{g}_{\mu\nu}$, which may be expressed as a functional of $\omega(x')$ and $g_{\mu\nu}(x')$. Then (3.144) can be functionally integrated between the integration limits $\omega = 1$ (the optical metric) and $\omega = \Omega$ (the physical metric). This procedure yields a relation between $\langle \overline{T}^{\mu}_{\nu} \rangle$ in the physical metric and $\langle T^{\mu}_{\nu} \rangle$ in the optical metric. Page actually finds this relation by looking for a generalization of the known energy-momentum tensor in conformally flat spacetimes that is conserved and has the correct trace. The result is (Page 1982), in our metric convention,

$$\overline{T}^{\mu}_{\nu} = \Omega^{-4} T^{\mu}_{\nu} - 8\alpha_s \Omega^{-4} \left\{ -(C^{\alpha\mu}{}_{\beta\nu} \ln \Omega)^{;\beta}_{;\alpha} + \frac{1}{2} R^{\beta}_{\alpha} C^{\alpha\mu}{}_{\beta\nu} \ln \Omega \right\}$$
$$+ \beta_s \left\{ \left(4\overline{R}^{\beta}_{\alpha} \overline{C}^{\alpha\mu}{}_{\beta\nu} - 2\overline{H}^{\mu}_{\nu} \right) - \Omega^{-4} \left(4 R^{\beta}_{\alpha} C^{\alpha\mu}{}_{\beta\nu} - 2 H^{\mu}_{\nu} \right) \right\}$$
$$- \frac{1}{6} \gamma_s \left\{ \overline{I}^{\mu}_{\nu} - \Omega^{-4} I^{\mu}_{\nu} \right\}. \tag{3.145}$$

Here, $C^{\alpha}{}_{\beta\gamma\delta}$ is the Weyl tensor, and for scalar fields

$$\alpha_s = -\frac{3}{2(2880\pi^2)}, \quad \beta_s = \frac{1}{2(2880\pi^2)}, \quad \gamma_s = \frac{1}{2880\pi^2}. \tag{3.146}$$

$H_{\mu\nu}$ and $I_{\mu\nu}$ in (3.145) are the local geometric tensors

$$H_{\mu\nu} = R^{\rho}_{\mu} R_{\rho\nu} - \frac{2}{3} R R_{\mu\nu} - \frac{1}{2} R_{\rho\sigma} R^{\rho\sigma} g_{\mu\nu} + \frac{1}{4} R^2 g_{\mu\nu}, \tag{3.147}$$

$$I_{\mu\nu} = 2 R_{;\mu\nu} - 2 g_{\mu\nu} \Box R - \frac{1}{2} g_{\mu\nu} R^2 + 2 R R_{\mu\nu}. \tag{3.148}$$

We now consider the application of the above results to two metrics of interest, namely the de Sitter and Schwarzschild metrics. The de Sitter metric, which is a solution of the Einstein equations with a cosmological constant, can be expressed as a static metric in the form

$$d\bar{s}^2 = -\left(1 - \frac{\Lambda r^2}{3}\right)$$
$$\times \left\{ d\tau^2 + \frac{dr^2}{\left(1 - \frac{\Lambda r^2}{3}\right)^2} + \frac{r^2}{\left(1 - \frac{\Lambda r^2}{3}\right)} (d\theta^2 + \sin^2\theta d\phi^2) \right\}. \tag{3.149}$$

The de Sitter metric is maximally symmetric and therefore invariant under the isometries defined by the $SO(4,1)$ group.[9] The thermal state that shares the full $SO(4,1)$ symmetry of the de Sitter metric has an inverse temperature of (Gibbons and Hawking 1977)

$$\beta = \left(\frac{12\pi^2}{\Lambda}\right)^{1/2}. \tag{3.150}$$

Equations (3.133) and (3.135), for this metric, give

---

[9] $SO(4,1)$ holds for the case of real, as opposed to imaginary, time.

$$\beta_{\text{loc}} = \beta \left(1 - \frac{\Lambda r^2}{3}\right)^{1/2}, \tag{3.151}$$

$$\beta_{\text{acc}} = \frac{6\pi}{\Lambda r} \left(1 - \frac{\Lambda r^2}{3}\right)^{1/2}. \tag{3.152}$$

Equations (3.142) and (3.145) then yield

$$\langle \bar{\phi}^2 \rangle_{\text{Gauss}} = -\frac{\Lambda}{72\pi^2}, \tag{3.153}$$

$$\langle \bar{T}^\mu_\nu \rangle_{\text{Gauss}} = -\frac{\Lambda^2}{8640\,\pi^2}\,\delta^\mu_\nu, \tag{3.154}$$

which are *exact* results for the de Sitter-invariant thermal state.

The Schwarzschild metric, in usual Schwarzschild coordinates, with $z \equiv 2M/r$, is

$$d\bar{s}^2 = -(1-z)\left\{d\tau^2 + (1-z)^{-2}dr^2 + (1-z)^{-1}r^2(d\theta^2 + \sin^2\theta d\phi^2)\right\}. \tag{3.155}$$

The Gaussian-approximated value for $\langle \bar{\phi}^2 \rangle$, found from (3.142), (3.133), and (3.135), is

$$\langle \bar{\phi}^2 \rangle_{\text{Gauss}} = (12\beta^2)^{-1}(1 + z + z^2 + z^3). \tag{3.156}$$

The above formula, found earlier by Whiting as an analytical fit to the numerical results of Fawcett and Whiting (1982), agrees with the exact result (Candelas 1980) on the horizon ($z = 1$), and reduces to the flat-space result as $r \to \infty$ ($z \to 0$). For intermediate values of the r coordinate, it gives very good agreement with the numerical results of Fawcett and Whiting (1982).

The Gaussian approximation for the expectation value of the energy-momentum tensor is

$$\langle \bar{T}^\mu_\nu \rangle_{\text{Gauss}} = \frac{\pi^2}{90}(8\pi M)^{-4}\left\{\frac{1-(4-3z)^2 z^6}{(1-z)^2}\left(\delta^\mu_\nu - 4\delta^\mu_0 \delta^0_\nu\right)\right.$$
$$\left. + 24z^6 \left(3\delta^\mu_0 \delta^0_\mu + \delta^\mu_1 \delta^1_\mu\right)\right\}. \tag{3.157}$$

It may be verified that the above expression is conserved and has the correct trace. It is finite at the horizon ($r = 2M$), as expected in the Hartle–Hawking state. For $r \gg 2M$, it agrees with the energy-momentum tensor for thermal radiation in flat space. Page (1982) also shows that the Gaussian-approximated energy-momentum tensor gives good agreement with the exact energy-momentum tensor on the horizon.

Following Page's work on the Gaussian approximation, there have been a number of studies of the energy-momentum tensor in the Hartle–Hawking state in black hole spacetimes. Howard (1984) found that $\langle T^\mu_\nu \rangle$ separates naturally into the sum of two terms, the first of which is Page's approximate energy-momentum tensor while the second is a remainder that may be evaluated numerically. He finds that the total energy-momentum tensor is in good qualitative agreement with Page's approximation everywhere outside the horizon. Extensions of Page's work to spin-1/2 and spin-1 fields have been carried out by Brown et al. (1986).

Another analytic approximation for the energy-momentum tensor in black hole spacetimes, which we do not discuss in detail here, has been given by Frolov and Zel'nikov (1987). Their approximation is based on constructing the most general energy-momentum tensor (for conformally invariant fields) from the timelike Killing vector field, the Riemann tensor, and their derivatives, such that it is conserved and possesses the correct trace anomaly. This approximation has been extended to non-conformally coupled fields by Anderson et al. (1995), who also give a numerical scheme for computing the stress tensor for fields of arbitrary mass and curvature coupling in arbitrary spherically symmetric spacetimes.

3.6 R-summed form of propagator

Comparison of (3.84) for $G_F(x,x')$ with (3.99), which defines the kernel, $\langle x,s|x',0\rangle$, in any spacetime dimension, n, motivates us to write the kernel in the form

$$\langle x,s|x',0\rangle = i(4\pi is)^{-n/2}\Delta^{1/2}(x,x')\exp\left[\frac{\sigma(x,x')}{2is}\right]F(x,x';is). \tag{3.158}$$

The expansion of $F(x,x';is)$ in powers of s is given in (3.85) in terms of coefficients $a_i(x,x')$, where i is the corresponding power of s in the expansion. The coincidence limits, $a_i(x) \equiv a_i(x,x)$, for $i=0,1,2$ are given in (3.86)–(3.88). The coincidence limit $a_3(x)$, as found by Sakai (1971) and Gilkey (1975, 1979), has 28 terms:

$$\begin{aligned}a_3(x) = -\frac{1}{7!}\Big(&-18\Box^2 R + 17R_{;\alpha}R^{;\alpha} - 2R_{\alpha\beta;\gamma}R^{\alpha\beta;\gamma} - 4R_{\alpha\beta;\gamma}R^{\alpha\gamma;\beta}\\
&+ 9R_{\alpha\beta\gamma\delta;\epsilon}R^{\alpha\beta\gamma\delta;\epsilon} + 28R\Box R - 8R_{\alpha\beta}\Box R^{\alpha\beta} + 24R_\alpha{}^\beta R^{\alpha\gamma}{}_{;\beta\gamma}\\
&+ 12R_{\alpha\beta\gamma\delta}\Box R^{\alpha\beta\gamma\delta} - \frac{35}{9}R^3 + \frac{14}{3}RR_{\alpha\beta}R^{\alpha\beta}\\
&- \frac{14}{3}RR_{\alpha\beta\gamma\delta}R^{\alpha\beta\gamma\delta} + \frac{208}{9}R_{\alpha\beta}R^{\alpha\gamma}R^\beta{}_\gamma\\
&- \frac{64}{3}R^{\alpha\beta}R^{\gamma\delta}R_{\alpha\gamma\beta\delta} + \frac{16}{3}R_{\alpha\beta}R^\alpha{}_{\gamma\delta\epsilon}R^{\beta\gamma\delta\epsilon}\\
&- \frac{44}{9}R_{\alpha\beta\gamma\delta}R^{\alpha\beta\epsilon\phi}R^{\gamma\delta}{}_{\epsilon\phi} - \frac{80}{9}R_{\alpha\beta\gamma\delta}R^{\alpha\epsilon\gamma\phi}R^\beta{}_\epsilon{}^\delta{}_\phi\Big)\\
-\frac{1}{360}\Big(&6\xi\Box^2 R + 60\xi^2 R\Box R + 30\xi^2 R_{;\alpha}R^{;\alpha} + 60\xi^3 R^3\\
&- 22\xi R\Box R - 4\xi R^{\alpha\beta}R_{;\alpha\beta} - 12\xi R_{;\alpha}R^{;\alpha} - 30\xi^2 R^3\\
&+ 5\xi R^3 - 2\xi RR_{\alpha\beta}R^{\alpha\beta} + 2\xi RR_{\alpha\beta\gamma\delta}R^{\alpha\beta\gamma\delta}\Big). \end{aligned} \tag{3.159}$$

Using the fact that $\sigma(x,x) = 0$ and $\Delta(x,x) = 1$, we find that the coincidence limit, $\langle x, s|x, 0\rangle$, of the propagator can be written in the form

$$\langle x, s|x, 0\rangle = i(4\pi is)^{-n/2} F(x, x; is). \tag{3.160}$$

The form of the Gaussian approximation of (3.101) leads us to ask what happens when we write the coincidence limit, $\langle x, s|x, 0\rangle$, of the exact propagator in the following form:

$$\langle x, s|x, 0\rangle = i(4\pi is)^{-d/2} \bar{F}(x, x; is) \exp\left[-is(\xi - \frac{1}{6})R(x)\right]. \tag{3.161}$$

Following Parker and Toms (1985a), let us compare the expansion of $\bar{F}(x, x; is)$ in powers of s with that of $F(x, x; is)$. The proper-time series for $\bar{F}(x, x; is)$ can be written as

$$\bar{F}(x, x; is) \sim \bar{a}_0(x) + \bar{a}_1(x)(is) + \bar{a}_2(x)(is)^2 + \cdots. \tag{3.162}$$

From (3.160) and (3.161), we have

$$\bar{F}(x, x; is) \exp\left[-is\left(\xi - \frac{1}{6}\right)R\right] = F(x, x; is). \tag{3.163}$$

Expanding the exponential, equating the coefficients of equal powers of s, we obtain a set of equations that recursively define the $\bar{a}_i(x)$ in terms of the known coefficients $a_i(x)$ of (3.85).

$$\bar{a}_0 + is\bar{a}_1 + \cdots = [a_0 + isa_1 + \cdots][1 + is(\xi - 1/6)R + \cdots]. \tag{3.164}$$

The resulting expressions for the $\bar{a}_i(x)$ are (Parker and Toms 1985a)

$$\bar{a}_0(x) = 1, \tag{3.165}$$

$$\bar{a}_1(x) = 0, \tag{3.166}$$

$$\bar{a}_2(x) = \frac{1}{180} R_{\alpha\beta\gamma\delta} R^{\alpha\beta\gamma\delta} - \frac{1}{180} R^{\alpha\beta} R_{\alpha\beta}$$
$$- \frac{1}{6}\left(\frac{1}{5} - \xi\right)\Box R, \tag{3.167}$$

and

$$\bar{a}_3(x) = -\frac{1}{7!}\Big(-18\Box^2 R + 17 R_{;\alpha} R^{;\alpha} - 2 R_{\alpha\beta;\gamma} R^{\alpha\beta;\gamma} - 4 R_{\alpha\beta;\gamma} R^{\alpha\gamma;\beta}$$
$$+ 9 R_{\alpha\beta\gamma\delta;\epsilon} R^{\alpha\beta\gamma\delta;\epsilon} - 8 R_{\alpha\beta} \Box R^{\alpha\beta} + 24 R_\alpha{}^\beta R^{\alpha\gamma}{}_{;\beta\gamma}$$

$$+ 12R_{\alpha\beta\gamma\delta}\Box R^{\alpha\beta\gamma\delta} + \frac{208}{9}R_{\alpha\beta}R^{\alpha\gamma}R^{\beta}{}_{\gamma} - \frac{64}{3}R^{\alpha\beta}R^{\gamma\delta}R_{\alpha\gamma\beta\delta}$$
$$+ \frac{16}{3}R_{\alpha\beta}R^{\alpha}{}_{\gamma\delta\epsilon}R^{\beta\gamma\delta\epsilon} - \frac{44}{9}R_{\alpha\beta\gamma\delta}R^{\alpha\beta\epsilon\phi}R^{\gamma\delta}{}_{\epsilon\phi}$$
$$- \frac{80}{9}R_{\alpha\beta\gamma\delta}R^{\alpha\epsilon\gamma\phi}R^{\beta}{}_{\epsilon}{}^{\delta}{}_{\phi}\Big) - \frac{1}{360}\Big(6\xi\Box^2 R + 30\xi^2 R_{;\alpha}R^{;\alpha}$$
$$- 4\xi R^{\alpha\beta}R_{;\alpha\beta} - 12\xi R_{;\alpha}R^{;\alpha}\Big). \tag{3.168}$$

Notice that these expressions are the same as the ones for the $a_i(x)$, except that they contain no terms that involve at least one factor of $R(x)$. There are 11 such terms in $a_3(x)$ that are absent in $\bar{a}_3(x)$. (Terms with derivatives of R can appear, provided they contain no factor of R.)

Based on this observation, Parker and Toms (1985a) conjectured the following theorem.

Theorem: The coefficients $\bar{a}_i(x)$ in the proper-time series for $\bar{F}(x,x;is)$ of (3.159), when expressed in terms of sums and products and contractions of $R^{\alpha}{}_{\beta\gamma\delta}$ and its covariant derivatives, contain no terms which vanish when R (but not its covariant derivative) is replaced by zero.

The form of the $a_i(x)$ and hence of the $\bar{a}_i(x)$ derived by iteration are the same for any value of the dimension n of spacetime. There are particular relations that hold in spacetimes of a definite number of dimensions that could be used, in a spacetime having a given value of n, to rewrite some terms that do not involve R as a set of terms that do involve R. Therefore, it should be understood that the theorem is stated for the form of the coefficients $\bar{a}_i(x)$ that is dimensionally invariant (i.e., the same in all dimensions). The above theorem was proved to all orders by Jack and Parker (1985, 1987) by means of induction. Because of the above theorem, (3.161) is called the partially summed, or R-summed, form of the amplitude (or heat kernel) $\langle x, s|x, 0\rangle$.

3.7 *R*-summed action and cosmic acceleration

The energy-momentum tensor of the quantized field can be expressed in terms of the coincidence limits of the Feynman propagator G_F and its derivatives. Of particular interest are the physical consequences of the nonperturbative factor $\exp[-is(\xi - \frac{1}{6})R(x)]$ in (3.161). These were considered by Parker and Toms (1985b,c), and with respect to the observed cosmological acceleration by Parker and Raval (1999a,b, 2000, 2001), Parker et al., (2003), and Parker and Vanzella (2004). They calculated an effective action W, from which the energy-momentum tensor can be calculated by varying with respect to the metric, as in (2.9) or (2.11), with the action S replaced by the effective action W. The effective action, in the present case, is a function of the metric (and its derivatives) that includes, in addition to the classical Einstein action, terms that give the effect of the quantized field in its vacuum state on the metric. The most complete form of this effective action and the effective Einstein equations is given in Parker and Vanzella (2004), where they also do numerical calculations showing that the evolution to the accelerated expansion of the universe from an earlier classical expansion takes place in a stable fashion.

In calculating the effective action in Parker and Raval (1999a), derivatives of the curvature tensors were dropped on the grounds that they are small in the recent universe compared to the curvature tensors. Although that is true on the average, those terms would introduce higher derivatives of the metric in the effective Einstein equations that could lead to runaway solutions or instabilities. In Parker and Vanzella (2004) all terms in the effective action, including the ones with higher derivatives of the metric, were considered. They also considered the asymptotic form of the heat kernel in order to arrive at a somewhat more general action. The effective Einstein equations were integrated numerically. Before the transition to constant Ricci scalar curvature occurs during the expansion of the universe, the vacuum effects coming from the corrections to the classical Einstein equations are negligible, so that the classical cosmology (with or without a cosmological constant) is valid to excellent approximation. Therefore, a standard classical evolution during that period can be assumed. (The cosmological constant was taken to be zero in the calculations.) It was shown that numerical solution of the Einstein equations with the quantum correction terms (and zero cosmological constant) gave rise to a transition from the classical expansion of the universe to an accelerated expansion having constant scalar curvature. There were no runaway or unstable solutions unless initial conditions far from the classical solution before the transition were assumed. The accelerating solution approaches the de Sitter universe at late times just as does the acceleration caused by a cosmological constant. However, the acceleration is somewhat larger than that caused by a cosmological constant, except at late times. The role of the cosmological constant is replaced in this theory by that of the mass squared of the low-mass particle, which determines both the time of transition and the magnitude of the acceleration.

A more general class of cosmological transitions, in which a linear combination of curvature invariants become constant (Tkachev 1992), were studied by Caldwell et al. (2006). A transition to constant scalar curvature R is included in the class of models they considered. It was shown that, of the class of transitions under consideration, only the transition to a constant R gives an acceptable fit to the observational data with a zero cosmological constant. Caldwell et al. (2006) also found that a sufficiently small non-zero vacuum expectation value of a low-mass scalar field (which acts like a cosmological constant) would still allow the transition to constant R and could agree better with the data than a cosmological constant alone.

The transition to constant R is a key feature of the explanation based on an effective action, as described in the previous paragraphs. A small non-zero vacuum expectation value of the low-mass scalar field in that theory would give the small effective cosmological constant considered in Caldwell et al. (2006), where they also discussed metric perturbations in the effective action of Parker and Raval (1999a), including derivatives of curvature tensors, as presented in its complete form by Parker and Vanzella (2004). Although the equations involving metric perturbations studied by Caldwell et al. (2006) are quite complicated, they suggested that the transition to constant R may stiffen the spacetime in such a way as to suppress anisotropies in the CMB radiation's power spectrum for long wavelength perturbations. (Such a suppression appears to be observed, but has several possible explanations.)

The idea of the effective action was originally introduced to find the effect of a quantized spinor field (corresponding to electrons and positrons) in its vacuum state on a classical electromagnetic field (Heisenberg and Euler 1936; Schwinger 1951a). The existence of an imaginary part of the effective action when the electric field becomes sufficiently strong is taken to imply that real electron–positron pairs are produced (so to speak, pulled out of the vacuum where they are present momentarily as virtual pairs). The magnitude of the imaginary part gives the rate of pair production. We will consider this for scalar fields in Section 5.4.1 and spinor fields in Section 5.8.1. Similarly, we can define an effective action that describes the effect of a quantized field on a classical gravitational field. Effective actions, including background fields (DeWitt 1972; Jackiw 1974), are discussed more fully in Chapters 5–7.

Here we give the minimal effective action of Parker and Raval (1999a, 2000) that includes the effect of the non-perturbative factor $\exp[-is(\xi - \frac{1}{6})R(x)]$ in (3.161), and that gives the correct trace anomaly of the energy-momentum tensor:

$$W = \int d^4x \sqrt{-g}\, \kappa_0 R + \frac{\hbar}{64\pi^2} \int d^4x \sqrt{-g} \left[-m^4 \ln \left| \frac{M^2}{m^2} \right| \right.$$

$$\left. + m^2 \bar{\xi} R \left(1 - 2\ln\left|\frac{M^2}{m^2}\right|\right) - 2a_2 \ln\left|\frac{M^2}{m^2}\right| + \frac{3}{2}\bar{\xi}^2 R^2 \right]$$

$$+ \frac{i\hbar}{64\pi} \int d^4x \sqrt{-g}(M^4 + 2\bar{a}_2)\Theta(-M^2). \qquad (3.169)$$

Here $M^2 = m^2 + (\xi - 1/6)R$, where m is the inverse Compton wavelength of the scalar field, $\bar{\xi} = \xi - 1/6$, and $\kappa_0 = (16\pi G)^{-1}$ where G is Newton's constant. $\bar{a}_2$ was given earlier in (3.167). This effective action describes vacuum effects of the quantized scalar field under consideration on the classical gravitational field, including non-perturbative effects. The cosmological constant has been taken to be zero in the classical part of this action. We have already discussed the transition to constant curvature that occurs in this model and in the more complete action discussed in Parker and Vanzella (2004). (We mention in passing that in the effective action they obtain by considering the asymptotic form of the heat kernel, the parameter $(\xi - 1/6)$ is replaced by a constant of order 1 and derivatives of curvature tensors are included.) Therefore, we turn now to a discussion of the particle creation by the gravitational field in the above somewhat simplified form of the effective action.

The imaginary term in (3.169) implies that real pairs of these scalar particles are produced by the gravitational field when $(\xi - 1/6)R$ becomes sufficiently negative that $\Theta(-M^2)$ is non-zero. Parker and Raval (2000) find that the rate of particle production given by the imaginary term in W agrees with the rate found by other methods (in cases where they are applicable), and reduces to the approximation of Zeldovich and Starobinsky (1977) when $m = 0$ and $(\xi - 1/6)R < 0$.

Finally, we mention an interesting non-local cosmology that is considered by Deser and Woodard (2008), and may be of relevance to the recent acceleration of the universe. The non-local gravitational terms in this class of cosmological models give rise to a delayed response to the transition from a radiation to a matter-dominated expansion of the universe.

3.8 Normal coordinate momentum space

In a general curved spacetime, it is very useful to introduce a local momentum-space representation based on Riemann normal coordinates (Bunch and Parker 1979; Parker 1979b; Parker and Toms 1984). Using this momentum-space representation, we will carry out dimensional regularization in order to renormalize the quadratic expectation values considered earlier. Our sign conventions are opposite to those of Parker (1979b) and Bunch and Parker (1979), which accounts for any sign differences between their expressions and the ones we give here.

Riemann normal coordinates with respect to an origin at a point Q are constructed as follows (see, for example, Petrov 1969). Suppose that there exists a neighborhood of Q in which there is a unique geodesic joining any point of the neighborhood to Q. As mentioned earlier, such a neighborhood is called a normal neighborhood of Q. Let P be an arbitrary point in a normal neighborhood of Q. The Riemann coordinates y^μ of a point P are given by

$$y^\mu = \lambda \xi^\mu, \qquad (3.170)$$

where λ is the value at P of an affine parameter of the geodesic joining Q to P. The affine parameter is chosen such that $\lambda = 0$ at the origin Q. Furthermore, ξ^μ is the tangent to the geodesic at the point Q,

$$\xi^\mu = (dx^\mu/d\lambda)_Q. \qquad (3.171)$$

If the spacetime is n dimensional, then the ξ^μ can be thought of as a function of $n - 1$ "angles" which are constants along the geodesic, while λ is analogous to a radial coordinate which locates points along the geodesic.

Along any given geodesic through Q, ξ^μ is constant or independent of λ. Therefore, the equation of the geodesic is

$$\frac{d^2 y^\alpha}{d\lambda^2} = 0, \tag{3.172}$$

which implies that in Riemann coordinates

$$\Gamma^\alpha{}_{\beta\gamma}(y)\frac{dy^\beta}{d\lambda}\frac{dy^\gamma}{d\lambda} = \Gamma^\alpha{}_{\beta\gamma}(y)\xi^\beta(y)\xi^\gamma(y) = 0, \tag{3.173}$$

where $\xi^\beta(y)$ is the particular tangent vector at Q along the geodesic joining Q to the point at y^μ. Multiplying by λ^2, we have

$$\Gamma^\alpha{}_{\beta\gamma}(y) y^\beta y^\gamma = 0. \tag{3.174}$$

At point Q itself, we have $\Gamma^\alpha{}_{\beta\gamma}(Q)\xi^\beta\xi^\gamma = 0$ for ξ^μ pointing along any geodesic through Q. Hence, in these coordinates

$$\Gamma^\alpha{}_{\beta\gamma}(Q) = 0. \tag{3.175}$$

We are free to diagonalize and scale the metric at Q so that

$$g_{\mu\nu}(Q) = \eta_{\mu\nu}. \tag{3.176}$$

The coordinates are then called Riemann normal coordinates.

For non-null geodesics, we take λ to be the absolute value of the proper length τ along the geodesic measured from Q, where

$$\tau = \int_Q^P d\sigma \left(g_{\alpha\beta} \frac{dx^\alpha}{d\sigma} \frac{dx^\beta}{d\sigma} \right)^{1/2} \tag{3.177}$$

and σ is any parameter. With the choice $\lambda = |\tau|$, we have for non-null geodesics in Riemann normal coordinates with origin at Q:

$$\eta_{\mu\nu}\xi^\mu\xi^\nu = \begin{cases} 1 & \text{(timelike geodesic)} \\ -1 & \text{(spacelike geodesic)} \end{cases}.$$

This implies that

$$\eta_{\mu\nu} y^\mu y^\nu = \tau^2, \tag{3.178}$$

where the sign of τ^2 depends on the type of geodesic from Q to the point P having Riemann normal coordinates y^μ. (This is also valid for null geodesics, which have $\tau^2 = 0$.) Recall that $y^\mu = \lambda \xi^\mu \equiv |\tau|\xi^\mu$ for non-null geodesics.

Note that (3.172) with the boundary conditions that $y^\alpha(0) = 0$ and $(dy^\alpha/d\lambda)_0 = \xi^\alpha$ is equivalent to (3.170). It follows that (3.174) implies that the coordinate system y^μ is Riemannian. This can be translated into conditions on the $\Gamma^\alpha{}_{\beta\gamma}$ and their derivatives at the region Q. We expand about the point Q at $y^\alpha = 0$,

$$\Gamma^\alpha{}_{\beta\gamma}(y) = \Gamma^\alpha{}_{\beta\gamma}(0) + (\partial_\mu \Gamma^\alpha{}_{\beta\gamma})(0) y^\mu + \frac{1}{2!}(\partial_\mu \partial_\nu \Gamma^\alpha{}_{\beta\gamma})(0) y^\mu y^\nu + \cdots ,$$

where $\partial_\mu \equiv \partial/\partial y^\mu$. Then the condition for Riemann coordinates, that $\Gamma^\alpha{}_{\beta\gamma}(y) y^\beta y^\gamma = 0$ for all y^σ, is equivalent to

$$\Gamma^\alpha{}_{\beta\gamma}(0) = 0, \quad \partial_{(\mu} \Gamma^\alpha{}_{\beta\gamma)}(0) = 0, \quad \partial_{(\mu} \partial_\nu \Gamma^\alpha{}_{\beta\gamma)}(0) = 0, \ldots . \qquad (3.179)$$

Here we use the notation that

$$A_{(\alpha_1 \cdots \alpha_m)} = \frac{1}{m!} \sum_P A_{\alpha_1 \cdots \alpha_m},$$

where $\sum_P$ denotes the sum over all permutations of $\alpha_1 \cdots \alpha_m$.

From (3.179) and the expression for the Riemann tensor at the origin of Riemann coordinates,

$$R^\alpha{}_{\beta\gamma\delta}(0) = -\partial_\gamma \Gamma^\alpha{}_{\beta\delta}(0) + \partial_\delta \Gamma^\alpha{}_{\beta\gamma}(0),$$

we obtain the following results (Petrov (1969), p. 35),

$$\partial_{(\beta} \Gamma^\nu{}_{\alpha)\mu}(0) = -\frac{1}{3} R^\nu{}_{(\alpha\beta)\mu}(0),$$

$$\partial_{(\gamma} \partial_\beta \Gamma^\nu{}_{\alpha)\mu}(0) = \frac{1}{2} R_{\mu(\gamma}{}^\nu{}_{\beta;\alpha)}(0),$$

$$\partial_{(\delta} \partial_\gamma \partial_\beta \Gamma^\nu{}_{\alpha)\mu}(0) = -\frac{3}{5} \left(\frac{2}{9} R_{(\alpha}{}^\rho{}_\beta{}^\nu R_{\delta\rho\gamma)\mu} - R_{\mu(\delta}{}^\nu{}_{\gamma;\alpha\beta)} \right)\Big|_{y=0}.$$

If $W_{\alpha_1 \cdots \alpha_p}(y)$ is any tensor field with analytic components in a neighborhood of $y^\alpha = 0$, it can be Taylor expanded about the origin, and the ordinary derivatives of $W_{\alpha_1 \cdots \alpha_p}$ expressed in terms of covariant derivatives and affine connections. Symmetrization over groups of indices will occur because of the symmetric products of y^α's appearing in the expansion. Then, using the above relations, it can be shown that (Petrov (1969), p. 36)

$$W_{\alpha_1 \cdots \alpha_p}(y) = W_{\alpha_1 \cdots \alpha_p}(0) + W_{\alpha_1 \cdots \alpha_p;\mu}(0) y^\mu$$

$$+ \frac{1}{2!} \left[W_{\alpha_1 \cdots \alpha_p;\mu\omega} - \frac{1}{3} \sum_{k=1}^p R^\nu{}_{\mu\alpha_k\omega} W_{\alpha_1 \cdots \alpha_{k-1}\nu\alpha_{k+1} \cdots \alpha_p} \right]_0 y^\mu y^\omega$$

$$+ \frac{1}{3!} \left[W_{\alpha_1 \cdots \alpha_p;\mu\omega\sigma} - \sum_{k=1}^p R^\nu{}_{\mu\alpha_k\omega} W_{\alpha_1 \cdots \alpha_{k-1}\nu\alpha_{k+1} \cdots \alpha_p;\sigma} \right.$$

$$\left. - \frac{1}{2} \sum_{k=1}^p R^\nu{}_{\mu\alpha_k\omega;\sigma} W_{\alpha_1 \cdots \alpha_{k-1}\nu\alpha_{k+1} \cdots \alpha_p} \right]_0 y^\mu y^\omega y^\sigma + \cdots . \qquad (3.180)$$

For the metric, in particular, we have (including the fourth-order term):

$$g_{\alpha\beta}(y) = \eta_{\alpha\beta} - \frac{1}{3}R_{\alpha\mu\beta\lambda}(0)y^\mu y^\lambda - \frac{1}{3!}R_{\alpha\gamma\beta\lambda;\mu}(0)y^\lambda y^\mu y^\gamma$$
$$+ \frac{1}{5!}\left(-6R_{\alpha\delta\beta\gamma;\lambda\mu} + \frac{16}{3}R_{\lambda\beta\mu}{}^\rho R_{\gamma\alpha\delta\rho}\right)_0 y^\lambda y^\mu y^\gamma y^\delta + \cdots \quad (3.181)$$

If x^ν is an arbitrary coordinate system in which the origin Q of Riemann coordinates is at $x_0{}^\nu$, and in which the affine connections of Q have the values $\Gamma^\mu{}_{\nu\lambda}(x_0)$, a transformation giving x^ν in terms of Riemann coordinates y^ν is (Petrov 1969)

$$x^\nu = x_0{}^\nu + y^\nu - \Gamma^\nu{}_{(\alpha\beta)}(x_0)y^\alpha y^\beta$$
$$+ \frac{1}{3!}\left[2\Gamma^\nu{}_{\sigma(\alpha}\Gamma^\sigma{}_{\beta\gamma)} - \partial_{(\alpha}\Gamma^\nu{}_{\beta\gamma)}\right]_{x_0} y^\alpha y^\beta y^\gamma + \cdots, \quad (3.182)$$

where ∂_α here refers to $\partial/\partial x^\alpha$. We can further make $g_{\mu\nu}(y=0) = \eta_{\mu\nu}$ by an additional linear homogeneous transformation, which does not alter the Riemannian nature of the coordinates.

It is also true in Riemann normal coordinates that

$$g_{\mu\nu}(y)y^\nu = \eta_{\mu\nu}y^\nu. \quad (3.183)$$

This can be checked by contracting (3.181) with y^β, and noting that in every term but the first, there are contractions of two antisymmetric indices of the Riemann tensor with a symmetric product of two y's yielding zero.

Consider again the basic equation satisfied by the scalar Green functions,

$$(\Box_x + m^2 + \xi R(x))\, G(x,x') = -\delta(x,x'), \quad (3.184)$$

where $\delta(x,x') \equiv |g(x)|^{-1/2}\delta(x-x')$ is the biscalar Dirac δ-distribution. In order to define a momentum-space transform, which faithfully describes the behavior of $G(x,x')$ when x and x' are within a normal neighborhood, we introduce a Riemann normal coordinate system y. It is convenient to take the origin Q at the spacetime point x'. (The origin Q can be chosen to be anywhere in the normal neighborhood containing x and x'. If Q were not at x' then the Fourier transform would be taken with respect to the difference of the normal coordinates of x and x'.) With the origin at x', we have $|g(x')| = 1$ in normal coordinates as a consequence of (3.176). In these coordinates, some simplification will occur if we define a function $\overline{G}$ by

$$G(x,x') \equiv |g(x)|^{-1/4}\overline{G}(x,x'). \quad (3.185)$$

Note that we have only defined $\overline{G}$ in normal coordinates. The relation between G and $\overline{G}$ in arbitrary coordinates will be given in (3.205). Denoting the normal coordinate of x by y^μ and $\partial/\partial y^\mu$ by ∂_μ, and using the normal coordinate expansion of $g_{\mu\nu}$ in (3.181) and of $|g(x)|$ obtained from it, as well as the expansion of

$R(x)$ obtained from (3.180), after some calculation (Bunch and Parker 1979) it can be shown that $\overline{G}$ satisfies the equation

$$\delta(y) = -\eta^{\mu\nu}\partial_\mu\partial_\nu\overline{G} - \left[m^2 + \left(\xi - \frac{1}{6}\right)R\right]\overline{G}$$
$$- \frac{1}{3}R_\alpha{}^\nu y^\alpha \partial_\nu\overline{G} + \frac{1}{3}R^\mu{}_\alpha{}^\nu{}_\beta y^\alpha y^\beta \partial_\mu\partial_\nu\overline{G}$$
$$- \left(\xi - \frac{1}{6}\right)R_{;\alpha}y^\alpha\overline{G} + \left(-\frac{1}{3}R_\alpha{}^\nu{}_{;\beta} + \frac{1}{6}R_{\alpha\beta}{}^{;\nu}\right)y^\alpha y^\beta \partial_\nu\overline{G}$$
$$+ \frac{1}{6}R^\mu{}_\alpha{}^\nu{}_{\beta;\gamma}y^\alpha y^\beta y^\gamma \partial_\mu\partial_\nu\overline{G} - \frac{1}{2}\left(\xi - \frac{1}{6}\right)R_{;\alpha\beta}y^\alpha y^\beta\overline{G}$$
$$+ \left(\frac{1}{30}R_\alpha^\lambda R_{\lambda\beta} - \frac{1}{60}R^\kappa{}_\alpha{}^\lambda{}_\beta R_{\kappa\lambda} - \frac{1}{60}R^{\lambda\mu\kappa}{}_\alpha R_{\lambda\mu\kappa\beta}\right.$$
$$\left. - \frac{1}{120}R_{;\alpha\beta} + \frac{1}{40}\Box R_{\alpha\beta}\right)y^\alpha y^\beta \overline{G}$$
$$+ \left(-\frac{3}{20}R^\nu{}_{\alpha;\beta\gamma} + \frac{1}{10}R_{\alpha\beta}{}^{;\nu}{}_\gamma + \frac{1}{60}R^\kappa{}_\alpha{}^\nu{}_\beta R_{\kappa\gamma}\right.$$
$$\left. - \frac{1}{15}R^\kappa{}_{\alpha\lambda\beta}R_\kappa{}^\nu{}_\gamma{}^\lambda\right)y^\alpha y^\beta y^\gamma \partial_\nu\overline{G}$$
$$+ \left(\frac{1}{20}R^\mu{}_\alpha{}^\nu{}_{\beta;\gamma\delta} - \frac{1}{15}R^\mu{}_{\alpha\lambda\beta}R^\lambda{}_\gamma{}^\nu{}_\delta\right)y^\alpha y^\beta y^\gamma y^\delta \partial_\mu\partial_\nu\overline{G}. \quad (3.186)$$

Here we have retained only terms with coefficients involving four or fewer derivatives of the metric (i.e., terms up to fourth adiabatic order). Thus, for example, a term $R_{;\alpha\beta\gamma}y^\alpha y^\beta y^\gamma$, which is of fifth adiabatic order, has been dropped. It is important to note that all the coefficients are evaluated at the origin Q of the normal coordinate system (in this case at the point x'). Thus, the complicated looking terms formed from the Riemann tensor and its covariant derivatives have no dependence on y and are, in effect, constants. It will be sufficient in treating UV divergences in four dimensions to work to fourth adiabatic order.

The momentum space associated with the point x' ($y = 0$) is introduced by making the generalized Fourier transformation (the negative sign on the right is convenient):

$$\overline{G}(x, x') = -\int \frac{d^n k}{(2\pi)^n} e^{iky}\overline{G}(k), \quad (3.187)$$

where $ky = k_\alpha y^\alpha$ and $\overline{G}(k)$ is a function of x' as well as k. If the origin Q were at an arbitrary point we would write

$$\overline{G}(x, x') = \overline{G}\left(X + \frac{u}{2}, X - \frac{u}{2}\right) = -\int \frac{d^n k}{(2\pi)^n} e^{iku}\overline{G}(k; X),$$

where $u = y - y'$ and $X = \frac{1}{2}(y + y')$ with y and y' being the Riemann normal coordinates of x and x', respectively (Bunch 1981a; Calzetta et al., 1988). This generalized Fourier transform is sufficient for describing the singularities which arise when x approaches x' in the Green functions. (It should also give a good approximation when x and x' are in a single normal neighborhood.) We work with the n-dimensional Fourier transform to facilitate the use of dimensional regularization later.

We obtain the adiabatic expansion of $\overline{G}(k)$ by substituting (3.187) into (3.186) and writing

$$\overline{G}(k) = \overline{G}_0(k) + \overline{G}_1(k) + \overline{G}_2(k) + \cdots, \tag{3.188}$$

where $\overline{G}_i(k)$ is of adiabatic order i, meaning that it has a geometrical coefficient involving i derivatives of the metric (evaluated at the origin x'). To lowest adiabatic order, (3.186) becomes in momentum space

$$(-\eta^{\mu\nu} k_\mu k_\nu + m^2)\overline{G}_0(k) = 1,$$

with the solution

$$\overline{G}_0(k) = (-k^2 + m^2)^{-1}, \tag{3.189}$$

where $k^2 = \eta^{\mu\nu} k_\mu k_\nu$. In first adiabatic order, we find that

$$\overline{G}_1(k) = 0. \tag{3.190}$$

The second adiabatic order momentum-space equation is

$$(-\eta^{\mu\nu} k_\mu k_\nu + m^2)\overline{G}_2(k) + \left(\xi - \frac{1}{6}\right) R\overline{G}_0(k)$$

$$+ \frac{1}{3} R_\alpha{}^\nu (ik_\nu) \left(i\frac{\partial}{\partial k_\alpha}\right) \overline{G}_0(k)$$

$$- \frac{1}{3} R^\mu{}_\alpha{}^\nu{}_\beta (ik_\mu)(ik_\nu) \left(i\frac{\partial}{\partial k_\alpha}\right) \left(i\frac{\partial}{\partial k_\beta}\right) \overline{G}_0(k) = 0,$$

where we have used the fact that in going to momentum space, $\partial/\partial y^\alpha \to ik_\alpha$ and $y^\alpha \to i\partial/\partial k_\alpha$ acting on $\overline{G}_i(k)$ (this follows from integration by parts, using the fact that $\overline{G}_i(k)$ vanishes for large k^2). Using (3.189) for $\overline{G}_0(k)$, we find that the last two terms sum to zero. Solving for $\overline{G}_2(k)$ then gives

$$\overline{G}_2(k) = \left(\frac{1}{6} - \xi\right) R(-k^2 + m^2)^{-2}. \tag{3.191}$$

Continuing the iteration, we obtain the solution to fourth adiabatic order:

$$\overline{G}(k) = (-k^2 + m^2)^{-1} + \left(\frac{1}{6} - \xi\right) R(-k^2 + m^2)^{-2}$$

$$+ i\left(\frac{1}{6} - \xi\right) R_{;\alpha}(-k^2 + m^2)^{-1} \frac{\partial}{\partial k_\alpha} (-k^2 + m^2)^{-1}$$

$$+ \left(\frac{1}{6} - \xi\right)^2 R^2 (-k^2 + m^2)^{-3}$$

$$+ a_{\alpha\beta}(-k^2 + m^2)^{-1} \frac{\partial}{\partial k_\alpha} \frac{\partial}{\partial k_\beta} (-k^2 + m^2)^{-1}, \tag{3.192}$$

where

$$a_{\alpha\beta} = \frac{1}{2}\left(\xi - \frac{1}{6}\right)R_{;\alpha\beta} + \frac{1}{120}R_{;\alpha\beta} - \frac{1}{40}\Box R_{\alpha\beta}$$
$$- \frac{1}{30}R\alpha^{\lambda}R_{\lambda\beta} + \frac{1}{60}R_{\kappa\alpha\lambda\beta}R^{\kappa\lambda} + \frac{1}{60}R^{\lambda\mu\kappa}{}_{\alpha}R_{\lambda\mu\kappa\beta}. \quad (3.193)$$

All geometrical quantities are evaluated at the origin of the Riemann normal coordinate system (the point x'). Then the expansion of $\overline{G}(x,x')$ is given by (3.187) with $\overline{G}(k)$ of (3.192). The momentum-space representation of $G(x,x')$ may also readily be obtained from $\overline{G}(k)$ by using (3.185) and (3.187), expressing the factor of $g^{-1/4}(x)$ as a polynomial in y^{α}, and integrating by parts to replace each y^{α} by $i\partial/\partial k_{\alpha}$ acting on $\overline{G}(k)$ in momentum space.

Using the identities

$$(-k^2 + m^2)^{-1}\frac{\partial}{\partial k_{\alpha}}(-k^2 + m^2)^{-1} = \frac{1}{2}\frac{\partial}{\partial k_{\alpha}}(-k^2 + m^2)^{-2},$$
$$(-k^2 + m^2)^{-1}\frac{\partial}{\partial k_{\alpha}}\frac{\partial}{\partial k_{\beta}}(-k^2 + m^2)^{-1} = \frac{1}{3}\frac{\partial}{\partial k_{\alpha}}\frac{\partial}{\partial k_{\beta}}(-k^2 + m^2)^{-2}$$
$$+ \frac{2}{3}\eta^{\alpha\beta}(-k^2 + m^2)^{-3},$$

the previous expression for $\overline{G}(k)$ may be rewritten as

$$\overline{G}(k) = (-k^2 + m^2)^{-1} + \left(\frac{1}{6} - \xi\right)R(-k^2 + m^2)^{-2}$$
$$+ \frac{1}{2}i\left(\frac{1}{6} - \xi\right)R_{;\alpha}\frac{\partial}{\partial k_{\alpha}}(-k^2 + m^2)^{-2}$$
$$+ \frac{1}{3}a_{\alpha\beta}\frac{\partial}{\partial k_{\alpha}}\frac{\partial}{\partial k_{\beta}}(-k^2 + m^2)^{-2}$$
$$+ \left[\left(\frac{1}{6} - \xi\right)^2 R^2 + \frac{2}{3}a^{\alpha}{}_{\alpha}\right](-k^2 + m^2)^{-3}. \quad (3.194)$$

Note that $G_0(k)$ in (3.189) has the same form as in Minkowski spacetime. The higher-order terms follow by iteration from $G_0(k)$, so that boundary conditions placed on $G_0(k)$ determine those on the higher-order terms. It is natural to define the curved spacetime analogues of the various Minkowski spacetime Green functions by displacing the poles in $G_0(k)$ at $k_0 = \pm(\vec{k}^2 + m^2)^{1/2}$ in the same way as in the corresponding Minkowski spacetime Green functions. In particular, the Feynman Green function is defined by displacing the pole at $k_0 = -(\vec{k}^2 + m^2)^{1/2}$ to lie above the real k_0-axis and the pole at $k_0 = (\vec{k}^2 + m^2)^{1/2}$ to lie below the real k_0-axis. (This is equivalent to replacing m^2 by $m^2 - i\epsilon$ where ϵ is a positive infinitesimal.)

Thus, it is clear in the Riemann normal coordinate momentum-space representation how to define the momentum-space expansions of each of the possible Green functions. (The same is true, as well, in the massless case.) Except for the way in which the poles on the real k_0-axis are displaced, all the Green functions have the same momentum-space expansion. Consequently, they all give rise to the same UV divergences. Dimensional regularization can be applied in the momentum-space representation in curved spacetime much as it is in flat spacetime (Bunch and Parker 1979). We will discuss a version of how this works for an interacting scalar field in Section 6.7.

The trace anomaly is readily obtained from the momentum-space adiabatic expansion of $\overline{G}(k)$. Formally, we have with $\xi = 1/6$ in the massless limit,

$$\langle T_\mu{}^\mu(x)\rangle = \lim_{m\to 0} m^2 \langle \phi(x)^2\rangle$$
$$= \lim_{m\to 0} im^2 G(x,x).$$

Here we have defined

$$\langle \phi(x')^2\rangle = \lim_{x\to x'} \langle T\phi(x)\phi(x')\rangle \equiv iG(x',x'),$$

where G satisfies (3.184). The time-ordered product appears here because in taking the limit as $x \to x'$, $\phi(x)$ and $\phi(x')$ do not commute when x and x' are connected by a timelike geodesic. The time-ordered product provides a generally covariant way of specifying the ordering of $\phi(x)$ and $\phi(x')$, and of insuring that a unique limit is approached. The expectation value can be taken in any state satisfying the adiabatic condition that the first term in the momentum-space adiabatic expansion of G is given by (3.189). This is the momentum-space form of the adiabatic condition. Then the adiabatic expansion of $\overline{G}(k)$ is given by (3.194).

As we know from Section 3.1, to obtain the physical expectation value of $T_\mu{}^\mu(x)$ we must subtract to fourth order in the adiabatic expansion. Thus

$$\langle T_\mu{}^\mu(x)\rangle_{\text{phys}} = \lim_{m\to 0} im^2\{G(x,x) - G_0(x,x) - G_1(x,x) - G_2(x,x)$$
$$- G_3(x,x) - G_4(x,x)\},$$
$$= \lim_{m\to 0} im^2\{\langle \phi^2(x)\rangle_{\text{phys}} - G_3(x,x) - G_4(x,x)\}$$
$$= -\lim_{m\to 0} im^2\{G_3(x,x) + G_4(x,x)\}, \qquad (3.195)$$

where $G_i(x,x)$ is the term of adiabatic order i in the adiabatic expansion of $G(x,x)$. From (3.185), when $x = x'$ we have $G(x,x) = \overline{G}(x,x)$. Then from (3.187) with $y = 0$ and (3.194), we have

$$G_3(x,x) = -\int \frac{d^n k}{(2\pi)^2} \overline{G}_3(k)$$
$$= \frac{i}{2}\left(\frac{1}{6} - \xi\right) R_{;\alpha} \int \frac{d^n k}{(2\pi)^n} \frac{\partial}{\partial k_\alpha}(-k^2 + m^2)^{-2}$$
$$= 0,$$

since in the present case we have $\xi = 1/6$, or in general because $(-k^2 + m^2)^{-2}$ vanishes at the limits of integration, or because the integrand is an odd function of k. Similarly, (3.194) gives, with $\xi = 1/6$,

$$G_4(x,x) = -\int \frac{d^n k}{(2\pi)^n} \overline{G}_4(k) = -\frac{2}{3} a^\alpha{}_\alpha(x) \int \frac{d^4 k}{(2\pi)^4} (-k^2 + m^2)^{-3},$$

where we have dropped a term which vanishes at the limits of integration, and we have taken the limit $n \to 4$, as there is no pole at $n = 4$. The integration yields

$$G_4(x,x) = -\frac{2}{3} a^\alpha{}_\alpha(x) \left(\frac{i\pi^2}{2m^2(2\pi)^4}\right), \quad \left(\xi = \frac{1}{6}\right).$$

From (3.193) with $\xi = 1/6$,

$$a^\alpha{}_\alpha = -\frac{1}{60}\Box R - \frac{1}{60} R_{\alpha\beta} R^{\alpha\beta} + \frac{1}{60} R^{\alpha\beta\gamma\delta} R_{\alpha\beta\gamma\delta}$$

$$= 3a_2(x), \quad \left(\xi = \frac{1}{6}\right) \tag{3.196}$$

with $a_2(x)$ given in (3.92). Hence

$$G_4(x,x) = -\frac{i\pi^2}{m^2(2\pi)^4} a_2(x) \tag{3.197}$$

and (3.195) yields the trace anomaly

$$\langle T_\mu{}^\mu(x) \rangle_{\text{phys}} = -\frac{1}{16\pi^2} a_2(x), \tag{3.198}$$

in agreement with our earlier derivation in (3.91).

Finally, following Bunch and Parker (1979), we show how to derive the DeWitt–Schwinger or heat kernel expansion from the Riemann normal coordinate momentum-space expansion (ξ and m are again arbitrary). Substitute (3.194) into (3.187). Then integrate by parts so that factors of $\partial/\partial k^\alpha$ act on $\exp(iky)$, thus being transmuted into factors of $-iy^\alpha$. Finally, write $(j!)(-k^2 + m^2)^{-j-1} = (-\partial/\partial m^2)^j (-k^2 + m^2)^{-1}$. The result is (for any of the inhomogeneous Green functions)

$$\overline{G}(x,x') = -\int \frac{d^n k}{(2\pi)^n} e^{iky} [1 + a_1(x,x')(-\partial/\partial m^2)$$
$$+ a_2(x,x')(\partial/\partial m^2)^2 + \cdots](-k^2 + m^2)^{-1}, \tag{3.199}$$

where (in these Riemann normal coordinates with origin at x')

$$a_1(x,x') = \left(\frac{1}{6} - \xi\right) R(x') + \frac{1}{2}\left(\frac{1}{6} - \xi\right) R_{;\alpha}(x') y^\alpha - \frac{1}{3} a_{\alpha\beta}(x') y^\alpha y^\beta, \tag{3.200}$$

$$a_2(x,x') = \frac{1}{2}\left(\frac{1}{6} - \xi\right)^2 R(x')^2 + \frac{1}{3} a^\alpha{}_\alpha(x'). \tag{3.201}$$

3.8 Normal coordinate momentum space

We emphasize that (3.199) is valid for *any* of the inhomogeneous Green functions. This can be put into the DeWitt–Schwinger or heat kernel form *only* when the Feynman contour of integration in the complex k_0-plane is chosen, or equivalently, the poles are moved off the real k_0-plane by replacing m^2 by $m^2 - i\epsilon$ with ϵ a positive infinitesimal (meaning it is taken to zero in the end). With $m^2 \to m^2 - i\epsilon$ understood, we can write

$$(-k^2 + m^2)^{-1} = \int_0^\infty ids \exp[-is(-k^2 + m^2)]$$

and

$$\int \frac{d^n k}{(2\pi)^n} \exp[-is(-k^2 + m^2) + iky]$$
$$= \frac{i}{(4\pi)^{n/2}} (is)^{-n/2} \exp\left(-im^2 s + \frac{\sigma(x, x')}{2is}\right),$$

where $\sigma(x, x') = \frac{1}{2}\tau^2 = \frac{1}{2}\eta_{\alpha\beta} y^\alpha y^\beta$ is half the square of the geodesic distance between x and x'. We can also make the replacement of y^α with $\sigma(x, x')^{;\alpha} = g^{\alpha\beta}(x)\partial\sigma(x, x')/\partial x^\beta$ in the expressions for $a_1(x, x')$ and $a_2(x, x')$. Then, we obtain from (3.199)

$$\overline{G}(x, x') = -\frac{i}{(4\pi)^{n/2}} \int_0^\infty \frac{ids}{(is)^{n/2}} \exp\left(-im^2 s + \frac{\sigma}{2is}\right) F(x, x'; is), \quad (3.202)$$

where

$$F(x, x'; is) = 1 + a_1(x, x') is + a_2(x, x') (is)^2 + \cdots \quad (3.203)$$

As defined above, $\overline{G}(x, x')$ is manifestly a biscalar in arbitrary coordinates. To obtain $G(x, x')$ in arbitrary coordinates, we need only find the biscalar which reduces to $|g(x)|^{-1/4}$ in Riemann normal coordinates with origin at x'. Then (3.185) will give a manifestly covariant expression for $G(x, x')$.

The biscalar Van Vleck–Morette determinant was defined in (3.82) as

$$\Delta(x, x') = -|g(x)|^{-1/2} \det[-\partial_\mu \partial_{\nu'} \sigma(x, x')] |g(x')|^{-1/2}.$$

In Riemann normal coordinates with origin at x', we have $\sigma(x, x') = \frac{1}{2}\tau^2 = \frac{1}{2}\eta_{\mu\nu} y^\mu y^\nu$ where the y^μ are the coordinates of point x. If we do not set the affine parameter to zero at x', but call it λ' instead, then we would find that $\sigma(x, x') = \frac{1}{2}\eta_{\mu\nu}(y^\mu - y'^\mu)(y^\nu - y'^\nu)$ where $y'^\mu = \lambda' \xi^\mu$, and ξ^μ is the tangent at x' to the geodesics joining x' to x. Then it is clear that $\partial_\mu \partial_{\nu'} \sigma(x, x') = -\eta_{\mu\nu}$, and

$$\Delta(x, x') = |g(x)|^{-1/2} \quad (3.204)$$

in Riemann normal coordinates with origin at x'. Thus (3.185) yields

$$G(x, x') = \Delta^{1/2}(x, x') \overline{G}(x, x'). \quad (3.205)$$

With $\overline{G}(x, x')$ written in the manifestly covariant form of (3.202), this gives the generally covariant expression for $G(x, x')$. From the momentum-space expansion

3.9 Chiral current anomaly caused by spacetime curvature

A free massless Dirac field in flat spacetime has a conserved vector current J_V and a conserved axial vector, or chiral, current J_A. If the field has mass, then the chiral current is not conserved. We will show that as a consequence, $\langle \nabla_\mu J_A^\mu \rangle$ is non-zero for the massless Dirac field in curved spacetime. Our derivation of this local state-independent anomaly is analogous to our earlier derivation of the trace anomaly $\langle T_\mu^\mu \rangle$ for the massless conformal scalar field. We will first summarize the theory of the Dirac field in curved spacetime, which was developed by Fock (1929), Schrödinger (1932), and Bargmann (1932). A different treatment of spinors in curved spacetime of arbitrary dimension will be given in Section 5.6.

In flat spacetime, the Dirac equation (1.117) is obtained from the Lagrangian (1.115), and the γ-matrices satisfy (1.116) and the hermiticity properties described there. In curved spacetime, (1.116) is generalized to

$$\underline{\gamma}^\mu(x)\underline{\gamma}^\nu(x) + \underline{\gamma}^\nu(x)\underline{\gamma}^\mu(x) = 2g^{\mu\nu}(x), \tag{3.206}$$

where the underline is used to distinguish these spacetime-dependent γ-matrices from the constant γ-matrices defined in (1.116). The spinorial affine connections $\Gamma_\mu(x)$ are matrices defined by the vanishing of the covariant derivative of the $\underline{\gamma}$-matrices:

$$\nabla_\mu \underline{\gamma}_\nu \equiv \partial_\mu \underline{\gamma}_\nu - \Gamma^\lambda{}_{\mu\nu}\underline{\gamma}_\lambda - \Gamma_\mu \underline{\gamma}_\nu + \underline{\gamma}_\nu \Gamma_\mu = 0, \tag{3.207}$$

where $\underline{\gamma}_\nu = g_{\nu\mu}\underline{\gamma}^\mu$. The covariant derivative acting on a spinor field ψ is

$$\nabla_\mu \psi \equiv (\partial_\mu - \Gamma_\mu)\psi. \tag{3.208}$$

Then, the generally covariant Dirac equation is

$$\left(i\underline{\gamma}^\mu(x)\nabla_\mu - m\right)\psi(x) = 0. \tag{3.209}$$

The quantity $\psi^\dagger \epsilon \psi$ can be shown to transform as a scalar if the matrix ϵ satisfies the equations

$$\epsilon - \epsilon^\dagger = 0,$$

$$\epsilon \underline{\gamma}^\mu - \underline{\gamma}^{\mu\dagger}\epsilon = 0,$$

$$\nabla_\mu \epsilon \equiv \partial_\mu \epsilon + \Gamma_\mu^\dagger \epsilon + \epsilon \Gamma_\mu = 0. \tag{3.210}$$

(Note that in general $\underline{\gamma}^0$ is not Hermitian, so that $\epsilon \neq \underline{\gamma}^0$.) Defining

$$\bar{\psi} = \psi^\dagger \epsilon, \tag{3.211}$$

3.9 Chiral current anomaly caused by spacetime curvature

the scalar density of weight 1/2 that generalizes the Lagrangian (1.115) is

$$\mathcal{L} = |g|^{1/2} \overline{\psi}(i\underline{\gamma}^\mu \nabla_\mu - m)\psi. \tag{3.212}$$

We can show that the current

$$J_V^\mu \equiv \overline{\psi}\underline{\gamma}^\mu \psi \tag{3.213}$$

is Hermitian and transforms as a vector. From the generalized Dirac equation (3.209) and its conjugate, which upon multiplication on the right by ϵ takes the form

$$\overline{\psi}(i\underline{\gamma}^\mu \overleftarrow{\nabla}_\mu + m) = 0, \tag{3.214}$$

it follows that

$$\nabla_\mu J_V^\mu = 0. \tag{3.215}$$

(Here $\overleftarrow{\nabla}$ denotes the covariant derivative operating to the left.)

An explicit representation for the $\underline{\gamma}_\mu(x)$ matrices in terms of the flat spacetime γ_μ matrices is obtained by introducing a vierbein $b^\alpha{}_\mu(x)$ of vector fields, defined by

$$g_{\mu\nu}(x) = \eta_{\alpha\beta} b^\alpha{}_\mu(x) b^\beta{}_\nu(x). \tag{3.216}$$

Under transformations of the coordinates, indices $\mu, \nu, \cdots$ are regarded as tensor indices, while indices $\alpha, \beta, \cdots$ act merely as labels (thus the $b^\alpha{}_\mu$ constitute four different vector fields). In addition to the covariance under general coordinate transformation acting on the spacetime indices μ, ν, the formalism is covariant under Lorentz transformations applied to the vierbein indices α, β. Vierbein indices are lowered with $\eta_{\alpha\beta}$, while spacetime indices are lowered with the metric $g_{\mu\nu}$. In terms of the special relativistic matrices γ^μ of (1.116), the matrices $\underline{\gamma}^\mu(x)$ of (3.206) can be written in the form

$$\underline{\gamma}^\mu(x) = b_\alpha{}^\mu(x) \gamma^\alpha. \tag{3.217}$$

A set of $\Gamma_\mu(x)$ satisfying (3.207), and reducing to zero in Minkowski spacetime, are given by

$$\Gamma_\mu(x) = -\frac{1}{4} \gamma_\alpha \gamma_\beta b^{\alpha\lambda}(x) \nabla_\mu b^\beta{}_\lambda(x), \tag{3.218}$$

where $\nabla_\mu b^\beta{}_\lambda = \partial_\mu b^\beta{}_\lambda - \Gamma^\sigma{}_{\mu\lambda} b^\beta{}_\sigma$. Note that $b^{\alpha\lambda} \nabla_\mu b^\beta{}_\lambda$ is antisymmetric under exchange of α and β because $b^{\alpha\lambda} b^\beta{}_\lambda = \eta^{\alpha\beta}$. It can be further proved (Parker (1980), Eq. (2.16)) that a suitable choice of ϵ satisfying (3.210) is

$$\epsilon = \gamma^0, \tag{3.219}$$

where γ^0 is the constant special relativistic matrix. The vierbein formalism will be considered in more depth in Section 5.6.

In the case when $m = 0$, the curved space Dirac Lagrangian is symmetric under the unitary "chiral" transformation defined by

$$\psi \to \psi' = e^{i\alpha\gamma_5}\psi. \tag{3.220}$$

This corresponds to the infinitesimal change $\delta\psi = i\alpha\gamma_5\psi$. Here

$$\gamma_5 \equiv i\gamma^0\gamma^1\gamma^2\gamma^3, \tag{3.221}$$

and satisfies $\gamma^{5\dagger} = \gamma_5$, $\gamma_5\gamma_5 = 1$, and $\gamma_5\gamma^\mu = -\gamma^\mu\gamma_5$ for $\mu = 0, 1, 2, 3$. It follows that

$$\gamma_5\underline{\gamma}^\mu(x) = -\underline{\gamma}^\mu(x)\gamma_5. \tag{3.222}$$

In accordance with the result obtained earlier from the Schwinger action principle, the symmetry of $\mathcal{L}$ implies (before renormalization) that the quantity

$$-\frac{\partial \mathcal{L}}{\partial(\partial_\mu\psi)}\alpha^{-1}\delta\psi = |g|^{1/2}\overline{\psi}\underline{\gamma}^\mu\gamma_5\psi \equiv |g|^{1/2}J^\mu{}_A$$

satisfies $\partial_\mu(|g|^{1/2}J_A{}^\mu) = 0$, or

$$\nabla_\mu J_A{}^\mu = 0 \quad (m = 0), \tag{3.223}$$

where

$$J_A{}^\mu \equiv \overline{\psi}\underline{\gamma}^\mu\gamma_5\psi \tag{3.224}$$

is the generalization of the axial vector current to curved spacetime. One readily finds that $J_A{}^\mu$ is Hermitian. (Here we are treating the components of ψ as ordinary numbers, rather than anticommuting Grassmann numbers. The alternative method of treating them as Grassmann numbers is discussed in Chapter 5, and does not alter the results obtained here.) Furthermore, from the requirement that $J_A{}^\mu$ transforms as a vector under infinitesimal coordinate transformations, we can show that for consistency the covariant derivative of γ_5 should have the form

$$\nabla_\mu\gamma_5 = \partial_\mu\gamma_5 - \Gamma_\mu\gamma_5 + \gamma_5\Gamma_\mu.$$

Then from (3.218) and the fact that γ_5 is a constant matrix, it follows immediately that

$$\nabla_\mu\gamma_5 = 0. \tag{3.225}$$

When the mass is not zero, we find from the curved space Dirac equation that

$$\nabla_\mu J_A{}^\mu = 2im\overline{\psi}\gamma_5\psi. \tag{3.226}$$

The form of (3.226) is analogous to (3.54) for $T^\mu{}_\mu$, suggesting that in the limit $m \to 0$, $\langle \nabla_\mu J_A{}^\mu \rangle$ may not vanish, but will be given by a local function quadratic in the Riemann tensor. We now proceed to calculate this chiral current anomaly for the massless Dirac field.

3.9 Chiral current anomaly caused by spacetime curvature

Formally, from (3.226) we have

$$\langle \nabla_\mu J_A^\mu(x) \rangle = 2im \langle \bar\psi(x)\gamma_5 \psi(x) \rangle. \tag{3.227}$$

As in the case of $\langle T^\mu{}_\mu \rangle$ we must subtract up to fourth adiabatic order because of the degree of divergence in the individual terms such as $\langle \nabla_1 J_A^1 \rangle$. As in the scalar case, we define

$$\langle \bar\psi_b(x)\psi_a(x) \rangle \equiv - \lim_{x \to x'} \langle T\psi_a(x)\bar\psi_b(x') \rangle,$$

where

$$T\psi_a(x)\bar\psi_b(x') = \Theta(t-t')\psi_a(x)\bar\psi_b(x') - \Theta(t'-t)\bar\psi_b(x')\psi_a(x)$$

and

$$\Theta(u) = \int_{-\infty}^{u} \delta(s)ds = \begin{cases} 1 \text{ if } u > 0 \\ 0 \text{ if } u < 0 \end{cases}.$$

Here the subscripts a, b are spinor indices. The state in (3.227), as for $\langle T^\mu{}_\mu \rangle$, may be any normalized state in the Fock space based on an adiabatic vacuum state. We remind the reader that an adiabatic vacuum state in the present context may be defined as one such that the *leading* term in the adiabatic expansion of the inhomogeneous Green functions has the same form at the origin of Riemann normal coordinates as it does in flat spacetime. Although the adiabatic vacuum state is not unique, the vector space spanned by the Fock space built from it is unique.

We now define a Feynman Green function as

$$S_{ab}(x, x') \equiv -i \langle T\psi_a(x)\bar\psi_b(x') \rangle. \tag{3.228}$$

Because the choice of the state vector is not unique, the Feynman Green function is not unique. However, its adiabatic expansion is unique. (As mentioned earlier, the adiabatic expansion is generally asymptotic and not convergent.)

The canonical momentum conjugate to ψ is

$$\pi = \frac{\partial \mathcal{L}}{\partial(\partial_0 \psi)} = i|g|^{1/2}\bar\psi\gamma^0.$$

Then

$$\{\psi_a(\vec{x}, t), \pi_b(\vec{x}', t)\} = i\delta(\vec{x} - \vec{x}')\delta_{ab}$$

yields the anticommutator

$$\{\psi_a(\vec{x}, t), (\bar\psi(\vec{x}', t)\gamma^0(\vec{x}', t))_b\} = |g(\vec{x}', t)|^{-1/2}\delta(\vec{x} - \vec{x}')\delta_{ab}. \tag{3.229}$$

Denoting the inverse of $\gamma^0(x)$ by $\gamma^0(x)^{-1}$, this can be written as

$$\{\psi_a(\vec{x}, t), \bar\psi_b(\vec{x}', t)\} = |g(\vec{x}, t)|^{-1/2}\delta(\vec{x} - \vec{x}')\gamma^0(\vec{x}, t)^{-1}_{ab}. \tag{3.230}$$

It now follows that
$$\partial_t T \psi_a(x) \overline{\psi}_b(x') = \delta(x,x') \underline{\gamma}^0(x)^{-1} + T \partial_t \psi_a(x) \overline{\psi}_b(x'),$$
where
$$\delta(x,x') \equiv |g(x)|^{-1/2} \delta(x-x').$$
Hence, in matrix notation, we have
$$(i\underline{\gamma}^\mu \nabla_\mu - m) S(x,x') = \delta(x,x') \mathbf{1}, \qquad (3.231)$$
where $\mathbf{1} = \underline{\gamma}^0(x)\underline{\gamma}^0(x)^{-1}$ is the unit matrix.

Acting on an arbitrary spinor ψ, noting (3.207), we can write
$$\underline{\gamma}^\mu \nabla_\mu \left(\underline{\gamma}^\nu \nabla_\nu \psi \right) = \left(\frac{1}{2}\{\underline{\gamma}^\mu, \underline{\gamma}^\nu\} \nabla_\mu \nabla_\nu + \frac{1}{2}[\underline{\gamma}^\mu, \underline{\gamma}^\nu] \nabla_\mu \nabla_\nu \right) \psi.$$
Clearly, from (3.206)
$$\frac{1}{2}\{\underline{\gamma}^\mu, \underline{\gamma}^\nu\} \nabla_\mu \nabla_\nu \psi = g^{\mu\nu} \nabla_\mu \nabla_\nu \psi,$$
and upon appropriate index relabeling,
$$\frac{1}{2}[\underline{\gamma}^\mu, \underline{\gamma}^\nu] \nabla_\mu \nabla_\nu \psi = \frac{1}{4}[\underline{\gamma}^\mu, \underline{\gamma}^\nu][\nabla_\mu, \nabla_\nu]\psi.$$
Furthermore, it will be shown in Section 5.6.1 that
$$[\nabla_\mu, \nabla_\nu]\psi = \frac{1}{4} \underline{\gamma}^\lambda \underline{\gamma}^\sigma R_{\mu\nu\lambda\sigma} \psi,$$
and that
$$\underline{\gamma}^\mu \underline{\gamma}^\nu \underline{\gamma}^\lambda \underline{\gamma}^\sigma R_{\mu\nu\lambda\sigma} = -2R.$$
From these results, we readily obtain
$$\frac{1}{2}[\underline{\gamma}^\mu, \underline{\gamma}^\nu][\nabla_\mu, \nabla_\nu]\psi = -\frac{1}{4} R\psi.$$
Therefore, we have the identity
$$\underline{\gamma}^\mu \nabla_\mu \left(\underline{\gamma}^\nu \nabla_\nu \psi \right) = \left(g^{\mu\nu} \nabla_\mu \nabla_\nu - \frac{1}{4} R \right) \psi. \qquad (3.232)$$
It follows that if we define a bispinor $\mathcal{G}(x,x')$ as
$$S(x,x') \equiv (i\underline{\gamma}^\mu \nabla_\mu + m) \mathcal{G}(x,x'), \qquad (3.233)$$
and apply the operator $(i\underline{\gamma}^\nu \nabla_\mu - m)$, which appears in (3.231), then
$$(i\underline{\gamma}^\mu \nabla_\mu - m)(i\underline{\gamma}^\nu \nabla_\nu + m) \mathcal{G}(x,x') = \delta(x,x') \mathbf{1},$$
or using (3.232),
$$\left(g^{\mu\nu} \nabla_\mu \nabla_\nu - \frac{1}{4} R + m^2 \right) \mathcal{G}(x,x') = -\delta(x,x') \mathbf{1}, \qquad (3.234)$$

3.9 Chiral current anomaly caused by spacetime curvature

where the covariant derivatives act on $\mathcal{G}$ as a spinor at x (as in $S(x,x')$). $\mathcal{G}(x,x')$ also transforms as a spinor at x', as is clear from the definition of $S(x,x')$ and the definition (3.233).

We can make a local momentum-space expansion in Riemann normal coordinates with origin at x', just as in the scalar case. The result (Bunch and Parker 1979) in Riemann normal coordinates is

$$\mathcal{G}(x,x') = |g(x)|^{-1/4} \int \frac{d^n k}{(2\pi)^n} e^{iky} \mathcal{G}(k), \tag{3.235}$$

where

$$\mathcal{G}(k) = -\left[1 + \left(A_1(x') + iA_{1\alpha}(x')\frac{\partial}{\partial k_\alpha} - A_{1\alpha\beta}(x')\frac{\partial}{\partial k_\alpha}\frac{\partial}{\partial k_\beta}\right)\left(-\frac{\partial}{\partial m^2}\right)\right.$$
$$\left. + A_2(x')\left(\frac{\partial}{\partial m^2}\right)^2\right](-k^2 + m^2)^{-1}. \tag{3.236}$$

The coefficients A_1, $A_{1\alpha\beta}$, and A_2 were obtained first using other methods in DeWitt (1964a) and Christensen (1978). Here m^2 is understood to be $m^2 - i\epsilon$ as in the scalar Feynman propagator, and

$$A_1 = \frac{1}{12}R\mathbf{1}, \tag{3.237}$$

$$A_2 = \left(+\frac{1}{120}R_{;\mu}{}^\mu + \frac{1}{288}R^2 - \frac{1}{180}R_{\mu\nu}R^{\mu\nu} + \frac{1}{180}R_{\mu\nu\sigma\tau}R^{\mu\nu\sigma\tau}\right)\mathbf{1}$$
$$+ \frac{1}{48}\Sigma_{[\alpha\beta]}\Sigma_{[\gamma\delta]}R^{\alpha\beta\lambda\chi}R^{\gamma\delta}{}_{\lambda\chi}, \tag{3.238}$$

$$A_{1\mu} = -\frac{1}{24}R_{;\mu}\mathbf{1} - \frac{1}{12}\Sigma_{[\alpha\beta]}R^{\alpha\beta}{}_\mu{}^\lambda{}_{;\lambda}, \tag{3.239}$$

$$A_{1\mu\nu} = \frac{1}{30}\left(\frac{1}{4}R_{\mu\nu;\lambda}{}^\lambda - \frac{1}{2}R_{;\mu\nu} + \frac{1}{3}R_{\mu\lambda}R^\lambda{}_\nu\right.$$
$$\left.- \frac{1}{6}R^{\lambda\chi}R_{\lambda\mu\chi\nu} - \frac{1}{6}R^{\lambda\chi\sigma}{}_\mu R_{\lambda\chi\sigma\nu}\right)\mathbf{1}$$
$$+ \frac{1}{48}\Sigma_{[\alpha\beta]}\left(RR^{\alpha\beta}{}_{\mu\nu} + R^{\alpha\beta\lambda}{}_{\mu;\lambda\nu} + R^{\alpha\beta\lambda}{}_{\nu;\lambda\mu}\right)$$
$$- \frac{1}{96}\Sigma_{[\alpha\beta]}\Sigma_{[\gamma\delta]}\left(R^{\alpha\beta\lambda}{}_\mu R^{\gamma\delta}{}_{\lambda\nu} + R^{\alpha\beta\lambda}{}_\nu R^{\gamma\delta}{}_{\lambda\mu}\right), \tag{3.240}$$

where

$$\Sigma_{[\alpha\beta]} \equiv \frac{1}{4}(\gamma_\alpha\gamma_\beta - \gamma_\beta\gamma_\alpha). \tag{3.241}$$

At the origin of these normal coordinates we have taken $b^\alpha{}_\sigma(x') = \delta^\alpha{}_\sigma$, so spacetime and vierbein indices need not be distinguished on the above Riemann tensors.

Then from (3.233) and (3.235),

$$S(x, x') = (i\underline{\gamma}^\mu(x)\partial_\mu + m)\mathcal{G}(x, x') - i\underline{\gamma}^\mu(x)\Gamma_\mu(x)\mathcal{G}(x, x')$$
$$= |g(x)|^{-1/4} \int \frac{d^n k}{(2\pi)^n} e^{iky}(-\underline{\gamma}^\mu(y)k_\mu + m)\mathcal{G}(k)$$
$$- i\underline{\gamma}^\mu(x)\Gamma_\mu(x)\mathcal{G}(x, x').$$

When $x \to x'$, we can use $\Gamma_\mu(x') = 0$, $|g(x')| = 1$, $\partial_\mu|g(x)|\big|_{x=x'} = 0$, and $\underline{\gamma}^\mu(x') = \gamma^\mu$, so

$$S(x', x') = \int \frac{d^n k}{(2\pi)^n}(-\gamma^\mu k_\mu + m)\mathcal{G}(k). \tag{3.242}$$

From (3.227) and (3.228), we have formally

$$\langle \nabla_\mu J_A^\mu(x) \rangle = 2m\gamma_{5\,ab} S_{ba}(x, x)$$
$$= 2m\,\text{tr}[\gamma_5 S(x, x)]. \tag{3.243}$$

Because individual terms like $\langle \nabla_1 J_A^1 \rangle$ diverge to fourth adiabatic order, we must subtract terms up to fourth adiabatic order to regularize the expression:

$$\langle \nabla_\mu J_A^\mu(k) \rangle_{\text{phys}} = \lim_{m \to 0} 2m\,\text{tr}\left[\gamma_5 \left(S(x, x) - \sum_{i=0}^{4} S_i(x, x)\right)\right]. \tag{3.244}$$

$S_i(x, x)$ is the term of ith adiabatic order in the adiabatic expansion of $S(x, x)$. If we substitute the adiabatic expansion of (3.236) into (3.242), then we see that the term involving $A_{1\alpha\beta}$ is zero upon integration by parts. The remaining term of fourth adiabatic order gives

$$S_4(x, x) = -2\int \frac{d^4 x}{(2\pi)^4}(-\gamma^\mu k_\mu + m)A_2(x)(-k^2 + m^2)^{-3}, \tag{3.245}$$

where we have put $n = 4$ because this term has no pole at $n = 4$. Then

$$\langle \nabla_\mu J_A^\mu(x) \rangle_{\text{phys}} = \lim_{m \to 0} 2m\,\text{tr}[\gamma_5 S_{\text{phys}}(x, x) - \gamma_5 S_4(x, x)]$$
$$= -\lim_{m \to 0} 2m\,\text{tr}[\gamma_5 S_4(x, x)], \tag{3.246}$$

where $S_{\text{phys}}(x, x)$ is the renormalized propagator having the divergent terms subtracted away. It is well behaved as m approaches zero, so the limit of $mS_{\text{phys}}(x, x)$ is zero. Thus

$$\langle \nabla_\mu J_A^\mu(x) \rangle = +\lim_{m \to 0} 4m \int \frac{d^4 k}{(2\pi)^4} \{-\text{tr}[\gamma_5 \gamma^\mu A_2(x)]k_\mu$$
$$+ \text{tr}[\gamma_5 A_2(x)]m\}(-k^2 + m^2)^{-3}. \tag{3.247}$$

Referring to the form of $A_2(x)$ given in (3.238), we can evaluate the traces which appear in (3.247). We use[10]

$$\text{tr}(\gamma_5\gamma^\mu) = -\text{tr}(\gamma^\mu\gamma_5) = \text{tr}(\gamma^\mu\gamma_5) = 0,$$
$$\text{tr}(\gamma_5\gamma^\mu\Sigma_{[\alpha\beta]}\Sigma_{[\gamma\delta]}) = 0,$$

because $\gamma_5\gamma^\mu$ is a product of three γ^λ matrices (with $\lambda = 0, \ldots, 3$), and the trace of a product of an odd number of γ^λ matrices is zero. Hence

$$\text{tr}[\gamma_5\gamma^\mu A_2(x)] = 0.$$

Furthermore,

$$\text{tr }\gamma_5 = 0,$$

so

$$\text{tr}(\gamma_5 A_2) = \frac{1}{48}\text{tr}(\gamma_5\Sigma_{[\alpha\beta]}\Sigma_{[\gamma\delta]})R^{\alpha\beta\lambda\chi}R^{\gamma\delta}{}_{\lambda\chi}$$
$$= \left(\frac{1}{48}\right)\left(\frac{1}{4}\right)\text{tr}(\gamma_5\gamma^\alpha\gamma^\beta\gamma^\gamma\gamma^\delta)R_{\alpha\beta}{}^{\lambda\chi}R_{\gamma\delta\lambda\chi}.$$

Now[11]

$$\text{tr}(\gamma_5\gamma^\alpha\gamma^\beta\gamma^\gamma\gamma^\delta) = 4i\epsilon^{\alpha\beta\gamma\delta}.$$

Hence,

$$\text{tr}(\gamma_5 A_2) = \frac{i}{48}\epsilon^{\alpha\beta\gamma\delta}R_{\alpha\beta}{}^{\lambda\chi}R_{\gamma\delta\lambda\chi},$$

and

$$\langle\nabla_\mu J_A^\mu\rangle = -\lim_{m\to 0} 4m^2 \frac{i}{48}\epsilon^{\alpha\beta\gamma\delta}R_{\alpha\beta}{}^{\lambda\chi}R_{\gamma\delta\lambda\chi}\int\frac{d^4k}{(2\pi)^4}(k^2-m^2)^{-3}. \quad (3.248)$$

We have, with $m^2 \to m^2 - i\epsilon$ understood, that

$$\int\frac{d^4k}{(2\pi)^4}(k^2-m^2)^{-3} = \frac{i\pi^2}{2m^2(2\pi)^4}, \quad (3.249)$$

so finally

$$\langle\nabla_\mu J_A^\mu\rangle = \frac{1}{384\pi^2}\epsilon^{\alpha\beta\gamma\delta}R_{\alpha\beta}{}^{\lambda\chi}R_{\gamma\delta\lambda\chi}. \quad (3.250)$$

This is the gravitational axial vector, or chiral, current anomaly. It was first calculated by Kimura (1969). For further discussion, see (Gibbons 1979, p. 661). In Section 5.9 we will show how to generalize this anomaly to a spacetime whose dimension may differ from 4.

---

[10] These are special cases of more general results that we will show in Section 5.9.
[11] See, for example, Bjorken and Drell (1964 p. 105).

4
Particle creation by black holes

4.1 Introduction

We have already seen how time-dependent gravitational fields, such as occur in cosmology, can give rise to the creation of elementary particles. The time development of a quantum field induces a redefinition of creation and annihilation operators that correspond to physical particles. For example, the annihilation operators of particles at late times can be expressed as superpositions of annihilation and creation operators of particles at early times. This superposition is known as a Bogolubov transformation. The coefficients in the superposition determine the probability of creation of pairs of elementary particles by the gravitational field.

This method of using Bogolubov transformations to analyze particle creation by gravitational fields was first introduced by Parker (1965, 1968, 1969, 1971) in the cosmological context. Fulling (1972, 1973) found that the creation and annihilation operators defined in an accelerated coordinate system (known as Rindler coordinates (Rindler 1966)) in Minkowski spacetime were related to the usual Minkowski creation and annihilation operators through a Bogolubov transformation. This was shown by Unruh (1976) to imply that an accelerated particle detector in empty Minkowski spacetime would respond as if it were observing a flux of particles. The particle spectrum corresponding to the Bogolubov transformation of Fulling was a thermal spectrum with a temperature determined by the acceleration (Davies 1975).

Several investigations of particle creation by black holes were also in progress at this time. Parker and Tiomno (1972) (see text near footnote 1) pointed out that a Reissner–Nordström (i.e., charged) black hole of sufficiently small mass has an electric field that would create electron–positron pairs through the Heisenberg–Euler–Schwinger process. Gibbons (1975) independently investigated this process in complete detail for charged black holes. Progress was also being made on particle creation by the gravitational fields of rotating (Starobinsky 1973, Unruh

1974) and non-rotating (Fulling and Parker 1973) black holes, when Hawking (1974); Hawking (1975) published his now classic papers showing that a black hole will create particles having a thermal spectrum with a temperature determined by the surface gravity (i.e., the gravitational acceleration at the event horizon) of the black hole.

Hawking considered a quantum field in the classical spacetime background of a dust cloud that collapses to form a black hole. Using the Bogolubov transformation method, he found that if no particles of the quantum field are present at early times, a distant observer at late times will detect an outgoing flux of particles having a thermal spectrum. In calculating the Bogolubov coefficients, the back reaction of the particle creation on the spacetime is neglected. The thermal flux of particles is caused in this case by the formation of an event horizon that occurs as the body collapses. The flux of energy emitted by the black hole implies that its mass will decrease. During this process of black hole evaporation, the surface gravity and temperature of the black hole increase, so the end stage of black hole evaporation is explosive. There are still many open questions about the end state of black hole evaporation. The process of emission of a thermal spectrum of particles by a black hole is connected in a deep way with the generalized second law of thermodynamics, according to which a black hole has entropy proportional to the surface area of its event horizon (Bekenstein 1973).

Hawking's original calculation (Hawking 1974; Hawking 1975) showed that the average number of outgoing particles in each mode is distributed in accordance with a thermal spectrum. Later it was shown by Hawking (1976), Parker (1975a), and Wald (1975) that the full probability distribution, and not just the average number, is that of thermal radiation.

In this chapter, we consider a quantum field in the spacetime of a body that collapses to form a black hole and obtain the thermal spectrum of created particles, first in the case of a non-rotating (Schwarzschild) black hole, and then for a rotating (Kerr) black hole.

4.2 Classical considerations

Imagine an isolated collapsing body that at early times is widely dispersed and of sufficiently low density that the entire spacetime is nearly flat. Suppose that in a finite proper time, as measured with respect to clocks co-moving with the particles of the body, it collapses to form a black hole. The black hole at late times is characterized by its mass, angular momentum, and charge. When the angular momentum and charge are zero, we have a spherically symmetric Schwarzschild black hole. When only the charge is zero, we have a Kerr black hole. A comprehensive discussion of classical black holes may be found in Chandrasekhar (1983).

The Schwarzschild black hole is described by the line element (in Schwarzschild coordinates, and in units with $G = c = 1$):

$$ds^2 = \left(1 - \frac{2M}{r}\right) dt^2 - \left(1 - \frac{2M}{r}\right)^{-1} dr^2$$
$$- r^2 d\theta^2 - r^2 \sin^2\theta d\phi^2. \tag{4.1}$$

There is an event horizon at $r = 2M$. Outgoing light-like, or null, geodesics cannot cross from inside to outside the event horizon. The event horizon is generated by the radially outgoing null geodesics at $r = 2M$. There is a coordinate singularity at $r = 2M$ in Schwarzschild coordinates, but other coordinate systems can be defined that are well behaved at the event horizon.

The Kerr black hole is described by the line element (in Boyer–Lindquist coordinates):

$$ds^2 = \frac{\Delta}{\rho^2}\left[dt - a\sin^2\theta d\phi\right]^2 - \frac{\sin^2\theta}{\rho^2}\left[(r^2+a^2)d\phi - adt\right]^2$$
$$-\frac{\rho^2}{\Delta} dr^2 - \rho^2 d\theta^2, \tag{4.2}$$

with

$$\Delta \equiv r^2 - 2Mr + a^2, \tag{4.3}$$

and

$$\rho^2 \equiv r^2 + a^2 \cos^2\theta. \tag{4.4}$$

Notice that $\Delta = (r - r_+)(r - r_-)$ has zeroes at

$$r_+ = M + (M^2 - a^2)^{1/2} \tag{4.5}$$

and

$$r_- = M - (M^2 - a^2)^{1/2}. \tag{4.6}$$

We will assume that $a^2 < M^2$.

The event horizon of the Kerr black hole is at $r = r_+$. In Boyer–Lindquist coordinates, there is a coordinate singularity at $r = r_+$ that can be removed by a suitable coordinate transformation. The constant a is the angular momentum per unit mass of the black hole. The Kerr and Schwarzschild geometries have many fascinating properties that are elucidated in Chandrasekhar (1983), Misner et al. (1973), and Wald (1984).

As we shall see, the properties of the massless quanta created by the gravitational field of a black hole are governed at late times by the null geodesics that begin far outside the collapsing body at early times, move inward through the body to become outgoing null geodesics, and escape from the collapsing body just before it collapses within the black hole event horizon. These outgoing null geodesics reach future null infinity ($\mathcal{I}^+$) at arbitrarily late times.

Therefore, we consider null geodesics in the spacetimes of (4.1) and (4.2). In the Schwarzschild geometry, the relevant class of geodesics are the radially outgoing

4.2 Classical considerations

null geodesics that begin just outside the event horizon. In the Kerr geometry, the null generators of the event horizon have angular velocity

$$d\phi/dt = \Omega_H \equiv \frac{a}{a^2 + r_+^2}. \tag{4.7}$$

Therefore, the relevant outgoing geodesics have an angular velocity that approaches the value Ω_H arbitrarily close to the event horizon.

Geodesics $x^\mu(\lambda)$ may be affinely parameterized, meaning that in an infinitesimal neighborhood of the origin of a locally inertial coordinate system y^μ, the geodesic path is a linear function of λ. In general coordinates this implies that

$$\frac{D}{d\lambda}\left(\frac{dx^\mu}{d\lambda}\right) = 0, \tag{4.8}$$

where $D/d\lambda$ is the covariant derivative along the geodesic with respect to affine parameter λ. The problem of finding geodesics is conveniently stated as the variational problem of extremizing $\int_a^b \mathcal{L} d\lambda$, where λ is an affine parameter and

$$\mathcal{L} = \frac{1}{2} g_{\mu\nu} \frac{dx^\mu}{d\lambda} \frac{dx^\nu}{d\lambda}. \tag{4.9}$$

We may view $\mathcal{L}$ as a Lagrangian of a classical particle with coordinates x^μ and velocities $dx^\mu/d\lambda$. The Euler–Lagrange equations imply that if $g_{\mu\nu}$ and hence $\mathcal{L}$ is independent of the coordinate x^μ, then the conjugate momentum

$$p_\mu \equiv \frac{\partial \mathcal{L}}{\partial(dx^\mu/d\lambda)} = g_\mu \frac{dx^\nu}{d\lambda}, \tag{4.10}$$

is constant along the geodesic. In Schwarzschild and Kerr the metric depends only on r and θ, so there are constant momenta conjugate to the (cyclic) coordinates t and ϕ. In Schwarzschild, the geodesics lie in a plane, while in Kerr, there is a third constant of the motion given originally by Carter (1968). The most complete calculation of the geodesics is given in Chandrasekhar (1983) for the Schwarzschild and Kerr spacetimes.

4.2.1 Schwarzschild

One finds that in Schwarzschild coordinates, with the plane containing the geodesic taken at $\theta = \pi/2$, the constants of the motion given by (4.10) for the momenta p_t and p_ϕ are

$$\left(1 - \frac{2M}{r}\right)\frac{dt}{d\lambda} = E \tag{4.11}$$

and

$$r^2 d\phi/d\lambda = L. \tag{4.12}$$

The vanishing of the expression for $\mathcal{L}$ in (4.9) along null geodesics leads to the further equation

$$\left(\frac{dr}{d\lambda}\right)^2 + \frac{L^2}{r^2}\left(1 - \frac{2M}{r}\right) = E^2. \tag{4.13}$$

The radial null geodesics are those with $L = 0$. They are characterized by (4.11) and

$$dr/d\lambda = \pm E, \tag{4.14}$$

as well as $d\phi/d\lambda = 0$. The upper sign corresponds to outgoing geodesics (for $r > 2M$) and the lower sign to incoming geodesics.

From (4.11) and (4.14) it follows that

$$\frac{dt}{d\lambda} \mp \left(1 - \frac{2M}{r}\right)^{-1} \frac{dr}{d\lambda} = 0 \tag{4.15}$$

or

$$\frac{d}{d\lambda}(t \mp r^*) = 0, \tag{4.16}$$

where r^* is defined by

$$\frac{dr^*}{dr} = \left(1 - \frac{2M}{r}\right)^{-1}. \tag{4.17}$$

As r approaches $2M$ from above, r^* approaches $-\infty$. Along any outgoing radial null geodesic, the null coordinate

$$u \equiv t - r^* \tag{4.18}$$

is constant. Along any incoming radial null geodesic, the null coordinate

$$v \equiv t + r^* \tag{4.19}$$

is constant.

Let $\mathcal{C}$ be an incoming radial null geodesic defined by $v = v_1$ for some v_1 that passes through the event horizon of the Schwarzschild black hole. Let λ be an affine parameter along this geodesic. The null coordinate u is given along $\mathcal{C}$ by some function $u(\lambda)$. It is the form of this function just outside the event horizon that will determine the spectrum of the particles created by the black hole.

Along the null geodesic $\mathcal{C}$,

$$\frac{du}{d\lambda} = \frac{dt}{d\lambda} - \frac{dr^*}{d\lambda},$$

where $dt/d\lambda$ is given by (4.11), and from (4.14) with $-E$, we have

$$\frac{dr^*}{d\lambda} = \frac{dr^*}{dr}\frac{dr}{d\lambda} = -\left(1 - \frac{2M}{r}\right)^{-1} E.$$

Thus

$$\frac{du}{d\lambda} = 2\left(1 - \frac{2M}{r}\right)^{-1} E. \tag{4.20}$$

4.2 Classical considerations

From (4.14), $dr/d\lambda = -E$ along $\mathcal{C}$, so

$$r - 2M = -E\lambda, \tag{4.21}$$

where λ is taken as zero at the event horizon. For $r > 2M$, the affine parameter λ is negative. It follows that

$$\left(1 - \frac{2M}{r}\right)^{-1} = 1 - \frac{2M}{E\lambda}, \tag{4.22}$$

and

$$\frac{du}{d\lambda} = 2E - \frac{4M}{\lambda}. \tag{4.23}$$

Therefore, on the incoming null geodesic, $\mathcal{C}$,

$$u = 2E\lambda - 4M \ln(\lambda/K_1), \tag{4.24}$$

where K_1 is a (negative) constant. Far from the event horizon, $u \approx 2E\lambda$, while near the event horizon ($\lambda = 0$),

$$u \approx -4M \ln(\lambda/K_1). \tag{4.25}$$

The null coordinate u is $-\infty$ at past null infinity $\mathcal{I}^-$ and $+\infty$ at the event horizon.

A convenient representation of the spacetime of the spherical collapsing body is a Penrose conformal diagram (Fig 4.1), in which radial null geodesics are represented by lines at $\pm 45°$ angles and a conformal transformation is made, so infinity can be represented on the diagram.

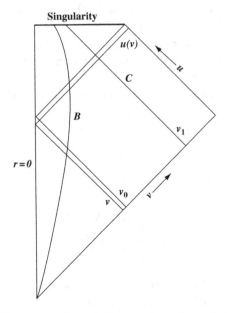

Fig 4.1. The Penrose conformal diagram for collapse in Schwarzschild.

The timelike curve B that begins at the point representing past timelike infinity is the outer boundary of the collapsing body. Past null infinity, $\mathcal{I}^-$, from which incoming null geodesics originate, is labeled by the arrow showing the direction in which the null coordinate v increases along $\mathcal{I}^-$. Future null infinity, $\mathcal{I}^+$, where outgoing null geodesics terminate, is labeled by the arrow showing the direction in which the null coordinate u increases along $\mathcal{I}^+$. The ray $\mathcal{C}$ is an incoming null geodesic, $v = v_1$, that enters the black hole through the event horizon, H, which is the unlabeled null ray running from $r = 0$ to the point at which the singularity and $\mathcal{I}^+$ seem to intersect in the figure. (That intersection point is future timelike infinity, where timelike geodesics go if they do not end on the singularity.) The null ray with constant incoming null coordinate v originates on $\mathcal{I}^-$, passes through the center of the collapsing body, and becomes the null ray having constant outgoing null coordinate, $u = u(v)$. The incoming ray, $v = v_0$, is the last one that passes through the center of the body and reaches $\mathcal{I}^+$. Incoming null rays with $v > v_0$ enter the black hole through H, and run into the singularity.

The affine parameter λ along all radially incoming null geodesics, such as $\mathcal{C}$, that pass through H can be chosen so that (4.25) relates u to λ near the event horizon. Then the affine parameter distance between the outgoing rays $u(v_0)$ and $u(v)$ is constant along the entire length of the geodesics, as measured by the change of the affine parameter λ along any incoming null ray, such as $\mathcal{C}$, that intersects the two outgoing null rays. (One can think of the "rays" in the figure as representing incoming or outgoing spherical waves.) Moving backward along these outgoing geodesics through the collapsing body, they become the incoming geodesics that originate on $\mathcal{I}^-$ at v_0 and v, respectively. The affine separation along the null direction between these geodesics can be chosen to remain constant along their entire length, as they go from $\mathcal{I}^-$ to $\mathcal{I}^+$. Therefore, the affine separation between v and v_0 at $\mathcal{I}^-$ is the same as that between $u(v)$ and $u(v_0)$ at $\mathcal{I}^+$. Because $\lambda = 0$ on H, the affine separation between $u(v)$ and $u(v_0)$ at $\mathcal{I}^+$ has the value λ that satisfies (4.25) with u having the value $u(v)$.

Because $\mathcal{I}^-$ is far from the collapsing body, the coordinate v is itself an affine parameter along $\mathcal{I}^-$. Therefore, $v - v_0$ must be related to the affine separation λ between $u(v)$ and $u(v_0)$ on $\mathcal{I}^+$ by

$$v_0 - v = K_2 \lambda, \tag{4.26}$$

where K_2 is a negative constant. Hence

$$\begin{aligned} u(v) &= -4M \ln(\lambda/K_1) \\ &= -4M \ln\left(\frac{v_0 - v}{K_1 K_2}\right), \end{aligned} \tag{4.27}$$

with the product $K_1 K_2$ being a positive constant. This is the relation that will determine the spectrum of created particles when we consider quantum fields.

4.2.2 Null geodesics in Kerr spacetime

In the Kerr metric of (4.2), equations governing geodesic motion are obtained from the constants of the motion, $p_t = E$, $-p_\phi = L_z$, and from a third constant of the motion, the Carter constant K_C, that exists for Kerr (Carter 1968). These constants of the motion, together with $\mathcal{L} = 0$ for null geodesics, lead to the following equations (Chandrasekhar (1983), pp. 342–344):

$$\rho^4 (dr/d\lambda)^2 = [(r^2 + a^2)E - aL_z]^2 - K_C \Delta, \qquad (4.28)$$

$$\rho^4 (d\theta/d\lambda)^2 = -(aE \sin\theta - L_z \operatorname{cosec}\theta)^2 + K_C, \qquad (4.29)$$

$$\rho^2 (dt/d\lambda) = \Delta^{-1}\{[(r^2 + a^2)^2 - \Delta a^2 \sin^2\theta]E - 2aMrL_z\}, \qquad (4.30)$$

and

$$\rho^2 (d\phi/d\lambda) = \Delta^{-1}[2aMrE + (\rho^2 - 2Mr)L_z \operatorname{cosec}^2\theta]. \qquad (4.31)$$

The generalization of the radial null geodesics that we considered in the Schwarzschild spacetime are a set of null geodesics in the Kerr spacetime that move on the (curved) surfaces of constant θ. From (4.29), we see that a constant solution $\theta(\lambda) = \theta_0$ is possible if we take

$$K_C = 0 \qquad (4.32)$$

and

$$L_z = aE \sin^2 \theta_0. \qquad (4.33)$$

Then (4.28) becomes

$$dr/d\lambda = \pm E, \qquad (4.34)$$

with (4.30) and (4.31) simplifying to

$$dt/d\lambda = (r^2 + a^2)E/\Delta, \qquad (4.35)$$

and

$$d\phi/d\lambda = aE/\Delta. \qquad (4.36)$$

In simplifying (4.30), we have used the identity

$$(r^2 + a^2)^2 - \Delta a^2 \sin^2 \theta = \rho^2 (r^2 + a^2) + 2a^2 Mr \sin^2 \theta. \qquad (4.37)$$

For the case $a = 0$, these equations of motion reduce to those of radial geodesics in the Schwarzschild spacetime found in Section 4.2.1. The geodesics specified by (4.34)–(4.36) are called the principal null congruence in the Kerr spacetime.

It is clear from (4.35) and (4.36) that as $r \to r_+$, we have $t \to \infty$ and $\phi \to \infty$. In Boyer–Lindquist coordinates, which are the generalization of Schwarzschild coordinates to the Kerr spacetime, not only does an incoming principal null geodesic take an infinite coordinate time to reach the event horizon, but it also winds an infinite number of times around the z axis. As in the Schwarzschild spacetime, the geodesic passes through the event horizon at a finite value of the affine parameter, λ. The generalization of the Schwarzschild null coordinates considered earlier are

$$u \equiv t - r^* \tag{4.38}$$

and

$$v = t + r^*, \tag{4.39}$$

with r^* defined by

$$\frac{dr^*}{dr} = \frac{r^2 + a^2}{\Delta}. \tag{4.40}$$

In addition, we define an angular coordinate $\tilde{\phi}_+$ that is well behaved as an incoming null geodesic falls into the Kerr event horizon at $r = r_+$:

$$\tilde{\phi}_+ \equiv \phi - \frac{a}{r_+^2 + a^2} t = \phi - \Omega_H t. \tag{4.41}$$

Then along incoming and outgoing geodesics of the principal null congruence,

$$\begin{aligned}\frac{d\tilde{\phi}_+}{d\lambda} &= \frac{aE}{\Delta} - \frac{a}{r_+^2 + a^2}\frac{r^2 + a^2}{\Delta}E \\ &= -\Omega_H E(r + r_+)/(r - r_-),\end{aligned} \tag{4.42}$$

which is well behaved at $r = r_+$. One can further define coordinates $U(u)$ and $V(v)$ that, with azimuthal angular coordinate $\tilde{\phi}_+$, give a form of the Kerr metric having no coordinate singularities as $r \to r_+$ (Chandrasekhar 1983, pp. 314–315). The coordinate $\tilde{\phi}_+$ of (4.41) is the azimuthal angle in a coordinate system rotating about the z-axis relative to Boyer–Lindquist coordinates at the angular velocity Ω_H of the Kerr event horizon.

The spectrum of the particles created by a Kerr black hole will be determined by the function $u(\lambda)$, where λ is an affine parameter along an incoming geodesic, $\mathcal{C}$, of the principal null congruence. Along such a null geodesic

$$\begin{aligned}\frac{du}{d\lambda} &= \frac{dt}{d\lambda} - \frac{dr^*}{dr}\frac{dr}{d\lambda} \\ &= 2E\frac{r^2 + a^2}{\Delta},\end{aligned} \tag{4.43}$$

where we used (4.34), (4.35), and (4.40). From (4.34), along the incoming geodesic

$$r - r_+ = -E\lambda, \tag{4.44}$$

where we have taken $\lambda = 0$ at $r = r_+$. The affine parameter λ is negative for $r > r_+$.

Then from (4.43) and (4.44) we find

$$\frac{du}{d(E\lambda)} = 2\frac{(r_+ - E\lambda)^2 + a^2}{E\lambda[E\lambda - (r_+ - r_-)]}. \tag{4.45}$$

It follows that along the null geodesic $\mathcal{C}$,

$$u = 2E\lambda - (1/\alpha_+)\ln(E\lambda/K_1) + (1/\alpha_-)\ln\{[E\lambda - (r_+ - r_-)]/K'_1\} \tag{4.46}$$

where K_1 and K'_1 are constants, and

$$\alpha_+ \equiv \frac{1}{2}(r_+ - r_-)/(r_+^2 + a^2), \tag{4.47}$$

$$\alpha_- \equiv \frac{1}{2}(r_+ - r_-)/(r_-^2 + a^2). \tag{4.48}$$

When $a \to 0$, we have $r_- \to 0$ and $r_+ \to 2M$. Then $\alpha_+ \to (4M)^{-1}$, $\alpha_- \to \infty$, and we recover the results found in the Schwarzschild spacetime. Just as $(4M)^{-1}$ is the surface gravity κ at the event horizon of the Schwarzschild black hole, the quantity α_+ of (4.47) is the surface gravity κ at the event horizon ($r = r_+$) of the Kerr black hole. We will find that the temperature of a black hole is proportional to its surface gravity.

Far outside the event horizon of the Kerr black hole, $u \approx 2E\lambda$, with u approaching $-\infty$ at $\mathcal{I}^-$. As $r \to r_+$, we have $u \to \infty$, with

$$u \approx -(1/\alpha_+)\ln(E\lambda/K_1). \tag{4.49}$$

As we continue inward along the null geodesic $\mathcal{C}$, the affine parameter goes through 0 at $r = r_+$ and becomes positive. As we approach $r = r_-$, where $E\lambda = r_+ - r_-$, the second logarithm in (4.46) causes u to approach $-\infty$. The behavior of $u(\lambda)$ near $r = r_-$ suggests that the (Cauchy) horizon at $r = r_-$ has a temperature proportional to α_-, but that temperature would not be observable in the region of spacetime from which the rotating body collapsed to form the Kerr black hole.

In the spacetime of the rotating body that undergoes gravitational collapse, consider two outgoing null geodesics, $\mathcal{C}_1$ and $\mathcal{C}_2$, that at late times belong to the principal null congruence. Along these outgoing geodesics, u is constant. Let $\mathcal{C}_1$ and $\mathcal{C}_2$ intersect the *incoming* null geodesic $\mathcal{C}$ where its affine parameter λ has values λ_1 and λ_2, respectively. Let these values of λ be negative and of sufficiently small magnitude that (4.49) is valid. The null geodesics $\mathcal{C}_1$ and $\mathcal{C}_2$ originate at $\mathcal{I}^-$ as incoming null geodesics at $v = v_1$ and $v = v_2$, respectively. As $\lambda_1 \to 0^-$, we have $v_1 \to v_0$, where v_0 is the value of v on the last incoming null geodesic that starts from $\mathcal{I}^-$, passes through the rotating collapsing body, and reaches $\mathcal{I}^+$ on the outgoing null geodesic $u = u(v_0)$. Similarly $\mathcal{C}_2$ originates at a value $v_2 < v_0$ on $\mathcal{I}^-$ and reaches $\mathcal{I}^+$ at $u = u(v_2)$. The affine separation λ along null geodesics between $\mathcal{C}_1$ and $\mathcal{C}_2$ is given by (4.49) with $u = u(v_2)$. At $\mathcal{I}^-$, v is an affine parameter. As in the Schwarzschild spacetime, we have

$$v_0 - v = K_2\lambda, \tag{4.50}$$

where K_2 is a negative constant, and we have dropped the subscript on v_2. It follows from (4.49) that

$$u(v) \approx -(1/\alpha_+)\ln[(v_0 - v)/K], \tag{4.51}$$

where K is a positive constant. This expression will determine the spectrum of the outgoing particles created by the Kerr black hole.

4.3 Quantum aspects

The spectrum of created particles observed at late times far from the black hole is determined by the relations in (4.27) and (4.51) that give the affine separation between two outgoing null geodesics that escape from the collapsing body just before formation of the event horizon. In calculating the spectrum, we neglect the effect of the particle creation on the mass of the evaporating black hole. This appears to be an excellent approximation until the mass of the evaporating black hole approaches within a few orders of magnitude of the Planck mass (10^{-8} kg).

We consider a quantized massless Hermitian scalar field satisfying the generally covariant wave equation $\Box\phi = 0$, or

$$(-g)^{-\frac{1}{2}}\partial_\mu\left[(-g)^{\frac{1}{2}}g^{\mu\nu}\partial_\nu\phi\right] = 0. \tag{4.52}$$

In the Schwarzschild and Kerr spacetimes, an additional term of the form $\xi R\phi$ in (4.52) would be zero, and thus would not affect the late time spectrum in the collapse spacetime. The canonical quantization of this field has already been described in Chapter 2. The conserved scalar product (ϕ_1, ϕ_2) was defined in (2.48).

The created particles observed at late times are created at a short affine distance from the event horizon. Their spectrum is not affected by the regions, such as that inside the collapsing body, where the metric is not stationary. In the spacetime of a body that collapses to form a Schwarzschild or Kerr black hole, we can write the field in the entire spacetime in the form

$$\phi = \int d\omega(a_\omega f_\omega + a_\omega^\dagger f_\omega^*), \tag{4.53}$$

where the f_ω and f_ω^* are a complete set of solutions of (4.52) having normalization

$$(f_{\omega_1}, f_{\omega_2}) = \delta(\omega_1 - \omega_2), \tag{4.54}$$

and the a_ω are time-independent operators. Then the canonical commutation relations imply that the a_ω are annihilation operators and $a_\omega^\dagger$ are creation operators obeying

$$[a_{\omega_1}, a^\dagger_{\omega_2}] = \delta(\omega_1 - \omega_2), \qquad (4.55)$$
$$[a_{\omega_1}, a_{\omega_2}] = 0 = [a^\dagger_{\omega_1}, a^\dagger_{\omega_2}]. \qquad (4.56)$$

The physical interpretation of the a_ω depends on the choice of the complete set of solutions f_ω.

Far outside the collapsing body at early times, the definition of physical particles that would be detected by inertial observers, or equivalently of positive frequency solutions of the wave equation (4.52), is unambiguous. Let the f_ω be chosen such that at early times and large distances they form a complete set of incoming positive frequency solutions of energy ω. Their asymptotic form on past null infinity, $\mathcal{I}^-$, is

$$f_\omega \sim \omega^{-1/2} r^{-1} \exp(-i\omega v) S(\theta, \phi), \qquad (4.57)$$

where discrete quantum numbers (l, n) have been suppressed, and $v = t + r$ is the incoming null coordinate at $\mathcal{I}^-$. The factor of $\omega^{-1/2}$ is required by the normalization of the scalar product. Then the operators a_ω annihilate particles on $\mathcal{I}^-$.

The expansion of the field in terms of the f_ω and f^*_ω, given in (4.53), is analogous to the corresponding expansion of the field in terms of early time positive frequency solutions in the cosmological spacetimes considered in Chapter 3. As in the cosmological examples, we will be able to calculate the spectrum of created particles by making an expansion of the field in terms of late time positive frequency solutions. In the black hole spacetimes, solutions of (4.52) outside the black hole are uniquely determined by giving data *both* on the event horizon and on future null infinity, $\mathcal{I}^+$.

On $\mathcal{I}^+$, just as on $\mathcal{I}^-$, the definition of positive frequency solutions is unambiguous. Let the p_ω be the solutions of (4.52) that have zero Cauchy data on the event horizon and are asymptotically outgoing and positive frequency at $\mathcal{I}^+$. Assume that p_ω and p^*_ω form a complete set of solutions on $\mathcal{I}^+$ satisfying the normalization condition

$$(p_{\omega_1}, p_{\omega_2}) = \delta(\omega_1 - \omega_2). \qquad (4.58)$$

The asymptotic form of p_ω on $\mathcal{I}^+$ is

$$p_\omega \sim \omega^{-1/2} r^{-1} \exp(-i\omega u) S(\theta, \phi), \qquad (4.59)$$

where again quantum numbers (l, m) are suppressed, and $u = t - r$ is the outgoing null coordinate at $\mathcal{I}^+$. A wave packet formed by a superposition of the p_ω is outgoing and localized at large r at late times.

The most general solution of the wave equation will have a part that is incoming at the event horizon at late times. Therefore, we must introduce a set of solutions q_ω such that a superposition of them at late times is localized near the event horizon and has zero Cauchy data on $\mathcal{I}^+$. The precise form of the q_ω will not affect observations on $\mathcal{I}^+$, since those observations can only depend on the p_ω. Let the q_ω and q_ω^* form a complete set on the horizon, with normalization

$$(q_{\omega_1}, q_{\omega_2}) = \delta(\omega_1 - \omega_2). \tag{4.60}$$

Since wave packets formed from the p_ω and q_ω are in disjoint regions at late times, their conserved scalar product must vanish:

$$(q_{\omega_1}, p_{\omega_2}) = 0. \tag{4.61}$$

We also have $(q_{\omega_1}, q_{\omega_2}^*) = 0$, $(q_{\omega_1}, p_{\omega_2}^*) = 0$, and $(p_{\omega_1}, p_{\omega_2}^*) = 0$. In calculating the probability distribution of created particles, it is useful to specify the q_ω more precisely (Wald 1975; Parker 1975a; Hawking 1976). However, we do not require their precise form to calculate the spectrum observed at late times on $\mathcal{I}^+$.

We can expand ϕ in the entire spacetime in the form

$$\phi = \int d\omega \{ b_\omega p_\omega + c_\omega q_\omega + b_\omega^\dagger p_\omega^* + c_\omega^\dagger q_\omega^* \}, \tag{4.62}$$

where the b_ω are annihilation operators for particles outgoing at late times at infinity. We have

$$[b_{\omega_1}, b_{\omega_2}^\dagger] = \delta(\omega_1 - \omega_2)$$
$$[c_{\omega_1}, c_{\omega_2}^\dagger] = \delta(\omega_1 - \omega_2), \tag{4.63}$$

with all other commutation relations between the b_{ω_1} and c_{ω_2} (such as $[b_{\omega_1}, c_{\omega_2}] = 0$) and their Hermitian conjugates vanishing.

We are working in the Heisenberg picture, in which the state vector is independent of time. Let the state vector, $|0\rangle$, be chosen to have no particles of the field incoming from $\mathcal{I}^-$. Thus, $|0\rangle$ is annihilated by the a_ω corresponding to particles incoming from $\mathcal{I}^-$:

$$a_\omega |0\rangle = 0 \tag{4.64}$$

for all ω.

As in the cosmological case, the spectrum of outgoing particles is determined by the coefficients of the Bogolubov transformation relating b_ω to $a_{\omega'}$ and $a_{\omega'}^\dagger$. The f_ω and f_ω^* are a complete set for expanding any solution of the field equation, so we can write

$$p_\omega = \int d\omega' (\alpha_{\omega\omega'} f_{\omega'} + \beta_{\omega\omega'} f_{\omega'}^*), \tag{4.65}$$

where the $\alpha_{\omega\omega'}$ and $\beta_{\omega\omega'}$ are complex numbers, independent of the coordinates. From (4.58), (4.61), and (4.62), we have

$$b_\omega = (p_\omega, \phi). \tag{4.66}$$

4.3 Quantum aspects

Expressing ϕ and p_ω in terms of $f_{\omega'}$ and $f^*_{\omega'}$ using (4.53) and (4.65), we obtain

$$b_\omega = \int d\omega' (\alpha^*_{\omega\omega'} a_{\omega'} - \beta^*_{\omega\omega'} a^\dagger_{\omega'}), \qquad (4.67)$$

where we used $(f^*_{\omega'}, f^*_{\omega''}) = -\delta(\omega' - \omega'')$. Furthermore, using (4.65) it follows that

$$(p_{\omega_1}, p_{\omega_2}) = \int d\omega' (\alpha^*_{\omega_1\omega'} \alpha_{\omega_2\omega'} - \beta^*_{\omega_1\omega'} \beta_{\omega_2\omega'}). \qquad (4.68)$$

We also note that the coefficients in (4.65) can be expressed as

$$\beta_{\omega\omega'} = -(f^*_{\omega'}, p_\omega), \qquad (4.69)$$

and

$$\alpha_{\omega\omega'} = (f_{\omega'}, p_\omega). \qquad (4.70)$$

For simplicity, we consider first the spacetime of a non-rotating collapsing body and then that of a rotating collapsing body.

4.3.1 Schwarzschild

A non-rotating body undergoing gravitational collapse forms a Schwarzschild black hole. Here we calculate the coefficients $\alpha_{\omega\omega'}$ and $\beta_{\omega\omega'}$ and find the spectrum of created particles observed at late times far from this non-rotating black hole. The relevant classical geodesics were discussed in Section 4.2.1. The spacetime of a body collapsing to form a Schwarzschild black hole is pictured in Fig. 4.1.

We can form a wave packet from a superposition of the p_ω for a range of frequencies near a given value, ω. The coefficients in the superposition can be chosen so that the outgoing wave packet approaches $\mathcal{I}^+$ along a null geodesic characterized by a large constant value of u (i.e., at late times). The components p_ω of this wave packet can be expressed in terms of the $f_{\omega'}$ and $f^*_{\omega'}$ by means of (4.65). Imagine propagating this wave packet backward in time. Part of it will be scattered back toward infinity by the curved spacetime, and will reach $\mathcal{I}^-$ as a superposition of the $f_{\omega'}$ with frequencies near the original frequency ω. Another part of the wave packet will pass through the center of the collapsing body (ignoring interaction with the matter of the collapsing body, or assuming that the interaction is negligible at sufficiently high frequencies) and reach $\mathcal{I}^-$ as a superposition of the $f_{\omega'}$ and $f^*_{\omega'}$ having highly blueshifted values of $\omega' \gg \omega$.

Therefore, the p_ω in this latter part of the wave packet can be expressed in terms of the $f_{\omega'}$ and $f^*_{\omega'}$ by means of (4.65) with coefficients $\alpha_{\omega\omega'}$ and $\beta_{\omega\omega'}$ having $\omega' \gg \omega$. Furthermore, the relevant values of ω' become arbitrarily large at sufficiently late times (i.e., as $u \to \infty$). Thus, the late time spectrum of outgoing particles is determined by the asymptotic form of the coefficients $\beta_{\omega\omega'}$ for arbitrarily large ω'.

To determine those coefficients, trace this latter part of p_ω back in time along an outgoing geodesic having a very large value u. The geodesic passes through the center of the collapsing body just before the event horizon has formed, and emerges as an incoming geodesic characterized by a value of v close to v_0 (see Fig. 4.1). The value of v at which the packet reaches $\mathcal{I}^-$ is related to the value of u that it had at $\mathcal{I}^+$ by (4.27), derived in Section 4.2:

$$u(v) = -4M \ln\left(\frac{v_0 - v}{K}\right). \tag{4.71}$$

Here $K(=K_1 K_2)$ is a positive constant characterizing the affine parameterizations of the geodesic when it is near $\mathcal{I}^+$ and $\mathcal{I}^-$. The asymptotic form of p_ω near $\mathcal{I}^+$ is given by (4.59). The location of the center of this wave packet formed from p_ω with a small range of frequencies near the value of ω is determined by the principle of stationary phase. It follows that at early times, the components p_ω forming the part of the wave packet that passes back through the collapsing body and reaches $\mathcal{I}^-$ at v have (to within a normalization constant) the form on $\mathcal{I}^-$ (with $v < v_0$):

$$p_\omega \sim \omega^{-1/2} r^{-1} \exp(-i\omega\, u(v)) S(\theta, \phi), \tag{4.72}$$

with $u(v)$ given by (4.71). The $f_{\omega'}$ in the expansion (4.65) of p_ω have asymptotic form near $\mathcal{I}^-$ given by (4.57) with $v < v_0$. (This part of the wave packet cannot reach $\mathcal{I}^-$ at $v > v_0$.)

Using these early time asymptotic forms for p_ω and $f_{\omega'}$, Fourier's theorem may be used to show that

$$\alpha_{\omega\omega'} = C \int_{-\infty}^{v_0} dv \left(\frac{\omega'}{\omega}\right)^{\frac{1}{2}} e^{i\omega' v} e^{-i\omega u(v)}, \tag{4.73}$$

and

$$\beta_{\omega\omega'} = C \int_{-\infty}^{v_0} dv \left(\frac{\omega'}{\omega}\right)^{\frac{1}{2}} e^{-i\omega' v} e^{-i\omega u(v)}, \tag{4.74}$$

where C is a constant. Substitute (4.71) for $u(v)$, and let $s \equiv v_0 - v$ in (4.73) and let $s \equiv v - v_0$ in (4.74). Then

$$\alpha_{\omega\omega'} = -C \int_\infty^0 ds \left(\frac{\omega'}{\omega}\right)^{\frac{1}{2}} e^{-i\omega' s} e^{i\omega' v_0} \exp\left[i\omega 4M \ln\left(\frac{s}{K}\right)\right], \tag{4.75}$$

and

$$\beta_{\omega\omega'} = C \int_{-\infty}^0 ds \left(\frac{\omega'}{\omega}\right)^{\frac{1}{2}} e^{-i\omega' s} e^{-i\omega' v_0} \exp\left[i\omega 4M \ln\left(-\frac{s}{K}\right)\right]. \tag{4.76}$$

In (4.75), the contour of integration along the real axis from 0 to ∞ in the complex s-plane can be joined by a quarter circle at infinity to the contour along the imaginary axis from $-i\infty$ to 0. Because there are no poles of the integrand in the quadrant enclosed by the contour, and the integrand vanishes on the boundary at infinity, the integral from 0 to ∞ along the real s-axis equals the integral from $-i\infty$ to 0 along the imaginary s-axis. Thus, putting $s \equiv is'$, we have

$$\alpha_{\omega\omega'} = -iC \int_{-\infty}^{0} ds' \left(\frac{\omega'}{\omega}\right)^{\frac{1}{2}} e^{\omega' s'} e^{i\omega' v_0} \exp\left[i\omega 4M \ln\left(\frac{is'}{K}\right)\right]. \tag{4.77}$$

Similarly, in (4.76), the integral along the real axis in the complex s-plane from $-\infty$ to 0 can be joined by a quarter circle at infinity to the contour along the imaginary s-axis from $-i\infty$ to 0. As before, the integrals are equal. Therefore, putting $s \equiv is'$, we have

$$\beta_{\omega\omega'} = iC \int_{-\infty}^{0} ds' \left(\frac{\omega'}{\omega}\right)^{\frac{1}{2}} e^{\omega' s'} e^{-i\omega' v_0} \exp\left[i\omega 4M \ln\left(\frac{-is'}{K}\right)\right]. \tag{4.78}$$

Taking the cut in the complex plane along the negative real axis to define a single-valued natural logarithm function, we find that for $s' < 0$ (as in these integrands),

$$\ln(is'/K) = \ln(-i|s'|/K) = -i(\pi/2) + \ln(|s'|/K), \tag{4.79}$$

and

$$\ln(-is'/K) = \ln(i|s'|/K) = i(\pi/2) + \ln(|s'|/K). \tag{4.80}$$

Then

$$\alpha_{\omega\omega'} = -iC\, e^{i\omega' v_0}\, e^{2\pi\omega M} \int_{-\infty}^{0} ds' \left(\frac{\omega'}{\omega}\right)^{\frac{1}{2}} e^{\omega' s'} \exp\left[i\omega 4M \ln\left(\frac{|s'|}{K}\right)\right], \tag{4.81}$$

and

$$\beta_{\omega\omega'} = iC\, e^{-i\omega' v_0}\, e^{-2\pi\omega M} \int_{-\infty}^{0} ds' \left(\frac{\omega'}{\omega}\right)^{\frac{1}{2}} e^{\omega' s'} \exp\left[i\omega 4M \ln\left(\frac{|s'|}{K}\right)\right]. \tag{4.82}$$

Hence, it follows that

$$|\alpha_{\omega\omega'}|^2 = \exp(8\pi M\omega)|\beta_{\omega\omega'}|^2, \tag{4.83}$$

for the part of the wave packet that was propagated back in time through the collapsing body just before it formed a black hole.

For the components p_ω of this part of the wave packet, we have the scalar product,

$$(p_{\omega_1}, p_{\omega_2}) = \Gamma(\omega_1)\delta(\omega_1 - \omega_2), \tag{4.84}$$

where $\Gamma(\omega_1)$ is the fraction of an outgoing packet of frequency ω_1 at $\mathcal{I}^+$ that would propagate backward in time through the collapsing body to $\mathcal{I}^-$. To see this, let $p_\omega^{(2)}$ denote the component of this part of the wave packet, and let $p_\omega^{(1)}$ denote the component of the part of the wave packet that if propagated backward in time would be scattered from the spacetime outside the collapsing body and would travel back in time to past null infinity, reaching $\mathcal{I}^-$ with the same frequency, ω, as it had when it started from $\mathcal{I}^+$ (since it "climbs" away from the black hole while remaining at all times in the exterior region). Because $p_\omega^{(1)}$ and $p_\omega^{(2)}$ propagate to disjoint regions on $\mathcal{I}^-$ (i.e., $v > v_0$ and $v < v_0$, respectively), they are orthogonal, and with $p_\omega = p_\omega^{(1)} + p_\omega^{(2)}$, we have

$$(p_{\omega_1}, p_{\omega_2}) = (p_{\omega_1}^{(1)}, p_{\omega_2}^{(1)}) + (p_{\omega_1}^{(2)}, p_{\omega_2}^{(2)}). \tag{4.85}$$

Then from (4.58), it follows that

$$(p_{\omega_1}^{(2)}, p_{\omega_2}^{(2)}) = \Gamma(\omega_1)\delta(\omega_1 - \omega_2), \tag{4.86}$$

and

$$(p_{\omega_1}^{(1)}, p_{\omega_2}^{(1)}) = (1 - \Gamma(\omega_1))\delta(\omega_1 - \omega_2), \tag{4.87}$$

where $\Gamma(\omega_1)$ is the fraction of the packet of frequency ω_1 at $\mathcal{I}^+$ that would propagate back through the collapsing body to reach $\mathcal{I}^-$.

It follows from (4.68) and (4.85) that

$$\Gamma(\omega_1)\delta(\omega_1 - \omega_2) = \int d\omega' (\alpha^*_{\omega_1\omega'}\alpha_{\omega_2\omega'} - \beta^*_{\omega_1\omega'}\beta_{\omega_2\omega'}), \tag{4.88}$$

where $\alpha_{\omega\omega'}$ and $\beta_{\omega\omega'}$ now refer to the coefficients in the expansion of $p_\omega^{(2)}$ in terms of the $f_{\omega'}$ and $f^*_{\omega'}$, as in (4.65). The part of b_ω in (4.66) that is of interest to us is

$$b_\omega^{(2)} = (p_\omega^{(2)}, \phi). \tag{4.89}$$

For brevity of notation, we will let b_ω refer only to $b_\omega^{(2)}$ below.

The information about the particles that are created in the collapse of the body to form a black hole should be contained in b_ω, but we encounter an infinity if we straightforwardly try to evaluate

$$\langle 0 | b_\omega^\dagger b_\omega | 0 \rangle = \int d\omega' |\beta_{\omega\omega'}|^2. \tag{4.90}$$

This infinity is a consequence of the $\delta(\omega_1 - \omega_2)$ that appears in (4.88). We expect that $\langle 0 | b_\omega^\dagger b_\omega | 0 \rangle$ is the total number of created particles per unit frequency that reach $\mathcal{I}^+$ at late times in the wave $p_\omega^{(2)}$. This total number is infinite (neglecting the change in mass of the black hole) because there is a steady flux of particles reaching $\mathcal{I}^+$ at late times.

One heuristic way to see this is to replace $\delta(\omega_1 - \omega_2)$ in (4.88) by

$$\delta(\omega_1 - \omega_2) = \lim_{T \to \infty} \frac{1}{2\pi} \int_{-T/2}^{T/2} dt \, \exp[i(\omega_1 - \omega_2)t]. \tag{4.91}$$

4.3 Quantum aspects

Then, formally, we can write (4.88) when $\omega_1 = \omega_2 = \omega$, as

$$\lim_{T\to\infty} \Gamma(\omega)(T/2\pi) = \int d\omega' (|\alpha_{\omega\omega'}|^2 - |\beta_{\omega\omega'}|^2)$$

$$= [\exp(8\pi M\omega) - 1] \int d\omega' |\beta_{\omega\omega'}|^2, \qquad (4.92)$$

where we have used (4.83). Hence,

$$\langle 0| b_\omega^\dagger b_\omega |0\rangle = \lim_{T\to\infty} (T/2\pi)\Gamma(\omega)[\exp(8\pi M\omega) - 1]^{-1}. \qquad (4.93)$$

The interpretation of this is that at late times, the number of created particles per unit angular frequency and per unit time that pass through a surface $r = R$ (where R is much larger than the circumference of the black hole event horizon) is

$$(2\pi)^{-1}\Gamma(\omega)[\exp(8\pi M\omega) - 1]^{-1}. \qquad (4.94)$$

(The number per unit *frequency* per unit time has no factor of $(2\pi)^{-1}$.)

Recall that the quantity $\Gamma(\omega)$ is the fraction of a purely outgoing wave packet that when propagated from $\mathcal{I}^+$ backward in time would enter the collapsing body just before it had formed a black hole. At sufficiently late times this fraction is the same as the fraction of the wave packet that would enter the black hole past event horizon if the collapsing body were replaced in the spacetime by the analytic extension of the black hole spacetime. This means that $\Gamma_{lm}(\omega)$ is also the probability that a purely incoming wave packet that starts from $\mathcal{I}^-$ at late times will enter the black hole event horizon, that is, will be absorbed by the black hole. Therefore, (4.94) implies that a Schwarzschild black hole emits and absorbs radiation exactly like a gray body of absorptivity $\Gamma(\omega)$ and temperature T given by

$$k_B T = (8\pi M)^{-1}$$
$$= (2\pi)^{-1}\kappa, \qquad (4.95)$$

where k_B is Boltzmann's constant, and $\kappa = (4M)^{-1}$ is the surface gravity of the Schwarzschild black hole.

The same conclusion follows by considering a wave packet

$$p = \int d\omega_1\, A(\omega_1)\, e^{i\gamma(\omega_1)} p_{\omega_1}, \qquad (4.96)$$

where $A(\omega_1)$ and $\gamma(\omega_1)$ are real. Let

$$\int d\omega_1\, A^2(\omega_1) = 1, \qquad (4.97)$$

so

$$(p,p) = \int d\omega_1 \int d\omega_2\, A(\omega_1)A(\omega_2)\, e^{i[\gamma(\omega_1)-\gamma(\omega_2)]}(p_{\omega_1}, p_{\omega_2})$$
$$= 1.$$

Furthermore, let $A(\omega_1)$ be non-zero only for ω_1 in a narrow range $\Delta\omega$ about the value $\omega_1 = \omega$. Then, if $p^{(2)}$ is the part of the wave packet that when propagated back from $\mathcal{I}^+$ would enter the collapsing body before formation of the event horizon and would pass through the body to reach $\mathcal{I}^-$, it follows from (4.86) that

$$(p^{(2)}, p^{(2)}) = \Gamma(\omega), \qquad (4.98)$$

assuming that $\Gamma(\omega_1)$ is nearly constant for the range $\Delta\omega$ about the value $\omega_1 = \omega$.

The annihilation operator corresponding to particles described by the wave packet $p^{(2)}$ is

$$b \equiv (p^{(2)}, \phi). \qquad (4.99)$$

The number of outgoing particles observed at $\mathcal{I}^+$ having wave function $p^{(2)}$ is

$$\langle N \rangle = \langle 0 \mid b^\dagger b \mid 0 \rangle$$
$$= \int d\omega' |\beta_{p\omega'}|^2, \qquad (4.100)$$

with

$$\beta_{p\omega'} \equiv \int d\omega_1 \, A(\omega_1) \, e^{i\gamma(\omega_1)} \beta_{\omega_1 \omega'}. \qquad (4.101)$$

Here the $\beta_{\omega_1 \omega'}$ are coefficients in the expansion of $p_\omega^{(2)}$ in terms of the $f_{\omega'}$ and $f_{\omega'}^*$, as in (4.65).

The wave packet p near $\mathcal{I}^+$ has the asymptotic form (see (4.59)):

$$p \sim \int d\omega_1 \, A(\omega_1) \, e^{i\gamma(\omega_1)} \omega_1^{-1/2} r^{-1} e^{-i\omega_1 u} S(\theta, \phi). \qquad (4.102)$$

The position of the center of the packet is determined by the principle of stationary phase, as

$$u = (d\gamma/d\omega_1)_\omega. \qquad (4.103)$$

Let $(d\gamma/d\omega_1)_\omega$ be very large with respect to the circumference of the black hole event horizon. Then the wave packet reaches large values of r long after the black hole has formed (i.e., at late times). In that case, the $\beta_{\omega_1 \omega'}$ in (4.101) satisfy (4.83). Then, since

$$\beta_{p\omega'} \approx \beta_{\omega\omega'} \int d\omega_1 \, A(\omega_1) \, e^{i\gamma(\omega_1)},$$

it follows that to good approximation

$$|\alpha_{p\omega'}|^2 = \exp(8\pi M\omega)|\beta_{p\omega'}|^2, \qquad (4.104)$$

where $\alpha_{p\omega'}$ is defined as in (4.101).

4.3 Quantum aspects

As a consequence of (4.98), we have (see (4.68)):

$$\Gamma(\omega) = \int d\omega' (|\alpha_{p\omega'}|^2 - |\beta_{p\omega'}|^2)$$

$$= [\exp(8\pi M\omega) - 1] \int d\omega' |\beta_{p\omega}|^2. \qquad (4.105)$$

Hence,

$$\langle N \rangle = \Gamma(\omega)[\exp(8\pi M\omega) - 1]^{-1} \qquad (4.106)$$

is the number of created particles observed at late times in the wave packet mode p. This confirms that the black hole appears as a gray body of temperature $T = (2\pi k_B)^{-1}\kappa$ and absorptivity $\Gamma(\omega)$.

4.3.2 Kerr

A rotating body that collapses to form a Kerr black hole may be treated in essentially the same way as in the non-rotating case, with two basic changes. First, the radial geodesics in the Schwarzschild spacetime are replaced by the principal null congruence of geodesics in the Kerr spacetime that were derived in Section 4.2. This means that as we trace back in time from $\mathcal{I}^+$ to $\mathcal{I}^-$ the part of an outgoing wave packet that passes through the collapsing body just before the event horizon has formed, the value of u that the packet has on $\mathcal{I}^+$ is related to the value of v that it has on $\mathcal{I}^-$ by (4.51), which can be written as

$$u(v) \approx -(1/\kappa) \ln[(v - v_0)/K], \qquad (4.107)$$

where

$$\kappa = \alpha_+ = \frac{1}{2}(r_+ - r_-)/(r_+^2 + a^2) \qquad (4.108)$$

is the surface gravity of the Kerr black hole given in (4.48). This expression for $u(v)$ replaces (4.71) that was used in the non-rotating case.

The second difference is that the event horizon of the Kerr black hole has angular velocity, $d\phi/dt = \Omega_H$, given by (4.7). As we approach arbitrarily close to the null generators of the event horizon at r_+, both ϕ and t diverge, but (see (4.41)) the angular coordinate $\tilde{\phi}_+ = \phi - \Omega_H t$ is well behaved in the vicinity of r_+. In tracing an outgoing wave packet with components $p_\omega^{(2)}$ back in time into the collapsing body just before it has fallen within the event horizon, the angular coordinate $\tilde{\phi}_+$ is appropriate as the wave packet passes into the collapsing body. The result is that if $p_\omega^{(2)}$ has the form $\exp[-i\omega u + im\phi]$ at $\mathcal{I}^+$, then it has the form $\exp[-i(\omega - m\Omega_H)u(v) + im\phi']$ at $\mathcal{I}^-$, where ϕ' is the azimuthal angular coordinate in an inertial coordinate system far outside the collapsing body at early times. Here m is the azimuthal quantum number, which may have either sign.

As a consequence of these two differences between the non-rotating and rotating cases, the quantity ω on the right-hand sides of (4.72) through (4.81) is replaced by the quantity $\omega - m\Omega_H$, and $u(v)$ in (4.72) through (4.74) is given by (4.107); so $4M$ in (4.75)–(4.81) is replaced by κ^{-1}. Hence, for the rotating black hole we find

$$|\alpha_{\omega\omega'}|^2 = \exp[2\pi\kappa^{-1}(\omega - m\Omega_H)]|\beta_{\omega\omega'}|^2 \qquad (4.109)$$

in place of (4.83). It then follows, as in the previous section, that the average number of particles created in a wave packet that reaches $\mathcal{I}^+$ with energy ω and angular momentum quantum numbers l, m is

$$\langle N_{\omega lm} \rangle = \Gamma_{lm}(\omega) \left\{ \exp[2\pi\kappa^{-1}(\omega - m\Omega_H)] - 1 \right\}^{-1}, \qquad (4.110)$$

where the surface gravity κ is given by (4.108), and $\Gamma_{lm}(\omega)$ is the same as the fraction of a similar wave packet incident on a Kerr black hole that would be absorbed by the black hole. Thus, the Kerr black hole acts like a gray body of temperature

$$T_H = (2\pi k_B)^{-1}\kappa. \qquad (4.111)$$

If (4.110) is to make sense with $\langle N_{\omega lm}\rangle > 0$, it must be true that $\Gamma_{lm}(\omega)$ is negative when $\omega < m\Omega_H$. This means that when an incoming wave packet with $\omega < m\Omega_H$ is sent toward a Kerr black hole, the backscattered part of the wave packet returns with a larger amplitude than the original incoming packet. This phenomenon, known as superradiant scattering, was investigated shortly before the discovery that black holes radiate as gray bodies. One can think of superradiant scattering as stimulated pair production caused by the incoming boson. For fermions, the result in (4.110) differs in that the -1 in curly brackets is replaced by $+1$. In that case, $\Gamma_{lm}(\omega)$ remains positive at all frequencies. Fermions do not exhibit superradiant scattering because of the Pauli exclusion principle. An extensive discussion of superradiance can be found in Chandrasekhar (1983).

4.3.3 Entropy

If there is an entropy associated with equilibrium states of a black hole, then the Hawking temperature of the black hole should serve as an integrating factor leading to an expression for the entropy.

A Kerr black hole having the metric of (4.2), with $M^2 > a^2$, has angular momentum

$$J = Ma. \qquad (4.112)$$

The event horizon at r_+ of (4.5) is rotating with angular velocity Ω given by (4.7). Furthermore, it has a temperature T given by (4.111). Imagine a macroscopically infinitesimal process in which the black hole emits neutral bosons of zero rest mass, changing its mass M by dM and its angular momentum J by dJ. The work done by the rotating black hole when its angular momentum changes by dJ is

$$dW = -\Omega dJ, \qquad (4.113)$$

4.3 Quantum aspects

as for any body rotating[1] at angular momentum Ω. The change in internal energy of the black hole is dM, so the first law of thermodynamics gives

$$dQ = dM - \Omega dJ, \qquad (4.114)$$

where dQ is the heat energy flowing into the black hole in this infinitesimal process. Emission of a particle in mode (ω, l, m) carries to $\mathcal{I}^+$ energy $\hbar\omega$ and angular momentum component $\hbar m$ with respect to the axis of rotational symmetry. In such an emission process, $dM - \Omega dJ$ is a positive integer multiple of $-\hbar\omega + \Omega\hbar m$. Therefore dQ is negative for $\omega > m\Omega$ and positive for $\omega < m\Omega$. Thus emission of a boson in a mode with $\omega < m\Omega$ causes energy to flow into the black hole. This is the result of superradiant scattering, in which more energy is scattered back into the black hole than is emitted to $\mathcal{I}^+$ because $\Gamma_{lm}(\omega)$ is negative (as discussed in connection with (4.110)).

Now imagine further that the macroscopically infinitesimal emission process can be made to occur quasi-statically between two equilibrium states. (This would require some care because the black hole has negative heat capacity.) If a unique entropy S can be associated with each equilibrium state of the black hole, then $dQ = TdS$ and (4.114) gives

$$dS = \frac{1}{T}dM - \frac{1}{T}\Omega dJ. \qquad (4.115)$$

Since we already know T, J, and Ω in terms of the mass M and the angular momentum per unit mass a of the Kerr black hole, we can try to integrate (4.115) to find its entropy S.

With the help of (4.5)–(4.7), (4.108), (4.111), and (4.112), we find from (4.115) that (in units with $k_B = 1$):

$$dS = \frac{4\pi M \left(M + \sqrt{M^2 - a^2}\right)}{\sqrt{M^2 - a^2}} \left\{ dM - \left[\frac{a}{2M\left(M + \sqrt{M^2 - a^2}\right)}\right] d(aM) \right\}$$

$$= \frac{4\pi M^2}{\sqrt{M^2 - a^2}} dM + 4\pi M dM - \frac{2\pi a}{\sqrt{M^2 - a^2}} d(aM)$$

$$= d\left(2\pi M^2 + 2\pi M \sqrt{M^2 - a^2}\right). \qquad (4.116)$$

Hence, the entropy of a Kerr black hole is

$$S = 2\pi M \left(M + \sqrt{M^2 - a^2}\right), \qquad (4.117)$$

assuming that S vanishes when both M and a are zero. The area of the event horizon is

$$A \equiv \int_{r=r_+} \sqrt{g_{\theta\theta} g_{\phi\phi}} \, d\theta \, d\phi,$$

---

[1] See Landau and Lifshitz (1958).

which yields

$$A = 4\pi(r_+^2 + a^2)$$
$$= 8\pi M \left(M + \sqrt{M^2 - a^2}\right). \tag{4.118}$$

Therefore, the entropy of (4.117) is

$$S = \frac{1}{4}A. \tag{4.119}$$

These results are, of course, also valid for the Schwarzschild black hole (for which $a = 0$).

It is noteworthy that the temperature and entropy of a black hole involve Planck's constant. In conventional units (with $\hbar, c, G$ restored), we have $k_B T = (2\pi)^{-1}(\hbar/c)\kappa$ and $S = (1/4)k_B c^3 (G\hbar)^{-1} A$. The appearance of $\hbar$ in the expressions for the temperature and entropy of a black hole, which is a classical object from the point of view of the theory of general relativity, suggests that the study of black holes may lead to a deeper understanding of how gravitation and quantum theory are interrelated.

If the black hole is also charged, it has an electrostatic potential Φ. Therefore, the work done by the black hole when its charge changes by an infinitesimal amount, dq, is $-\Phi dq$, and this term must be added to the right-hand side of (4.114). It can be shown (Hawking 1975) that the average number of particles of charge e emitted in mode ω, l, m has the form of (4.110), but with $\omega - m\Omega_H - e\Phi$ appearing in the exponential and with the κ and Γ appropriate to the rotating charged black hole. Similarly, the entropy has the form of (4.119), with the appropriate event horizon area A.

The assertion that a black hole entropy proportional to A entered into a generalized second law of thermodynamics was first made by Bekenstein (1972, 1973, 1974). The generalized second law states that the total entropy of a system of black holes and their environment does not decrease in any natural process. Such processes include the absorption by a black hole of subsystems containing entropy, as well as the evaporation of a black hole through Hawking radiation.

4.4 Energy-momentum tensor with Hawking flux

An important first step in the analysis of the back reaction of the Hawking flux on a black hole is the calculation of the expectation value of the energy-momentum tensor in a black hole spacetime. In Chapter 3, we discussed an approximation due to Page for the expectation value of the energy-momentum tensor in the Hartle–Hawking state. This state corresponded to a thermal equilibrium state of particles in the black hole spacetime. However, the Hawking radiation is a non-equilibrium process and it is therefore clear that the Hartle–Hawking state is not the appropriate state in which the Hawking radiation flux appears. Instead, we seek, as in

Hawking's original treatment of the problem, a vacuum state that corresponds to no incoming particles from past null infinity ($\mathcal{I}^-$) and a flux of particles into future null infinity ($\mathcal{I}^+$). Earlier in this chapter, we defined the vacuum state in exactly this way and obtained the Hawking flux for Schwarzschild (non-rotating) and Kerr (rotating) black holes. The black holes in our earlier calculations were formed by collapsing matter, and hence had no past event horizon.

As shown by Unruh (1976), we can obtain the same result for the Hawking flux in the spacetime of an eternal black hole. Because a black hole formed by collapsing matter has no past event horizon, one of the conditions that defined a Hartle–Hawking state in the spacetime of an eternal black hole, namely, regularity of expectation values of observables at the past event horizon, is no longer necessary. In the spacetime of an eternal black hole, the vacuum state with respect to which the Hawking flux is observed is known as the Unruh vacuum. A precise definition of this state may be given in terms of the Kruskal null coordinates U and V introduced earlier. The Unruh vacuum is defined by expanding the field operator in mode functions that have the following two properties:

(a) modes that are incoming from $\mathcal{I}^-$ are positive frequency with respect to t, the usual Schwarzschild time coordinate;
(b) modes that are outgoing from the past event horizon are positive frequency with respect to U.

The annihilation operators in this expansion of the field operator annihilate the Unruh vacuum state. Conditions (a) and (b) can be shown to imply that the expectation value of the energy-momentum tensor in the Unruh vacuum is regular on the future event horizon but diverges on the past event horizon.

The problem of calculating the expectation value of the energy-momentum tensor in the Unruh vacuum may be split into two parts: the calculation of the expectation value of the energy-momentum tensor in the Hartle–Hawking state, and the calculation of the difference of the energy-momentum tensors in the Unruh and Hartle–Hawking states (denoted by $T^U_{\mu\nu}$ and $T^{HH}_{\mu\nu}$ respectively). For the first part of the problem, there exist analytic approximations as well as numerical schemes, as outlined in Chapter 3. For the second part of the problem, we follow the work of Elster (1983) for a massless conformally coupled scalar field in the Schwarzschild black hole spacetime. No renormalization is necessary in the calculation of the difference $T^U_{\mu\nu} - T^{HH}_{\mu\nu}$ because the ultraviolet (short-distance) divergences in each energy-momentum tensor are state-independent. These divergences therefore cancel out in the difference, so $T^U_{\mu\nu} - T^{HH}_{\mu\nu}$ is a finite quantity. To compute this difference, we construct the two-point Hadamard functions

$$G(x, x') = \frac{1}{2}\langle 0|\phi(x)\phi(x') + \phi(x')\phi(x)|0\rangle, \qquad (4.120)$$

where $|0\rangle$ denotes either the Unruh or the Hartle–Hawking state. We expand the fields in the mode functions appropriate to each vacuum state in order to compute each Hadamard function. The difference between the Hadamard functions in the two states is found to be (Christensen and Fulling 1977; Candelas 1980; Hawking 1981; Elster 1983)

$$G^U(x,x') - G^{HH}(x,x') = -(4\pi rr')^{-1} \int_0^\infty d\omega\, \omega^{-1} e^{-i\omega(t-t')} \frac{1}{e^{8\pi M\omega} - 1}$$
$$\times \sum_{m,n} Y_n^m(\theta,\phi) Y_n^{m*}(\theta',\phi') R_{\omega n}(r) R_{\omega n}^*(r') + \text{c.c.},$$
(4.121)

where t, r, θ, and ϕ are the usual Schwarzschild coordinates, M is the mass of the black hole, and c.c. denotes complex conjugate[2]. The radial function $R_{\omega n}(r)$ is a complex solution of the differential equation

$$\frac{d^2R}{d\bar{r}^2} + \left[\omega^2 - n(n+1)r^{-3}(r-2M) - 2Mr^{-4}(r-2M)\right]R = 0, \qquad (4.122)$$

where $\bar{r}$ is the Regge–Wheeler coordinate

$$\bar{r} = r + 2M\ln(r/(2M) - 1). \qquad (4.123)$$

Note that (4.122) is analogous to a one-dimensional Schrödinger equation with a potential barrier and that its solution is analogous to the solution of an over-barrier transmission problem in quantum mechanics. In the range $2M < r < 4M$, (4.122) has the series solution (Elster 1983)

$$R_{\omega n}(r) = \tau_{\omega n} e^{-i\omega\bar{r}} \sum_{k=0}^\infty c_k (r/(2M) - 1)^k, \qquad (4.124)$$

where $\tau_{\omega n}$ is the transmission coefficient for a wave of unit amplitude incident from $r = \infty$, which may be computed by numerical integration of (4.122). The coefficients c_k are obtained recursively as follows:

$$c_0 = 1,$$
$$c_k = -(6k(k - 4iM\omega))^{-1} \sum_{l=0}^{k-1} (-1)^{k-l}$$
$$\times \left[(k-l)(k-l+1)(k+2l+3n^2+3n+2) + 6(k-l+1)l^2 - 24ilM\omega\right] c_l, \quad \text{for } k \geq 1. \qquad (4.125)$$

The difference of the expectation values of the stress tensor in the Unruh and Hartle–Hawking states is found by taking appropriate derivatives of the difference of Hadamard functions in (4.121) and then taking the coincidence limit $x' \to x$. Defining the radial pressure $p_r \equiv -\langle 0 \mid T_r^r \mid 0 \rangle$, the tangential pressure

---

[2] Units with $\hbar = c = G = 1$ are used.

4.4 Energy-momentum tensor with Hawking flux

$p_t \equiv -\langle 0 | T_\theta^\theta | 0 \rangle = -\langle 0 | T_\phi^\phi | 0 \rangle$, the energy density $\mu \equiv \langle 0 | T_t^t | 0 \rangle$, and the energy current $s \equiv -\langle 0 | T_{rt} | 0 \rangle$, we find (Elster 1983)

$$p_r^U(r) - p_r^{HH}(r) = \frac{1}{96\pi^2} \int_0^\infty \frac{d\omega}{\omega} \left(e^{8\pi M \omega} - 1\right)^{-1}$$
$$\times \sum_{n=0}^\infty (2n+1) f_{\omega n}(r), \qquad (4.126)$$

$$p_t^U(r) - p_t^{HH}(r) = \frac{1}{96\pi^2} \int_0^\infty \frac{d\omega}{\omega} \left(e^{8\pi M \omega} - 1\right)^{-1}$$
$$\times \sum_{n=0}^\infty (2n+1) g_{\omega n}(r), \qquad (4.127)$$

$$\mu^U(r) - \mu^{HH}(r) = \frac{1}{96\pi^2} \int_0^\infty \frac{d\omega}{\omega} \left(e^{8\pi M \omega} - 1\right)^{-1}$$
$$\times \sum_{n=0}^\infty (2n+1) h_{\omega n}(r), \qquad (4.128)$$

$$s^U(r) - s^{HH}(r) = \frac{L}{4\pi r(r-2M)}, \qquad (4.129)$$

where

$$f_{\omega n}(r) = 2\frac{(\gamma_2 - 3\delta)}{r(r-2M)} + 2\gamma_1 \frac{(r-3M)}{r^3(r-2M)}$$
$$+ 2\left[n(n+1) + \frac{r-4M}{r} - \omega^2 \frac{r^3}{r-2M}\right] \frac{\epsilon}{r^4}, \qquad (4.130)$$

$$g_{\omega n}(r) = 2\frac{\delta}{r(r-2M)}$$
$$- 2\left[2n(n+1) + \frac{r-2M}{r} + \omega^2 \frac{r^3}{r-2M}\right] \frac{\epsilon}{r^4}, \qquad (4.131)$$

$$h_{\omega n}(r) = -2\frac{\delta}{r(r-2M)} + 2\gamma_1 \frac{(r-3M)}{r^3(r-2M)}$$
$$- 2\left[n(n+1) + \frac{r-4M}{r} + 5\omega^2 \frac{r^3}{r-2M}\right] \frac{\epsilon}{r^4}, \qquad (4.132)$$

and

$$\epsilon = |R_{\omega n}|^2, \quad \delta = \left|\frac{dR_{\omega n}}{d\bar{r}}\right|^2, \quad \gamma_i = R_{\omega n}^* \frac{d^i R_{\omega n}}{d\bar{r}^i} + \text{c.c.} \qquad (4.133)$$

The expression for the constant L appearing in (4.129) has the form of an integral over ω that is given in (4.134) and (4.135) below.

To obtain the pressures, energy density, and current in the Unruh vacuum state, we add to (4.126)–(4.129) the corresponding values in the Hartle–Hawking state obtained, for example, from Page's approximation. Because of the invariance under time-reversal of the Hartle–Hawking state, $s^{HH} = 0$. Therefore (4.129) represents the energy current in the Unruh vacuum. It follows that L is the luminosity of the black hole. The explicit expression for L is (DeWitt 1975; Christensen and Fulling 1977; Elster 1983)

$$L = \int_0^\infty d\omega\, L_\omega, \qquad (4.134)$$

with

$$L_\omega = \frac{\omega}{2\pi} \left(e^{8\pi M\omega} - 1\right)^{-1} \sum_{n=0}^\infty (2n+1)|\tau_{\omega n}|^2, \qquad (4.135)$$

where $\tau_{\omega n}$ is defined by (4.124). We may use the values of $\tau_{\omega n}$ obtained by numerical integration of (4.122) to find the luminosity of the black hole. Elster (1983) gives the result

$$L = 7.44 \times 10^{-5} M^{-2}. \qquad (4.136)$$

A calculation similar to the one outlined above, carried out by Page (1976) for neutrinos, photons, and gravitons gives $L_\nu = 16.36 \times 10^{-5} M^{-2}$, $L_\gamma = 3.37 \times 10^{-5} M^{-2}$ and $L_g = 0.38 \times 10^{-5} M^{-2}$ respectively.

4.5 Back reaction to black hole evaporation

The discussion of the previous sections showed that a black hole emits Hawking radiation at a temperature that is inversely proportional to the mass of the hole, and a luminosity that is inversely proportional to the square of its mass (for a Schwarzschild black hole). Therefore, according to the law of conservation of energy, the mass of the black hole must decrease as it emits Hawking radiation. Furthermore, it is expected from energy conservation that the rate of loss of mass must be proportional to the luminosity, that is,

$$\frac{dM}{dt} = -\frac{C}{M^2}, \qquad (4.137)$$

where C is a positive constant that depends on the number of quantized matter fields that couple to gravity.

The above heuristic derivation of (4.137) brings into question its regime of validity. Clearly, there are at least two assumptions on which (4.137) is based; namely,

(a) that the right-hand side of (4.137), based on the derivation of the Hawking luminosity for a *constant* mass black hole, continues to be a good approximation when the black hole is a dynamical object with a time-dependent mass; and
(b) that the black hole is large enough that the dominant components of the Riemann tensor at the horizon are small with respect to the Planckian curvature, that is, that $M \gg M_{\text{Planck}}$.

Assumption (b) breaks down when the black hole, in the course of its evaporation, reaches a mass of the order of the Planck mass. At this point, we would expect that the gravitational field itself must be quantized, and the entire semiclassical derivation of the Hawking flux becomes invalid. The question of what happens to the black hole when it reaches Planck scale dimensions can therefore be satisfactorily answered only within the context of quantum gravity (e.g., loop quantum gravity or string theory). The resolution of this question remains unclear.

The range of validity of assumption (a), however, seems to be testable, at least in principle, within the semiclassical theory of gravity. It involves the computation of the Hawking flux in a sufficiently general *dynamical* black hole metric and the calculation of the back reaction of this flux on the black hole itself via the semiclassical Einstein equations. This is a mathematically formidable problem because the simpler computation of the Hawking flux in a static black hole metric is itself not amenable to an analytic solution, but must be investigated numerically. We describe here briefly a few approaches toward the full dynamical problem.

Hajicek and Israel (1980) showed that there exist black hole geometries whose mass diminishes according to (4.137). Subsequently, Bardeen (1981) (see also York 1983) showed that if we assume that at a large distance from the horizon there is an outgoing flux of radiation (with an integrated luminosity L) and that the energy-momentum tensor is regular on the future event horizon, then the mass of the black hole must decrease according to the law $dM/dt = -L$. Bardeen's analysis is valid for arbitrary L and his form for the dynamical metric is

$$ds^2 = e^{2\psi}\left(1 - \frac{2M(r,v)}{r}\right)dv^2 + 2e^{\psi}dvdr + r^2 d\Omega^2, \qquad (4.138)$$

where v is a null coordinate that when M is constant and $\psi = 0$ reduces to the incoming Eddington–Finkelstein coordinate. In general, the functional dependence of $M(r,v)$ on r and v is fixed by the form of the energy-momentum tensor of the outgoing flux of radiation, as is the functional dependence of ψ on r. Note that the above general form of the metric is invariant under an arbitrary reparameterization of v (i.e., the transformation $v \to f(v)$) and correspondingly, ψ is defined only up to the addition of an arbitrary function of v.

Bardeen's geometry must be valid at early times in the evaporation of a black hole of large mass because at early times the energy-momentum tensor must certainly be close to the energy-momentum tensor in the Unruh vacuum in the absence of back reaction. This latter energy-momentum tensor satisfies the flux and regularity assumptions of Bardeen and therefore the geometry at early times must be the geometry described by (4.138). However, at late times, when the black hole has lost a significant fraction of its mass, the energy-momentum tensor could in principle differ significantly from that calculated without back reaction and may not satisfy the assumptions of Bardeen. If this were the case (4.137) and (4.138) may not be valid.

To investigate this issue, Parentani and Piran (1994) carried out a numerical integration of the coupled Einstein and quantized scalar field equations in a simple model in which the expectation value of the energy-momentum tensor was given by its two-dimensional value (Davies et al., 1976) multiplied by a factor of $(4\pi r^2)^{-1}$ so as to be conserved in four dimensions (this is sometimes referred to as the s-wave approximation to the stress tensor). Their result is that the geometry is indeed that of Bardeen, with a luminosity that is proportional to M^{-2}. Also, Massar (1995) has considered the coupled Einstein and quantized scalar field equations in Bardeen's metric and argued that (4.137) holds at all times for which $M \gg M_{\text{Planck}}$.

4.6 Trans-Planckian physics in Hawking radiation and cosmology

After Hawking published his paper deriving the thermal spectrum of the radiation created by a black hole (Hawking 1974; Hawking 1975), questions were raised about the use of paths from $\mathcal{I}^-$ to $\mathcal{I}^+$; the frequencies of massless particles receive arbitrarily large redshifts along such paths as they pass through the collapsing dust cloud just prior to formation of the event horizon. Then the range of frequencies that can be seen by distant observers at late times would have had to originate at $\mathcal{I}^-$ with ultrahigh frequencies, including frequencies above the Planck scale. Local Lorentz invariance would be violated if such frequencies were arbitrarily cut off. Would the Hawking thermal spectrum nevertheless survive the breaking of local Lorentz invariance?

Much work has been done on classifying the ways that Lorentz invariance can be violated, and in devising experimental tests of Lorentz violation; see Kostelecky (2004) and references given there.

In the context of black holes, Unruh (1995) considered a definite model of sound waves propagating in a moving fluid that simulates the behavior of the event horizon a black hole. See Unruh (1995) and references cited there. By numerical methods he found that despite the breaking of Lorentz invariance in his fluid model, the sonic black hole nevertheless produced a spectrum of sound waves that was very close to a thermal spectrum. He demonstrated that the ultrahigh frequencies are not responsible for the thermal spectrum produced by a sonic black hole. This supports the viewpoint that the ultrahigh frequencies that appear in the derivation of the Hawking thermal spectrum in black hole evaporation are not necessarily essential for obtaining the thermal spectrum. In this context, related models with dispersion relations that break Lorentz invariance have been considered by, for example, Jacobson (1991). In addition, Jacobson and Mattingly (2001) have considered gravity with a preferred frame corresponding to a dynamical vector field.

In Agullo et al. (2008a) the Hadamard form of the two-point correlation function of the field at very short distances characterized by an *invariant* Planck length was altered. The invariance of the Planck length appearing in the two-point function is enforced by means of a non-linear physical realization of the Lorentz group, in analogy with the idea applied to dispersion relations in Magueijo and Smolin (2002). They showed that this alteration of the Hadamard form at the invariant Planck scale has negligible effect on the thermal spectrum of Hawking radiation. This conclusion extends to spectral frequencies much higher than the energy scale set by the Hawking temperature of the black hole. Thus, the thermal spectrum of an evaporating black hole of radius above the Planck scale appears to be insensitive to such changes in physics near the Planck scale. Similarly, for a uniformly accelerated detector in Minkowski spacetime, changing the Hadamard form in this way has an entirely negligible effect on the thermal response of the detector (Agullo et al., 2008b). A nice discussion of various models of black hole evaporation, including ones that incorporate the back reaction of the Hawking radiation on the black hole, is given in Fabbri and Navarro–Salas (2005).

In Deser and Levin (1998, 1999), the spacetime of a four-dimensional black hole is embedded in a six-dimensional Minkowski spacetime in a global way, in the sense that the embedding in the six-dimensional flat spacetime covers the usual Kruskal maximal extension of Schwarzschild spacetime without encountering a coordinate singularity at the Schwarzschild radius of the black hole. In this embedding, a detector held at rest at constant Schwarzschild radius, r, is mapped to a detector moving at constant acceleration in the six-dimensional Minkowski spacetime. They show that the temperature, $a/(2\pi)$, of the thermal spectrum measured by this uniformly accelerated detector is the temperature that the detector at constant r outside the black hole would detect as a result of the Hawking radiation. This correspondence makes no use of trans-Planckian frequencies and thus supports the view that they are not essential to the thermal spectrum of Hawking radiation.

The string theory derivation of the Hawking radiation for a nearly extremal supersymmetric black hole (Maldacena 1997) makes use of the Minkowski spacetime limit of the black hole in terms of D-branes and oppositely moving string excitations that interact and produce the Hawking thermal spectrum of radiation, including the gray-body factor, without appealing to large red- or blueshifts. (This string model also gives the correct Bekenstein–Hawking entropy of the black hole from the counting of microstates.) This again suggests that the thermal spectrum is not dependent on very high frequency modes of the radiation field.

Cosmological aspects of trans-Planckian physics, including its possible influence on the power spectrum of the CMB radiation and on dark matter and dark energy have been considered in many interesting papers (e.g., Chung et al., 1999, 2000; Mersini et al., 2001; Brandenberger and Martin 2002; Danielsson 2002; Easther et al., 2002; Armendariz-Picon and Lim 2003; Bastero-Gil et al., 2003; Bozza et al., 2003; López Nacir and Mazzitelli 2007; Holman et al., 2008). It is entirely possible that as the relevant cosmological observations are refined, and new results of experiments in the Large Hadron Collider are announced in the coming years, the influence of Planckian and trans-Planckian scale physics will be shown to be important.

4.7 Further topics: Closed timelike curves; closed-time-path integral

Here we briefly discuss some topics that are of interest, but that we have not had space to cover in detail.

4.7.1 Closed timelike curves

The work of Morris and Thorne (1988) and Morris et al. (1988) showed that if a traversable wormhole connecting two locations in relative motion in our universe could be constructed, then it would be possible for an object to pass through the wormhole moving forward in its own proper time, while exiting the wormhole at a time earlier than the time at which it entered, as measured by synchronized clocks in our universe. Using such a wormhole, a closed timelike curve in spacetime could be formed along which an object moving forward in proper time could return to a given location at a time earlier than the time at which it left. The question naturally arose as to the possibility in our universe of constructing such a wormhole. With sufficient negative-energy, such as appears in the Casimir effect, could we hope to hold open such a wormhole? Does quantum field theory in curved or flat spacetime make it impossible to construct such a wormhole? The weight of evidence seems to show that there are insurmountable obstacles to constructing such wormholes and to defining a viable quantum field theory in a spacetime with closed timelike curves (Friedman *et al.*, 1990; Hawking 1992; Friedman *et al.*, 1992; Ford and Roman 1996).

4.7.2 Closed-time-path functional integral in curved spacetime

The functional Feynman path integral is taken over a path in function space that runs from an early time field configuration to a late time field configuration. The result is the amplitude to go from the initial state, defined by the initial field configuration or a set of possible initial configurations, to the final state. (This is analogous to the quantum mechanical Feynman path integral in which the amplitude to go from an initial state of a particle to a final state is obtained by summing over all paths of the particle that go from the initial state to the final state.) The amplitude obtained from the functional Feynman path integral gives a direct measure of the particle creation between the initial and final states of the quantum field system. This is true in flat or curved spacetimes (Schwinger 1951a; DeWitt 1975; Parker 1979b).

On the other hand, we often wish to write the semiclassical Einstein gravitational field equations in order to evolve the classical background gravitational

4.7 Further topics: Closed timelike curves; closed-time-path integral

field under the influence of quantized particle fields (and possible additional classical matter). What can we do to represent the influence of the quantized fields in the Einstein equations? In cases where there is sufficient symmetry, then it is justified to use the expectation value of the energy-momentum tensor of the quantized fields. This allows us to evolve the system forward in time, while taking into account the averaged effect of the quantized field (Parker and Fulling 1973). The expectation value that we must use in such a case is that taken at a given time, not the one between an initial and a final time. This is commonly referred to as the "in-in" expectation value, as opposed to the "in-out" expectation value. The transformation from the "in-out" expectation value to the "in-in" expectation value can be expressed in terms of the Bogolubov coefficients that relate the early time and late time creation and annihilation operators. Such a transformation from an in-state to an out-state of the quantized field vacuum is given for a Robertson–Walker universe in Parker (1969). In the "in-out" expectation value, we can use the expression giving the out-state in terms of the in-state to convert the "in-out" expectation value to the "out-out" expectation value. So that gives one way to obtain the "in-in" expectation value that appears in the Einstein equations from the "in-out" expectation value that comes from the functional Feynman path integral.

A direct way to obtain "in-in" expectation values was introduced by Schwinger (1961) using closed-time-paths that move forward in time and then return to the initial time, thus yielding an amplitude between two states specified at the same time. From this "in-in" amplitude, including a source term in the action, Schwinger was able to obtain "in-in" expectation values directly. This technique was further developed by Keldysh (1964), Korenman (1966), and others, and is reviewed in Chou et al. (1985). In curved spacetime, the Schwinger–Keldysh closed-time-path formalism has been further developed by Jordan (1986) and in a series of papers by Calzetta and Hu (1987, 1988, 1989, 1994). Recently, Weinberg (2005, 2006) has extended the "in-in" formalism and used it to calculate higher-order Gaussian and non-Gaussian correlations in cosmology. These correlations depend on the whole history of inflation, not just on the behavior of fields when they exit from the horizon. Therefore, they would carry more detailed information about inflation. Weinberg's results show that within a perturbative framework these correlations are never large.

5
The one-loop effective action

5.1 Introduction

The main purpose of this chapter is to provide a link between the methods used in previous chapters and the more general methods contained in the following chapters necessary to study interacting fields. These more general methods, which may be applied to a wide class of theories, are based on the background field approach to the effective action. In this chapter we will concentrate on free fields, or fields interacting with background, or external, fields which are not quantized, as we did in the previous four chapters. We will defer the quantization of gauge fields to Chapter 7.

We begin this chapter by presenting the relation between the Schwinger action principle and the Feynman functional, or path, integral for the basic ⟨out|in⟩ transition amplitude. Regularization of the one-loop effective action, which is simply related to the in-out transition amplitude, is discussed using a number of popular methods. (Cut-off, dimensional, and ζ-function regularization are presented to complement our earlier treatment of regularization in Chapter 3.) Two explicit scalar field examples are given: the Schwinger effective Lagrangian for a constant electromagnetic field in flat Minkowski spacetime and the effective potential for a constant gauge field background in the spacetime $\mathbb{R}^{n-1} \times S^1$. In these two cases it is possible to calculate an exact result for the effective action. The conformal anomaly for a scalar field, considered earlier in four spacetime dimensions, is analyzed from the Feynman path integral viewpoint.

A substantial part of this chapter is devoted to spinor fields in spacetimes of arbitrary dimension, reinforcing our earlier four-dimensional presentation. We analyze the special features found in even- and odd-dimensional spacetimes, including representations of the Dirac γ-matrices. Majorana and Weyl, as well as Dirac spinors are defined. A brief discussion of some of the global considerations in defining spinors in general spacetimes is presented, and illustrated by means of a simple example. Properties of Grassmann variables, which are needed to describe spinor fields, are given, including integration of Gaussians. These results will prove useful in later chapters as well as in the present one.

Having presented enough background to deal with spinor fields on a general spacetime, we define the path integral expression for the one-loop effective action for spinor fields, and discuss some of the new features not present in the scalar field case. We show how the conformal anomaly for spinor fields arises in the functional integral approach, and present the spinor versions of the two scalar field examples given earlier where the effective action can be computed exactly. The axial, or chiral, anomaly in a spacetime of general dimension is calculated in terms of the coefficients appearing in the asymptotic expansion of the kernel for the Dirac operator. Some explicit results are presented. We also show how the two-dimensional axial anomaly may be used to arrive at the radiatively induced Chern–Simons theory in three spacetime dimensions.

5.2 Preliminary definition of the effective action

In this section we wish to discuss an approach to quantum field theory which uses the effective action. We will concentrate on free fields, or fields interacting with background, or external, fields which are not quantized. The intention of this section is to provide a heuristic derivation of the lowest order, one-loop, correction to the classical action functional arising from quantized fields. The more general case of quantized interacting fields will be deferred until Chapter 6, where the results of the present section will be derived again in a much more systematic manner.

The Schwinger action principle reads

$$\delta\langle \text{out}|\text{in}\rangle = \frac{i}{\hbar}\langle \text{out}|\delta S|\text{in}\rangle \qquad (5.1)$$

where $|\text{in}\rangle$ and $\langle\text{out}|$ represent any states, and we have chosen to leave $\hbar$ in this expression, rather than choose units with $\hbar = 1$. As we will see in Chapter 6, retaining the explicit factor of $\hbar$ proves to be a convenient method to keep track of the order in the perturbative expansion of the full effective action. δS represents the variation of the action operator S with respect to any variations of the fields, coordinates, or other parameters that appear in S as discussed in Chapters 1 and 2. (For a more complete discussion of the Schwinger action principle with applications, see Toms (2007).) It proves convenient to define an object W by

$$\langle \text{out}|\text{in}\rangle = \exp\left(\frac{i}{\hbar}W\right). \qquad (5.2)$$

Then from (5.1), it follows that

$$\delta W = \frac{\langle \text{out}|\delta S|\text{in}\rangle}{\langle \text{out}|\text{in}\rangle}. \qquad (5.3)$$

For concreteness, consider the specific case of a set of real scalar fields $\varphi^i(x)$ where $i = 1, \ldots, N$, whose action functional is

$$S[\varphi] = \frac{1}{2} \int dv_x \left[\delta_{ij} \partial^\mu \varphi^i \partial_\mu \varphi^j - M_{ij}^2(x) \varphi^i \varphi^j \right]. \tag{5.4}$$

$M_{ij}^2(x)$ is some given function of x, which is symmetric in i and j. (For example, we might have $M_{ij}^2(x) = (m^2 + \xi R)\delta_{ij}$, in which case the theory consists of N massive, non-minimally coupled (if $\xi \neq 0$) scalar fields.) Here dv_x is the invariant spacetime volume element defined in terms of the metric and coordinate differentials by $dv_x = |g(x)|^{1/2} d^n x$ with $g(x) = \det[g_{\mu\nu}(x)]$ and n the spacetime dimension.

We can choose the variation in (5.3) to be with respect to $M_{ij}^2(x)$. This variation will be denoted by $\delta M_{ij}^2(x)$. Then (5.3) gives

$$\delta W = -\frac{1}{2} \int dv_x \left(\delta M_{ij}^2(x)\right) \frac{\langle \text{out}|\varphi^i(x)\varphi^j(x)|\text{in}\rangle}{\langle \text{out}|\text{in}\rangle}, \tag{5.5}$$

using the action of (5.4). At this stage we will choose the states $|\text{in}\rangle$ and $\langle \text{out}|$ to be the vacuum states in the in and out regions of the spacetime. The dynamic spacetime models of Chapter 2 can be kept in mind. We will define

$$\langle \text{out}|\varphi^i(x)\varphi^j(x)|\text{in}\rangle = \lim_{x' \to x} \langle \text{out}|T\left(\varphi^i(x)\varphi^j(x')\right)|\text{in}\rangle, \tag{5.6}$$

which allows us to make contact with the Feynman Green function defined by

$$\left[\delta_{ij}\Box_x + M_{ij}^2(x)\right] G^{jk}(x, x') = \delta_i^k \delta(x, x'). \tag{5.7}$$

$\delta(x, x')$ represents the biscalar Dirac δ-distribution.[1] It then follows that

$$\frac{\langle \text{out}|\varphi^i(x)\varphi^j(x)|\text{in}\rangle}{\langle \text{out}|\text{in}\rangle} = -i\hbar G^{ij}(x, x). \tag{5.8}$$

(A factor of $\hbar$ has been placed in (5.8) to serve as an indicator of the order in the perturbative expansion, as mentioned at the start of this section. See Chapter 6 for full details.) Equation (5.5) now reads

$$\delta W = \frac{i\hbar}{2} \int dv_x \left(\delta M_{ij}^2(x)\right) G^{ij}(x, x). \tag{5.9}$$

At this stage, in order to facilitate the formal manipulations we are about to perform, it proves convenient to adopt a condensed notation used by DeWitt (1964a,b). This notation will be used extensively in Chapter 6, and it may be helpful to see it in action in a simpler context first. With this notation, we use the indices on any field or operator to also include the spacetime arguments. In

---

[1] By biscalar we mean that the object transforms as a scalar under independent coordinate changes in its two arguments. In some cases it may be more convenient to define a related distribution that transforms as a scalar density in one or both arguments.

5.2 Preliminary definition of the effective action

place of the conventional expression $\varphi^i(x)$ for the field we would simply write φ^i with the understanding that now i stands for both the field label and the spacetime argument of the conventional expression. Similarly, for the Green function $G^{ij}(x, x')$ we will write $G^{ij'}$ in condensed notation, where the prime on the index j signifies that we are using it to include the spacetime argument x'. With this condensed notation, we can now use the summation convention for any repeated index to include an integration over the associated spacetime argument.

To enable the use of condensed notation, suppose that we now define

$$D_{ij}(x, x') = \left[\delta_{ij}\Box_x + M_{ij}^2(x)\right]\delta(x, x'). \tag{5.10}$$

Then (5.7), which defines the Green function, can be rewritten as (in conventional notation)

$$\int dv_{x'}\, D_{ij}(x, x') G^{jk}(x', x'') = \delta_i^k \delta(x, x''). \tag{5.11}$$

If we now rewrite this same equation using condensed notation, it reads

$$D_{ij'} G^{j'k''} = \delta_i^{k''}, \tag{5.12}$$

where $\delta_i^{k''}$ is the condensed version of $\delta_i^k \delta(x, x'')$. (Note that the placement of the primes on the right-hand side of (5.12) is unambiguous given their placement on the left; both sides of the equation must have the same behavior under a general coordinate transformation.) The result in (5.12) may be interpreted, by analogy with the case of finite dimensional matrices, as stating that the Green function $G^{j'k''}$ is the inverse of the operator $D_{ij'}$. We are therefore justified in writing

$$G^{ij'} = (D^{-1})^{ij'}. \tag{5.13}$$

In keeping with the adoption of condensed notation, we will introduce

$$\delta M_{ij}^2(x, x') = \delta M_{ij}^2(x)\delta(x, x'), \tag{5.14}$$

with $\delta M_{ij'}^2$ representing the left-hand side in condensed notation. The result in (5.9) becomes

$$\delta W = \frac{i\hbar}{2} \int dv_x\, dv_{x'}\, \delta M_{ij}^2(x, x') G^{ji}(x', x) \tag{5.15}$$

if (5.14) is used. (Note that $\delta M_{ij}^2(x, x') = \delta M_{ji}^2(x', x)$ is symmetric, as is the Green function.) Using condensed notation, we can rewrite (5.15) as

$$\delta W = \frac{i\hbar}{2} \delta M_{ij'}^2 G^{j'i}$$
$$= \frac{i\hbar}{2} \mathrm{tr}[\delta M_{ij'}^2 G^{j'k''}]. \tag{5.16}$$

In the last line we have introduced the trace operator in an obvious way. If $A_i{}^{j'}$ is the condensed notation version of $A_i{}^j(x,x')$, then the trace is defined by

$$\operatorname{tr} A_i{}^{j'} = \int dv_x\, A_i{}^i(x,x). \tag{5.17}$$

(The right-hand side can equally well be written as $A_i{}^i$ in condensed notation.)

Because the metric is not being varied here, from (5.10) and (5.14) it is clear that

$$\delta D_{ij}(x,x') = \delta M_{ij}^2(x,x'). \tag{5.18}$$

Using the condensed notation version of (5.18), along with (5.13) in (5.16), leads to

$$\begin{aligned}\delta W &= \frac{i\hbar}{2}\operatorname{tr}\left[\delta D_{ij'}(D^{-1})^{j'k''}\right]\\ &= \delta\left[\frac{i\hbar}{2}\operatorname{tr}\ln\left(\ell^2 D_{ij'}\right)\right],\end{aligned} \tag{5.19}$$

since formally $A^{-1}\delta A = \delta\ln(\ell^2 A)$ for any operator A. Here ℓ is an arbitrary constant with dimensions of length inserted so that the argument of the logarithm is dimensionless, as it must be. The role of ℓ will be examined when we discuss the renormalization of the theory.

The result in (5.19) suggests the definition

$$W = \frac{i\hbar}{2}\operatorname{tr}\ln\left(\ell^2 D_{ij'}\right). \tag{5.20}$$

This result will be derived again in Chapter 6 in a more systematic way using the functional integral approach. By extending the matrix identity $\operatorname{tr}\ln M = \ln\det M$ to operators such as $D_{ij'}$, we may write (5.20) in the equivalent form

$$W = \frac{i\hbar}{2}\ln\det\left(\ell^2 D_{ij'}\right). \tag{5.21}$$

The result in (5.20) or (5.21) will be referred to as the one-loop effective action. (The terminology 'one-loop' will become more transparent in Chapter 6.) Either expression may be used to evaluate W, and we will illustrate how this is done later in the present chapter. The transition amplitude $\langle\text{out}|\text{in}\rangle$ now follows from (5.2) as

$$\langle\text{out}|\text{in}\rangle = [\det\left(\ell^2 D_{ij'}\right)]^{-1/2}. \tag{5.22}$$

Although we will obtain the general link between the Schwinger action principle and the Feynman path integral in the next chapter, contact with the path integral in the simple case we have been considering can be made directly from (5.22). To obtain this, consider first of all a real symmetric, non-degenerate matrix A. Let

$$(x, Ax) = x^i A_{ij} x^j, \tag{5.23}$$

5.2 Preliminary definition of the effective action

where the summation over i and j extends over the dimension of the matrix. Consider the Gaussian integral

$$I(A) = \int d^n x \exp\left\{\frac{i}{2}(x, Ax)\right\}, \tag{5.24}$$

with the integration extending over all space. Because A is symmetric, it can be diagonalized by an orthogonal transformation. We can write $A = M^T L M$ where $L_{ij} = L_i \delta_{ij}$ (no sum on i) is the matrix of eigenvalues of A. By changing variables $x \to M^T x$ in (5.24), and noting that because M is orthogonal the Jacobian of the transformation is unity, it may be seen easily that

$$I(A) = \int d^n x \exp\left\{\frac{i}{2}(x, Lx)\right\}$$

$$= \prod_j \int dx \exp\left(\frac{i}{2} L_j x^2\right)$$

$$= \prod_j \left(\frac{2\pi i}{L_j}\right)^{1/2},$$

using the standard result $\int_{-\infty}^{\infty} \exp(iax^2)dx = \left(\frac{i\pi}{a}\right)^{1/2}$. Because $\prod_j L_j = \det L = \det A$, we therefore have

$$I(A) = (2\pi i)^{n/2} (\det A)^{-1/2}. \tag{5.25}$$

This leads to the central result

$$\int d\mu(x) \exp\left\{\frac{i}{2}(x, Ax)\right\} = (\det A)^{-1/2}, \tag{5.26}$$

where

$$d\mu(x) = \prod_j \frac{dx^j}{(2\pi i)^{1/2}}. \tag{5.27}$$

If the result in (5.26) is extended to the infinite-dimensional case, then we may write

$$[\det(\ell^2 D_{ij'})]^{-1/2} = \int d\mu[\varphi] \exp\left\{-\frac{i}{2\hbar}\varphi^i D_{ij'} \varphi^{j'}\right\}, \tag{5.28}$$

where

$$d\mu[\varphi] = \prod_j \frac{d\varphi^j}{\ell(-2\pi i\hbar)^{1/2}}. \tag{5.29}$$

We can express the argument of the exponential in (5.28) in normal notation as follows. Using (5.10) we have

$$\frac{1}{2}\varphi^i D_{ij'}\varphi^{j'} = \frac{1}{2}\int dv_x dv_{x'} \varphi^i(x) D_{ij}(x,x')\varphi^j(x')$$
$$= \frac{1}{2}\int dv_x \varphi^i(x)\left(\delta_{ij}\Box_x + M_{ij}^2(x)\right)\varphi^j(x)$$
$$= -S[\varphi],$$

where in obtaining the last equality we have performed an integration by parts and then used the expression (5.4) for the action. We may therefore write (5.22) as

$$\langle \text{out}|\text{in}\rangle = \int d\mu[\varphi]\exp\left\{\frac{i}{\hbar}S[\varphi]\right\}, \quad (5.30)$$

which gives the Feynman path integral expression for the transition amplitude. This equation is often adopted as the basic starting point for quantum field theory.

5.3 Regularization of the effective action

In the last section we obtained in (5.21) a result for the effective action which read

$$W = \frac{i\hbar}{2}\ln\det(\ell^2 D_{ij'}), \quad (5.31)$$

where $D_{ij'}$ was a differential operator. Suppose that we now try to evaluate this expression in the simple case of a massive scalar field in flat n-dimensional spacetime. In this case $D_{ij'} = (\Box_x + m^2)\delta(x,x')$. We are now faced with the task of evaluating the determinant of this differential operator.

If $D_{ij'}$ was a finite dimensional matrix, then we could define its determinant in a number of different ways. The most useful way for our present situation is to make use of its eigenvalues and use the fact that the determinant of any Hermitian matrix is invariant under a unitary transformation. We define the eigenvalue problem by

$$D_{ij'}f_n^{j'} = \lambda_n f_{ni}, \quad (5.32)$$

with f_n^i the normalized (possibly complex) eigenvector. We assume that the set of all eigenvectors forms a complete orthonormal set: $f_n^i f_{n'i}^* = \delta_{nn'}$. We can then write

$$\det D_{ij'} = \prod_n \lambda_n, \quad (5.33)$$

with the product extending over all eigenvalues. Using this definition in (5.31) above results in

$$W = \frac{i\hbar}{2}\sum_n \ln(\ell^2 \lambda_n). \quad (5.34)$$

5.3 Regularization of the effective action

For flat n-dimensional spacetime, we can impose box normalization by assuming that the eigenvectors obey periodic boundary conditions on the box walls. Let L_μ represent the length of the box in the direction specified by the spacetime coordinate x^μ. The eigenvalue problem (5.32) becomes

$$(\Box_x + m^2) f_n(x) = \lambda_n f_n(x) \tag{5.35}$$

when transcribed from condensed notation. With our choice of box normalization, the normalized eigenvector is

$$f_n(x) = (L_0 L_1 \cdots L_{n-1})^{-1/2} \exp\left\{ 2\pi i \sum_{\nu=0}^{n-1} \left(\frac{n_\nu x^\nu}{L_\nu} \right) \right\}, \tag{5.36}$$

where $\mathbf{n} = (n_0, n_1, \ldots, n_{n-1})$ with $n_\nu = 0, \pm 1, \pm 2, \ldots$ by virtue of the periodic boundary conditions assumed. The eigenvalues λ_n appearing in (5.35) are given by

$$\lambda_\mathbf{n} = -\left(\frac{2\pi n_0}{L_0} \right)^2 + \sum_{j=1}^{n-1} \left(\frac{2\pi n_j}{L_j} \right)^2 + m^2. \tag{5.37}$$

If we take the large box limit ($L_\mu \to \infty$), then the sums over $n_0, \ldots, n_{n-1}$ which result when (5.37) is used in (5.34) may be approximated with integrals. By making the change of variables $n_\mu \to \frac{L_\mu}{2\pi} k_\mu$ in the resulting expression, we arrive at

$$W = \frac{i\hbar}{2} (L_0 L_1 \cdots L_{n-1}) \int \frac{d^n k}{(2\pi)^n} \ln[\ell^2(-k^2 + m^2)]. \tag{5.38}$$

The factors of $L_0 L_1 \cdots L_{n-1}$ may be associated with the spacetime volume $\int dv_x$, so that we may write W in the form

$$W = \frac{i\hbar}{2} \int dv_x \int \frac{d^n k}{(2\pi)^n} \ln[\ell^2(-k^2 + m^2)]. \tag{5.39}$$

The result just obtained is still not satisfactory since any attempt to evaluate it results in an infinite answer. There are two possible sources for the infinity. In the first place, there is no x-dependence in the last integral, and so $\int dv_x$ is infinite in the case of Minkowski spacetime. This infinity is easy to deal with by defining the effective potential density V by

$$W = -\int dv_x V. \tag{5.40}$$

The minus sign in this definition is for conformity with that in the classical action; namely, $S = \int dv_x (T - V)$, where T is the field kinetic energy density and V is the potential energy density. The effective potential density is then given by

$$V = -\frac{i\hbar}{2} \int \frac{d^n k}{(2\pi)^n} \ln[\ell^2(-k^2 + m^2)]. \tag{5.41}$$

This result for the effective potential density is still seen to be infinite because the integral over k diverges. This infinity must be dealt with by regularization. We will discuss several methods of regularization in this section.

In order to apply regularization to the effective action (5.31), it is convenient to write it in another form. If we change the operator D to $D + \delta D$, then the change in the effective action is (to first order in δD)

$$\delta W = \frac{i\hbar}{2} \mathrm{tr}(D^{-1} \delta D). \tag{5.42}$$

D is assumed to have a small negative imaginary part by virtue of Feynman boundary conditions. We may therefore write

$$D^{-1} = i \int_0^\infty d\tau \exp(-i\tau D). \tag{5.43}$$

Note that τ must have dimensions of length squared if the argument of the exponential is to be dimensionless. It is then seen from (5.42) that

$$\delta W = -\frac{i\hbar}{2} \delta \int_0^\infty \frac{d\tau}{\tau} \mathrm{tr} \exp(-i\tau D). \tag{5.44}$$

This suggests the definition

$$W = -\frac{i\hbar}{2} \int_0^\infty \frac{d\tau}{\tau} \mathrm{tr} \exp(-i\tau D). \tag{5.45}$$

Another way to obtain (5.45) is to begin with the identity

$$\ln \frac{\alpha}{\beta} = -\int_0^\infty \frac{d\tau}{\tau} [\exp(-i\tau\alpha) - \exp(-i\tau\beta)], \tag{5.46}$$

which assumes α, β to have small negative imaginary parts. The operator version of this is

$$\ln\left(\frac{D}{D_0}\right) = -\int_0^\infty \frac{d\tau}{\tau} [\exp(-i\tau D) - \exp(-i\tau D_0)] \tag{5.47}$$

for two operators D, D_0. Taking $D_0 = \ell^{-2} I$, where ℓ is an arbitrary constant with dimensions of length, (5.45) is regained provided that the term independent of D is dropped. This was also done in integrating (5.44) to get (5.45).

By evaluating W as expressed in (5.45) in a particular way, we can make contact with a very powerful method of analysis originally introduced into quantum field theory by Schwinger (1951a) in a special example, and later developed extensively by DeWitt (1964a).

Because the exponential of any operator is defined through its Taylor series expansion, if we make use of the eigenvalue problem (5.32) it is clear that

$$\mathrm{tr}\left(\exp(-i\tau D)\right) = \sum_n e^{-i\tau \lambda_n}. \tag{5.48}$$

5.3 Regularization of the effective action

Suppose that we now define

$$K^i{}_j(\tau;x,x') = \sum_n e^{-i\tau\lambda_n} f_n{}^i(x) f^*_{nj}(x'), \qquad (5.49)$$

where we assume the eigenfunctions are normalized by

$$\int dv_x f_n{}^i(x) f^*_{n'i}(x) = \delta_{nn'}. \qquad (5.50)$$

(This latter equation corresponds to $f_n{}^i f^*_{n'i} = \delta_{nn'}$ in condensed notation written down below (5.32).) Then from (5.48) and (5.49) we find

$$\operatorname{tr}(\exp(-i\tau D)) = \int dv_x K^i{}_i(\tau;x,x). \qquad (5.51)$$

If (5.51) is now used in (5.45), we have

$$W = -\frac{i\hbar}{2} \int dv_x \int_0^\infty \frac{d\tau}{\tau} \operatorname{tr} K(\tau;x,x), \qquad (5.52)$$

where the trace on the right-hand side is over the field indices i and j carried by $K^i{}_j(\tau;x,x')$ defined in (5.49).

The importance of $K^i{}_j(\tau;x,x')$ resides in the fact that it obeys the equation

$$i\frac{\partial}{\partial \tau} K^i{}_k(\tau;x,x') = D^i{}_j K^j{}_k(\tau;x,x'), \qquad (5.53)$$

which resembles a Schrödinger equation with $D^i{}_j$ playing the role of the Hamiltonian operator. (This follows very simply from the definition in (5.49) and the eigenvalue equation (5.32).) By letting $\tau \to 0$ in (5.49), and making the assumption that the eigenfunctions form a complete set, it follows that

$$K^i{}_j(\tau=0;x,x') = \delta^i{}_j \delta(x,x'). \qquad (5.54)$$

This may be regarded as the boundary condition for (5.53). $K^i{}_j(\tau;x,x')$ is often referred to as the heat kernel, since if we analytically continue τ to imaginary values the Schrödinger equation becomes the heat equation.[2]

The importance of the kernel $K(\tau;x,x')$ lies in the fact that for small τ it has an asymptotic expansion of the form

$$K(\tau;x,x) \sim i(4\pi i\tau)^{-n/2} \sum_{k=0}^\infty (i\tau)^k E_k(x), \qquad (5.55)$$

where $E_k(x)$ are local quantities constructed from a knowledge of the operator D_x. For any operator of the form

$$D_x = g^{\mu\nu}(x) \nabla_\mu \nabla_\nu + Q(x), \qquad (5.56)$$

---

[2] See (5.79) and (5.80) later in this chapter.

where ∇_μ is any covariant derivative (which includes gauge potentials or connections due to spacetime curvature) and $Q(x)$ does not contain any derivatives, the first few coefficients are[3]

$$E_0(x) = I, \tag{5.57}$$

$$E_1(x) = \frac{1}{6}RI - Q, \tag{5.58}$$

$$E_2(x) = \left(-\frac{1}{30}\Box R + \frac{1}{72}R^2 - \frac{1}{180}R^{\mu\nu}R_{\mu\nu} + \frac{1}{180}R^{\mu\nu\rho\sigma}R_{\mu\nu\rho\sigma}\right)I$$
$$+ \frac{1}{12}W^{\mu\nu}W_{\mu\nu} + \frac{1}{2}Q^2 - \frac{1}{6}RQ + \frac{1}{6}\Box Q, \tag{5.59}$$

where

$$W_{\mu\nu} = [\nabla_\mu, \nabla_\nu]. \tag{5.60}$$

For the case of spin-0 and spin-1/2 fields an equivalent expansion of the Green function (which we will see later is related to $K(\tau; x, x')$ in a simple way) was given in Chapter 3. Additional terms may arise in the case where boundaries are present.[4]

Armed with these results, we can now return to analyze the divergences of the effective action expressed in (5.52). It is evident from the simple flat spacetime example, considered at the beginning of this section, that the interesting divergences are contained in the integration over τ in (5.52). For large τ, because D has a negative imaginary part, $K(\tau; x, x)$ is expected to decay exponentially fast. It would then be expected that any divergences should come only from the $\tau = 0$ limit in the integral. If **divp** is used to denote the divergent part of any expression, then

$$\mathbf{divp}\, W = -\frac{i\hbar}{2} \int dv_x\, \mathbf{divp} \int_0^{\tau_0} \frac{d\tau}{\tau} \mathrm{tr} K(\tau; x, x) \tag{5.61}$$

for some constant τ_0. Because the divergence comes only from $\tau = 0$, we may take τ_0 to be as small as we like, in which case we can use the asymptotic expansion of the kernel given in (5.55). This leads to

$$\mathbf{divp}W = \frac{i\hbar}{2}(4\pi)^{-n/2} \int dv_x \sum_{k=0}^{\infty} \mathrm{tr} E_k(x)\, \mathbf{divp} \int_0^{\tau_0} d\tau (i\tau)^{k-1-n/2}. \tag{5.62}$$

The divergent part of W has been isolated in a very simple integral. The only terms in the summation which can contain divergences are those with $k \leq \frac{n}{2}$. In order to obtain an explicit form for the divergent part of W, it is necessary to adopt a regularization technique. We will now discuss some such techniques.

---

[3] See DeWitt (1964a); Gilkey (1975, 1979) for example.
[4] See Kirsten (2002) and references therein for a review.

5.3.1 Regularization by a cut-off

Because the divergent part of W comes from $\tau = 0$, one simple method of regularization consists of cutting off the integration over τ at some lower limit $\tau = \tau_c$. The divergence will occur if we try to take the limit $\tau_c \to 0$. Provided that $k \neq n/2$,

$$\int_{\tau_c}^{\tau_0} d\tau (i\tau)^{k-1-n/2} = (i)^{k-1-n/2} \frac{\tau_0^{k-n/2} - \tau_c^{k-n/2}}{k - n/2}. \tag{5.63}$$

If $k = n/2$, then

$$\int_{\tau_c}^{\tau_0} d\tau (i\tau)^{-1} = -i \ln(\tau_0/\tau_c). \tag{5.64}$$

If $k \geq n/2$ then the limit $\tau_c \to 0$ may be taken in (5.63). We therefore conclude, from (5.63) that for even spacetime dimensions (n even),

$$\mathbf{divp}W = -\frac{\hbar}{2}(4\pi)^{-n/2} \int dv_x \sum_{k=0}^{\frac{n}{2}-1} \mathrm{tr} E_k(x)(i\ell^2)^{k-\frac{n}{2}} \frac{\tau_c^{k-\frac{n}{2}}}{k-\frac{n}{2}}$$

$$- \frac{\hbar}{2}(4\pi)^{-n/2} \int dv_x \mathrm{tr} E_{n/2}(x) \ln \tau_c. \tag{5.65}$$

For odd spacetime dimensions (n odd),

$$\mathbf{divp}W = -\frac{\hbar}{2}(4\pi)^{-n/2} \int dv_x \sum_{k=0}^{[\frac{n}{2}]} \mathrm{tr} E_k(x)(i\ell^2)^{k-\frac{n}{2}} \frac{\tau_c^{k-\frac{n}{2}}}{k-\frac{n}{2}}, \tag{5.66}$$

where $[x]$ is the largest integer less than or equal to x. In both cases, the divergent part of W has been isolated in a finite number of terms, all of which may be found for a given operator D. The difference between the cases of even-dimensional and odd-dimensional spacetimes is seen to reside in the presence or absence of the factor of $\ln \tau_c$. The significance of this will be discussed later. The relation with the adiabatic subtraction method described in Section 3.1 in the case $n = 4$ should be evident, given that the heat kernel coefficient E_n is of adiabatic order $2n$.

5.3.2 Dimensional regularization

Another method of regularization consists of regarding the spacetime dimension n as a complex variable, evaluating a given expression in a region of the complex n-plane where the expression is finite, and then performing an analytic continuation of n back to the physical spacetime dimension. The divergences of the original expression are isolated as poles in the complex n-plane. This method, called dimensional regularization, was first discussed by Ashmore (1972); Bollini and Giambiagi (1972), and most extensively by 't Hooft and Veltman (1972).

We will take the extra dimensions to be flat. The fact that this choice is completely arbitrary is sometimes regarded as a weak point in the application of dimensional regularization to general spacetimes. Our view is that the purpose of any regularization technique is to make things finite in as simple a way as possible, consistent with any symmetries manifest in the formulation of the theory. (In some cases it is not possible to preserve all of the symmetries, as we will see later in the chapter.)

If we take $\Re(n) < 0$, then

$$\int_0^{\tau_0} d\tau (i\tau)^{k-1-n/2} = (i)^{k-1-n/2} \frac{\tau_0^{k-n/2}}{k-n/2}. \tag{5.67}$$

We can analytically continue n to $\Re(n) \geq 0$ since (5.67) only has simple poles at $n = 2k, k = 0, 1, 2, \ldots$. It follows that if n is odd, then $\mathbf{divp}W = 0$; whereas if n is even,

$$\mathbf{divp}W = -\frac{\hbar}{\epsilon}(4\pi)^{-n/2} \int dv_x \mathrm{tr} E_{n/2}(x), \tag{5.68}$$

where we have replaced n with $n + \epsilon$, with $\epsilon \to 0$ understood. The divergent part of W in this case is seen to take the form of a simple pole.

Comparison of (5.68) with (5.65) shows that the coefficient of the pole term in dimensional regularization is the same as that of $\ln \tau_c^{1/2}$ in the cut-off method. In the case of odd-dimensional spacetimes there is no dependence on $\ln \tau_c^{1/2}$ in the cut-off method and no pole term present in dimensional regularization. The actual form taken for the divergences in the two methods of regularization is seen to differ.

5.3.3 ζ-function regularization

A view is often taken that quantum field theory should be defined on a Riemannian rather than a pseudo-Riemannian spacetime. This view is based on mathematical rather than physical grounds, and is advocated because many of the operators encountered become elliptic rather than hyperbolic. Elliptic operators on Riemannian manifolds have the desirable property of having a spectrum which is bounded from below. For example, in flat Euclidean space, the Laplacian $-\nabla_\mu \nabla^\mu$ acting on scalars has eigenvalues $k^2 \geq 0$. In flat Minkowski spacetime, $\Box = \nabla_\mu \nabla^\mu$ has eigenvalues $-(k^0)^2 + \mathbf{k} \cdot \mathbf{k}$ which have no definite sign. The Feynman boundary conditions adopted in Minkowski spacetime quantum field theory ensure that the Wick rotation $k^0 \to ik^0$ may be performed leading to a result which is equivalent to using elliptic operators in a Euclidean spacetime. A rigorous proof of the equivalence between Minkowski and Euclidean field theory may be given.[5] Unfortunately the situation in curved spacetime is more complicated.

---

[5] See Osterwalder and Schrader (1973, 1975) for example.

5.3 Regularization of the effective action

If the flat spacetime Wick rotation $k^0 \to ik^0$ is rephrased in terms of coordinates, it becomes $t \to -it$. This may not lead to a Riemannian metric in general. (A simple example where it does not is $ds^2 = -dt^2 + t^2 dx^2 + dy^2 + dz^2$.) In addition, viewing the Wick rotation as $t \to -it$ is a coordinate-dependent statement. If the spacetime is static, then there always exists a local coordinate system in which the metric is time-independent with $ds^2 = g_{00}(\mathbf{x})dt^2 + g_{ij}(\mathbf{x})dx^i dx^j$. In this restricted case, $t \to -it$ leads to a Riemannian metric. There is always a Wick rotation for stationary spacetimes which leads to a Riemannian metric. For this section we will assume either that attention is restricted to spacetimes which lead to a Riemannian metric when a Wick rotation is performed, or that the quantum field theory is formulated in a Riemannian spacetime from the start.

Suppose that the theory is described by a real scalar field $\varphi(x)$ whose action is

$$S = \frac{1}{2}\int dv_x \varphi(x) D_x \varphi(x), \tag{5.69}$$

where $D_x = -\Box_x + M^2(x)$. This is the Riemannian counterpart of the theory considered in the previous section. (The metric is chosen to be positive definite in this section. This means that because of our signature in the Lorentzian spacetime of $(+ - \cdots -)$, we must replace the d'Alembertian with its negative.) $M^2(x)$ is assumed to be non-negative. In Riemannian space we define the effective action W by

$$\exp\left(-\frac{1}{\hbar}W\right) = \int d\mu[\varphi] \exp\left(-\frac{1}{\hbar}S\right), \tag{5.70}$$

which gives

$$W = \frac{\hbar}{2} \ln \det(\ell^2 D). \tag{5.71}$$

The choice of signs in (5.70) follows without ambiguity from the Wick rotation of the spacetime to one with a positive definite Riemannian signature, and ensures that the expression for the effective action agrees with the one found in a Lorentzian spacetime.

We may regard $\det(\ell^2 D)$ as being defined in terms of the eigenvalues of D as mentioned above. If λ_N denotes the eigenvalues of D, then

$$W = \frac{\hbar}{2} \sum_N \ln(\ell^2 \lambda_N). \tag{5.72}$$

In order to give meaning to this expression, consider the ζ-function for D defined by

$$\zeta(s) = \sum_N \lambda_N^{-s}. \tag{5.73}$$

Then $\zeta'(s) = -\sum_N \lambda_N^{-s} \ln \lambda_N$. This suggests defining W by

$$W = \frac{\hbar}{2} \lim_{s \to 0} \sum_N \lambda_N^{-s} \ln(\ell^2 \lambda_N)$$

$$= \frac{\hbar}{2}\{-\zeta'(0) + \zeta(0) \ln \ell^2\}, \tag{5.74}$$

provided that $\zeta'(s)$ and $\zeta(s)$ are finite at $s = 0$. The use of generalized ζ-functions to regularize functional determinants was given by Ray and Singer (1973). It was first used in quantum field theory calculations in different versions by Dowker and Critchley (1976) and Hawking (1977).

To study the behavior of $\zeta(s)$ we proceed as follows. Write

$$\lambda_N^{-s} = \frac{1}{\Gamma(s)} \int_0^\infty \tau^{s-1} \exp(-\tau \lambda_N) d\tau,$$

which is valid for $\Re(s) > 0$ using the integral representation of the Γ-function. Then

$$\zeta(s) = \frac{1}{\Gamma(s)} \int_0^\infty \tau^{s-1} \sum_N \exp(-\tau \lambda_N) d\tau. \tag{5.75}$$

Let $\varphi_N(x)$ denote the eigenfunctions of D normalized by

$$\int dv_x \varphi_N^*(x) \varphi_{N'}(x) = \delta_{NN'}. \tag{5.76}$$

Then

$$\sum_N \exp(-\tau \lambda_N) = \operatorname{tr} \exp(-\tau D)$$

$$= \int dv_x \operatorname{tr} K(\tau; x, x), \tag{5.77}$$

with

$$K(\tau; x, x') = \sum_N \exp(-\tau \lambda_N) \varphi_N(x) \varphi_N^*(x'). \tag{5.78}$$

This result gives the Riemannian counterpart of the earlier result in Lorentzian space in (5.49), and we will refer to $K(\tau; x, x')$ as the kernel for the operator D. The kernel satisfies

$$-\frac{\partial}{\partial \tau} K(\tau; x, x') = D_x K(\tau; x, x') \tag{5.79}$$

in place of (5.53). This equation is analogous to the diffusion, or heat, equation, and for this reason $K(\tau; x, x')$ is sometimes called the heat kernel. The boundary condition for (5.79) is

$$K(\tau = 0; x, x') = \delta(x, x'). \tag{5.80}$$

5.3 Regularization of the effective action

A solution for the heat kernel was written down in (5.78). The boundary condition (5.80) follows from the assumption that the eigenfunctions form a complete set. These results are the Riemannian counterparts of (5.48)–(5.51), (5.53) and (5.54).

We can now use these results to analyze the effective action W. For large τ we have
$$\sum_N \exp(-\tau \lambda_N) \sim \exp(-\tau \lambda_0).$$

If we split up the integration over τ in (5.75), and define
$$F_1(s) = \int_0^{\tau_0} \tau^{s-1} \sum_N \exp(-\tau \lambda_N) d\tau, \qquad (5.81)$$

$$F_2(s) = \int_{\tau_0}^{\infty} \tau^{s-1} \sum_N \exp(-\tau \lambda_N) d\tau, \qquad (5.82)$$

it follows that $F_2(s)$ is an analytic function of s for any $\tau_0 > 0$ provided that $\lambda_0 > 0$, which we will assume here. By taking τ_0 sufficiently small, we can use the asymptotic expansion
$$K(\tau; x, x') \sim (4\pi\tau)^{-n/2} \sum_{k=0}^{\infty} \tau^k E_k(x), \qquad (5.83)$$

which is the Riemannian counterpart to (5.55). If we assume $\Re(s) > n/2$, then it is easily seen from (5.78) that
$$F_1(s) = (4\pi)^{-n/2} \sum_{k=0}^{\infty} \frac{\tau_0^{s+k-n/2}}{s+k-n/2} \int dv_x \, \mathrm{tr}\, E_k(x). \qquad (5.84)$$

This result allows us to deduce that $\zeta(s)$ is an analytic function of s except at $s = n/2 - k$ for $k = 0, 1, 2, \ldots$ where simple poles occur. We may therefore define $\zeta(s)$ for $\Re(s) < n/2$ by analytic continuation. For s in a neighborhood of $s = -k$, with $k = 0, 1, 2, \ldots$, we have the expansion $\Gamma(s) = \frac{(-1)^k}{k!}(s+k)^{-1} + \cdots$, where the ellipsis indicates terms which are finite at $s = -k$. This result is easily confirmed by repeatedly using the recursion relation $\Gamma(s) = \Gamma(s+1)/s$. We find $(k = 0, 1, \ldots)$
$$\zeta(-k) = (-1)^k k! (4\pi)^{-n/2} \int dv_x \, \mathrm{tr}\, E_{k+n/2}(x). \qquad (5.85)$$

Note that the presence of the Γ-function in (5.75) ensures that all terms which are analytic functions of s in the higher-order expansion of the heat kernel vanish. In particular, if the spacetime dimension n is an odd integer, we find that $\zeta(-k) = 0$. (If there is a boundary to the spacetime then boundary terms will be present in $\zeta(-k)$.) By taking $k = 0$, we have an evaluation of $\zeta(0)$ which multiplies the $\ln \ell^2$ term in W. There is no analogously simple result for $\zeta'(0)$. It can be observed that $\zeta(0)$ is the analogue of the simple pole term in W found using dimensional regularization.

5.4 Effective action for scalar fields: Some examples

5.4.1 Scalar field in a constant background electromagnetic field

In order to consider a practical use for the kernel, we will look at an example where it is possible to obtain an explicit expression for it. The example we will consider is a spin-0 particle in a constant electromagnetic field background. This was first evaluated by Schwinger (1951a), and the method we use is a slight modification of his.

The operator representing a spin-0 field in an electromagnetic background described by the vector potential A_μ is $\eta^{\mu\nu} D_\mu D_\nu + m^2$ where $D_\mu = \partial_\mu - ieA_\mu$ is the gauge-covariant derivative. The field strength tensor is $F_{\mu\nu} = \partial_\mu A_\nu - \partial_\nu A_\mu$, and we will assume that $F_{\mu\nu}$ is constant here. The spacetime will be taken to be flat n-dimensional Minkowski spacetime. We will take the vector potential describing the constant background field to be

$$A_\mu = -\frac{1}{2} F_{\mu\nu} x^\nu + a_\mu, \tag{5.86}$$

where a_μ is an arbitrary constant. In Minkowski spacetime, the constant a_μ may always be removed by a gauge transformation. (This will be discussed in the following section.) However, for later reasons, we will leave it in at this stage.

We have previously expressed the effective action in terms of the kernel $K(\tau; x, x')$ associated with a differential operator. (See Section 5.3.) The kernel obeys the differential equation

$$i\frac{\partial}{\partial \tau} K(\tau; x, x') = (D^\mu D_\mu + m^2) K(\tau; x, x'), \tag{5.87}$$

in the present case with the boundary condition

$$K(\tau = 0; x, x') = \delta(x, x'). \tag{5.88}$$

As we remarked in Section 5.3, the differential equation (5.87) resembles a Schrödinger equation with the operator, in this case $D^\mu D_\mu + m^2$ playing the role of the Hamiltonian operator, and τ playing the role of the time. τ is sometimes (Schwinger 1951a) called the proper time. The similarity between (5.87) and the Schrödinger equation leads to the interpretation of $K(\tau; x, x')$ as the probability amplitude for a particle located at position x'^μ at proper time $\tau = 0$ to propagate to position x^μ at proper time τ. This interpretation was exploited to great effect by Schwinger (1951a), who evaluated the transition amplitude by setting up and solving the Heisenberg equations of motion for the quantum mechanical problem. We will adopt a different method here.

The standard way of constructing the quantum mechanical Hamiltonian operator from a classical Hamiltonian, in the Schrödinger picture, consists of the replacement of the classical momentum p_μ with the differential operator $-i\partial_\mu$. The classical Hamiltonian corresponding to the Hamiltonian operator on the right-hand side of (5.87) can be seen to be

5.4 Effective action for scalar fields: Some examples

$$H = [-\eta^{\mu\nu}(p_\mu - eA_\mu)(p_\nu - eA_\nu) + m^2]. \tag{5.89}$$

The classical Hamiltonian equation of motion $\dot{x}^\mu = \frac{\partial H}{\partial p_\mu}$ results in

$$p^\mu = -\frac{1}{2}\dot{x}^\mu + eA^\mu.$$

The Lagrangian then follows from

$$L = p_\mu \dot{x}^\mu - H$$
$$= -\frac{1}{4}\eta_{\mu\nu}\dot{x}^\mu \dot{x}^\nu + eA_\mu \dot{x}^\mu - m^2. \tag{5.90}$$

The action functional for this Lagrangian is then

$$S[x] = \int_0^\tau d\tau' L(\tau'), \tag{5.91}$$

where L is regarded as a function of proper time through $x^\mu(\tau)$ in (5.90).

At this stage we will make a departure from Schwinger's approach and instead use the fact that because the kernel $K(\tau; x, x')$ may be interpreted as a transition amplitude, it may be expressed as a Feynman path integral (Feynman and Hibbs 1965). The integration extends over all paths which begin at x'^μ and end at x^μ:

$$K(\tau; x, x') = \int_{x^\mu(0)=x'^\mu}^{x^\mu(\tau)=x^\mu} [dx] \exp(iS[x]). \tag{5.92}$$

(We set $\hbar = 1$ here.) Now let $x_{cl}(\tau')$ denote the solution to the classical equations of motion for the theory whose Lagrangian is given by (5.90). We may write $x^\mu(\tau') = x_{cl}^\mu(\tau') + z^\mu(\tau')$ where $z^\mu(0) = z^\mu(\tau) = 0$. The important feature of the constant electromagnetic field is that L in (5.90) is only quadratic in the coordinates and velocities. (See (5.86).) This means that when we expand $S[x]$ about a classical solution,

$$S[x] = S[x_{cl}] + S_2[z],$$

where $S_2[z]$ depends quadratically on z; the term of first order in z vanishes because x_{cl} is a stationary point of the action. We then have

$$K(\tau; x, x') = \exp(iS_{cl}) \int_{z^\mu(0)=0}^{z^\mu(\tau)=0} [dz] \exp(iS_2[z]),$$

where $S_{cl} = S[x_{cl}]$ is the action evaluated for the classical path. The remaining functional integral extends over all paths $z^\mu(\tau')$ with $z^\mu(0) = z^\mu(\tau) = 0$. It is possible to actually evaluate this path integral, but a simpler approach (Feynman and Hibbs 1965) is to note that whatever the path integral is, because of the boundary conditions on $z^\mu(\tau')$, it cannot depend on the initial or final points x'^μ

and x^μ, but only on τ and parameters appearing in L. The kernel function must therefore take the form

$$K(\tau; x, x') = f(\tau) \exp(iS_{cl}) \tag{5.93}$$

for some function $f(\tau)$. The function $f(\tau)$ can be determined from the knowledge that the kernel satisfies the Schrödinger equation (5.87) with boundary condition (5.88).

The first task is to find the solution to the Euler–Lagrange equations of motion. These equations follow from (5.90) as

$$\ddot{x}^\mu = -2eF^\mu{}_\nu \dot{x}^\nu. \tag{5.94}$$

Although (5.94) is true for general $F^\mu{}_\nu$, because $F^\mu{}_\nu$ is constant in our case it is possible to solve (5.94) exactly. To do this, it is useful to adopt a matrix notation in which $x(\tau')$ represents a column vector with components $x^\mu(\tau')$, and F represents a matrix with components $F^\mu{}_\nu$. Then (5.94) may be written in matrix form as

$$\ddot{x}(\tau') = -2eF\dot{x}(\tau'). \tag{5.95}$$

With this matrix notation, $F^T = -F$ and x^T has components x_μ. The solution to (5.95) is

$$x_{cl}(\tau') = -\frac{1}{2e} F^{-1} \exp(-2e\tau' F) u + v, \tag{5.96}$$

where u and v are constant vectors which may be determined from the boundary conditions $x(0) = x'$, $x(\tau) = x$ to be

$$u = (\exp(-2e\tau F) - 1)^{-1} 2eF(x' - x), \tag{5.97}$$
$$v = (1 - \exp(2e\tau F))^{-1} (x' - \exp(2e\tau F)x). \tag{5.98}$$

It is then straightforward to show that the classical action is given by

$$S_{cl} = \frac{e}{2}(x' - x)^T F \left(\exp(2e\tau F) - 1\right)^{-1} \left(\exp(2e\tau F)x - x'\right)$$
$$+ ea^T(x - x') - m^2 \tau. \tag{5.99}$$

The requirement that (5.93) satisfy (5.87) leads to, after some calculation,

$$\frac{\partial}{\partial \tau} \ln f(\tau) = -e\,\mathrm{tr}\left[F\left(1 - \exp(-2e\tau F)\right)^{-1}\right].$$

Because[6] $\mathrm{tr}F = 0$, the right-hand side of this equation may be rewritten giving

$$\frac{\partial}{\partial \tau} \ln f(\tau) = -\frac{1}{2}\frac{\partial}{\partial \tau}\mathrm{tr}\ln\sinh(e\tau F),$$

---

[6] Note that $F_{\mu\nu} = -F_{\nu\mu}$.

5.4 Effective action for scalar fields: Some examples

from which it follows that

$$f(\tau) = C \exp\left(-\frac{1}{2}\mathrm{tr}\ln\sinh(e\tau F)\right) \tag{5.100}$$

for some constant C. This may be rewritten in the alternate form

$$f(\tau) = C\left[\det\left(\sinh(e\tau F)\right)\right]^{-1/2} \tag{5.101}$$

if the identity $\mathrm{tr}\ln = \ln\det$ is used.

We now have an expression for the kernel determined up to an overall multiplicative constant, and this constant may be calculated from the boundary condition (5.88). Because $\lim_{\tau\to 0} K(\tau; x, x') = \delta(x, x')$ we must have

$$1 = \lim_{\tau\to 0} \int d^n x' K(\tau; x, x').$$

Making the change of variable $x'^\mu = x^\mu + (4\tau)^{1/2} z^\mu$ in this integral and using (5.99) and (5.101) lead to

$$1 = 2^n C \left[\det(eF)\right]^{-1/2} \int d^n z \exp(-i\eta_{\mu\nu} z^\mu z^\nu) \tag{5.102}$$

when the limit $\tau \to 0$ is taken. Using

$$\int_{-\infty}^{+\infty} dx \exp(\pm i x^2) = \pi^{1/2} \exp\left(\pm i\frac{\pi}{4}\right), \tag{5.103}$$

which may be proven either by contour integration or by inserting the convergence factor $\exp(-\epsilon x^2)$ in the integrand with $\epsilon \to 0$ taken at the end, it follows that

$$\int d^n z \exp(-i\eta_{\mu\nu} z^\mu z^\nu) = -i\pi^{n/2} \exp\left(i\frac{\pi n}{4}\right). \tag{5.104}$$

We therefore see that

$$C = i\left[\det\left(\frac{eF}{4\pi i}\right)\right]^{1/2}, \tag{5.105}$$

if we place the factors of π and i which arise from the use of (5.104) inside of the determinant. This determines $f(\tau)$ in (5.101) to be

$$f(\tau) = i\left[\det\left(\frac{\left(\frac{eF}{4\pi i}\right)}{\sinh(e\tau F)}\right)\right]^{1/2}, \tag{5.106}$$

and the calculation of the kernel is complete.

Before calculating the one-loop effective action for a spin-0 field in an electromagnetic field background, we will examine first the simpler case when there is no electromagnetic field present. Letting $F \to 0$ in (5.106) leads to

$$f(\tau) \to i(4\pi i \tau)^{-n/2}. \tag{5.107}$$

Taking $F \to 0, a \to 0$ in the expression for the classical action (5.99) results in

$$S_{cl} \to -\frac{1}{4\tau}\eta_{\mu\nu}(x^\mu - x'^\mu)(x^\nu - x'^\nu) - m^2\tau. \tag{5.108}$$

The kernel for a free scalar field in n-dimensions is then

$$K_{\text{free}}(\tau; x, x') = i(4\pi i\tau)^{-n/2} \exp\left(-\frac{i}{4\tau}(x-x')^2 - im^2\tau\right), \tag{5.109}$$

where $(x-x')^2 = \eta_{\mu\nu}(x^\mu - x'^\mu)(x^\nu - x'^\nu)$. We will now show that this leads to the standard expression for the Feynman Green function.

The Green function is given in terms of the kernel by

$$G(x, x') = i\int_0^\infty d\tau\, K_{\text{free}}(\tau; x, x'). \tag{5.110}$$

The standard momentum-space expression is obtained by using the identity

$$\exp\left(-\frac{i}{4\tau}(x-x')^2\right) = -i(4\pi i\tau)^{n/2}\int\frac{d^n k}{(2\pi)^n}\exp\left[i\tau k^2 + ik_\mu(x^\mu - x'^\mu)\right]. \tag{5.111}$$

(This result is proved by completing the square in the argument of the exponential on the right-hand side, making a change of variables, and using (5.104).) The integration over proper time τ may then be performed using

$$\int_0^\infty d\tau \exp\left(i\tau(k^2 - m^2)\right) = \frac{i}{(k^2 - m^2)}, \tag{5.112}$$

which is valid since m^2 is understood to have a small negative imaginary part. We then obtain

$$G(x, x') = -\int \frac{d^n k}{(2\pi)^n}\frac{\exp(ik_\mu(x^\mu - x'^\mu))}{(k^2 - m^2)}, \tag{5.113}$$

which is the standard expression for the Feynman Green function. (The minus sign on the right-hand side of (5.113) is because we have defined $G(x,x')$ to satisfy $(\Box_x + m^2)G(x,x') = \delta(x,x')$.)

We now return to the problem of evaluating the one-loop effective action for the spin-0 particle in a constant background electromagnetic field. The one-loop effective action is determined from (see (5.52))

$$W = -i\hbar \int dv_x \int_0^\infty \frac{d\tau}{\tau} K(\tau; x, x). \tag{5.114}$$

There is a factor of 2 difference with our previous result for the effective action because our earlier consideration was for a real scalar field, whereas here we have

5.4 Effective action for scalar fields: Some examples

a charged or complex scalar field which has two real degrees of freedom. From our expression for the kernel in (5.99) and (5.106),

$$K(\tau; x, x) = i \left[\det \left(\frac{\left(\frac{eF}{4\pi i} \right)}{\sinh(e\tau F)} \right) \right]^{1/2} \exp(-im^2\tau). \tag{5.115}$$

We will first show that this expression really does have the asymptotic form claimed earlier.

First of all, inverting the Taylor expansion of $\sinh \theta$ shows that

$$\frac{\theta}{\sinh \theta} = 1 - \frac{1}{6}\theta^2 + \frac{7}{360}\theta^4 + \cdots. \tag{5.116}$$

Next, using the result

$$\det(I + M) = \exp\left(\operatorname{tr} \ln(I + M) \right), \tag{5.117}$$

along with

$$\ln(I + M) = M - \frac{1}{2}M^2 + \frac{1}{3}M^3 + \cdots, \tag{5.118}$$

it may be seen that

$$\det(I + M) = 1 + \operatorname{tr} M + \frac{1}{2}(\operatorname{tr} M)^2 - \frac{1}{2}\operatorname{tr} M^2 + \cdots, \tag{5.119}$$

where M is any matrix. Taking the determinant of (5.116), with θ regarded as a matrix, and using (5.119) gives

$$\det\left(\frac{\theta}{\sinh \theta} \right) = 1 - \frac{1}{6}\operatorname{tr}\theta^2 + \frac{1}{72}\left(\operatorname{tr}\theta^2\right)^2 + \frac{1}{180}\operatorname{tr}\theta^4 + \cdots. \tag{5.120}$$

If only the term of order F^2 is kept, it may be seen from using (5.120) in (5.115) that

$$K(\tau; x, x) \sim i(4\pi i \tau)^{-n/2}\left\{1 - im^2\tau \right.$$
$$\left. + \tau^2\left[\frac{1}{12}e^2 F_{\mu\nu}F^{\mu\nu} - \frac{1}{2}m^4\right] + \cdots \right\}, \tag{5.121}$$

if we expand the kernel in powers of τ. (Note that $\operatorname{tr} F^2 = F^\mu{}_\nu F^\nu{}_\mu = -F_{\mu\nu}F^{\mu\nu}$.) This result is consistent with the general asymptotic expansion of (5.55) with (5.57)–(5.59) for the operator $D^2 + m^2$ in flat spacetime, since in the present case $Q = m^2$ and $W_{\mu\nu} = [D_\mu, D_\nu] = -ieF_{\mu\nu}$. The exact result obtained for the kernel is seen to contain considerable information about higher-order terms in the general asymptotic expansion of the kernel.

The one-loop effective Lagrangian may be defined in terms of the one-loop effective action by

$$W = \int dv_x \mathcal{L}^{(1)}. \tag{5.122}$$

Then

$$\mathcal{L}^{(1)} = \hbar \int_0^\infty \frac{d\tau}{\tau} \exp(-im^2\tau) \left[\det \left(\frac{\left(\frac{eF}{4\pi i}\right)}{\sinh(e\tau F)} \right) \right]^{1/2}. \qquad (5.123)$$

The integral over τ will be divergent unless we adopt a regularization technique. We will choose dimensional regularization here, in which there are ϵ extra flat spatial dimensions. This modifies the kernel function with an additional factor of $(4\pi i\tau)^{-\epsilon/2}$. (See (5.109).) Another way to see how this factor arises is to take the background electromagnetic field to only be non-zero in the physical spacetime dimensions. The determinant over the ϵ extra dimensions gives rise to the factor of $(4\pi i\tau)^{-\epsilon/2}$ just as in the free scalar field case discussed above.

If, as in Schwinger (1951a), we concentrate on the case of a four-dimensional spacetime, then the regularized one-loop effective Lagrangian is

$$\mathcal{L}^{(1)} = \hbar \ell^\epsilon \int_0^\infty \frac{d\tau}{\tau} \exp(-im^2\tau)(4\pi i\tau)^{-\epsilon/2} \left[\det \left(\frac{\left(\frac{eF}{4\pi i}\right)}{\sinh(e\tau F)} \right) \right]^{1/2}. \qquad (5.124)$$

(The factor of ℓ^ϵ in front of the integral may be understood as arising from the volume associated with the ϵ extra dimensions which arise in going from W in (5.122) to $\mathcal{L}^{(1)}$ if we use dimensional regularization.) For large τ the integrand vanishes exponentially fast due to the fact that m^2 is understood to have a small negative imaginary part. Any divergences can come only from small τ.

In order to calculate the renormalized effective Lagrangian, first rewrite $\mathcal{L}^{(1)}$ above in the form

$$\mathcal{L}^{(1)} = \hbar \ell^\epsilon \int_0^\infty \frac{d\tau}{\tau} \exp(-im^2\tau)(4\pi i\tau)^{-2-\epsilon/2} \left[\det \left(\frac{(e\tau F)}{\sinh(e\tau F)} \right) \right]^{1/2}. \qquad (5.125)$$

Note that the determinant in (5.125) is four-dimensional. In order to evaluate (5.125), we follow Schwinger (1951a) and rotate the contour of integration from the positive real axis to lie along the negative imaginary axis. This may be accomplished by replacing $\tau \to -i\tau$. (We will see later that there may be poles along the imaginary axis which must be avoided.) The choice of which way to rotate the contour is dictated by the requirement that there be no contribution from the portion of the contour at infinity. Equation (5.125) then becomes

$$\mathcal{L}^{(1)} = \hbar \ell^\epsilon \int_0^\infty \frac{d\tau}{\tau} \exp(-m^2\tau)(4\pi\tau)^{-2-\epsilon/2} \left[\det \left(\frac{(e\tau F)}{\sin(e\tau F)} \right) \right]^{1/2}. \qquad (5.126)$$

The determinant in (5.126) may now be evaluated by calculating the eigenvalues of the matrix F as described by Schwinger (1951a).

5.4 Effective action for scalar fields: Some examples

The components of $F_{\mu\nu}$ are given in terms of the electric and magnetic fields by

$$F_{\mu\nu} = \begin{pmatrix} 0 & E_1 & E_2 & E_3 \\ -E_1 & 0 & -B_3 & B_2 \\ -E_2 & B_3 & 0 & -B_1 \\ -E_3 & -B_2 & B_1 & 0 \end{pmatrix}. \tag{5.127}$$

Define the dual tensor $\tilde{F}_{\mu\nu}$ by

$$\tilde{F}_{\mu\nu} = \frac{1}{2}\epsilon_{\mu\nu\rho\sigma}F^{\rho\sigma}, \tag{5.128}$$

where $\epsilon_{\mu\nu\rho\sigma}$ is the totally antisymmetric Levi–Civita tensor, normalized by $\epsilon_{0123} = 1$. It is easy to see that $\tilde{F}_{\mu\nu}$ has components

$$\tilde{F}_{\mu\nu} = \begin{pmatrix} 0 & -B_1 & -B_2 & -B_3 \\ B_1 & 0 & -E_3 & E_2 \\ B_2 & E_3 & 0 & -E_1 \\ B_3 & -E_2 & E_1 & 0 \end{pmatrix}. \tag{5.129}$$

Define the two invariants $\mathcal{F}$ and $\mathcal{G}$ by

$$\mathcal{F} = \frac{1}{4}F_{\mu\nu}F^{\mu\nu}, \tag{5.130}$$

$$\mathcal{G} = \frac{1}{4}\tilde{F}_{\mu\nu}F^{\mu\nu}. \tag{5.131}$$

In terms of $\vec{E}$ and $\vec{B}$, these two invariants are $\mathcal{F} = \frac{1}{2}(\vec{B}\cdot\vec{B} - \vec{E}\cdot\vec{E})$ and $\mathcal{G} = \vec{E}\cdot\vec{B}$. It may then be shown that

$$\tilde{F}^\mu{}_\lambda F^\lambda{}_\nu = -\mathcal{G}\delta^\mu_\nu, \tag{5.132}$$

$$\tilde{F}^\mu{}_\lambda \tilde{F}^\lambda{}_\nu - F^\mu{}_\lambda F^\lambda{}_\nu = 2\mathcal{F}\delta^\mu_\nu. \tag{5.133}$$

We need the determinant of expressions that involve F. To deal with this, set up the eigenvalue equation

$$F^\mu{}_\nu \psi^\nu = f\psi^\mu. \tag{5.134}$$

Operating on both sides of (5.134) with $\tilde{F}^\lambda{}_\mu$, and using (5.132) leads to

$$\tilde{F}^\lambda{}_\mu \psi^\mu = -\frac{1}{f}\mathcal{G}\psi^\lambda. \tag{5.135}$$

If we now operate on ψ^ν with (5.133), and use (5.134) and (5.135), we obtain

$$\left(-f^{-1}\mathcal{G}\right)^2 - f^2 = 2\mathcal{F},$$

or

$$f^4 + 2\mathcal{F}f^2 - \mathcal{G}^2 = 0. \tag{5.136}$$

The four roots of this biquadratic equation are $\pm f_+$ and $\pm f_-$ where

$$f_\pm = \frac{i}{2}(\psi^* \pm \psi), \tag{5.137}$$

with

$$\psi = \sqrt{2}(\mathcal{F} + i\mathcal{G})^{1/2}. \tag{5.138}$$

We then find

$$\left[\det \frac{(e\tau F)}{\sin(e\tau F)}\right]^{1/2} = \frac{(e\tau)^2 f_+ f_-}{\sin(e\tau f_+)\sin(e\tau f_-)}$$

$$= \frac{(e\tau)^2 \mathcal{G}}{\Im[\cosh(e\tau\psi)]}, \tag{5.139}$$

using (5.137) and (5.138). Here $\Im$ denotes the imaginary part of the expression. The effective Lagrangian becomes

$$\mathcal{L}^{(1)} = \hbar\ell^\epsilon \int_0^\infty \frac{d\tau}{\tau} \exp(-m^2\tau)(4\pi\tau)^{-2-\epsilon/2} \frac{(e\tau)^2 \mathcal{G}}{\Im[\cosh(e\tau\psi)]}. \tag{5.140}$$

In order to identify the precise manner in which (5.140) diverges as $\epsilon \to 0$, we will expand the integrand in powers of τ. (As we discussed in Section 5.3, the divergences show up from the small τ behavior of the kernel.) Expansion of (5.139) in powers of τ leads to

$$\mathcal{L}^{(1)} = \hbar\ell^\epsilon \int_0^\infty \frac{d\tau}{\tau} \exp(-m^2\tau)(4\pi\tau)^{-2-\epsilon/2} \left\{ 1 - \frac{1}{3}(e\tau)^2 \mathcal{F} \right.$$

$$\left. + \frac{1}{90}(e\tau)^4 (7\mathcal{F}^2 + \mathcal{G}^2) + \cdots \right\}. \tag{5.141}$$

We may make use of the integral representation of the Γ-function,

$$\Gamma(z) = \int_0^\infty d\tau\, \tau^{z-1} \exp(-\tau),$$

valid for $\Re(z) > 1$ to perform the integrations in (5.141). The result of this is

$$\mathcal{L}^{(1)} = \frac{\hbar}{16\pi^2} \left(\frac{m^2 \ell^2}{4\pi}\right)^{\epsilon/2} \left\{ m^4 \Gamma(-2 - \epsilon/2) - \frac{1}{3}\Gamma(-\epsilon/2) e^2 \mathcal{F} \right.$$

$$\left. + \frac{e^4}{90 m^4} \Gamma(2 - \epsilon/2)(7\mathcal{F}^2 + \mathcal{G}^2) + \cdots \right\}. \tag{5.142}$$

Recalling that $\Gamma(z)$ has poles only for $z = 0, -1, -2, \ldots$, it is observed that only the first two terms on the right-hand side of (5.142) diverge as $\epsilon \to 0$. To deal with these divergences, we add to $\mathcal{L}^{(1)}$ the classical Lagrangian for an electromagnetic field

$$\mathcal{L}^{(0)} = -\frac{1}{4} F^B_{\mu\nu} F^{B\,\mu\nu} - c_B, \tag{5.143}$$

where the 'B' denotes a bare, unrenormalized quantity. Here c_B denotes a renormalization of the vacuum energy density, which may also be interpreted as a renormalization of the cosmological constant. Define the renormalized gauge field A_μ in terms of the bare one A_μ^B by

$$A_\mu^B = Z_A^{1/2} A_\mu, \tag{5.144}$$

where $Z_A = 1 + \delta Z_A$ does not depend on the spacetime coordinates x^μ, with δZ_A allowed to diverge as $\epsilon \to 0$. We then have

$$\mathcal{L}^{(0)} = -\frac{1}{4} Z_A F_{\mu\nu} F^{\mu\nu} - c_B$$
$$= -Z_A \mathcal{F} - c_B. \tag{5.145}$$

These two terms are seen to be sufficient to deal with that part of $\mathcal{L}^{(1)}$ which diverges as $\epsilon \to 0$. Specifically, we will define

$$\delta Z_A = -\frac{\hbar e^2}{48\pi^2} \Gamma(-\epsilon/2) \left(\frac{m^2 \ell^2}{4\pi}\right)^{\epsilon/2}, \tag{5.146}$$

$$c_B = \frac{\hbar m^4}{16\pi^2} \Gamma(-2-\epsilon/2) \left(\frac{m^2 \ell^2}{4\pi}\right)^{\epsilon/2}. \tag{5.147}$$

The renormalized one-loop effective Lagrangian, which is defined to be $\mathcal{L} = \mathcal{L}^{(0)} + \mathcal{L}^{(1)}$, is therefore given by

$$\mathcal{L} = -\mathcal{F} + \frac{\hbar}{16\pi^2} \int_0^\infty \frac{d\tau}{\tau^3} \exp(-m^2 \tau) \left\{ \frac{(e\tau)^2 \mathcal{G}}{\Im[\cosh(e\tau\psi)]} - 1 + \frac{1}{3}(e\tau)^2 \mathcal{F} \right\}. \tag{5.148}$$

(The $\epsilon \to 0$ limit may be taken since the renormalization has removed all terms that diverge.) The same result could be obtained by subtracting off all terms up to adiabatic order 4 as described in a different problem in Chapter 3.

If the electromagnetic field is weak, the lowest order correction to the classical theory results in

$$\mathcal{L} = -\mathcal{F} + \frac{\hbar e^4}{1440\pi^2 m^4}(7\mathcal{F}^2 + \mathcal{G}^2)$$
$$= -\frac{1}{2}(\vec{B}^2 - \vec{E}^2) + \frac{\hbar e^4}{1440\pi^2 m^4}\left[(\vec{E}\cdot\vec{B})^2 + \frac{7}{4}(\vec{B}^2 - \vec{E}^2)^2\right]. \tag{5.149}$$

This leads to a non-linear modification of Maxwell's theory of electromagnetism.[7]

One of the consequences we do want to discuss concerns the special case of an external field which is purely electric or purely magnetic. Either case is seen to correspond to $\mathcal{G} = 0$. For a pure magnetic field, $\psi = (\mathcal{F})^{1/2} = |\vec{B}|$ is real. For a

---

[7] Some of the physical consequences are discussed in Schwinger (1989).

pure electric field, $\psi = (2\mathcal{F})^{1/2} = i|\vec{E}|$ is imaginary. Taking the limit $\vec{E} \to 0$ in (5.148), with $\vec{B} \neq 0$, leads to

$$\mathcal{L}(B) = -\frac{1}{2}B^2 + \frac{\hbar}{16\pi^2}\int_0^\infty \frac{d\tau}{\tau^3}\exp(-m^2\tau)\left\{\frac{(e\tau B)}{\sinh(e\tau B)} - 1 + \frac{1}{6}(e\tau B)^2\right\}, \tag{5.150}$$

as the result for a pure magnetic field of magnitude $B = |\vec{B}|$. For a pure electric field of non-zero magnitude $E = |\vec{E}|$ we have, upon letting $\vec{B} \to 0$ in (5.148),

$$\mathcal{L}(E) = \frac{1}{2}E^2 + \frac{\hbar}{16\pi^2}\int_0^\infty \frac{d\tau}{\tau^3}\exp(-m^2\tau)\left\{\frac{(e\tau E)}{\sin(e\tau E)} - 1 - \frac{1}{6}(e\tau E)^2\right\}. \tag{5.151}$$

In this case, the path of integration is seen to run through poles at $\tau = \tau_n = (\pi n)/(eE)$, for $n = 1, 2, \ldots$. Recall that we have rotated the original path of integration from the positive real axis onto the negative imaginary axis. This of course must be done in such a way that any poles are avoided. This means that the integration in (5.151) should be understood as being deformed above the poles at $\tau = \tau_n$ that now lie along the positive real axis in the complex plane of the present integration variable τ. By taking the deformation about the $\tau = \tau_n$ pole to be a semicircle whose radius tends to zero, it may be seen that $\mathcal{L}(E)$ has an imaginary part given by

$$\Im[\mathcal{L}(E)] = \frac{\hbar e^2 E^2}{16\pi^3}\sum_{n=1}^\infty \frac{(-1)^{n+1}}{n^2}\exp\left(-\frac{n\pi m^2}{eE}\right), \tag{5.152}$$

where we used $\sin(n\pi + r\exp(i\theta)) = (-1)^n r\exp(i\theta)$ to first order in r. The interpretation of this imaginary part is as follows. Recall that $\exp(iW) = \langle\text{out}|\text{in}\rangle$ is the vacuum persistence amplitude. If the effective action is allowed to be complex, then

$$|\langle\text{out}|\text{in}\rangle|^2 = \exp(-2\Im(W)).$$

If $\Im(W) \neq 0$, this indicates that the in-vacuum does not remain as the vacuum state. Physically, this means that the electric field has become strong enough to create particle–antiparticle pairs. Therefore, (5.152) may be interpreted as the number of particles per unit volume and per unit time that appear in the particle–antiparticle pairs created by a constant electric field $\vec{E}$.

5.4.2 Scalar field in a constant gauge field background in $\mathbb{R}^{n-1} \times \mathbb{S}^1$

Suppose that we look at a spacetime of the form $\mathbb{R}^{n-1} \times \mathbb{S}^1$, where one of the spatial dimensions is periodically identified to form the circle $\mathbb{S}^1$. Take $F_{\mu\nu} = 0$, so that there is no background electromagnetic field. Then (5.86) shows that $A_\mu = a_\mu$ where a_μ is a constant. Because $A_\mu \to A'_\mu = A_\mu + \partial_\mu\theta$ under a gauge transformation, where θ is some scalar function, it might be thought that by choosing $\theta = -a_\mu x^\mu$ we could always transform any constant gauge field to zero.

5.4 Effective action for scalar fields: Some examples

This is true in Minkowski spacetime; however, if one (or more) of the spatial dimensions is compact this may no longer be true. For $\mathbb{R}^{n-1} \times \mathbb{S}^1$, the gauge transformations must be periodic around the identified spatial coordinate, and $\theta = -a_\mu x^\mu$ clearly is not. We can however transform all components of a_μ which lie in the $\mathbb{R}^{n-1}$ directions to zero, leaving only a single non-zero component, say a, which lies in the circular direction. We will take the background gauge field to be

$$A^\mu = a\delta_1^\mu, \tag{5.153}$$

where we have chosen x^1 to be the coordinate on $\mathbb{S}^1$. We will restrict $0 \leq x^1 \leq L$, so that L may be thought of as the circumference of $\mathbb{S}^1$.

Because one of the spatial dimensions is compact, it is necessary to specify some sort of boundary condition on the field $\Phi(\vec{x}, x^1)$ when x^1 goes through one period. We will use $\vec{x}$ to denote all coordinates other than the coordinate on $\mathbb{S}^1$. ($\vec{x}$ represents the coordinates on $\mathbb{R}^{n-1}$, which is $(n-1)$-dimensional Minkowski spacetime.) We will require the Lagrangian, or Hamiltonian, to be single-valued:

$$\mathcal{L}(\vec{x}, x^1 + L) = \mathcal{L}(\vec{x}, x^1). \tag{5.154}$$

Because $\mathcal{L}$ involves Φ in terms such as $\Phi^\dagger\Phi$, this condition is met if

$$\Phi(\vec{x}, x^1 + L) = \exp(2\pi i\delta)\Phi(\vec{x}, x^1), \tag{5.155}$$

where the phase δ lies in the range $0 \leq \delta < 1$, but is otherwise arbitrary. This includes the special cases of periodic ($\delta = 0$) and antiperiodic ($\delta = 1/2$) boundary conditions.

The kernel $K(\tau; x, x')$ for this problem must behave like $\Phi(x)\Phi^\dagger(x')$ under any transformation. This may be most easily seen from its expansion in terms of mode functions (5.49) or the analogous result in Riemannian space (5.78). It follows that we must have

$$K(\tau; \vec{x}, x^1 + L, \vec{x}', x'^1) = \exp(2\pi i\delta)K(\tau; \vec{x}, x^1, \vec{x}', x'^1), \tag{5.156}$$
$$K(\tau; \vec{x}, x^1, \vec{x}', x'^1 + L) = \exp(-2\pi i\delta)K(\tau; \vec{x}, x^1, \vec{x}', x'^1). \tag{5.157}$$

We can expand the kernel in a Fourier series in the periodic coordinates x^1 and x'^1 as

$$K(\tau; \vec{x}, x^1, \vec{x}', x'^1) = \sum_{j=-\infty}^{\infty} \frac{1}{L} K_j(\tau; \vec{x}, \vec{x}') \exp\left\{\frac{2\pi i}{L}(j + \delta)(x^1 - x'^1)\right\}, \tag{5.158}$$

where the Fourier coefficients $K_j(\tau; \vec{x}, \vec{x}')$ have no dependence on the coordinates on the circle. The kernel must satisfy

$$\left(D^\mu D_\mu + m^2\right) K(\tau; x, x') = i\frac{\partial}{\partial \tau} K(\tau; x, x'), \tag{5.159}$$

where $D_\mu = \partial_\mu - ieA_\mu$. Using $A^\mu = a\delta_1^\mu$, it is easily seen that

$$D^\mu D_\mu + m^2 = \Box_{n-1} - \left(\frac{\partial}{\partial x^1} + iea\right)^2 + m^2, \tag{5.160}$$

where $\Box_{n-1}$ is the d'Alembertian on $\mathbb{R}^{n-1}$. Substitution of the Fourier expansion (5.158) into (5.159) shows that the Fourier coefficients K_j satisfy

$$\left(\Box_{n-1} + m_j^2\right) K_j(\tau; \vec{x}, \vec{x}') = i\frac{\partial}{\partial \tau} K_j(\tau; \vec{x}, \vec{x}'), \tag{5.161}$$

where

$$m_j^2 = m^2 + \left(\frac{2\pi}{L}\right)^2 \left(j + \delta + \frac{eaL}{2\pi}\right)^2. \tag{5.162}$$

In this form, K_j is seen to be the kernel for a free scalar field of mass m_j in an $(n-1)$-dimensional Minkowski spacetime. From (5.109), we have

$$K_j(\tau; \vec{x}, \vec{x}') = i(4\pi i\tau)^{-(n-1)/2} \exp\left\{-\frac{i}{4\tau}|\vec{x} - \vec{x}'|^2 - im_j^2 \tau\right\}, \tag{5.163}$$

where $|\vec{x} - \vec{x}'|^2 = \eta_{\mu\nu}(\vec{x}^\mu - \vec{x}'^\mu)(\vec{x}^\nu - \vec{x}'^\nu)$. We know from the preceding section that

$$K_j(\tau; \vec{x}, \vec{x}') \to \delta^{(n-1)}(\vec{x} - \vec{x}'), \tag{5.164}$$

as $\tau \to 0$. Therefore, as $\tau \to 0$ in (5.158),

$$K(\tau; x, x') \to \delta^{(n-1)}(\vec{x} - \vec{x}') \sum_{j=-\infty}^{\infty} \frac{1}{L} \exp\left\{\frac{2\pi i}{L}(j + \delta)(x^1 - x'^1)\right\}. \tag{5.165}$$

If we can show that

$$\sum_{j=-\infty}^{\infty} \frac{1}{L} \exp\left\{\frac{2\pi i}{L}(j + \delta)(x^1 - x'^1)\right\} = \delta(x^1 - x'^1), \tag{5.166}$$

then we have found the complete kernel. The validity of this result is demonstrated as follows.

Consider any function $f(z)$ which is periodic in z: $f(z + L) = f(z)$. We can expand $f(z)$ in a Fourier series:

$$f(z) = \sum_{j=-\infty}^{\infty} c_j \exp\left(\frac{2\pi i}{L}jz\right), \tag{5.167}$$

where the Fourier coefficients c_j are given by

$$c_j = \frac{1}{L}\int_0^L dz' f(z') \exp\left(-\frac{2\pi i}{L}jz'\right). \tag{5.168}$$

Substitution of this result for c_j back into the Fourier series (5.167), and interchanging the order of summation and integration gives

$$f(z) = \int_0^L dz' \left\{\sum_{j=-\infty}^{\infty} \frac{1}{L} \exp\left(\frac{2\pi i}{L}j(z - z')\right)\right\} f(z'), \tag{5.169}$$

5.4 Effective action for scalar fields: Some examples

from which we see that

$$\delta(z-z') = \sum_{j=-\infty}^{\infty} \frac{1}{L} \exp\left(\frac{2\pi i}{L} j(z-z')\right) \tag{5.170}$$

is the periodic δ-function. The result in (5.166) follows immediately.

The one-loop effective action is given in terms of the kernel function by

$$W = -i\hbar \int dv_x \int_0^\infty \frac{d\tau}{\tau} K(\tau; \vec{x}, x^1, \vec{x}, x^1). \tag{5.171}$$

Because the coincidence limit of the kernel involves no spatial dependence, we can concentrate on the one-loop effective potential V defined by $W = -\ell^\epsilon \int dv_x V$. (We again adopt dimensional regularization here, and ℓ^ϵ is associated with the volume of the extra spatial dimensions.) It is straightforward to show that

$$V = -\frac{\hbar}{L}\Gamma\left(\frac{1-n}{2}\right)\ell^\epsilon (4\pi)^{-(n-1)/2}\left(\frac{2\pi}{L}\right)^{n-1} F\left(\frac{1-n}{2}; \frac{ml}{2\pi}, \delta + \frac{eaL}{2\pi}\right), \tag{5.172}$$

where we define the function

$$F(\lambda; \alpha, \beta) = \sum_{j=-\infty}^{\infty} \left[(j+\beta)^2 + \alpha^2\right]^{-\lambda}. \tag{5.173}$$

The summation defined in (5.173) converges for $\Re(\lambda) > 1/2$, and may be evaluated by contour integral methods to be[8]

$$F(\lambda; \alpha, \beta) = \pi^{1/2}\frac{\Gamma(\lambda-1/2)}{\Gamma(\lambda)}\alpha^{1-2\lambda} + 4\sin\pi\lambda\, f_\lambda(\alpha, \beta), \tag{5.174}$$

where

$$f_\lambda(\alpha, \beta) = \Re\left(\int_\alpha^\infty dx \frac{(x^2-\alpha^2)^{-\lambda}}{[\exp(2\pi x + 2\pi i\beta) - 1]}\right). \tag{5.175}$$

The explicit form for the potential, which follows from (5.172) and (5.174), is

$$V = -\hbar\Gamma(-n/2)\ell^\epsilon \left(\frac{m^2}{4\pi}\right)^{n/2} - \frac{4\hbar}{L}\ell^\epsilon \left(\frac{\pi}{L^2}\right)^{(n-1)/2}\Gamma\left(\frac{1-n}{2}\right)$$
$$\times \cos\left(\frac{\pi n}{2}\right) f_{(1-n)/2}\left(\frac{mL}{2\pi}, \delta + \frac{eaL}{2\pi}\right). \tag{5.176}$$

The first term on the right-hand side is just the contribution which is found in flat n-dimensional spacetime, and is not very interesting. We will discard it in what

---

[8] See Ford (1980) for the original derivation, or Toms (2007) for a review.

follows, which amounts to renormalizing the vacuum energy of flat Minkowski spacetime to zero. We therefore take

$$V_{\text{ren}} = -\frac{4\hbar}{L}\left(\frac{\pi}{L^2}\right)^{(n-1)/2} \Gamma\left(\frac{1-n}{2}\right) \cos\left(\frac{\pi n}{2}\right) f_{(1-n)/2}\left(\frac{mL}{2\pi}, \delta + \frac{eaL}{2\pi}\right) \quad (5.177)$$

as the renormalized vacuum energy density. It is understood that we have taken $\epsilon \to 0$ here, so that n takes its physical value.

Note that $F(\lambda; \alpha, \beta+z) = F(\lambda; \alpha, \beta)$ for any integer z. (This may be seen either from the summation (5.173), or from the analytic continuation of the summation defined in (5.174).) Thus, the resulting effective potential is periodic in δ' where

$$\delta' = \delta + \frac{eaL}{2\pi}. \quad (5.178)$$

The result (5.172) for the effective potential may be used to obtain a number of interesting results. The first is found by setting $a = 0$, so that the potential depends only on the phase δ which entered the boundary condition for the scalar field. We might then demand that the phase, and hence the boundary conditions, be determined dynamically by requiring the potential to be minimized. If we call δ_{min} the value of δ that minimizes V, then $V(\delta = \delta_{\text{min}})$ will give the energy density for the field of lowest energy. In effect, our result allows us to compare the energy for fields satisfying different boundary conditions.

A second problem which we might try to analyze is to regard δ as fixed, and use the potential to determine the value of a that minimizes the potential. For the classical theory, without any quantum effects considered, any real value for a is allowed. Because of the periodic nature of the one-loop effective potential that we have found, demanding that a be determined by the minimum of the potential will restrict a to a discrete set of values, removing some of the classical degeneracy.

The two problems mentioned above are clearly related, since δ and a only enter the potential through the combination δ' defined in (5.178). Either problem requires a knowledge of the stationary points of $f_\lambda(\alpha, \beta)$ as a function of β with α fixed. A straightforward differentiation of (5.175) with respect to β gives[9]

$$\frac{\partial}{\partial \beta} f_\lambda(\alpha, \beta) = -2\pi \Im\left\{\int_\alpha^\infty dx\, (x^2 - \alpha^2)^{-\lambda} \frac{\exp(2\pi x + 2\pi i \beta)}{[\exp(2\pi x + 2\pi i \beta) - 1]^2}\right\}. \quad (5.179)$$

We may assume λ to have been continued to negative real values in this expression, since $\lambda = (1-n)/2$ in the case we are interested in. It is then clear that the integral has a non-vanishing imaginary part unless $\exp(2\pi i \beta)$ is real. This allows us to conclude that $\frac{\partial}{\partial \beta} f_\lambda(\alpha, \beta) = 0$ when $\beta = k/2$ for any integer k. Computing the second derivative of (5.179), and evaluating it at $\beta = k/2$, leads to

---

[9] Recall that $\Im$ defines the imaginary part.

5.4 Effective action for scalar fields: Some examples

$$\text{sign}\left\{\frac{\partial^2}{\partial\beta^2}f_\lambda(\alpha,\beta)\right\} = (-1)^k. \tag{5.180}$$

This proves that $f_\lambda(\alpha,\beta)$ has a minimum when $\beta = \varpi$, and a maximum when $\beta = \varpi + 1/2$, where ϖ is any integer.

From the discussion of the minima of $f_\lambda(\alpha,\beta)$ above, it follows that V_{ren} has stationary points at $\delta' = k/2$, where k is an integer. It is easy to show that the coefficient of $f_{(1-n)/2}(mL/2\pi, \delta')$ is negative regardless of the spacetime dimension n, so that

$$\text{sign}\left[\frac{\partial^2}{\partial\delta'^2}V_{\text{ren}}\right] = (-1)^{k+1}, \tag{5.181}$$

at $\delta' = k/2$. The effective potential is then minimized if $\delta' = \varpi$, and maximized if $\delta' = \varpi + 1/2$, where ϖ is an integer. If we restrict attention to one period of the potential, then we may restrict $0 \leq \delta' < 1$. In this case $\delta' = 0$ corresponds to a minimum, and $\delta' = 1/2$ corresponds to a maximum.

If we consider $a = 0$, so that the theory consists of a single complex scalar field with no coupling to a background gauge field, then our results show that if the field satisfies a periodic boundary condition ($\delta' = \delta = 0$), its vacuum energy density is lower than if it satisfies any other type of boundary condition. The antiperiodic field ($\delta' = \delta = 1/2$) has the largest vacuum energy density.

The result for the Casimir effect for a complex scalar field which satisfies periodic or antiperiodic boundary conditions can be obtained from the results derived here. We will take $n = 4$ and $m = 0$, although the results given above are general enough to relax either of these conditions. The basic integral is $f_{(1-n)/2}(0, \delta)$ where $\delta = 0$ or $1/2$. Both cases can be evaluated in terms of the Riemann ζ-function. In particular,

$$f_{(1-n)/2}(0,0) = (2\pi)^{-n}\Gamma(n)\zeta(n),$$
$$f_{(1-n)/2}(0,1/2) = (2\pi)^{-n}\Gamma(n)(2^{1-n} - 1)\zeta(n).$$

It is then easy to show that the renormalized vacuum energy density becomes, for $n = 4$,

$$V(\delta = 0) = -\frac{\pi^2\hbar}{45L^4}, \tag{5.182}$$

$$V(\delta = 1/2) = \frac{7\pi^2\hbar}{360L^4}. \tag{5.183}$$

(The results for a real scalar field are simply half of those found here.) The result for a massless complex scalar field which satisfies vanishing boundary conditions at $x^1 = 0$ and $x^1 = L$ may be inferred from our results as well, by replacing L with $2L$. The factor of 2 follows here if it is noted that for periodic boundary conditions the mode functions involve $\sin(2\pi n x^1/L)$, whereas those for vanishing boundary conditions involve $\sin(\pi n x^1/L)$. This conclusion is substantiated by the explicit calculation in Chapter 1.

5.5 The conformal anomaly and the functional integral

We have already discussed the conformal anomaly in Chapter 3. The present chapter has adopted a functional integral approach to quantum field theory, and it is instructive to re-examine the conformal anomaly from the functional integral point of view.

The effective action has been defined by the functional integral

$$\exp\left(\frac{i}{\hbar}W\right) = \int d\mu[\varphi] \exp\left(\frac{i}{\hbar}S\right), \tag{5.184}$$

where

$$S = -\frac{1}{2}\int dv_x \varphi^i(x)\left(\delta_{ij}\Box_x + M_{ij}^2(x)\right)\varphi^j(x) \tag{5.185}$$

is the classical action. Suppose that S is invariant under a conformal transformation

$$g_{\mu\nu}(x) \to \tilde{g}_{\mu\nu}(x) = \Omega^2(x)g_{\mu\nu}(x) \tag{5.186}$$

of the spacetime metric. Under this transformation,

$$\det \tilde{g}_{\mu\nu}(x) = \Omega^{2n}(x)\det g_{\mu\nu}(x),$$

so that

$$dv_x \to d\tilde{v}_x = \Omega^n(x)dv_x. \tag{5.187}$$

We keep the spacetime dimension n general here. A necessary condition for the action S above to be invariant is that the fields $\varphi^i(x)$ transform like

$$\varphi^i(x) \to \tilde{\varphi}^i(x) = \Omega^{1-n/2}(x)\varphi^i(x) \tag{5.188}$$

under a conformal transformation. If $M_{ij}^2 = \frac{1}{4}\left(\frac{n-2}{n-1}\right)\delta_{ij}R$, (5.188) is also a sufficient condition for conformal invariance of the classical action. (Note that when $n = 4$ the coefficient of R required for conformal invariance becomes $1/6$ as in Section 2.2.) We will now consider the case of only one field.

Because the classical action S is invariant under a conformal transformation, if we assume that the functional measure $d\mu[\varphi]$ is invariant, we will end up with the effective action W invariant as well. But we know from the discussion of the conformal (or trace) anomaly of the energy momentum tensor in Chapter 3 that quantum effects lead to a non-zero value for the expectation value of $T^\mu{}_\mu$. This would conflict with the invariance of W which should contain all of the information on quantum effects, including the trace anomaly. It must therefore follow that the functional measure cannot be invariant, and it should be possible to extract the conformal anomaly from the measure. A way to do this was first given by Fujikawa (1980a, 1981). Here we will follow the treatment of Toms (1987).

5.5 The conformal anomaly and the functional integral

Suppose that we have a set of functions $\{f_N(x)\}$ that obeys

$$\left[\Box_x + \frac{1}{4}\left(\frac{n-2}{n-1}\right)R(x)\right]f_N(x) = \lambda_N f_N(x), \quad (5.189)$$

and that this set is orthonormal:

$$\int dv_x f_N^*(x) f_{N'}(x) = \ell^2 \delta_{NN'}, \quad (5.190)$$

and complete:

$$\sum_N f_N^*(x) f_N(x') = \ell^2 \delta(x, x'). \quad (5.191)$$

We choose $f_N(x)$ to have the same dimensions and transformation properties as the scalar field $\varphi(x)$; ℓ is a constant with dimensions of length which is introduced so that the dimensions on both sides of (5.191) agree. (Note that the dimensions of f_N are $L^{1-n/2}$ where L is a length, and the dimensions of the Dirac δ-distribution are L^{-n}.) Because the set $\{f_N(x)\}$ is complete, we may expand the scalar field

$$\varphi(x) = \sum_N \varphi_N f_N(x), \quad (5.192)$$

where the expansion coefficients φ_N are dimensionless constants. It then follows that

$$S = -\frac{1}{2}\sum_N (\ell^2 \lambda_N)\varphi_N^2. \quad (5.193)$$

If we define the functional measure formally by

$$d\mu[\varphi] = \prod_N \frac{d\varphi_N}{(-2\pi i\hbar)^{1/2}}, \quad (5.194)$$

then

$$\exp\left(\frac{i}{\hbar}W\right) = \prod_N \int \frac{d\varphi_N}{(-2\pi i\hbar)^{1/2}}\exp\left\{-\frac{i}{2\hbar}(\ell^2\lambda_N)\varphi_N^2\right\}$$
$$= \prod_N (\ell^2 \lambda_N)^{-1/2}$$
$$= \left\{\det \ell^2\left[\Box + \frac{1}{4}\left(\frac{n-2}{n-1}\right)R\right]\right\}^{-1/2}, \quad (5.195)$$

in agreement with what we already know. This justifies the choice of measure made above.

In the conformally related spacetime we must have

$$\tilde{\varphi}(x) = \sum_N \varphi_N \tilde{f}_N(x), \quad (5.196)$$

where

$$\tilde{f}_N(x) = \Omega^{1-n/2}(x) f_N(x), \quad (5.197)$$

since $\varphi(x)$ and $f_N(x)$ behave in the same way under the conformal transformation (5.188). We therefore have

$$\int d\tilde{v}_x \tilde{f}_N^*(x)\tilde{f}_{N'}(x) = \int dv_x \Omega^2(x) f_N^*(x) f_{N'}(x) \tag{5.198}$$

and

$$\sum_N \tilde{f}_N^*(x)\tilde{f}_N(x') = \Omega(x)\Omega(x')\ell^2 \delta_c(x,x'), \tag{5.199}$$

where

$$\delta_c(x,x') = [\Omega(x)]^{-1-n/2}[\Omega(x')]^{1-n/2}\delta(x,x'). \tag{5.200}$$

(The factors of Ω on the right-hand side can be simplified to read $[\Omega(x)]^{-n}$ if desired by making use of the properties of the Dirac δ-distribution.) The conformally related basis $\{\tilde{f}_N(x)\}$ is therefore neither orthonormal nor complete. However, the derivation of (5.195) relied on both completeness and orthonormality. We must therefore perform a further transformation from $\{\tilde{f}_N(x)\}$ to a new basis $\{g_N(x)\}$ which is complete and orthonormal in the conformally related spacetime. It is clear from (5.198) and (5.199) that we should take

$$\tilde{f}_N(x) = \Omega(x)g_N(x). \tag{5.201}$$

However, this transformation of basis may result in a non-trivial Jacobian, since we do not know if the transformation results in $\det(\partial\varphi_N/\partial\tilde{\varphi}_{N'}) = 1$.

To compute the Jacobian arising from the transformation (5.201), write

$$\tilde{\varphi}(x) = \sum_N \tilde{\varphi}_N g_N(x). \tag{5.202}$$

We therefore have, equating (5.196)–(5.202),

$$\sum_N \varphi_N \tilde{f}_N(x) = \sum_N \tilde{\varphi}_N g_N(x) \tag{5.203}$$

giving (from (5.201))

$$\sum_N \varphi_N g_N(x) = \sum_N \Omega^{-1}(x)\tilde{\varphi}_N g_N(x). \tag{5.204}$$

It then follows using the orthonormality of the $g_N(x)$ that

$$\varphi_N = \sum_{N'} C_{NN'} \tilde{\varphi}_{N'} \tag{5.205}$$

where

$$C_{NN'} = \ell^{-2} \int d\tilde{v}_x \Omega^{-1}(x) g_N^*(x) g_{N'}(x), \tag{5.206}$$

or equivalently,

$$C_{NN'} = \ell^{-2} \int dv_x \Omega^{-1}(x) f_N^*(x) f_{N'}(x). \tag{5.207}$$

This leads to
$$\prod_N d\varphi_N = (\det C_{NN'}) \prod_N d\tilde{\varphi}_N. \tag{5.208}$$

The results in (5.206) and (5.207) show that the Jacobian of the transformation may be evaluated in the original space, or in the conformally related space. If $\det C_{NN'} = 1$ then the functional measure will be conformally invariant resulting in conformal invariance of the one-loop effective action.

The effective action $\tilde{W}$ for the conformally related theory is related to the original effective action W by
$$\tilde{W} = W + i\hbar \ln \det C_{NN'}. \tag{5.209}$$

Suppose that we look at infinitesimal conformal transformations only. (As discussed in Chapter 2, there is no loss of generality in this.) Take
$$\Omega(x) = 1 + \delta\omega(x). \tag{5.210}$$

Then
$$C_{NN'} = \ell^{-2} \int dv_x \, (1 - \delta\omega(x)) \, f_N^*(x) f_{N'}(x)$$
$$= \delta_{NN'} - \ell^{-2} \int dv_x \delta\omega(x) f_N^*(x) f_{N'}(x),$$

and so
$$\ln \det C_{NN'} = -\ell^{-2} \sum_N \int dv_x \delta\omega(x) f_N^*(x) f_N(x). \tag{5.211}$$

We have used the fact that $\ln \det C_{NN'} = \operatorname{tr} \ln C_{NN'}$, and expanded the logarithm to first order in the infinitesimal conformal parameter. The change in the effective action $\delta W = \tilde{W} - W$ is therefore
$$\delta W = -i\hbar \ell^{-2} \sum_N \int dv_x \delta\omega(x) f_N^*(x) f_N(x). \tag{5.212}$$

The expression for δW in (5.212) requires regularization. We will use ζ-function regularization here. Define (assuming no zero modes)
$$\zeta(s; x, x') = \sum_N \lambda_N^{-s} f_N^*(x) f_N(x') \ell^{-2}. \tag{5.213}$$

The generalized ζ-function defined earlier in Section 5.3.3 as $\zeta(s) = \sum_N \lambda_N^{-s}$ is obtained from $\zeta(s; x, x')$ by
$$\zeta(s) = \int dv_x \zeta(s; x, x). \tag{5.214}$$

We will therefore define
$$\delta W = -i\hbar \int dv_x \delta\omega(x) \zeta(0; x, x). \tag{5.215}$$

This definition is a natural one, since by simply setting $s = 0$ in (5.213) we have $\sum_N f_N^*(x) f_N(x') \ell^{-2} = \zeta(0; x, x')$. Of course, the regularization consists of defining $\zeta(0; x, x')$ by analytic continuation of $\zeta(s; x, x')$ to $s = 0$. $\zeta(s; x, x')$ is easily seen to be related to the kernel $K(\tau; x, x')$ defined earlier by

$$\zeta(s; x, x') = \frac{i^{-s}}{\Gamma(s)} \int_0^\infty \tau^{s-1} K(\tau; x, x') d\tau. \tag{5.216}$$

Using the properties of the kernel obtained in Section 5.3 leads to

$$\zeta(0; x, x) = i(4\pi)^{-n/2} E_{n/2}(x). \tag{5.217}$$

Thus

$$\delta W = \hbar(4\pi)^{-n/2} \int dv_x \delta\omega(x) E_{n/2}(x) \tag{5.218}$$

gives the change in the one-loop effective action under an infinitesimal conformal transformation. (We have assumed no boundary terms in the spacetime manifold, although the modification is easy to work out.)

The result in (5.218) may be used to evaluate the trace of the stress-energy tensor. From its definition in Section 5.2,

$$\delta W = -i\hbar Z^{-1} \delta Z, \tag{5.219}$$

where

$$Z = \int d\mu[\varphi] \exp\left(\frac{i}{\hbar} S\right).$$

But

$$\delta Z = \frac{i}{\hbar} \int d\mu[\varphi] \delta S \exp\left(\frac{i}{\hbar} S\right),$$

so that

$$\delta W = \langle \delta S \rangle, \tag{5.220}$$

where

$$\langle \delta S \rangle = \frac{\int d\mu[\varphi] \delta S \exp\left(\frac{i}{\hbar} S\right)}{\int d\mu[\varphi] \exp\left(\frac{i}{\hbar} S\right)}. \tag{5.221}$$

The basic definition of the stress-energy tensor is that if the spacetime metric is varied while holding all matter fields fixed, the change in the classical action functional is

$$\delta S = \frac{1}{2} \int dv_x T_{\mu\nu} \delta g^{\mu\nu}. \tag{5.222}$$

This leads to, from (5.220),

$$\delta W = \frac{1}{2} \int dv_x \langle T_{\mu\nu} \rangle \delta g^{\mu\nu}. \tag{5.223}$$

Under the infinitesimal conformal transformation (5.210) we have $\delta g^{\mu\nu} = -2\delta\omega(x)g^{\mu\nu}$, resulting in

$$\delta W = -\int dv_x \langle T^\mu{}_\mu(x)\rangle \delta\omega(x). \tag{5.224}$$

Comparison of this result with that of (5.218) shows that

$$\langle T^\mu{}_\mu(x)\rangle = -\hbar(4\pi)^{-n/2} E_{n/2}(x) \tag{5.225}$$

gives the anomalous trace of the stress-energy tensor. If we now specialize to four spacetime dimensions ($n = 4$) and use (5.59) with $Q = R/6$ we recover the result of Chapter 3.

5.6 Spinors in curved spacetime

5.6.1 The n-bein formalism

In Chapter 3 we outlined how to deal with spinor fields in four-dimensional curved spacetime. This section reconsiders this approach in a spacetime of arbitrary dimension. We will refer to the method as the n-bein formalism, as opposed to the vierbein formalism in four spacetime dimensions.

If $\{x^\mu\}$ are local coordinates on spacetime M, then $\{dx^\mu\}$ is a set of coordinate basis one-forms for the cotangent space. (Recall that the cotangent space T^*M is the vector space dual of the tangent space TM.) The line element for the n-dimensional spacetime M is expressed in local coordinates as

$$ds^2 = g_{\mu\nu}(x)dx^\mu dx^\nu. \tag{5.226}$$

Instead of working in a coordinate basis, it is often more convenient to work in a local orthonormal frame which has as basis one-forms $\{\omega^a(x)\}$ and as line element

$$ds^2 = \eta_{ab}\omega^a(x)\omega^b(x). \tag{5.227}$$

(Here $\eta_{ab} = \text{diag}(+1, -1, -1, \ldots, -1)$ is the metric tensor for Minkowski spacetime in Cartesian coordinates. We will use Latin indices to distinguish components of tensors in the local orthonormal frame.) Because both $\{dx^\mu\}$ and $\{\omega^a(x)\}$ span T^*M, it must be possible to express

$$\omega^a(x) = e^a{}_\mu(x)dx^\mu \tag{5.228}$$

for some coefficients $e^a{}_\mu(x)$. We will call $e^a{}_\mu(x)$ the n-bein, analogously to the terminology vierbein used in four spacetime dimensions. $e^a{}_\mu(x)$ can be thought of as the components of a set of covariant vector fields $\mathbf{e}^a$, labeled by the orthonormal frame index a. Substitution of (5.228) into (5.227), and comparison of the result with (5.226), shows that

$$g_{\mu\nu}(x) = e^a{}_\mu(x)e^b{}_\nu(x)\eta_{ab}. \tag{5.229}$$

The inverse, or dual, of the n-bein will be denoted by $e_a{}^\mu(x)$, and satisfies

$$e_a{}^\mu(x) e^a{}_\nu(x) = \delta^\mu_\nu, \qquad (5.230)$$

$$e_a{}^\mu(x) e^b{}_\mu(x) = \delta^a_b. \qquad (5.231)$$

We then have

$$dx^\mu = e_a{}^\mu(x) \omega^a(x), \qquad (5.232)$$

$$\eta_{ab} = e_a{}^\mu(x) e_b{}^\nu(x) g_{\mu\nu}(x). \qquad (5.233)$$

It is easily seen from (5.229) or (5.233) that the n-bein is not uniquely determined by the metric tensor $g_{\mu\nu}(x)$. Rather, $e^a{}_\mu(x)$ and $e'^a{}_\mu(x)$ both satisfy (5.229) if

$$e'^a{}_\mu(x) = L^a{}_b(x) e^b{}_\mu(x), \qquad (5.234)$$

where

$$L^a{}_c(x) L^b{}_d(x) \eta_{ab} = \eta_{cd}. \qquad (5.235)$$

$L^a{}_b(x)$ is a matrix representing a local Lorentz transformation. (Here local means that the parameters representing the Lorentz transformation are functions of the spacetime coordinates, and may be arbitrarily specified at different points in spacetime.) Because $e^a{}_\mu(x)$ and $e'^a{}_\mu(x)$ given in (5.234) determine the same spacetime metric, the action, or field equations, for some given field should be independent of whether $e^a{}_\mu$ or $e'^a{}_\mu$ is chosen. This means that the action for the field should be invariant under a local Lorentz transformation.

Consider for example a vector field $A^\mu(x)$. We may use the n-bein formalism to relate the components to the local orthonormal frame by defining

$$A^a(x) = e^a{}_\mu(x) A^\mu(x). \qquad (5.236)$$

Under a local Lorentz transformation,

$$A^a(x) \to A'^a(x) = L^a{}_b(x) A^b(x), \qquad (5.237)$$

as a consequence of (5.234). Under a general coordinate transformation $x \to x'$, there is no reason to suppose that the basis vectors for the local orthonormal frame are unaffected. A change of coordinates will be taken to induce a local Lorentz transformation of the tangent space. This means that under a general coordinate transformation, $e^a{}_\mu(x)$ must transform like a covariant vector up to a local Lorentz transformation, and $A^a(x)$ must transform like a set of scalars, again up to a local Lorentz transformation:

$$A^a(x) \to A'^a(x') = L^a{}_b(x) A^b(x). \qquad (5.238)$$

The covariant derivative $\nabla_\mu A^a$ should be defined to transform like

$$\nabla'_\mu A'^a(x) = L^a{}_b(x) \nabla_\mu A^b(x) \qquad (5.239)$$

under a local Lorentz transformation, and

$$\nabla'_\mu A'^a(x') = \frac{\partial x^\nu}{\partial x'^\mu} L^a{}_b(x) \nabla_\nu A^b(x) \tag{5.240}$$

under a general coordinate transformation. Introduce a connection with components $\omega_\mu{}^a{}_b(x)$ defined so that

$$\nabla_\mu A^a = \partial_\mu A^a + \omega_\mu{}^a{}_b A^b. \tag{5.241}$$

The properties (5.239) and (5.240) are satisfied if $\omega_\mu{}^a{}_b(x)$ transforms like

$$\omega'_\mu{}^a{}_b = \frac{\partial x^\nu}{\partial x'^\mu} \{ L^a{}_c \omega_\nu{}^c{}_d (L^{-1})^d{}_b - (\partial_\nu L^a{}_c)(L^{-1})^c{}_b \} \tag{5.242}$$

under a general coordinate transformation. (The prefactor of $\partial x^\nu/\partial x'^\mu$ is replaced with δ^ν_μ for just a local Lorentz transformation.)

Consider the covariant derivative of the n-bein $e^a{}_\mu$. Because it transforms like (5.234) under a local Lorentz transformation, and like a covariant vector under a general coordinate transformation, we should define

$$\nabla_\mu e^a{}_\nu = \partial_\mu e^a{}_\nu - \Gamma^\lambda_{\nu\mu} e^a{}_\lambda + \omega_\mu{}^a{}_b e^b{}_\nu. \tag{5.243}$$

Another way to motivate this is to demand that covariant differentiation obey the product (Liebniz) rule, as in normal tensor calculus. Then (5.236) gives

$$\nabla_\mu A^a = (\nabla_\mu e^a{}_\nu) A^\nu + e^a{}_\nu (\nabla_\mu A^\nu). \tag{5.244}$$

Using $\nabla_\mu A^\nu = \partial_\mu A^\nu + \Gamma^\nu_{\mu\lambda} A^\lambda$, and (5.241), the result in (5.243) is regained.

The condition that the connection does not contain torsion fixes the covariant derivative of the n-bein to vanish. This is most easily seen using Cartan's approach to differential geometry. (See Misner et al., 1973 or Eguchi et al., 1980 for example.) From $\omega_\mu{}^a{}_b$, define a one-form

$$\omega^a{}_b = \omega_\mu{}^a{}_b dx^\mu. \tag{5.245}$$

Using (5.228) we have

$$d\omega^a = e^a{}_{[\nu,\mu]} dx^\mu \wedge dx^\nu, \tag{5.246}$$

where d is the exterior derivative and $\wedge$ is the wedge product. We have used the notation

$$e^a{}_{[\nu,\mu]} = \frac{1}{2}(e^a{}_{\nu,\mu} - e^a{}_{\mu,\nu}). \tag{5.247}$$

The condition that the connection be torsion-free is

$$0 = d\omega^a + \omega^a{}_b \wedge \omega^b. \tag{5.248}$$

(Torsion is described by replacing 0 on the left-hand side of this equation with the torsion two-form Eguchi et al. (1980).) From (5.245) and (5.246) we can infer that

$$\omega_\mu{}^a{}_b = -e_b{}^\nu (\partial_\mu e^a{}_\nu - \Gamma^\lambda_{\mu\nu} e^a{}_\lambda). \tag{5.249}$$

The condition that the Christoffel connection be torsion-free is equivalent to the requirement that $\Gamma^\lambda_{\nu\mu} = \Gamma^\lambda_{\mu\nu}$, and has been used here. If the index a in (5.249) is lowered with the Minkowski metric, we have

$$\omega_{\mu b a} = -\omega_{\mu a b} \tag{5.250}$$

by virtue of (5.231). If we now use (5.249) in (5.243), it follows that

$$\nabla_\mu e^a{}_\nu = 0. \tag{5.251}$$

This result shows that covariant differentiation commutes with the conversion of tensors to and from the local orthonormal frame. An alternate approach is to impose (5.251) and use it to prove that the connection is torsion-free.

For later use we will develop the Cartan approach further by defining the curvature two-form $\mathcal{R}^a{}_b$ by

$$\mathcal{R}^a{}_b = d\omega^a{}_b + \omega^a{}_c \wedge \omega^c{}_b. \tag{5.252}$$

Using (5.245), it is straightforward to show that

$$\mathcal{R}^a{}_b = \frac{1}{2}[\partial_\mu \omega_\nu{}^a{}_b - \partial_\nu \omega_\mu{}^a{}_b + \omega_\mu{}^a{}_c \omega_\nu{}^c{}_b - \omega_\nu{}^a{}_c \omega_\mu{}^c{}_b] dx^\mu \wedge dx^\nu. \tag{5.253}$$

The components of the curvature two-form will be defined by

$$\mathcal{R}^a{}_b = -\frac{1}{2} R_{\mu\nu}{}^a{}_b dx^\mu \wedge dx^\nu. \tag{5.254}$$

It is straightforward to see that

$$R_{\mu\nu}{}^a{}_b = \partial_\nu \omega_\mu{}^a{}_b - \partial_\mu \omega_\nu{}^a{}_b + \omega_\nu{}^a{}_c \omega_\mu{}^c{}_b - \omega_\mu{}^a{}_c \omega_\nu{}^c{}_b, \tag{5.255}$$

and that $R_{\mu\nu}{}^a{}_b = e^a{}_\lambda e_b{}^\sigma R^\lambda{}_{\sigma\mu\nu}$ where $R^\lambda{}_{\sigma\mu\nu}$ is the Riemann curvature tensor.

Suppose that we consider a field $\Psi(x)$ which transforms under a local Lorentz transformation like

$$\Psi(x) \to \Psi'(x) = S(L(x))\Psi(x), \tag{5.256}$$

where $S(L(x))$ denotes some representation of the Lorentz group. (As we will discuss later, in the case of spinors $S(L(x))$ really gives a representation of the double covering of the Lorentz group.) We will assume that $\Psi(x)$ transform like a scalar under a general coordinate transformation, which for tensor fields can always be done by using the n-bein formalism to relate the components to a local orthonormal frame. (The case of a vector field was considered in (5.236). The generalization to arbitrary tensor fields should be obvious from this.) We will suppress the indices on Ψ here and use a matrix notation. An infinitesimal local Lorentz transformation is characterized by

$$L^a{}_b(x) = \delta^a{}_b + \delta\epsilon^a{}_b(x), \tag{5.257}$$

where $\delta\epsilon_{ab} = \eta_{ac}\delta\epsilon^c{}_b = -\delta\epsilon_{ba}$. (The antisymmetry of the infinitesimal parameters follows immediately from (5.235).) We will write

$$S(1+\delta\epsilon) = 1 + i\delta\epsilon^{ab}\Sigma_{ab}, \tag{5.258}$$

where $\Sigma_{ab} = -\Sigma_{ba}$ provides a matrix representation for the Lie algebra of the Lorentz group $SO(1, n-1)$, and is therefore independent of x. The infinitesimal parameters $\delta\epsilon_{ab}$ do depend on x. (For a finite Lorentz transformation, we have $S(L(x)) = \exp(i\epsilon^{ab}\Sigma_{ab})$ for ϵ^{ab}, the parameters characterizing the Lorentz transformation.) Because $S(L)$ forms a representation of the Lorentz group, we must have

$$S(L)S(L')S(L^{-1}) = S(LL'L^{-1}) \tag{5.259}$$

for two arbitrary Lorentz matrices L, L'. Taking L' to have the infinitesimal form (5.257), using (5.258), and working to first order in $\delta\epsilon^{ab}$ give

$$S(L)\Sigma_{ab}S(L^{-1}) = \frac{1}{2}\left[L^c{}_a(L^{-1})_b{}^d - L^c{}_b(L^{-1})_a{}^d\right]\Sigma_{cd}. \tag{5.260}$$

(Note that care must be taken when comparing coefficients of $\delta\epsilon^{ab}$ to account for the antisymmetry in a, b.) Now taking $L^a{}_b$ to have the infinitesimal form (5.257), using (5.258), and working to first order in $\delta\epsilon^{ab}$ lead to

$$[\Sigma_{ab}, \Sigma_{cd}] = \frac{i}{2}\left(\eta_{ac}\Sigma_{bd} - \eta_{ad}\Sigma_{bc} - \eta_{bc}\Sigma_{ad} + \eta_{bd}\Sigma_{ac}\right). \tag{5.261}$$

This defines the Lie algebra for $SO(1, n-1)$.

Because the transformation (5.256) is local, the normal partial derivative of $\Psi(x)$ will not transform covariantly. Define the covariant derivative of $\Psi(x)$ by

$$\nabla_\mu \Psi(x) = \partial_\mu \Psi(x) + B_\mu(x)\Psi(x), \tag{5.262}$$

with the requirement that

$$\nabla'_\mu \Psi'(x) = S(L(x))\nabla_\mu \Psi(x) \tag{5.263}$$

under a local Lorentz transformation. $B_\mu(x)$ is called the spin connection and must transform like

$$B_\mu(x) \to B'_\mu(x) = S(L(x))B_\mu(x)S^{-1}(L(x)) - \partial_\mu S(L(x))S^{-1}(L(x)) \tag{5.264}$$

under a local Lorentz transformation. Using the infinitesimal form (5.258) for $S(L(x))$, and working to first order in $\delta\epsilon^{ab}$ give

$$B'_\mu(x) = B_\mu(x) + i\delta\epsilon^{ab}[\Sigma_{ab}, B_\mu(x)] - i\partial_\mu\delta\epsilon^{ab}\Sigma_{ab}. \tag{5.265}$$

The connection should take its value in the Lie algebra of $SO(1, n-1)$, so we may write

$$B_\mu(x) = B_\mu{}^{ab}(x)\Sigma_{ab}. \tag{5.266}$$

It is possible to add on a gauge potential representing an interaction with the electromagnetic field. At this stage we will only consider the interaction with gravity; the electromagnetic interaction will be given later. To first order in $\delta\epsilon$, (5.264) reads

$$B'_\mu{}^{ab} = B_\mu{}^{ab} + B_\mu{}^b{}_c \delta\epsilon^{ca} - B_\mu{}^a{}_c \delta\epsilon^{cb} - i\partial_\mu \delta\epsilon^{ab}. \tag{5.267}$$

This should be compared with the infinitesimal form of (5.242) which reads

$$\omega'_\mu{}^{ab} = \omega_\mu{}^{ab} + \omega_\mu{}^b{}_c \delta\epsilon^{ca} - \omega_\mu{}^a{}_c \delta\epsilon^{cb} - \partial_\mu \delta\epsilon^{ab}. \tag{5.268}$$

Comparison of the results in (5.267) and (5.268) shows that we may take

$$B_\mu{}^{ab} = i\omega_\mu{}^{ab}. \tag{5.269}$$

The covariant derivative of Ψ is therefore

$$\nabla_\mu \Psi = \partial_\mu \Psi + i\omega_\mu{}^{ab} \Sigma_{ab} \Psi. \tag{5.270}$$

For later use we record

$$[\nabla_\mu, \nabla_\nu]\Psi = -i R_{\mu\nu}{}^{ab} \Sigma_{ab} \Psi, \tag{5.271}$$

which follows from (5.261) and (5.255).

As a check on (5.270) consider the case when Ψ is a vector field. For reasons already explained, take the components of the vector with respect to a local orthonormal frame. Then

$$\nabla_\mu \Psi^a = \partial_\mu \Psi^a + i\omega_\mu{}^{cd} (\Sigma_{cd})^a{}_b \Psi^b, \tag{5.272}$$

where $(\Sigma_{cd})^a{}_b$ denotes the elements of the matrix Σ_{cd} appropriate to a vector representation of the Lorentz group. From (5.256) and (5.258), we have

$$\Psi'^a = [\delta^a{}_b + i\delta\epsilon^{cd} (\Sigma_{cd})^a{}_b] \Psi^b \tag{5.273}$$

under an infinitesimal Lorentz transformation. Because Ψ^a is a vector, we also have

$$\Psi'^a = (\delta^a{}_b + \delta\epsilon^a{}_b) \Psi^b, \tag{5.274}$$

which is the infinitesimal form of (5.237). Comparison of (5.273) with (5.274) shows that

$$(\Sigma_{cd})^a{}_b = -\frac{i}{2}(\delta^a_c \eta_{bd} - \delta^a_d \eta_{bc}). \tag{5.275}$$

(This may be easily shown to satisfy (5.261).) Using (5.275) in (5.272) leads to

$$\nabla_\mu \Psi^a = \partial_\mu \Psi^a + \omega_\mu{}^a{}_b \Psi^b$$

in agreement with our earlier result (5.241). The extension of the formalism to arbitrary tensor fields should be obvious.

5.6 Spinors in curved spacetime

We now want to consider the Dirac equation. In flat spacetime it reads

$$[i\gamma^a \partial_a - m]\Psi(x) = 0, \quad (5.276)$$

where the γ^a matrices satisfy

$$\{\gamma^a, \gamma^b\} = 2\eta^{ab}. \quad (5.277)$$

We will consider some of the properties of the Dirac γ-matrices in an arbitrary number of dimensions later.

The generalization of the Dirac equation to curved spacetime is obtained by replacing $\partial_a \to e_a{}^\mu \nabla_\mu$, where ∇_μ is an appropriate covariant derivative for a spinor representation of the Lorentz group. We then have

$$[i\gamma^a e_a{}^\mu \nabla_\mu - m]\Psi(x) = 0 \quad (5.278)$$

as the Dirac equation in curved spacetime. It is possible to regard

$$\underline{\gamma}^\mu = e_a{}^\mu \gamma^a \quad (5.279)$$

as spacetime-dependent γ-matrices as in Section 3.9, which satisfy[10]

$$\{\underline{\gamma}^\mu, \underline{\gamma}^\nu\} = 2g^{\mu\nu}(x), \quad (5.280)$$

in which case we may write

$$[i\underline{\gamma}^\mu \nabla_\mu - m]\Psi(x) = 0. \quad (5.281)$$

The only thing that remains is to find an explicit form for the spin connection.

Under a combined local Lorentz and general coordinate transformation we have

$$\Psi'(x') = S(L(x))\Psi(x),$$

$$e'_b{}^\nu(x') = \frac{\partial x'^\nu}{\partial x^\mu}(L^{-1})^a{}_b e_a{}^\mu(x).$$

Using these results in (5.278) leads to

$$[iS(L(x))\gamma^a S^{-1}(L(x)) L^b{}_a(x) e'_b{}^\nu(x') \nabla'_\nu - m]\Psi'(x') = 0.$$

For covariance of the Dirac equation we must therefore require

$$S(L(x))\gamma^a S^{-1}(L(x)) L^b{}_a(x) = \gamma^b. \quad (5.282)$$

This is identical to the requirement in flat spacetime (Bjorken and Drell 1964) that the Dirac equation be covariant under a Lorentz transformation, except that the transformation here is local. Using the infinitesimal form of the transformation given in (5.257) and (5.258) leads to

$$0 = \delta\epsilon^a{}_b \gamma^b + i\delta\epsilon_{cd}[\Sigma^{cd}, \gamma^a]. \quad (5.283)$$

---

[10] See (3.206).

This is satisfied if

$$\Sigma^{cd} = -\frac{i}{8}[\gamma^c, \gamma^d]. \tag{5.284}$$

(The Lie algebra of the Lorentz group (5.261) is also satisfied with this choice.) The covariant derivative of a Dirac spinor is therefore

$$\nabla_\mu \Psi = \partial_\mu \Psi + \frac{1}{8}\omega_\mu{}^{ab}[\gamma_a, \gamma_b]\Psi. \tag{5.285}$$

Equation (5.271) becomes

$$[\nabla_\mu, \nabla_\nu]\Psi = -\frac{1}{8}R_{\mu\nu}{}^{ab}[\gamma_a, \gamma_b]\Psi. \tag{5.286}$$

The transformation of the spacetime-dependent γ-matrices follows from (5.282) and the properties of the n-bein. We have

$$\underline{\gamma}'^\mu(x') = e'_a{}^\mu(x')\gamma^a$$

$$= \frac{\partial x'^\mu}{\partial x^\nu} S(L(x))\underline{\gamma}^\nu(x)S^{-1}(L(x)). \tag{5.287}$$

(This result may also be found from the covariance of (5.281).) In order to define the covariant derivative of an object which transforms like (5.287), consider first the case of a field $\Phi(x)$ which transforms like

$$\Phi'(x') = S(L(x))\Phi(x)S^{-1}(L(x)). \tag{5.288}$$

Such a field is said to transform under the adjoint representation. We want the covariant derivative of Φ to transform in the same way as Φ:

$$\nabla'_\mu \Phi'(x') = S(L(x))\nabla_\mu\Phi(x)S^{-1}(L(x)). \tag{5.289}$$

It is easily verified using (5.264) that

$$\nabla_\mu \Phi = \partial_\mu \Phi + [B_\mu, \Phi] \tag{5.290}$$

satisfies (5.289) with B_μ given in (5.266) and (5.269). The only difference between (5.287) and (5.288) is the presence of the additional factor of $\partial x'^\mu/\partial x^\nu$ which is associated with the vector index of $\underline{\gamma}^\mu$. Thus we should define

$$\nabla_\mu \underline{\gamma}^\nu = \partial_\mu \underline{\gamma}^\nu + \Gamma^\nu_{\mu\lambda}\underline{\gamma}^\lambda + [B_\mu, \underline{\gamma}^\nu], \tag{5.291}$$

which may be verified to transform covariantly under (5.287). Using $B_\mu = \frac{1}{8}\omega_{\mu ab}[\gamma^a, \gamma^b]$, which is the spin connection for a Dirac field, and (5.249), it may be shown that

$$\nabla_\mu \underline{\gamma}^\nu = 0. \tag{5.292}$$

This result also implies $\nabla_\mu \gamma^a = 0$, since the n-bein is covariantly constant.

Finally we use (5.286) and (5.292) to derive the important result

$$(\underline{\gamma}^\mu \nabla_\mu)^2 = \Box + \frac{1}{4}R. \tag{5.293}$$

Multiplication of both sides of (5.286) by $\underline{\gamma}^\mu \underline{\gamma}^\nu$ gives

$$\underline{\gamma}^\mu \underline{\gamma}^\nu (\nabla_\mu \nabla_\nu - \nabla_\nu \nabla_\mu)\Psi = -\frac{1}{4}\underline{\gamma}^\mu \underline{\gamma}^\nu \underline{\gamma}^\lambda \underline{\gamma}^\sigma R_{\mu\nu\lambda\sigma}\Psi. \qquad (5.294)$$

Using the cyclic identity,

$$R_{\mu\nu\lambda\sigma} + R_{\mu\lambda\sigma\nu} + R_{\mu\sigma\nu\lambda} = 0,$$

and the anticommutation relations (5.280) leads to

$$\underline{\gamma}^\mu \underline{\gamma}^\nu \underline{\gamma}^\lambda \underline{\gamma}^\sigma R_{\mu\nu\lambda\sigma} = -2R. \qquad (5.295)$$

(We have used $R_{\mu\nu} = R^\lambda{}_{\mu\lambda\nu}$ here.) The left-hand side of (5.294) is easily evaluated using (5.280) to obtain

$$\underline{\gamma}^\mu \underline{\gamma}^\nu (\nabla_\mu \nabla_\nu - \nabla_\nu \nabla_\mu) = (\underline{\gamma}^\mu \nabla_\mu)^2 - (2g^{\nu\mu} - \underline{\gamma}^\nu \underline{\gamma}^\mu)\nabla_\nu \nabla_\mu$$
$$= -2\Box + 2(\underline{\gamma}^\mu \nabla_\mu)^2. \qquad (5.296)$$

Combining (5.295) and (5.296) with (5.294) gives the result claimed in (5.293).

5.6.2 Dirac matrices in n dimensions

The Dirac matrices have already been defined as satisfying (5.277). The set of Dirac matrices, along with the unit matrix, is said to generate a Clifford algebra $\mathcal{C}$. $\mathcal{C}$ is defined to be the set of all linear combinations of all products formed from elements of the set $\{I, \gamma^0, \gamma^1, \ldots, \gamma^{n-1}\}$, where n is the spacetime dimension. A basis for $\mathcal{C}$ is

$$\{I, \gamma^a, \gamma^{a_1}\gamma^{a_2}, \ldots, \gamma^{a_1} \cdots \gamma^{a_r}, \ldots, \gamma^0 \gamma^1 \cdots \gamma^{n-1}\},$$

where $0 \leq a_1 < a_2 < \cdots \leq n-1$. The dimension of $\mathcal{C}$ is seen to be 2^n. Because $\mathcal{C}$ is finite dimensional, it follows that the basis elements may be chosen to be unitary matrices. From (5.277) we have $(\gamma^0)^2 = I$ and $(\gamma^i)^2 = -I$ for $i = 1, \ldots, n-1$. We may therefore choose $(\gamma^0)^\dagger = \gamma^0$ and $(\gamma^i)^\dagger = -\gamma^i$. These last results may be summarized as

$$(\gamma^a)^\dagger = \gamma^0 \gamma^a \gamma^0. \qquad (5.297)$$

In order to proceed, it is convenient to deal separately with the cases of even- and odd-dimensional spacetimes.[11]

---

[11] For more on the properties of the Dirac matrices, see Brauer and Weyl (1935) and Case (1955).

n even

It is possible to prove that there is a representation for γ^a in terms of $2^{n/2} \times 2^{n/2}$ complex matrices, and that this representation is unique up to a similarity transformation. That is, if γ^a and γ'^a give the Dirac matrices in two different representations of the same dimension, then there exists an invertible matrix M such that $\gamma'^a = M\gamma^a M^{-1}$.

By taking the complex conjugate of (5.277) it is clear that $-(\gamma^a)^*$ satisfies the same anticommutation relations as γ^a. By Schur's lemma, it follows that the same Clifford algebra is generated by $-(\gamma^a)^*$, and that there must be a non-singular matrix B such that

$$(\gamma^a)^* = -B\gamma^a B^{-1}. \tag{5.298}$$

By taking the transpose of (5.277) and applying a similar reasoning to that which led to (5.298), it follows that

$$(\gamma^a)^T = -C\gamma^a C^{-1} \tag{5.299}$$

for some matrix C. The overall scales and phases of B and C are not determined by (5.298) and (5.299). We will fix the scales by

$$|\det B| = |\det C| = 1. \tag{5.300}$$

Taking the complex conjugate of (5.298) gives $\gamma^a = -B^*(\gamma^a)^* B^{*-1}$. Using (5.298) again shows that $[B^*B, \gamma^a] = 0$. It follows from Schur's lemma that

$$B^*B = \epsilon I \tag{5.301}$$

for some real constant ϵ. If the scale of B is fixed as in (5.300), then $\epsilon = \pm 1$ are the only values allowed.

Taking the complex conjugate of (5.297) shows that

$$\begin{aligned}(\gamma^a)^T &= (\gamma^0)^*(\gamma^a)^*(\gamma^0)^* \\ &= -B(\gamma^a)^\dagger B^{-1}\end{aligned} \tag{5.302}$$

upon use of (5.298). Also

$$\begin{aligned}(\gamma^a)^T &= [(\gamma^a)^*]^\dagger \\ &= -(B^{-1})^\dagger (\gamma^a)^\dagger B^\dagger\end{aligned} \tag{5.303}$$

again upon use of (5.298). Comparison of (5.302) and (5.303) shows that $[B^\dagger B, \gamma^a] = 0$. Schur's lemma, combined with the normalization (5.300) is sufficient to prove that B must be a unitary matrix,

$$B^\dagger B = I. \tag{5.304}$$

Equation (5.301) and (5.304) may be combined to show that

$$B^T = \epsilon B, \tag{5.305}$$

proving that B is either symmetric or antisymmetric depending upon the sign of ϵ.

The matrix C defined in (5.232) may be easily related to B. From (5.297) and (5.302) we have

$$(\gamma^a)^T = -(B\gamma^0)\gamma^a(B\gamma^0)^{-1}. \tag{5.306}$$

Comparison of this result with (5.299) shows that $C = \epsilon' B\gamma^0$ for some constant ϵ'. The normalization (5.300) restricts $|\epsilon'| = 1$. We will choose the phase of C by fixing $\epsilon' = 1$, and so

$$C = B\gamma^0. \tag{5.307}$$

It is then easy to see that C is unitary and satisfies

$$C^T = -\epsilon C. \tag{5.308}$$

The sign of ϵ depends on the spacetime dimension in an interesting way. To determine this, consider $C\gamma^{a_1}\cdots\gamma^{a_r}C^{-1}$ where $\gamma^{a_1}\cdots\gamma^{a_r}$ is one of the basis elements of the Clifford algebra $\mathcal{C}$. Using (5.299), it may be seen that

$$C\gamma^{a_1}\cdots\gamma^{a_r}C^{-1} = (-1)^{\frac{1}{2}r(r+1)}(\gamma^{a_1}\cdots\gamma^{a_r})^T. \tag{5.309}$$

Also

$$(C\gamma^{a_1}\cdots\gamma^{a_r})^T = (\gamma^{a_1}\cdots\gamma^{a_r})^T C^T$$
$$= -\epsilon(-1)^{\frac{1}{2}r(r+1)} C\gamma^{a_1}\cdots\gamma^{a_r} \tag{5.310}$$

if (5.308) and (5.309) are used. This shows that $C\gamma^{a_1}\cdots\gamma^{a_r}$ is either symmetric or antisymmetric depending on the sign of $-\epsilon(-1)^{\frac{1}{2}r(r+1)}$. Symmetric matrices of the form $C\gamma^{a_1}\cdots\gamma^{a_r}$ form a basis for the set of all symmetric matrices of dimension $2^{n/2}\times 2^{n/2}$. There are a total of $\frac{1}{2}2^{n/2}(2^{n/2}+1)$ linearly independent symmetric matrices of this dimension. However, there are $\binom{n}{r}$ basis elements of the form $C\gamma^{a_1}\cdots\gamma^{a_r}$, where $r = 0, \ldots, n$. We must therefore have

$$\frac{1}{2}2^{n/2}(2^{n/2}+1) = \sum_{r=0}^{n}\binom{n}{r}\frac{1}{2}[1-\epsilon(-1)^{\frac{1}{2}r(r+1)}]. \tag{5.311}$$

The summation on the right-hand side is easily performed by noting that $(-1)^{\frac{1}{2}r(r+1)} = \sqrt{2}\cos\frac{\pi}{4}(2r+1)$. Then[12]

$$\sum_{r=0}^{n}\binom{n}{r}(-1)^{\frac{1}{2}r(r+1)} = \sqrt{2}\,\Re\left[\sum_{r=0}^{n}\binom{n}{r}\exp\{i\pi(2r+1)/4\}\right]$$
$$= \sqrt{2}\,\Re\left[\exp(i\pi/4)(1+\exp\{i\pi/2\})^n\right]$$
$$= 2^{\frac{n+1}{2}}\cos\frac{\pi}{4}(n+1).$$

Equation (5.311) then implies that

$$\epsilon = -\sqrt{2}\cos\frac{\pi}{4}(n+1). \tag{5.312}$$

---

[12] Here $\Re[\cdots]$ denotes the real part of any complex number.

Thus
$$\epsilon = \begin{cases} +1 & \text{if } n = 2, 4 \pmod 8 \\ -1 & \text{if } n = 0, 6 \pmod 8. \end{cases}$$

We will return to the significance of this later.

The special basis element $\gamma^0 \gamma^1 \cdots \gamma^{n-1}$ of the Clifford algebra plays an important role in physics. In the case $n = 4$ it is proportional to the γ_5 matrix. We will define
$$\gamma_{n+1} = \eta \gamma^0 \gamma^1 \cdots \gamma^{n-1}, \tag{5.313}$$
where η is chosen such that
$$(\gamma_{n+1})^2 = I, \tag{5.314}$$
$$(\gamma_{n+1})^\dagger = \gamma_{n+1}. \tag{5.315}$$

It is straightforward to show from (5.313) that
$$\eta^2 = (-1)^{\frac{n-2}{2}}, \tag{5.316}$$
which means that η is real if $n/2$ is odd, and η is imaginary if $n/2$ is even. From (5.277) and (5.313), we have
$$\{\gamma_{n+1}, \gamma^a\} = 0, \tag{5.317}$$
with $\{\,,\,\}$ the anticommutator. Note that all of the results in this section have been established in a representation-independent manner.

n odd

We may define γ_{n+1} as in (5.313)–(5.315). This time it may be seen that
$$\eta^2 = (-1)^{\frac{n-1}{2}}. \tag{5.318}$$

In place of (5.317), it is easily seen that
$$[\gamma_{n+1}, \gamma^a] = 0. \tag{5.319}$$

As a consequence of this result, γ_{n+1} commutes with all of the generators of the Clifford algebra $\mathcal{C}$. Thus, the algebra has a non-trivial centre generated by the set of two elements $\{I, \gamma_{n+1}\}$. (The centre is the set of all basis elements that commute with every basis element in the algebra.) Because γ_{n+1} commutes with all elements in the Clifford algebra, and $(\gamma_{n+1})^2 = I$, we can have $\gamma_{n+1} = \pm I$, proving that $\mathcal{C}$ has a centre isomorphic to Z_2 if the spacetime dimension is odd. (In contrast, the centre is trivial, consisting of only the identity element, if the spacetime dimension is even.)

Because of the non-trivial centre, there are two inequivalent irreducible representations for the γ-matrices corresponding to the two possibilities $\gamma_{n+1} =$

5.6 Spinors in curved spacetime

$\pm I$. Each representation is unique up to a similarity transformation, and has dimension $2^{\frac{n-1}{2}} \times 2^{\frac{n-1}{2}}$. Write

$$\gamma_{n+1} = \gamma I, \qquad (5.320)$$

where $\gamma = \pm 1$. We may solve for γ^{n-1} in terms of the other $(n-1)$ Dirac γ-matrices:

$$\gamma^{n-1} = -\gamma\eta\gamma^0\gamma^1\cdots\gamma^{n-2} \qquad (5.321)$$

using (5.313). As there is an irreducible representation in terms of $2^{\frac{n-1}{2}} \times 2^{\frac{n-1}{2}}$ matrices, we may choose $\gamma^0, \gamma^1, \ldots, \gamma^{n-2}$ to correspond to the irreducible representation in the even-dimensional spacetime of dimension $n-1$.

The matrices B and C may be defined as in (5.298) and (5.299). However, care must be taken to ensure that the representations provided by $\{(\gamma^a)^*\}$ and $\{(\gamma^a)^T\}$ are equivalent to that provided by $\{\gamma^a\}$. This will only be true if in the representation $\{\gamma'^a\}$ we have $\gamma'_{n+1} = \eta\gamma'^0\cdots\gamma'^{n-1} = \gamma I$ when $\gamma_{n+1} = \gamma I$.

Consider first the representation provided by $\{-\gamma^a\}$. In this case $\gamma'_{n+1} = -\gamma_{n+1}$, and the representation is always inequivalent to $\{\gamma^a\}$.

Next, consider the representation given by $\{(\gamma^a)^*\}$. Then

$$\gamma'_{n+1} = \eta(\gamma^0)^* \cdots (\gamma^{n-1})^*$$
$$= \gamma(-1)^{(n+1)/2} I$$

if (5.321) is used. This representation is only equivalent to that given by $\{\gamma^a\}$ if $n = 3 \pmod 4$. If $n = 1 \pmod 4$, then the representation is equivalent to that given by $\{-\gamma^a\}$.

Finally consider $\{(\gamma^a)^T\}$. It is again found that $\gamma'_{n+1} = \gamma(-1)^{(n+1)/2} I$, proving that if $n = 1 \pmod 4$ the representation is equivalent to $\{-\gamma^a\}$, and if $n = 3 \pmod 4$ the representation is equivalent to $\{\gamma^a\}$.

It therefore follows that if $n = 3 \pmod 4$, we may take

$$(\gamma^a)^* = -B_e\gamma^a B_e^{-1}, \qquad (5.322)$$
$$(\gamma^a)^T = -C_e\gamma^a C_e^{-1}, \qquad (5.323)$$

where B_e and C_e are the matrices found for the case of the even-dimensional spacetime. A direct computation using (5.321) shows that

$$(\gamma^{n-1})^* = -B_e\gamma^{n-1} B_e^{-1}, \qquad (5.324)$$
$$(\gamma^{n-1})^T = -C_e\gamma^{n-1} C_e^{-1}. \qquad (5.325)$$

When $n = 1 \pmod 4$, because the representations provided by $\{(\gamma^a)^*\}$ and $\{(\gamma^a)^T\}$ are equivalent to that given by $\{-\gamma^a\}$, we have

$$(\gamma^a)^* = B\gamma^a B^{-1}, \qquad (5.326)$$
$$(\gamma^a)^T = C\gamma^a C^{-1}. \qquad (5.327)$$

It may be seen that the only matrices satisfying these relations are

$$B = B_e \gamma^{n-1}, \tag{5.328}$$
$$C = C_e \gamma^{n-1}. \tag{5.329}$$

Explicit representations

In all of our considerations so far, we have not had to use an explicit form for the γ-matrices. Often an explicit form is useful. Representations for the γ-matrices in arbitrary-dimensional spacetimes may be constructed from those in lower dimensions by taking tensor products.

One useful representation for a spacetime of even dimension is called the Weyl representation. In this representation,

$$\gamma_{n+1} = \begin{pmatrix} I & 0 \\ 0 & -I \end{pmatrix}. \tag{5.330}$$

This representation always exists in an even-dimensional spacetime. Once we have found the Weyl representation in a spacetime of even dimension n, we can always find the representation in the odd-dimensional spacetime $n+1$ as described above. We will illustrate this construction explicitly here.

Consider first the case $n = 2$. The Pauli matrices may be defined by

$$\sigma_1 = \begin{pmatrix} 0 & 1 \\ 1 & 0 \end{pmatrix}, \sigma_2 = \begin{pmatrix} 0 & -i \\ i & 0 \end{pmatrix}, \sigma_3 = \begin{pmatrix} 1 & 0 \\ 0 & -1 \end{pmatrix}. \tag{5.331}$$

These matrices are Hermitian, and satisfy

$$\{\sigma_i, \sigma_j\} = 2\delta_{ij}, \tag{5.332}$$
$$\sigma_j \sigma_k = i\sigma_l \quad (j, k, l \text{ cyclic}). \tag{5.333}$$

It is then easy to see that the γ-matrices for $n = 2$ in the Weyl representation are

$$\gamma^0 = \sigma_1, \gamma^1 = i\sigma_2, \gamma_3 = \sigma_3. \tag{5.334}$$

For $n = 3$ they are then given by

$$\gamma^0 = \sigma_1, \gamma^1 = i\sigma_2, \gamma^2 = i\sigma_3. \tag{5.335}$$

Suppose that we have constructed the γ-matrices in the Weyl representation for a spacetime of even dimension $2n$. Denote the $2n$ γ-matrices by $\gamma^0, \gamma^1, \ldots, \gamma^{2n-1}$. γ_{2n+1} is given by (5.330). They are all $2^n \times 2^n$ dimensional matrices. Let $\Gamma^0, \Gamma^1, \ldots, \Gamma^{2n+1}$ be the γ-matrices in the Weyl representation in the spacetime of dimension $2n + 2$. It may easily be seen that the following tensor products satisfy the correct anticommutation relations:

$$\Gamma^0 = \sigma_1 \otimes I_{2n}, \Gamma^1 = i\sigma_2 \otimes \gamma^0, \Gamma^{j+1} = \sigma_2 \otimes \gamma^j \ (j = 1, \ldots, 2n - 1) \tag{5.336}$$

$$\Gamma^{2n+1} = i\sigma_2 \otimes \gamma_{2n+1}. \tag{5.337}$$

Here I_{2^n} denotes the $2^n \times 2^n$ dimensional unit matrix and $\otimes$ is the tensor product. (For the reader unfamiliar with tensor products of matrices, see Messiah (1961) for example.) More explicitly, we have

$$\Gamma^0 = \begin{pmatrix} 0 & I_{2^n} \\ I_{2^n} & 0 \end{pmatrix}, \Gamma^1 = \begin{pmatrix} 0 & \gamma^0 \\ -\gamma^0 & 0 \end{pmatrix}, \tag{5.338}$$

$$\Gamma^{j+1} = \begin{pmatrix} 0 & -i\gamma^j \\ i\gamma^j & 0 \end{pmatrix} \quad (j = 1, \ldots, 2n-1), \tag{5.339}$$

$$\Gamma^{2n+1} = \begin{pmatrix} 0 & \gamma_{2n+1} \\ -\gamma_{2n+1} & 0 \end{pmatrix}. \tag{5.340}$$

The results for $n = 4, 5$ are

$$\gamma^0 = \begin{pmatrix} 0 & I_2 \\ I_2 & 0 \end{pmatrix}, \gamma^j = \begin{pmatrix} 0 & \sigma_j \\ -\sigma_j & 0 \end{pmatrix} \quad (j = 1, 2, 3) \tag{5.341}$$

and

$$\gamma^0 = \begin{pmatrix} 0 & I_2 \\ I_2 & 0 \end{pmatrix}, \gamma^j = \begin{pmatrix} 0 & \sigma_j \\ -\sigma_j & 0 \end{pmatrix} \quad (j = 1, 2, 3), \gamma^4 = \begin{pmatrix} iI_2 & 0 \\ 0 & -iI_2 \end{pmatrix}, \tag{5.342}$$

respectively. It should be noted that (5.330) results from the the choice $\eta = -i^{n/2-1}$ for n even.

5.6.3 Dirac, Weyl, and Majorana spinors

The Dirac equation in curved spacetime has already been written down in (5.281). If we allow the Dirac field to be coupled to an electromagnetic field, we have

$$[i\gamma^\mu D_\mu - m]\Psi(x) = 0, \tag{5.343}$$

where $D_\mu = \nabla_\mu + ieA_\mu(x)$ is the gauge-covariant derivative. The Dirac matrices are given by an irreducible representation in terms of $2^{[n/2]} \times 2^{[n/2]}$ complex matrices. (Here $[n/2]$ denotes the greatest integer less than or equal to $n/2$.) This means that a Dirac spinor in n-dimensional spacetime has $2^{[n/2]}$ components.

We will initially assume that the spacetime dimension is even. Define the projection operators P_R and P_L by

$$P_R = \frac{1}{2}(I + \gamma_{n+1}), P_L = \frac{1}{2}(I - \gamma_{n+1}) \tag{5.344}$$

with the properties

$$P_R^2 = P_R, P_R P_L = P_L P_R = 0, P_L^2 = P_L. \tag{5.345}$$

These properties follow as a consequence of (5.314). (Note that γ_{n+1} in (5.344) is taken in the local orthonormal frame, and so is a constant matrix.) Using these projection operators, we may define a right-handed Weyl spinor by

$$P_R\Psi_R(x) = \Psi_R(x), P_L\Psi_R(x) = 0. \tag{5.346}$$

Using (5.344), it may be seen that

$$\gamma_{n+1}\Psi_R(x) = \Psi_R(x). \tag{5.347}$$

Similarly, a left-handed Weyl spinor may be defined by

$$P_L\Psi_L(x) = \Psi_L(x), P_R\Psi_L(x) = 0, \tag{5.348}$$

or

$$\gamma_{n+1}\Psi_L(x) = -\Psi_L(x). \tag{5.349}$$

Weyl spinors are also sometimes referred to as chiral spinors. Right-handed spinors are said to have negative chirality, and left-handed spinors to have positive chirality, the terminology coming from the eigenvalue of γ_{n+1}.

The chirality conditions (5.347) and (5.349) must be compatible with the Dirac equation. Assume that $\Psi(x)$ has a definite handedness and satisfies (5.343). Because $\{\gamma_{n+1},\gamma^\mu\} = 0$, it is easy to see that $\Psi_{L,R}$ must be massless, and hence uncharged. A Weyl spinor therefore satisfies

$$i\gamma^\mu\nabla_\mu\Psi(x) = 0, \tag{5.350}$$

in addition to (5.347) or (5.349).

If Ψ_D is a Dirac spinor, then we can always write

$$\Psi_D(x) = \Psi_L(x) + \Psi_R(x), \tag{5.351}$$

where $\Psi_{L,R}$ satisfy (5.347) and (5.349). (Note that $\Psi_{L,R}$ are not Weyl spinors unless $m = 0$ and $e = 0$.) If $\Psi_L = 0$ or $\Psi_R = 0$, then Ψ_D is a Weyl spinor, proving that a Weyl spinor has half the number of degrees of freedom of a Dirac spinor. If we choose the Weyl representation for the Dirac γ-matrices, then

$$P_R = \begin{pmatrix} I & 0 \\ 0 & 0 \end{pmatrix}, P_L = \begin{pmatrix} 0 & 0 \\ 0 & I \end{pmatrix} \tag{5.352}$$

and so

$$\Psi_D = \begin{pmatrix} \psi_R \\ \psi_L \end{pmatrix} \tag{5.353}$$

where $\psi_{L,R}$ have $\frac{1}{2}2^{[n/2]}$ components.

If the spacetime dimension is odd, then there is no element in the Clifford algebra which anticommutes with the other elements, and Weyl spinors cannot be defined. Another way to see this is to note that in odd-dimensional spacetimes letting $\gamma^a \to -\gamma^a$ changes $\gamma_{n+1} \to -\gamma_{n+1}$, which gives an inequivalent representation of the Clifford algebra. All physical consequences should be independent of which representation is chosen; however, if we define $\Psi_{L,R}$ as in (5.347) and (5.349), then the transformation $\gamma^a \to -\gamma^a$ interchanges Ψ_R and Ψ_L. Thus a consistent choice of chirality cannot be made in odd-dimensional spacetimes.

5.6 Spinors in curved spacetime

Suppose that we take the complex conjugate of the Dirac equation in an even-dimensional spacetime. This gives

$$[i(\gamma^\mu)^*(\nabla_\mu^* - ieA_\mu) + m]\Psi^*(x) = 0, \qquad (5.354)$$

where the complex conjugate on ∇_μ is necessary because of the presence of the spin connection in the spacetime covariant derivative. Using $(\gamma^\mu)^* = -B\gamma^\mu B^{-1}$, this equation may be written as

$$[i\gamma^\mu(\nabla_\mu - ieA_\mu) - m]B^{-1}\Psi^*(x) = 0. \qquad (5.355)$$

If we define

$$\Psi^c = B^{-1}\Psi^*, \qquad (5.356)$$

we see that Ψ^c satisfies the same equation as Ψ, but with $e \to -e$. Thus Ψ^c is the charge conjugate spinor representing the antiparticle of Ψ.

A spinor Ψ_M is called a Majorana spinor if

$$\Psi_M^c = \Psi_M. \qquad (5.357)$$

We will find later that this requirement can only be realized in certain spacetime dimensions. A Majorana spinor is its own antiparticle. It must therefore be uncharged. (A Majorana spinor may be thought of as the spinor analogue of a real scalar field.) We have

$$\Psi_M^* = B\Psi_M \qquad (5.358)$$

for a Majorana spinor Ψ_M. Taking the complex conjugate of (5.358) gives $\Psi_M = B^*\Psi_M^*$. Using this in (5.358) along with (5.301) gives

$$\Psi_M^* = BB^*\Psi_M^* = \epsilon\Psi_M^*. \qquad (5.359)$$

For consistency we must have $\epsilon = 1$, which means that for n even, Majorana spinors can only be defined if $n = 2, 4 \pmod 8$. The situation for $n = 3 \pmod 8$ is similar to that for $n = 2 \pmod 8$ since $(\gamma^\mu)^* = -B\gamma^\mu B^{-1}$ still holds. (See (5.322).) For other odd-dimensional spacetimes, the representation provided by $\{(\gamma^a)^*\}$ is not equivalent to that provided by $\{\gamma^a\}$, and Majorana spinors may not be defined. Thus, only in spacetimes of dimensionality $n = 2, 3, 4 \pmod 8$ may Majorana spinors be defined.

It may be shown (Gliozzi et al., 1977) that when Majorana spinors are defined, it is always possible to choose $B = I$. This means that the Majorana condition (5.357) becomes simply $\Psi^* = \Psi$, so that the spinor is real. This shows that a Majorana spinor has half the number of degrees of freedom of a Dirac spinor. The choice of representation of the γ-matrices with $B = I$ is called a Majorana representation. It is seen from (5.298) to correspond to a representation in which the γ-matrices are purely imaginary. The Dirac matrices in the Majorana representation for $n = 2, 3$ and 4 are respectively

$$n = 2 : \gamma^0 = \sigma_2, \gamma^1 = i\sigma_1 ;$$
$$n = 3 : \gamma^0 = \sigma_2, \gamma^1 = i\sigma_1, \gamma^2 = -i\sigma_3 ;$$
$$n = 4 : \gamma^0 = \sigma_2 \otimes \sigma_3 = \begin{pmatrix} 0 & i\sigma_3 \\ -i\sigma_3 & 0 \end{pmatrix}$$
$$\gamma^1 = i\sigma_1 \otimes I_2 = \begin{pmatrix} 0 & iI_2 \\ -iI_2 & 0 \end{pmatrix}$$
$$\gamma^3 = i\sigma_3 \otimes I_2 = \begin{pmatrix} iI_2 & 0 \\ 0 & -iI_2 \end{pmatrix}$$
$$\gamma^4 = i\sigma_2 \otimes \sigma_2 = \begin{pmatrix} 0 & \sigma_2 \\ -\sigma_2 & 0 \end{pmatrix}.$$

The Majorana representation is related to the Weyl representation by a similarity transformation.

Finally, we will consider when it is possible to define a spinor which is both Weyl and Majorana. First of all, n must be even in order to define a Weyl spinor. From the preceding discussion, $n = 2, 4 \pmod 8$ are the only potential dimensions for which Weyl–Majorana spinors are possible. The Weyl condition states that $\gamma_{n+1} \Psi = \pm \Psi$. Since $\Psi = B^{-1} \Psi^*$ is the Majorana condition, a Weyl–Majorana spinor must satisfy

$$B \gamma_{n+1} B^{-1} \Psi^* = \pm \Psi^*.$$

Taking the complex conjugate of the Weyl condition gives

$$\gamma_{n+1}^* \Psi^* = \pm \Psi^*.$$

Comparison of these last two results shows that

$$B \gamma_{n+1} B^{-1} = \gamma_{n+1}^*.$$

Using (5.313), it follows that

$$\gamma_{n+1}^* = \eta^* (\gamma^0)^* \cdots (\gamma^{n-1})^*$$
$$= \frac{\eta^*}{\eta} B \gamma_{n+1} B^{-1}.$$

Comparison of the two results for γ_{n+1}^* shows that η must be real. We had, from (5.316), $\eta^2 = (-1)^{n/2-1}$, so that η is real only if $n = 2 \pmod 4$. Therefore reality of η is only compatible with the Majorana condition if $n = 2 \pmod 8$. Hence Weyl–Majorana spinors may only be defined in spacetimes of dimensionality $n = 2 \pmod 8$. In particular, they do not exist in four spacetime dimensions.

5.6.4 Global considerations

A comprehensive treatment of this topic would take us rather far away from quantum theory into the realm of algebraic topology. Apart from a few introductory remarks, we will concentrate on issues of relevance for quantum theory. This section assumes a cursory knowledge of fiber bundle theory.[13]

The Dirac equation, as we have presented it, has used the n-bein formalism in a crucial way. In order that a Dirac spinor be globally defined, it is necessary that the n-bein exist globally. The n-bein may be regarded as the components of vector fields which are orthonormal. Thus, the global existence of a Dirac spinor can be related to the existence of a set of orthonormal vectors that, as a consequence of orthonormality, can never vanish. If the spacetime M admits such vector fields, it is called parallelizable, and spinor fields may be globally defined. (Another way to state this is that the bundle of orthonormal frames must be trivial, and hence have a global cross section. This cross section defines the local orthonormal frame.)

Of course not every spacetime M will be parallelizable. The classic example of a spacetime which is not parallelizable is $\mathbb{S}^2$. (This result is often picturesquely phrased as the statement that you cannot comb the hair on a billiard ball.) The only spheres which are parallelizable are $\mathbb{S}^1, \mathbb{S}^3$, and $\mathbb{S}^7$. If M is not parallelizable, it does not mean that spinor fields cannot be globally defined. The real determination is whether or not the second Steiffel–Whitney class of the tangent bundle vanishes. This characteristic class is determined by the cohomology group $H^2(M; \mathbb{Z}_2)$. It is necessary and sufficient that $H^2(M; \mathbb{Z}_2) = 0$ for the global existence of spinor fields. Geroch (1968) has shown that this holds automatically for four-dimensional non-compact spacetimes. An example of a space which does not possess a spinor structure is $\mathbb{C}P^{2n}$ (Milnor and Stasheff 1974). Because $H^2(\mathbb{S}^n; \mathbb{Z}_2) = 0$, all spheres possess a spinor structure, parallelizable or not. A spacetime M without a spinor structure is of no physical interest. Accordingly, we will assume that $H^2(M; \mathbb{Z}_2) = 0$. Even if the second Steiffel–Whitney class of the tangent bundle does not vanish, it may still be possible to define a generalized spin structure. The basic idea here is to replace $Spin(1, n-1)$, which is the double covering group of the Lorentz group, with some other group that does not lead to topological restrictions (or at least weaker topological restrictions than the vanishing of the second Steiffel–Whitney class). An extensive discussion of this, along with references to the literature, is contained in Avis and Isham (1980).

Given a spacetime M which admits a spinor structure, the next question which arises concerns the uniqueness of the structure.[14] The number of inequivalent spin structures is classified by the cohomology group $H^1(M; \mathbb{Z}_2)$. Because $H^1(M; \mathbb{Z}_2) = \text{Hom}(\pi_1(M), \mathbb{Z}_2)$, where Hom denotes the set of homomorphisms

---

[13] See Eguchi *et al.* (1980) for example.
[14] A more complete treatment has been given by Avis and Isham (1979a,b).

between the fundamental group $\pi_1(M)$ and $\mathbb{Z}_2$, it may be seen that if the spacetime M is simply connected, then there is a unique spin structure. For example, if $M = \mathbb{R}^n \times \mathbb{S}^m$ ($m > 2$), then $\pi_1(M) = \pi_1(\mathbb{S}^m) = 0$, and therefore M has a unique spin structure. If we take $M = \mathbb{R}^n \times \mathbb{S}^1$, then $\pi_1(M) = \pi_1(\mathbb{S}^1) = \mathbb{Z}$ and $\text{Hom}(\mathbb{Z}, \mathbb{Z}_2) = \mathbb{Z}_2$. Therefore, $\mathbb{R}^n \times \mathbb{S}^1$ has two inequivalent spin structures. As we will see, these two spin structures may be interpreted as the choice of periodic and antiperiodic boundary conditions around the circle.[15]

Another way of viewing inequivalent spin structures is as follows. A Dirac spinor has been defined as transforming under a spinor representation of $SO(1, n-1)$. What is really meant by this is that it transforms under a representation of the group $Spin(1, n-1)$, which is the simply connected (for $n > 2$) double covering group of $SO(1, n-1)$. Given a local Lorentz transformation $L^a{}_b(x)$, which may be viewed as the mapping $L : M \to SO(1, n-1)$, we need to construct the matrix $S(L(x))$ defined in (5.256). The situation may be summarized by the commutative diagram in Fig 5.1. Here Λ is the 2–1 covering map. (Note that $S(L)$ is not to be confused with the composition map $S \circ L$.) Given L and Λ, the problem amounts to the existence of the map $S(L)$ such that the diagram is commutative (i.e., such that $L = \Lambda \circ S(L)$). $S(L)$, if it exists, is called the lift of L. This is a standard problem in algebraic topology, and it may be shown that $S(L)$ exists if and only if $H^1(M; \mathbb{Z}_2) = 0$. Again, if $\pi_1(M) = 0$, then $S(L)$ exists, although $S(L)$ may also exist if $\pi_1(M) \neq 0$. A final remark is that in four spacetime dimensions we have the isomorphism $Spin(1,3) = SL(2,\mathbb{C})$. This is why it is possible to use two-component spinors in general relativity. The two components refer to $SL(2,\mathbb{C})$.

As a simple example where a non-trivial spin connection arises, consider the spacetime $M = \mathbb{R}(\text{time}) \times \mathbb{R}^2 \times \mathbb{S}^1$. This spacetime may be realized by regarding M as Minkowski spacetime with the x coordinate periodically identified on the interval $[0, L]$. L represents the circumference of the circle $\mathbb{S}^1$. The fundamental group of M is $\pi_1(M) = \pi_1(\mathbb{S}^1) = \mathbb{Z}$. We will regard $\mathbb{Z}$ as the set of integers, which forms a group under the operation of addition. Inequivalent spin connections

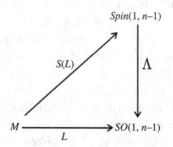

Fig 5.1. Commutative diagram.

---

[15] See Isham (1978a,b).

5.6 Spinors in curved spacetime

are classified by $H^1(M;\mathbb{Z}_2) = \text{Hom}(\pi_1(M), \mathbb{Z}_2) = \text{Hom}(\mathbb{Z}, \mathbb{Z}_2)$ in this case. Let $f : \mathbb{Z} \to \mathbb{Z}_2$ be a group homomorphism. Then

$$f(n+m) = f(n)f(m), \tag{5.360}$$

for any integers n and m. ($\mathbb{Z}_2$ may be regarded as the set $\{-1,+1\}$ with the group operation of multiplication.) Putting $n = m = 0$ in (5.360), we see that $f(0) = 1$. (Generally, a group homomorphism maps the identity element to the identity element.) Consider now $f(1)$. We can have $f(1) = 1$ or $f(1) = -1$. Because $f(n+1) = f(n)f(1)$, it can be seen that if $f(1) = 1$, then $f(n) = 1$ for all integral n. If $f(1) = -1$, then $f(n) = (-1)^n$. Therefore there are two distinct homomorphisms between $\mathbb{Z}$ and $\mathbb{Z}_2$, and it is expected that there will be two inequivalent spin connections.

Because M is just Minkowski spacetime with an identified spatial coordinate, the most obvious choice for the vierbein is $e^a{}_\mu = \delta^a_\mu$, which corresponds to $\omega_{\mu\,ab} = 0$. If $\Psi(x)$ is the spinor field, assume that $\Psi(x+L) = \exp(2\pi i \delta)\Psi(x)$ for some constant δ. (For example, we might choose $\delta = 0$, in which case the field is periodic around the circle.) However, there is another choice for the vierbein (DeWitt et al., 1979; Ford 1980). Let $L^a{}_b$ represent a rotation of $\theta = 2\pi x/L$ around the x-axis. Define the vierbein $e'^a{}_\mu$ by

$$e'^a{}_\mu = L^a{}_b e^b{}_\mu. \tag{5.361}$$

It is straightforward to show using (5.249) and (5.267) that this choice corresponds to a non-zero spin connection B'_μ given by

$$B'_\mu = \frac{\pi}{L}\delta^1_\mu \gamma^2 \gamma^3. \tag{5.362}$$

There is no reason to prefer one choice of vierbein and spin connection over the other. We will see later that the quantum theory based on the spin connection B'_μ differs from that based on $B_\mu = 0$. This may be more obvious if, following Chockalingham and Isham (1980), we transform away B'_μ. From the transformation of the spin connection (5.264), it may be seen that we want to find an S such that

$$B'_\mu = -(\partial_\mu S)S^{-1}. \tag{5.363}$$

Using the explicit result for the spin connection B'_μ given in (5.362), it is easily shown that

$$S(x) = \cos\left(\frac{\pi x}{L}\right) I - \sin\left(\frac{\pi x}{L}\right) \gamma^2 \gamma^3. \tag{5.364}$$

The spinor field transforms from $\Psi(x)$ to $\Psi'(x)$, where

$$\Psi'(x) = S(x)\Psi(x). \tag{5.365}$$

The new spinor field $\Psi'(x)$ may now be seen to obey different boundary conditions from $\Psi(x)$. From (5.364) we see that $S(x+L) = -S(x)$, and thus

$$\Psi'(x+L) = \exp\left\{2\pi i \left(\delta + \frac{1}{2}\right)\right\} \Psi'(x). \tag{5.366}$$

The conclusion is that although we can transform the non-trivial spin connection away, the boundary conditions on the spinor field are changed if we do so. In particular, if the original field $\Psi(x)$ is periodic, then the transformed field $\Psi'(x)$ is antiperiodic. The fact that the quantum theories based on the two inequivalent spin connections are different should therefore come as no surprise.

5.6.5 Grassmann variables

By taking the classical limit of the anticommutation relations for a Dirac spinor, it may be seen that spinor fields on a given spacelike hypersurface anticommute. (This can be seen from (1.120) if it is noted that there is actually a factor of $\hbar$ on the right-hand side which we have set to 1 by a choice of units. As stated there, the anticommutator of a Dirac spinor field with itself vanishes.) As a consequence, the square of a classical spinor field must vanish. If the field is treated as an ordinary number, then the field itself must obviously vanish. However, there is a well-defined mathematics which deals with numbers whose square vanishes, but for which the number itself does not vanish. Such numbers are called anticommuting numbers or sometimes just Grassmann numbers. They are the natural objects for discussing spinor fields.

Let $\{\xi^i\}$, where $i = 1, \ldots, n$, be a set of real quantities that satisfy

$$\xi^i \xi^j + \xi^j \xi^i = 0. \tag{5.367}$$

A particular consequence of this relation is $(\xi^i)^2 = 0$, showing that the ξ^i cannot be thought of as ordinary numbers or else we must have $\xi^i = 0$. We will call numbers which satisfy (5.367) a-type, where the "a" refers to anticommuting. Ordinary real or complex numbers will be referred to as c-type to denote that they are commuting. Note that the product of an even number of a-type factors is a c-type number, and the product of an odd number of factors is an a-type number. A c-type number is not necessarily an ordinary real or complex number. For a complete treatment of a-type numbers see Berezin (1966, 1987) or DeWitt (1984a).

If spinor fields are described by a-type numbers, then the functional integral approach to quantum field theory will involve integration over a-type variables. Berezin (1987) has provided a definition of integration for a-type numbers. The differential $d\xi^i$ of an a-type number is also defined to be an a-type number, and satisfies

$$d\xi^i \xi^j + \xi^j d\xi^i = 0, \tag{5.368}$$
$$d\xi^i d\xi^j + d\xi^j d\xi^i = 0. \tag{5.369}$$

5.6 Spinors in curved spacetime

Berezin integration is defined by the rules

$$\int d\xi^i = 0, \qquad (5.370)$$

$$\int d\xi^i \xi^j = \delta^{ij}, \qquad (5.371)$$

along with the assumption that the integral is linear over the integrand. (i.e., $\int d\xi^i (f+g) = \int d\xi^i f + \int d\xi^i g$.) Multiple Berezin integrals may be defined as iterated single integrals. We will define

$$d^n\xi = d\xi^1 \cdots d\xi^n. \qquad (5.372)$$

Note that unlike the case of integration over the real numbers, the order of factors in (5.372) is important.

An important feature of Berezin integration is the behavior of the measure (5.372) under a change of variables. Suppose that $\{\xi'^i\}$ satisfies (5.367)–(5.371) where

$$\xi'^i = L^i{}_j \xi^j, \qquad (5.373)$$

with $L^i{}_j$ an invertible $n \times n$ c-type matrix which is independent of all ξ^k. We then have

$$\delta^{ij} = \int d\xi'^i \xi'^j$$

$$= L^j{}_k \int d\xi'^i \xi^k. \qquad (5.374)$$

(Because $L^i{}_j$ is c-type and constant, it may be pulled outside the integration.) It follows from (5.374) that

$$d\xi'^i = (L^T)^{-1i}{}_j d\xi^j, \qquad (5.375)$$

where L^T is the transpose of the matrix L. Naively, from (5.373) it might have been expected that $d\xi'^i = L^i{}_j d\xi^j$, but it would have then been impossible to satisfy (5.367)–(5.371) for both sets $\{\xi^i\}$ and $\{\xi'^i\}$. We then have

$$d^n\xi' = d\xi'^1 \cdots d\xi'^n$$
$$= (L^T)^{-11}{}_{i_1} \cdots (L^T)^{-1n}{}_{i_n} d\xi^{i_1} \cdots d\xi^{i_n}. \qquad (5.376)$$

But because of (5.369) we have

$$d\xi^{i_1} \cdots d\xi^{i_n} = \sigma^{i_1 \cdots i_n} d\xi^1 \cdots d\xi^n, \qquad (5.377)$$

where $\sigma^{i_1 \cdots i_n}$ denotes the sign of the permutation of $(i_1 \cdots i_n)$. Equation (5.376) becomes, by definition of the determinant,

$$d^n\xi' = (\det(L^T)^{-1}) d^n\xi$$
$$= (\det L)^{-1} d^n\xi. \qquad (5.378)$$

A transformation of a-type variables is seen to result in the inverse of the usual Jacobian.

Let $\{\xi^i\}$ and $\{\eta^i\}$ for $i = 1,\ldots, n$ be two linearly independent sets of real a-type variables. We assume that $\xi^i \eta^j = -\eta^j \xi^i$ consistently with their a-type properties. Define

$$\zeta^j = \frac{1}{\sqrt{2}}(\xi^j + i\eta^j) \tag{5.379}$$

$$\bar{\zeta}^j = \frac{1}{\sqrt{2}}(\xi^j - i\eta^j). \tag{5.380}$$

This is completely analogous to defining an ordinary complex coordinate and its complex conjugate in terms of ordinary real variables. (Note that $(\zeta^i)^2 = 0$ and $(\bar{\zeta}^i)^2 = 0$, but $\bar{\zeta}^j \zeta^j = i\xi^j \eta^j \neq 0$.)

Let $A_{ij} = -A_{ji}$ be an $n \times n$ antisymmetric c-type matrix with no dependence on the Grassmann variables ξ^k, and consider the Gaussian Berezin integral

$$I(A) = \int d^n\zeta\, d^n\bar{\zeta}\, \exp(i\bar{\zeta}^i A_{ij} \zeta^j). \tag{5.381}$$

We may expand $\exp(i\bar{\zeta}^i A_{ij} \zeta^j)$ in a Taylor series. But because $(\zeta^i)^2 = 0 = (\bar{\zeta}^i)^2$, only the first n terms of the series will be non-zero:

$$\exp(i\bar{\zeta}^i A_{ij}\zeta^j) = \sum_{k=0}^{n} \frac{i^k}{k!}(\bar{\zeta}^i A_{ij}\zeta^j)^k. \tag{5.382}$$

Furthermore, because of the rules of Berezin integration, unless there are n factors of ζ and n factors of $\bar{\zeta}$ in the integrand of (5.381), the integral will vanish. Thus, from (5.381) and (5.382) we have

$$I(A) = \frac{i^n}{n!} \int d^n\zeta\, d^n\bar{\zeta}\, (\bar{\zeta}^i A_{ij}\zeta^j)^n \tag{5.383}$$

as only the nth term in the Taylor series (5.382) can be non-zero. We may write (5.383) as

$$I(A) = \frac{i^n}{n!} A_{i_1 j_1} \cdots A_{i_n j_n} \int d^n\zeta\, d^n\bar{\zeta}\, \bar{\zeta}^{i_1} \zeta^{j_1} \cdots \bar{\zeta}^{i_n} \zeta^{j_n}. \tag{5.384}$$

It is easy to show that

$$\bar{\zeta}^{i_1} \zeta^{j_1} \cdots \bar{\zeta}^{i_n} \zeta^{j_n} = (-1)^{\frac{1}{2}n(n-1)} \sigma^{i_1 i_2 \cdots i_n} \sigma^{j_1 j_2 \cdots j_n} \bar{\zeta}^n \bar{\zeta}^1 \cdots \zeta^n \zeta^1 \tag{5.385}$$

if the factors on the left-hand side are reordered. Using

$$\int d^n\zeta\, d^n\bar{\zeta}\, \bar{\zeta}^n \bar{\zeta}^1 \cdots \zeta^n \zeta^1 = 1, \tag{5.386}$$

it follows that

$$I(A) = i^n \det A. \tag{5.387}$$

If we define
$$d[\zeta,\bar{\zeta}] = i^{-n} d^n\zeta\, d^n\bar{\zeta}, \tag{5.388}$$
then we have the fundamental result
$$\int d[\zeta,\bar{\zeta}] \exp(i\bar{\zeta}^i A_{ij}\zeta^j) = \det A. \tag{5.389}$$

This should be contrasted with the result for integration over ordinary numbers which would have resulted in $(\det A)^{-1}$ on the right-hand side (along with some factors of π).

Finally, we may use (5.389) to deduce the result of performing Berezin integration of a Gaussian over real a-type variables. Because a-type variables transform with the inverse of the usual Jacobian, from (5.379) and (5.380) we have
$$d\zeta^j d\bar{\zeta}^j = i d\xi^j d\eta^j. \tag{5.390}$$

It follows from this result that
$$d^n\zeta\, d^n\bar{\zeta} = i^n d^n\xi\, d^n\eta. \tag{5.391}$$

Also, we have
$$\bar{\zeta}^i A_{ij}\zeta^j = \frac{1}{2}\xi^i A_{ij}\xi^j + \frac{1}{2}\eta^i A_{ij}\eta^j,$$
so that
$$\exp(i\bar{\zeta}^i A_{ij}\zeta^j) = \exp(\frac{i}{2}\xi^i A_{ij}\xi^j) \exp(\frac{i}{2}\eta^i A_{ij}\eta^j).$$

From (5.381) we have
$$I(A) = i^n \left[\int d^n\xi \exp\left(\frac{i}{2}\xi^i A_{ij}\xi^j\right)\right]^2$$
from which we can deduce that
$$\int d^n\xi \exp\left(\frac{i}{2}\xi^i A_{ij}\xi^j\right) = (\det A)^{1/2}. \tag{5.392}$$

Again this should be contrasted with the result $(\det A)^{-1/2}$ which would have been obtained if the ξ^i were ordinary real numbers.

The results in (5.389) and (5.392) may now be extended to the case of field theory, just as for the bosonic case.

5.7 The effective action for spinor fields

We will consider first the form for the classical action for a Dirac spinor in curved spacetime. Just as in flat spacetime, this requires the notion of the adjoint spinor $\bar{\Psi}$. Define
$$\bar{\Psi} = \Psi^\dagger \gamma^0. \tag{5.393}$$

Note that γ^0 here is the γ-matrix in a local orthonormal frame, and not the curved spacetime γ-matrix γ^0. The transformation of $\overline{\Psi}$ under local Lorentz and general coordinate transformations follows from that of Ψ which is given in (5.256) as $\Psi'(x') = S(L)\Psi(x)$. Because $(\gamma^0)^2 = I$, we can write

$$\overline{\Psi}'(x') = \overline{\Psi}(x)\gamma^0 S^\dagger(L)\gamma^0. \tag{5.394}$$

We had $\gamma_a^\dagger = \gamma^0 \gamma_a \gamma^0$ (see (5.297)) and $\Sigma_{ab} = -\frac{i}{8}[\gamma_a, \gamma_b]$ (see (5.284)), so that

$$\gamma^0 \Sigma_{ab}^\dagger \gamma^0 = \Sigma_{ab}. \tag{5.395}$$

Since $S(L) = \exp\left(i\epsilon^{ab}\Sigma_{ab}\right)$, it follows that

$$\gamma^0 S^\dagger(L)\gamma^0 = S^{-1}(L). \tag{5.396}$$

From (5.394), we therefore have

$$\overline{\Psi}'(x') = \overline{\Psi}(x) S^{-1}(L) \tag{5.397}$$

as the transformation property of the adjoint spinor. This is sufficient to prove that $\overline{\Psi}\Psi$ is generally coordinate invariant, as well as invariant under a local Lorentz transformation.

One of the basic properties of the classical action functional is that it must be real. It is expected to be bilinear in the fields (if its variation is to lead to linear field equations), and so to examine its reality we need to define the complex conjugate of a product of Grassmann variables. (It was discussed in the last section how Dirac spinors may be described by a-type Grassmann variables.) This will be defined by analogy with Hermitian conjugation of matrices. If A and B are two matrices whose product AB is defined, then the Hermitian conjugate of the product is $(AB)^\dagger = B^\dagger A^\dagger$. Matrix A is said to be Hermitian if $A^\dagger = A$. If A and B are both Hermitian, then their product is not Hermitian in general. (This follows from the fact that if $A^\dagger = A$ and $B^\dagger = B$, then $(AB)^\dagger = BA \neq AB$ for general matrices.)

If ζ and η are any two Grassmann numbers, we define the complex conjugate of their product by

$$(\zeta\eta)^* = \eta^*\zeta^*. \tag{5.398}$$

(More formally, complex conjugation is defined to be an involution of the Grassmann algebra Berezin (1966).) If either, or both, of the Grassmann numbers is c-type, then the order of factors on the right-hand side of (5.398) is not important. If both ζ and η are a-type, then the order of factors is important. A Grassmann number ζ is defined to be real if $\zeta^* = \zeta$ and imaginary if $\zeta^* = -\zeta$. It follows from these definitions that the product of two real a-type Grassmann numbers is imaginary.

The components Ψ_α of the Dirac spinor Ψ will be represented by a-type Grassmann numbers. The components of $\Psi^\dagger$ are Ψ_α^* where $*$ denotes complex conjugation as defined in (5.398). (We will use Greek letters from the beginning of the alphabet to run over the appropriate dimension of the spinor.) From (5.398), we have

$$(\overline{\Psi}\Psi)^* = (\Psi_\alpha^*(\gamma^0)^{\alpha\beta}\Psi_\beta)^*$$
$$= \Psi_\beta^*(\gamma^0)^{*\alpha\beta}\Psi_\alpha,$$

where $(\gamma^0)^{\alpha\beta}$ denotes the matrix elements of γ^0. Because γ^0 is Hermitian, $(\gamma^0)^{*\alpha\beta} = (\gamma^0)^{\beta\alpha}$. Thus,

$$(\overline{\Psi}\Psi)^* = \overline{\Psi}\Psi, \qquad (5.399)$$

proving that $\overline{\Psi}\Psi$ is real.

Consider $\overline{\Psi}\gamma^\mu \nabla_\mu \Psi$. From the properties of the covariant derivative, this is seen to be a scalar under a general coordinate transformation. To check reality, evaluate

$$(\overline{\Psi}\gamma^\mu \nabla_\mu \Psi)^* = (\nabla_\mu \Psi)^\dagger \gamma^0 \gamma^\mu \Psi$$

using $(\gamma^0)^2 = I$ and (5.297). The spin connection is[16] $B_\mu = i\omega_\mu{}^{ab}\Sigma_{ab}$ so that

$$(\nabla_\mu \Psi)^\dagger \gamma^0 = \partial_\mu \overline{\Psi} + \overline{\Psi}\gamma^0 B_\mu^\dagger \gamma^0$$
$$= \partial_\mu \overline{\Psi} - \overline{\Psi} B_\mu$$

using (5.395). This leads us to define

$$\nabla_\mu \overline{\Psi} = \partial_\mu \overline{\Psi} - \overline{\Psi} B_\mu, \qquad (5.400)$$

which can be seen to transform like $\overline{\Psi}$ under a local Lorentz transformation. We then have

$$(\overline{\Psi}\gamma^\mu \nabla_\mu \Psi)^* = (\nabla_\mu \overline{\Psi})\gamma^0 \gamma^\mu \Psi. \qquad (5.401)$$

The Dirac equation in curved spacetime is (see (5.281)) $i\gamma^\mu \nabla_\mu \Psi - m\Psi = 0$. (The factor of i in front of $\gamma^\mu \nabla_\mu$ is necessary for the Dirac operator $i\gamma^\mu \nabla_\mu - m$ to be self-adjoint with respect to the natural inner product $(\Psi_1, \Psi_2) = \int dv_x \overline{\Psi}_1 \Psi_2$ between two spinors Ψ_1 and Ψ_2. It also ensures that in flat spacetime, every solution of the Dirac equation is also a solution to the Klein–Gordon equation.) It is clear that under a general variation with respect to $\overline{\Psi}$, $i\overline{\Psi}\gamma^\mu \nabla_\mu \Psi - m\overline{\Psi}\Psi$ gives rise to the Dirac equation. But from (5.401), the first term is seen not to be real. Therefore, if $i\overline{\Psi}\gamma^\mu \nabla_\mu \Psi - m\overline{\Psi}\Psi$ is adopted as the Dirac Lagrangian, the resulting Lagrangian and Hamiltonian are not real. The problem is easily solved by taking the Dirac Lagrangian to be

$$\mathcal{L}_D = \frac{i}{2}\left(\overline{\Psi}\gamma^\mu \nabla_\mu \Psi - \nabla_\mu \overline{\Psi}\gamma^\mu \Psi\right) - m\overline{\Psi}\Psi \qquad (5.402)$$

$$= \frac{i}{2}\overline{\Psi}\gamma^\mu \overleftrightarrow{\nabla}_\mu \Psi - m\overline{\Psi}\Psi. \qquad (5.403)$$

---

[16] See (5.266) and (5.269).

Because $\nabla_\mu \underline{\gamma}^\mu = 0$, $\mathcal{L}_D$ is seen to differ from $i\overline{\Psi}\underline{\gamma}^\mu \nabla_\mu \Psi - m\overline{\Psi}\Psi$ by a total derivative. The action is

$$S_D = \int dv_x \mathcal{L}_D$$
$$= \int dv_x \overline{\Psi}(i\underline{\gamma}^\mu \nabla_\mu - m)\Psi \qquad (5.404)$$

if the integral of the above-mentioned total derivative is discarded. The Dirac equation results upon variation with respect to $\overline{\Psi}$, and the adjoint of the Dirac equation results upon variation with respect to Ψ.

If we adopt the functional integral representation for the effective action, then we have

$$\exp\left(\frac{i}{\hbar}W\right) = \int d\mu[\overline{\Psi}, \Psi] \exp\left(\frac{i}{\hbar}S_D\right). \qquad (5.405)$$

The integration is understood to be Berezin integration with $d\mu[\overline{\Psi}, \Psi]$ the functional measure analogue of $d[\zeta, \overline{\zeta}]$ in (5.388). The functional analogue of (5.389) results in

$$\exp\left(\frac{i}{\hbar}W\right) = \det(\ell D), \qquad (5.406)$$

where

$$D = i\underline{\gamma}^\mu \nabla_\mu - m \qquad (5.407)$$

is the Dirac operator. The length scale ℓ has been introduced in (5.406) to keep the argument of the determinant dimensionless, just as it was for bosonic fields. We therefore have

$$W = -i\hbar \ln \det(\ell D) \qquad (5.408)$$

as the one-loop effective action for a Dirac spinor. This should be compared with the result in Section 5.2 or Section 5.3 for a bosonic field. First of all, there is a factor of 2 different from the real scalar field result, which is due simply to the fact that the functional integral has been performed over a complex field here, rather than a real one. Second, the overall sign is different, which is due to the nature of Berezin integration. This in turn, is related to the fact that fermions satisfy anticommutation rather than commutation relations.

If we are interested in the one-loop divergences of W, then the results quoted earlier in the chapter (see Section 5.3) are not directly applicable because the Dirac operator is first, rather than second, order. In any case, if the spacetime dimension is odd, we would not expect any divergences in W since curvature invariants which can form possible counterterms are all of even dimension. We will restrict our attention to even-dimensional spacetimes for the remainder of this section.

Define

$$\tilde{D} = -i\underline{\gamma}^\mu \nabla_\mu - m, \qquad (5.409)$$

5.7 The effective action for spinor fields

which is obtained from D in (5.407) by replacing γ^μ with $-\gamma^\mu$. In an even-dimensional spacetime, we know from the discussion in Section 5.6 that $\{\gamma^\mu\}$ and $\{-\gamma^\mu\}$ form equivalent representations of the Clifford algebra. We therefore expect that $\det(\ell\tilde{D}) = \det(\ell D)$. A direct proof of this is obtained by using the fact that if the spacetime dimension is even, then there exists a Hermitian matrix γ_{n+1} that anticommutes with γ^μ and that satisfies $(\gamma_{n+1})^2 = I$. We then have

$$D = \gamma_{n+1} \tilde{D} \gamma_{n+1}, \tag{5.410}$$

from which it follows that $\det(\ell\tilde{D}) = \det(\ell D)$. This argument does not work in an odd-dimensional spacetime, since γ_{n+1} does not anticommute with γ^μ. (In an odd-dimensional spacetime, the representations of the Clifford algebra provided by $\{\gamma^\mu\}$ and $\{-\gamma^\mu\}$ are not equivalent.)

We can write W in (5.408) as

$$W = -\frac{i\hbar}{2} \ln[\det(\ell D)]^2$$
$$= -\frac{i\hbar}{2} \ln \det(\ell^2 D\tilde{D}) \tag{5.411}$$

if we use $\det(\ell\tilde{D}) = \det(\ell D)$. From (5.407) and (5.409),

$$D\tilde{D} = \tilde{D}D = (\gamma^\mu \nabla_\mu)^2 + m^2$$
$$= \Box + \frac{1}{4}R + m^2, \tag{5.412}$$

if (5.293) is used. This result establishes that

$$W = -\frac{i\hbar}{2} \ln \det\left[\ell^2 \left(\Box + \frac{1}{4}R + m^2\right)\right]. \tag{5.413}$$

This casts W into a form which involves a second-order differential operator, and the heat kernel results of Section 5.3 may be used. (See (5.68).) If we adopt dimensional regularization, with the physical spacetime dimension n replaced with $n + \epsilon$, then

$$\text{divp} W = \frac{\hbar}{\epsilon}(4\pi)^{-n/2} \int dv_x \text{tr} E_{n/2}(x). \tag{5.414}$$

(The overall sign is different from (5.68) because of the rules for Berezin integration.) In particular, in four spacetime dimensions, the differential operator appearing in (5.413) is of the general form given in (5.56) with

$$Q = \left(\frac{1}{4}R + m^2\right) I, \tag{5.415}$$

$$W_{\mu\nu} = -i R_{\mu\nu ab} \Sigma^{ab}. \tag{5.416}$$

If we use (5.59) we find

$$\mathrm{tr}\, E_2 = \frac{1}{30}\Box R + \frac{1}{72}R^2 - \frac{1}{45}R_{\mu\nu}R^{\mu\nu} - \frac{7}{360}R_{\mu\nu\rho\sigma}R^{\mu\nu\rho\sigma}$$
$$+ \frac{1}{3}m^2 R + 2m^4. \tag{5.417}$$

5.7.1 The conformal anomaly

We will now study the conformal anomaly for a spin-1/2 field using the Fujikawa method already applied to the spin-0 field in Section 5.5. Define, as before,

$$\tilde{g}_{\mu\nu}(x) = \Omega^2(x) g_{\mu\nu}(x). \tag{5.418}$$

The n-bein in the conformally related spacetime is defined in terms of the conformal metric $\tilde{g}_{\mu\nu}$ by

$$\tilde{g}_{\mu\nu}(x) = \tilde{e}^a{}_\mu(x)\tilde{e}^b{}_\nu(x)\eta_{ab}, \tag{5.419}$$

from which it follows that

$$\tilde{e}^a{}_\mu(x) = \Omega(x) e^a{}_\mu(x), \tag{5.420}$$

and hence

$$\tilde{e}_a{}^\mu(x) = \Omega^{-1}(x) e_a{}^\mu(x). \tag{5.421}$$

The connection components in the conformally related spacetime are given by (see (5.249))

$$\tilde{\omega}_\mu{}^a{}_b = -\tilde{e}_b{}^\nu \left(\partial_\mu \tilde{e}^a{}_\nu - \tilde{\Gamma}^\lambda_{\mu\nu} \tilde{e}^a{}_\lambda \right)$$
$$= \omega_\mu{}^a{}_b + \Omega^{-1}\Omega_{,\nu}\left(e^a{}_\mu e_b{}^\nu - e^{a\nu}e_{b\mu}\right). \tag{5.422}$$

The spacetime-dependent γ-matrices in the conformally related spacetime are given by

$$\underline{\tilde{\gamma}}^\mu = \tilde{e}_a{}^\mu(x)\gamma^a$$
$$= \Omega^{-1}\underline{\gamma}^\mu. \tag{5.423}$$

If we define

$$\tilde{\Psi}(x) = \Omega^p(x)\Psi(x), \tag{5.424}$$

for some p, then it is easy to show that

$$i\underline{\tilde{\gamma}}^\mu \tilde{\nabla}_\mu \tilde{\Psi} = i\underline{\tilde{\gamma}}^\mu \left(\partial_\mu + i\tilde{\omega}_{\mu ab}\Sigma^{ab} \right) \tilde{\Psi}$$
$$= \Omega^{p-1}\left\{ i\underline{\gamma}^\mu \nabla_\mu \Psi + i[p + \frac{1}{2}(n-1)]\Omega^{-1}\underline{\gamma}^\mu \Omega_{,\mu}\Psi \right\}. \tag{5.425}$$

(The result $\gamma_a \Sigma^{ab} = -\frac{i}{4}(n-1)\gamma^b$, which follows directly from the definition of the γ-matrices, has been used here.) This result shows that the Dirac equation is only conformally invariant if $m = 0$ (i.e., the spinors must be massless) and if p in (5.424) is chosen as

$$p = -\frac{1}{2}(n-1). \tag{5.426}$$

It is now easy to show that the classical action for a massless Dirac spinor is conformally invariant if

$$\tilde{\Psi} = \Omega^{-(n-1)/2}\Psi. \tag{5.427}$$

(Conformal invariance also holds for Weyl and massless Majorana spinors under this transformation.)

Define an eigenfunction decomposition of the massless Dirac operator as

$$D\psi_N = \lambda_N \psi_N, \tag{5.428}$$

where $D = i\gamma^\mu \nabla_\mu$. We treat λ_N and ψ_N as c-type. The set of all ψ_N is assumed to be complete. Expand the spinor field Ψ in terms of these eigenfunctions

$$\Psi = \sum_N a_N \psi_N, \tag{5.429}$$

where the expansion coefficients a_N are a-type Grassmann numbers. The expansion of the adjoint spinor is then

$$\overline{\Psi} = \sum_N a_N^* \overline{\psi}_N, \tag{5.430}$$

where $\overline{\psi}_N = \psi_N^\dagger \gamma^0$. Choose the normalization of the eigenfunctions ψ_N to be

$$\int dv_x \overline{\psi}_N \psi_{N'} = \ell \delta_{NN'}, \tag{5.431}$$

where the unit of length ℓ is introduced to balance dimensions on both sides. (ψ_N is chosen to have the same dimensions as Ψ, namely $\ell^{(1-n)/2}$. This makes the expansion coefficients a_N in (5.429) dimensionless.) The classical action S_D becomes

$$S_D = \sum_N (\ell \lambda_N) a_N^* a_N, \tag{5.432}$$

in terms of the expansion coefficients. The one-loop effective action becomes

$$\exp\left(\frac{i}{\hbar}W\right) = \int \left(\prod_N da_N\right)\left(\prod_N da_N^*\right) \exp\left\{\frac{i}{\hbar}\sum_N (\ell \lambda_N) a_N^* a_N\right\}. \tag{5.433}$$

Under a conformal transformation, defined by $\tilde{g}_{\mu\nu}(x) = \Omega^2(x) g_{\mu\nu}(x)$ with $\Omega(x)$ a general function of the spacetime coordinates, choose ψ_N to transform in the same way as Ψ (see (5.427)):

$$\tilde{\psi}_N(x) = \Omega^{-(n-1)/2} \psi_N. \tag{5.434}$$

Then
$$\int d\tilde{v}_x \overline{\tilde{\psi}}_N \tilde{\psi}_{N'} = \int dv_x \Omega \overline{\psi}_N \psi_{N'}. \qquad (5.435)$$

Just as in Section 5.5, the conformally related basis $\{\tilde{\psi}_N\}$ is neither complete nor orthonormal in general. Let $\{g_N(x)\}$ be a complete orthonormal basis in the conformally related spacetime. We will choose

$$\tilde{\psi}_N(x) = \Omega^{1/2}(x) g_N(x). \qquad (5.436)$$

If we write
$$\tilde{\Psi} = \sum_N \tilde{a}_N g_N \qquad (5.437)$$

for some new expansion coefficients $\tilde{a}_N$, it may be seen that

$$a_{N'} = \sum_N C_{N'N} \tilde{a}_N, \qquad (5.438)$$

where
$$C_{N'N} = \ell^{-1} \int d\tilde{v}_x \Omega^{-1/2} \overline{g}_{N'} g_N \qquad (5.439)$$
$$= \ell^{-1} \int dv_x \Omega^{-1/2} \overline{\psi}_{N'} \psi_N. \qquad (5.440)$$

Because of the rules for Berezin integration (specifically that a-type Grassmann variables transform with the inverse Jacobian), we have from (5.438)

$$\prod_N da_N = (\det C_{N'N})^{-1} \left(\prod_N d\tilde{a}_N \right). \qquad (5.441)$$

Since the effective action in the conformally related spacetime $\tilde{W}$ is defined by

$$\exp\left(\frac{i}{\hbar} \widetilde{W}\right) = \int \left(\prod_N d\tilde{a}_N\right) \left(\prod_N d\tilde{a}_N^*\right) \exp\left(\frac{i}{\hbar} \sum_N (\ell \lambda_N) \tilde{a}_N^* \tilde{a}_N\right), \qquad (5.442)$$

we have
$$\widetilde{W} = W - 2i\hbar \ln \det C_{NN'}. \qquad (5.443)$$

This should be compared with the analogous scalar field result in (5.209). The sign of the second term is changed due to the nature of Berezin integration, and there is a factor of 2 because we have considered a Dirac spinor, whereas (5.209) was for a real scalar field.

The remainder of the calculation proceeds as in Section 5.5 and results in

$$\delta W = -\hbar (4\pi)^{-n/2} \int dv_x \delta\omega(x) \mathrm{tr} E_{n/2}(x). \qquad (5.444)$$

($\delta W = \widetilde{W} - W$ and $\Omega = 1 + \delta\omega$.) The anomalous trace of the stress-energy tensor is

$$\langle T^\mu{}_\mu \rangle = \hbar(4\pi)^{-n/2} \text{tr} E_{n/2}(x). \tag{5.445}$$

In (5.444) and (5.445), $E_{n/2}$ is the relevant heat kernel coefficient for $\tilde{D}D = \Box + \frac{1}{4}R$.

5.8 Application of the effective action for spinor fields

The two examples considered in this section are the spin-1/2 versions of the examples in Section 5.4. Because many of the steps are similar, we will be brief.

5.8.1 Schwinger effective Lagrangian for spin-1/2

The one-loop effective action for a spinor field coupled to a background electromagnetic field is

$$W = -i\hbar \ln \det \ell(i\gamma^\mu D_\mu - m), \tag{5.446}$$

where $D_\mu = \partial_\mu - ieA_\mu$. The spacetime is taken to be flat n-dimensional Minkowski spacetime, so that the distinction between curved spacetime γ-matrices and those in the local orthonormal frame disappears (assuming Cartesian coordinates are used). The Feynman Green function for the Dirac field is defined by

$$(i\gamma^\mu D_\mu - m) S(x, x') = \delta(x, x'). \tag{5.447}$$

By differentiating (5.446) with respect to the mass, it may be shown that

$$\frac{\partial}{\partial m} W = i\hbar \int dv_x \text{tr} S(x, x), \tag{5.448}$$

where the trace is over spinor indices. (This result follows from using $\ln \det = \text{tr} \ln$ in (5.446) and noting that the Feynman Green function defined in (5.447) is the inverse of the differential operator that appears in (5.446).) In order to calculate W, define a new Green function $G(x, x')$ by

$$S(x, x') = (-i\gamma^\mu D_\mu - m) G(x, x'). \tag{5.449}$$

Substitution of this into (5.447) leads to

$$\left[D^2 + m^2 - \frac{i}{2} e F_{\mu\nu} \sigma^{\mu\nu} \right] G(x, x') = \delta(x, x'), \tag{5.450}$$

where we have set

$$\sigma^{\mu\nu} = \frac{1}{2} [\gamma^\mu, \gamma^\nu], \tag{5.451}$$

and used $(\gamma^\mu D_\mu)^2 = D^2 - \frac{i}{2} e F_{\mu\nu} \sigma^{\mu\nu}$.

If we define
$$H = -\eta^{\mu\nu}\Pi_\mu\Pi_\nu + m^2 - \frac{i}{2}eF_{\mu\nu}\sigma^{\mu\nu}, \tag{5.452}$$

where $\Pi_\mu = p_\mu - eA_\mu$, then $G(x,x') = \langle x|G|x'\rangle$, where

$$G = H^{-1} = i\int_0^\infty d\tau \exp(-i\tau H). \tag{5.453}$$

(As for the scalar field, assume $\Im(m^2) < 0$. We will set the renormalization length $\ell = 1$ here.) The kernel is defined by

$$K(\tau;x,x') = \langle x|\exp(-i\tau H)|x'\rangle, \tag{5.454}$$

in which case

$$G(x,x') = i\int_0^\infty d\tau\, K(\tau;x,x'), \tag{5.455}$$

and

$$S(x,x') = -i\int_0^\infty d\tau\,(i\gamma^\mu D_\mu + m)K(\tau;x,x') \tag{5.456}$$

gives the Dirac propagator.

The kernel $K(\tau;x,x')$ may be calculated exactly as in Section 5.4 by expressing it as a Feynman path integral. The Lagrangian corresponding to the Hamiltonian H in (5.452) is

$$L = -\frac{1}{4}\dot{x}^\mu\dot{x}_\mu + eA_\mu\dot{x}^\mu - m^2 + \frac{i}{2}eF_{\mu\nu}\sigma^{\mu\nu}. \tag{5.457}$$

The only essential difference between the spin-0 and spin-1/2 cases is the presence of the last term in (5.457), which represents a coupling between the spin of the Dirac field and the background electromagnetic field. This coupling corresponds to a $\vec{\sigma}\cdot\mathbf{B}$ interaction in non-relativistic quantum mechanics, where $\mathbf{B}$ is the magnetic field and $\vec{\sigma}$ is the vector whose components are the Pauli spin matrices. In fact, if $F_{\mu\nu}$ represents a magnetic field in the z direction in four-dimensional spacetime, then

$$-\frac{i}{2}eF_{\mu\nu}\sigma^{\mu\nu} = \frac{1}{2}e\begin{pmatrix}\vec{\sigma}\cdot\mathbf{B} & 0 \\ 0 & \vec{\sigma}\cdot\mathbf{B}\end{pmatrix}.$$

The classical action S_{cl} is obtained from that for the scalar field by making the replacement

$$m^2 \to m^2 - \frac{i}{2}eF_{\mu\nu}\sigma^{\mu\nu}.$$

5.8 Application of the effective action for spinor fields

It is then easy to prove that

$$\lim_{x \to x'} D_\mu K(\tau; x, x') = i \lim_{x \to x'} (\partial_\mu S_{cl} - eA_\mu) K(\tau; x, x')$$
$$= 0.$$

This means that

$$\mathrm{tr} S(x, x) = -im \int_0^\infty d\tau \, \mathrm{tr} K(\tau; x, x), \tag{5.458}$$

if we use (5.456). In an even-dimensional spacetime, this result could have been deduced immediately from the fact that the trace of an odd number of γ-matrices vanishes. The one-loop effective Lagrangian is

$$\mathcal{L}^{(1)} = -\frac{\hbar}{2} \int_0^\infty \frac{d\tau}{\tau} \exp(-im^2\tau) \left[\det\left(\frac{\left(\frac{eF}{4\pi i}\right)}{\sinh(e\tau F)} \right) \right]^{1/2}$$
$$\times \mathrm{tr} \exp\left(-\frac{1}{2} e\tau F_{\mu\nu} \sigma^{\mu\nu} \right). \tag{5.459}$$

As discussed above, the only essential difference between this result and that for spin-0 is the presence of $\mathrm{tr}\exp(-\frac{1}{2}e\tau F_{\mu\nu}\sigma^{\mu\nu})$, and an overall sign change due to Fermi–Dirac statistics.

In order to evaluate $\mathcal{L}^{(1)}$ explicitly, we will consider four-dimensional spacetime. Following Schwinger (1951a), we will work out the eigenvalues of $F_{\mu\nu}\sigma^{\mu\nu}$. This may be done by using the identity

$$\{\sigma^{\mu\nu}, \sigma^{\rho\sigma}\} = 2(\eta^{\mu\sigma}\eta^{\nu\rho} - \eta^{\mu\rho}\eta^{\nu\sigma})I - 2i\epsilon^{\mu\nu\rho\sigma}\gamma_5, \tag{5.460}$$

which may be proven using the explicit representation of the γ-matrices given in Section 5.6.2. Contraction of both sides of this identity with $\frac{1}{4}F_{\mu\nu}F_{\rho\sigma}$ leads to

$$\left(\frac{1}{2} F_{\mu\nu}\sigma^{\mu\nu} \right)^2 = -2(\mathcal{F} + i\mathcal{G}\gamma_5), \tag{5.461}$$

where $\mathcal{F}$ and $\mathcal{G}$ were defined in (5.130) and (5.131). We therefore have

$$\left(\frac{i}{2} F_{\mu\nu}\sigma^{\mu\nu} \right)^2 = \begin{pmatrix} \psi^2 I_2 & 0 \\ 0 & \psi^{2*} I_2 \end{pmatrix}, \tag{5.462}$$

where $\psi = 2^{1/2}(\mathcal{F} + i\mathcal{G})^{1/2}$ was defined in (5.138) and I_2 is the 2×2 identity matrix. This result allows us to deduce that $\frac{i}{2}F_{\mu\nu}\sigma^{\mu\nu}$ has eigenvalues $\pm\psi, \pm\psi^*$, and hence that

$$\mathrm{tr}\exp\left(-\frac{1}{2} e\tau F_{\mu\nu}\sigma^{\mu\nu} \right) = 4\Re\left[\cosh(e\tau\psi)\right]. \tag{5.463}$$

Using dimensional regularization, as in the scalar field case, we find (restoring the renormalization length ℓ and rotating $\tau \to -i\tau$)

$$\mathcal{L}^{(1)} = -2\hbar\ell^\epsilon \int_0^\infty \frac{d\tau}{\tau} \exp(-m^2\tau)(4\pi\tau)^{-2-\epsilon/2}(e\tau)^2 \mathcal{G} \frac{\Re[\cosh(e\tau\psi)]}{\Im[\cosh(e\tau\psi)]}. \quad (5.464)$$

The effective Lagrangian may be renormalized using the same method as in the scalar case. Take the bare Lagrangian to be given as in (5.145). The counterterms are

$$\delta Z_A = -\frac{4}{3}\hbar e^2 (4\pi)^{-2} \left(\frac{m^2\ell^2}{4\pi}\right)^{\epsilon/2} \Gamma(-\epsilon/2), \quad (5.465)$$

$$C_B = -2\hbar(4\pi)^{-2} \left(\frac{m^2\ell^2}{4\pi}\right)^{\epsilon/2} m^4 \Gamma(-2-\epsilon/2). \quad (5.466)$$

The renormalized one-loop effective Lagrangian is

$$\mathcal{L}_{\text{ren}} = -\mathcal{F} - \frac{\hbar}{8\pi^2} \int_0^\infty \frac{d\tau}{\tau^3} \exp(-m^2\tau) \left\{ (e\tau)^2 \mathcal{G} \frac{\Re(\cosh(e\tau\psi))}{\Im(\cosh(e\tau\psi))} \right.$$
$$\left. -1 - \frac{2}{3}(e\tau)^2 \mathcal{F} \right\}. \quad (5.467)$$

For weak fields,

$$\mathcal{L}_{\text{ren}} \approx \frac{1}{2}(\mathbf{E}^2 - \mathbf{B}^2) + \frac{\hbar e^4}{360\pi^2 m^4} \left[(\mathbf{E}^2 - \mathbf{B}^2)^2 + 7(\mathbf{E}\cdot\mathbf{B})^2\right]. \quad (5.468)$$

Finally, in the case of a pure electric field, $\mathcal{L}$ has an imaginary part, as for the scalar field, which may be calculated by carefully considering the rotation $\tau \to -i\tau$ of the integration contour. It is easily shown that

$$\Im(\mathcal{L}) = \frac{\hbar}{8\pi^3}(eE)^2 \sum_{n=1}^\infty \frac{1}{n^2} \exp\left(-\frac{\pi n m^2}{eE}\right). \quad (5.469)$$

The physical interpretation of having $\Im(\mathcal{L}) \neq 0$ is discussed at the end of Section 5.4.1.

5.8.2 Spinor field in constant gauge background in $\mathbb{R}^{n-1} \times \mathbb{S}^1$

Assume that the background vector potential is given by (5.153) and that the spinor field satisfies the same boundary condition (5.155). The kernel function, defined as in (5.454), may be expanded as in (5.158), and because $F_{\mu\nu} = 0$ the kernel still satisfies (5.159) with the understanding that it is now a multiple of the unit spinor. The one-loop effective action is given in terms of the Dirac propagator in (5.448). Because of the tracelessness of the Dirac matrices, we have

$$\text{tr} S(x, x') = -im \int_0^\infty d\tau\, \text{tr} K(\tau; x, x'). \quad (5.470)$$

Recalling that the effective potential is defined in terms of the effective action by $W = -\ell^\epsilon \int dv_x V_{1/2}$, where the subscript $1/2$ has been used to designate the fact that we are dealing with spin-$1/2$, we have

$$\frac{\partial}{\partial m} V_{1/2} = -m\hbar\ell^\epsilon \int_0^\infty d\tau \, \mathrm{tr} K(\tau; x, x). \tag{5.471}$$

(Note that the effective action is given by (5.448).) The m-dependence in the kernel is $\exp(-im^2\ell^2\tau)$ (see (5.163)), and so we have

$$mK(\tau; x, x) = \frac{i}{2\tau} \frac{\partial}{\partial m} K(\tau; x, x).$$

This gives

$$V_{1/2} = -\frac{i\hbar}{2} \ell^\epsilon \int_0^\infty \frac{d\tau}{\tau} \mathrm{tr} K(\tau; x, x). \tag{5.472}$$

Because the kernel function is obtained from that of the scalar field by multiplying with the unit spinor, we can relate the effective potential for the spin-$1/2$ field directly to that for the spin-0 field:

$$V_{1/2} = -\frac{1}{2} (\mathrm{tr} I) V_0, \tag{5.473}$$

where V_0 is the result found earlier for spin-0. (See (5.176) and (5.177).) The presence of the minus sign reverses conclusions found in the case of the scalar field. Minima found in the potential for the scalar field are now maxima and vice versa. In particular, a periodic spinor field has a lower vacuum energy density than an antiperiodic field.

5.9 The axial, or chiral, anomaly

Consider an even-dimensional spacetime. In such a case, there exists a matrix γ_{n+1} which anticommutes with all of the Dirac matrices. (See Section 5.6.2.) To save writing, we will use γ_5 in place of γ_{n+1} although we will not restrict the spacetime dimension to four. The specific case of the four-dimensional axial anomaly was dealt with in Section 3.9.

A chiral, or axial, transformation is defined by

$$\Psi'(x) = \exp\left(i\alpha(x)\gamma_5\right)\Psi(x), \tag{5.474}$$

where $\alpha(x)$ is an arbitrary function. The adjoint spinor $\bar{\Psi} = \Psi^\dagger \gamma^0$ transforms like

$$\bar{\Psi}(x) = \bar{\Psi}(x) \exp\left(i\alpha(x)\gamma_5\right) \tag{5.475}$$

if we use $\gamma_5^\dagger = \gamma_5$, and the fact that γ_5 anticommutes with γ^0. Because the spin connection $B_\mu(x)$ contains two γ-matrices, it follows that $[B_\mu, \gamma_5] = 0$. The spacetime covariant derivative of the Dirac spinor is seen to satisfy

$$\nabla_\mu \Psi' = (\partial_\mu + iB_\mu)\Psi'$$
$$= \exp(i\alpha(x)\gamma_5)\left(\nabla_\mu \Psi + i\partial_\mu \alpha \gamma_5 \Psi\right). \tag{5.476}$$

The covariant derivative of $\bar{\Psi}$ was defined in (5.400), and is easily seen to transform like

$$\nabla_\mu \bar{\Psi}' = \left(\nabla_\mu \bar{\Psi} + i\bar{\Psi}\partial_\mu \alpha \gamma_5\right)\exp(i\alpha\gamma_5). \tag{5.477}$$

The Dirac Lagrangian was defined in (5.402). Using the results found above, it may be seen that the Dirac action transforms like

$$S'_D = S_D - m\int dv_x \bar{\Psi}\left(\exp(2i\alpha\gamma_5) - 1\right)\Psi - \int dv_x \partial_\mu \alpha \bar{\Psi}\gamma^\mu \gamma_5 \Psi. \tag{5.478}$$

If the last term is integrated by parts, and the surface term assumed to vanish, this becomes

$$S'_D = S_D - m\int dv_x \bar{\Psi}\left(\exp(2i\alpha\gamma_5) - 1\right)\Psi + \int dv_x \alpha(x)\nabla_\mu J_A^\mu, \tag{5.479}$$

where

$$J_A^\mu = \bar{\Psi}\gamma^\mu \gamma_5 \Psi \tag{5.480}$$

defines the axial vector current. This contrasts with the vector current J_V^μ defined in (5.507).

Under a rigid, or global, chiral transformation (i.e., α constant), the Dirac equation and the classical action are both seen to be invariant if $m = 0$. If $m \neq 0$, then the chiral transformation is not a symmetry of the Dirac equation. Restricting $m = 0$, (5.479) becomes

$$S'_D = S_D + \int dv_x \alpha(x)\nabla_\mu J_A^\mu. \tag{5.481}$$

By taking the divergence of (5.480) and using the massless Dirac equation, it is easily seen that the axial current is conserved:

$$\nabla_\mu J_A^\mu = 0. \tag{5.482}$$

The classical action is invariant in this case. Conversely, demanding that the classical action be invariant under the chiral transformation results in (5.482).

We now wish to study what happens in the quantum theory. The four-dimensional theory was already discussed in Chapter 3. Our approach here will be slightly different, following Fujikawa (1979, 1980a,b).

Let ψ_N be defined as in (5.428), and expand Ψ as in (5.429). We may expand Ψ' in a similar way as

$$\Psi' = \sum_N a'_N \Psi_N. \tag{5.483}$$

5.9 The axial, or chiral, anomaly

It is then easily seen that the chiral transformation induces the transformation

$$a'_{N'} = \sum_N C_{N'N} a_N, \tag{5.484}$$

where

$$C_{N'N} = \ell^{-1} \int dv_x \bar{\psi}_{N'}(x) \exp\left(i\alpha(x)\gamma_5\right) \psi_N(x), \tag{5.485}$$

on the expansion coefficients. (The orthogonality condition (5.431) has been used.) Taking α to be infinitesimal, we have

$$C_{N'N} = \delta_{N'N} + i\ell^{-1} \int dv_x \alpha(x) \bar{\psi}_{N'}(x) \gamma_5 \psi_N(x), \tag{5.486}$$

and hence

$$\det C_{N'N} = 1 + i\ell^{-1} \sum_N \int dv_x \alpha(x) \bar{\psi}_N(x) \gamma_5 \psi_N(x). \tag{5.487}$$

Because the transformation of Grassmann variables involves the inverse Jacobian,[17] we have

$$d\mu[\bar{\Psi}', \Psi'] = \left(\prod_N da'_N\right)\left(\prod_N da'^*_N\right)$$
$$= \left(\det C_{N'N}\right)^{-2} d\mu[\bar{\Psi}, \Psi]. \tag{5.488}$$

Defining the transformed effective action W' by

$$\exp\left(\frac{i}{\hbar}W'\right) = \int d\mu[\bar{\Psi}', \Psi'] \exp\left(\frac{i}{\hbar}S'_D\right),$$

it may be seen using (5.481) and (5.488) that

$$\exp\left(\frac{i}{\hbar}(W'-W)\right) = 1 + \frac{i}{\hbar}\int dv_x \alpha(x)\langle \nabla_\mu J_A^\mu\rangle - 2i\ell^{-1}\sum_N \int dv_x \alpha \bar{\psi}_N \gamma_5 \psi_N,$$

if only first-order terms in α are retained. Demanding that the quantum theory be invariant under a chiral transformation (i.e., $W' = W$) forces

$$\langle \nabla_\mu J_A^\mu\rangle = 2\hbar \ell^{-1} \sum_N \bar{\psi}_N(x) \gamma_5 \psi_N(x). \tag{5.489}$$

In the classical limit ($\hbar \to 0$), we recover (5.482); however, it is possible that the axial current is not conserved in the quantum theory if the right-hand side of (5.489) is non-zero.

The right-hand side of (5.489) is not well defined and must be regulated. We will adopt a regularization that makes contact with the asymptotic expansion of

---
[17] See (5.378).

the kernel for the operator[18] $(-i\underline{\gamma}^\mu \nabla_\mu)(i\underline{\gamma}^\mu \nabla_\mu) = \Box + \frac{1}{4}R$, which has eigenvalues $-\lambda_N^2$. We will define

$$\sum_N \bar{\psi}_N(x) \gamma_5 \psi_N(x) = \lim_{\tau \to 0} \sum_N \bar{\psi}_N(x) \gamma_5 \exp(-i\tau \ell^2 \lambda_N^2) \psi_N(x). \quad (5.490)$$

Noting that $\bar{\psi}_N \gamma_5 \psi_N = \text{tr}(\gamma_5 \psi_N \bar{\psi}_N)$, where the trace is over spinor indices, we may write

$$\sum_N \bar{\psi}_N(x) \gamma_5 \psi_N(x) = \lim_{\tau \to 0} \text{tr}[\gamma_5 K(\tau; x, x)], \quad (5.491)$$

where

$$K_{\alpha\beta}(\tau; x, x') = \sum_N \exp(-i\ell^2 \tau \lambda_N^2) \psi_{N\alpha}(x) \bar{\psi}_{N\beta}(x'). \quad (5.492)$$

(α and β are spinor indices.) We therefore have

$$\langle \nabla_\mu J_A^\mu \rangle = 2\hbar \ell^{-1} \lim_{\tau \to 0} \text{tr}[\gamma_5 K(\tau; x, x)]. \quad (5.493)$$

The asymptotic expansion for the kernel may now be used, giving as $\tau \to 0$

$$\text{tr}[\gamma_5 K(\tau; x, x)] \sim i\ell(4\pi i\ell^2 \tau)^{-n/2} \sum_{k=0}^{\infty} (i\ell^2 \tau)^k \text{tr}[\gamma_5 E_k(x)]. \quad (5.494)$$

All terms in the sum with $k > n/2$ will clearly give a vanishing contribution when $\tau \to 0$. We may therefore restrict $k \le n/2$. We now claim (justification will be given later) that terms in the sum with $k < n/2$ vanish because of the properties of the Dirac γ-matrices. Accepting for the time being that this is the case results in

$$\langle \nabla_\mu J_A^\mu \rangle = 2i\hbar(4\pi)^{-n/2} \text{tr}[\gamma_5 E_{n/2}(x)], \quad (5.495)$$

which, as for the conformal anomaly, relates the anomaly directly to a coefficient in the asymptotic expansion of the kernel. Unlike the case for the conformal anomaly, we can proceed further, simply from a knowledge of the general form taken by the coefficient $E_{n/2}$. Before doing this, we will justify the claim made above that all terms in the sum (5.494) with $k < n/2$ vanish.

Consider $\text{tr}[\gamma_5 \gamma^{a_1} \cdots \gamma^{a_r}]$, with $r \le n$. (n is the spacetime dimension, assumed to be even.) First of all, because $\gamma_5 \gamma^a = -\gamma^a \gamma_5$, we have

$$\text{tr}[\gamma_5 \gamma^{a_1} \cdots \gamma^{a_r}] = (-1)^r \text{tr}[\gamma^{a_1} \cdots \gamma^{a_r} \gamma_5].$$

This is sufficient to prove that $\text{tr}[\gamma_5 \gamma^{a_1} \cdots \gamma^{a_r}] = 0$ if r is odd. Suppose now that r is even, but $r < n$. It is easy to see that if any of the r indices $a_1, \ldots, a_r$ are equal to each other, then the trace vanishes. However, the trace also vanishes if all the indices are unequal. This is true because $\gamma_5 \gamma^{a_1} \cdots \gamma^{a_r}$ consists, in this case, of a product of $(n-r)$ γ-matrices with unequal indices, and the trace of a

---

[18] See (5.293).

5.9 The axial, or chiral, anomaly

product of an even number of γ-matrices with unequal indices vanishes. (Consider $\mathrm{tr}[\gamma^{b_1}\cdots\gamma^{b_{2k}}] = (-1)^{2k-1}\mathrm{tr}[\gamma^{b_2}\cdots\gamma^{b_{2k}}\gamma^{b_1}]$ since $\gamma^{b_1}\gamma^{b_j} = -\gamma^{b_j}\gamma^{b_1}$ ($j \neq 1$) if $b_1 \neq b_2 \neq \cdots \neq b_{2k}$. This, and the cyclic property of the trace, proves the desired result.) The only possibility for a non-zero trace is if $r = n$, in which case $\gamma_5 \gamma^{a_1} \cdots \gamma^{a_r}$ is proportional to the identity matrix. We may therefore write[19]

$$\mathrm{tr}[\gamma_5 \gamma^{a_1} \cdots \gamma^{a_n}] = c\epsilon^{a_1 \cdots a_n} \qquad (5.496)$$

for some constant c. Using $\epsilon^{01\cdots(n-1)} = -1$, and $\gamma_5 = \eta \gamma^0 \cdots \gamma^{n-1}$, it follows that

$$c = -\frac{1}{\eta}\mathrm{tr}I = (-i)^{n/2-1} 2^{n/2}.$$

($\eta = -i^{n/2-1}$ has been used, and the reader should recall that the Dirac matrices are $2^{n/2} \times 2^{n/2}$ matrices for n even; thus, $\mathrm{tr}\,I = 2^{n/2}$ here.) Thus,

$$\mathrm{tr}[\gamma_5 \gamma^{a_1} \cdots \gamma^{a_n}] = (-i)^{n/2-1} 2^{n/2} \epsilon^{a_1 \cdots a_n}. \qquad (5.497)$$

Now consider how γ-matrices can enter E_k for $k \leq n/2$. Because we only need $\mathrm{tr}[\gamma_5 E_k]$, the trace vanishes unless E_k contains n factors of γ-matrices. E_k comes from the kernel for $\Box + \frac{1}{4}R$, so the only way γ-matrices can enter is through $[\nabla_\mu, \nabla_\nu] = W_{\mu\nu}$ where $W_{\mu\nu} = -iR_{\mu\nu ab}\Sigma^{ab}$. (See (5.416).) E_k has dimensions of (length)$^{-2k}$ and $W_{\mu\nu}$ has dimensions of (length)$^{-2}$. It therefore follows that E_k can contain at most k powers of Σ^{ab}, and hence $2k$ factors of γ-matrices. The trace identity established above may then be used to deduce that $\mathrm{tr}[\gamma_5 E_k] = 0$ for $k < n/2$. This justifies (5.495).

The only possible non-zero term in $E_{n/2}$ comes from $(W_{\mu\nu})^{n/2}$. Terms which involve derivatives of $W_{\mu\nu}$ will have too few factors of γ-matrices to survive the Dirac trace. Because of this, we only need the coefficient of $(W_{\mu\nu})^{n/2}$ in $E_{n/2}$, and this may be determined by treating $W_{\mu\nu}$ as constant. The relevant part of the kernel which gives rise to the anomaly may be inferred from the kernel found for a scalar field in a constant electromagnetic field background. We need only make the replacement $ieF_{\mu\nu} \to W_{\mu\nu} = -iR_{\mu\nu ab}\Sigma^{ab}$ in (5.115) giving (with $m = 0$)

$$K(\tau; x, x) = i\left\{\det\left[\frac{\left(\frac{1}{4\pi i}R_{\mu\nu ab}\Sigma^{ab}\right)}{\sinh(\tau R_{\mu\nu ab}\Sigma^{ab})}\right]\right\}^{1/2}. \qquad (5.498)$$

The determinant here is only over the spacetime indices, so that the result is regarded as a Dirac matrix. This may be rewritten as

$$K(\tau; x, x) = i(4\pi i\tau)^{-n/2}\left[\det\left(\frac{\theta}{\sinh\theta}\right)\right]^{1/2}, \qquad (5.499)$$

where $\theta = \tau R_{\mu\nu ab}\Sigma^{ab}$. All dependence on the γ-matrices comes from θ. $K(\tau; x, x)$ is seen to be invariant under the replacement of θ with $-\theta$, which means that

---

[19] $\epsilon^{a_1 \cdots a_n}$ is the Levi–Civita tensor that is antisymmetric in all of its indices.

only even powers of θ occur in the expansion of $K(\tau; x, x)$. (The first few terms in the expansion were given in (5.120).) This means that the number of factors of γ occurring is a multiple of 4, since θ contains two factors. From the trace identity above, this means that the gravitational part of the axial anomaly vanishes except possibly in spacetimes of dimension $n = 4k$, where $k = 1, 2, \ldots$. For example, in four spacetime dimensions,

$$K(\tau, x, x) = -i(4\pi\tau)^{-2}\left\{1 + \frac{1}{12}\tau^2 R_{\mu\nu ab}R^{\mu\nu}{}_{cd}\Sigma^{ab}\Sigma^{cd} + \cdots\right\}, \quad (5.500)$$

and thus,

$$\langle \nabla_\mu J_A^\mu \rangle = \frac{i\hbar}{1536\pi^2} R_{\mu\nu ab} R^{\mu\nu}{}_{cd} \mathrm{tr}\left(\gamma_5 \gamma^a \gamma^b \gamma^c \gamma^d\right)$$
$$= \frac{\hbar}{384\pi^2} \epsilon^{\lambda\tau\rho\sigma} R_{\mu\nu\lambda\tau} R^{\mu\nu}{}_{\rho\sigma}, \quad (5.501)$$

using (5.497). This agrees with our earlier result (3.250) in Chapter 3.

A similar procedure to the one just outlined may be used to obtain the axial anomaly in an electromagnetic field background. The results in (5.489), (5.493), and (5.495) still hold. The only difference is that $K(\tau; x, x')$ is the kernel for the operator $(-i\gamma^\mu D_\mu)(i\gamma^\mu D_\mu) = D^2 + \frac{1}{4}RI + 2eF_{\mu\nu}\Sigma^{\mu\nu}$, since $D_\mu = \nabla_\mu + ieA_\mu$. This time there are two possible sources for products of γ-matrices. One is through

$$[D_\mu, D_\nu] = W_{\mu\nu} = -iR_{\mu\nu ab}\Sigma^{ab} + ieF_{\mu\nu}I.$$

The other is through the explicit γ-matrices occurring in the term $2eF_{\mu\nu}\Sigma^{\mu\nu}$. The relevant terms in the anomaly can be inferred from a knowledge of the kernel for a spin-1/2 particle in a constant electromagnetic field. Rather than pursuing this general result, we will concentrate again on four spacetime dimensions, in which case we may use (5.495). This gives

$$\langle \nabla_\mu J_A^\mu \rangle = 2i\hbar(4\pi)^{-2}\mathrm{tr}[\gamma_5 E_2(x)]$$
$$= \frac{i\hbar}{8\pi^2}\mathrm{tr}\left[\gamma_5 \frac{1}{12} W_{\mu\nu}W^{\mu\nu} + \gamma_5 \frac{1}{2}Q^2\right].$$

Only the terms in E_2 with a non-zero trace have been retained. (See (5.19).) The first term on the right-hand side gives rise to the result in (5.501). Making use of $Q = \frac{1}{4}RI + 2eF_{\mu\nu}\Sigma^{\mu\nu}$, the second term involves

$$\mathrm{tr}(\gamma_5 Q^2) = 4e^2 F_{\mu\nu} F_{\rho\sigma} \mathrm{tr}(\gamma_5 \Sigma^{\mu\nu}\Sigma^{\rho\sigma})$$
$$= -\frac{1}{4} e^2 F_{\mu\nu} F_{\rho\sigma} \mathrm{tr}(\gamma_5 \gamma^\mu \gamma^\nu \gamma^\rho \gamma^\sigma)$$
$$= ie^2 \epsilon^{\mu\nu\rho\sigma} F_{\mu\nu} F_{\rho\sigma}.$$

We therefore have

$$\langle \nabla_\mu J_A^\mu \rangle = \frac{\hbar}{384\pi^2} \epsilon^{\mu\nu\rho\sigma} R_{\lambda\tau\mu\nu} R^{\lambda\tau}{}_{\rho\sigma} - \frac{\hbar}{16\pi^2} e^2 \epsilon^{\mu\nu\rho\sigma} F_{\mu\nu} F_{\rho\sigma}. \quad (5.502)$$

In two spacetime dimensions, we have (see (5.495))

$$\langle \nabla_\mu J_A^\mu \rangle = \frac{i\hbar}{2\pi} \operatorname{tr}[\gamma_5 E_1]$$
$$= -\frac{i\hbar}{2\pi} \operatorname{tr}(\gamma_5 Q)$$
$$= -\frac{\hbar e}{2\pi} \epsilon^{\mu\nu} F_{\mu\nu}. \tag{5.503}$$

There is no purely gravitational contribution to the anomaly, in agreement with the general conclusion above that they can only arise in spacetimes whose dimension is a multiple of 4.

5.9.1 Chern–Simons theory

There is a direct connection between the chiral anomaly and the so-called Chern–Simons gauge theory which we wish to consider. The effective action for a spinor field coupled to a background electromagnetic field is $W[A]$ defined by

$$\exp\left(\frac{i}{\hbar} W[A]\right) = \int d\mu[\bar{\Psi}, \Psi] \exp\left(\frac{i}{\hbar} S_D\right), \tag{5.504}$$

where S_D is

$$S_D = \int dv_x \bar{\Psi}\left(i\gamma^\mu \nabla_\mu + e\gamma_\mu A^\mu - m\right)\Psi. \tag{5.505}$$

Variation of both sides of (5.504) with respect to A_μ gives

$$\delta W[A] = \int dv_x \langle J_V^\mu(x) \rangle [A] \delta A_\mu(x), \tag{5.506}$$

where

$$J_V^\mu(x) = \bar{\Psi} \gamma^\mu \Psi \tag{5.507}$$

is the vector current, and

$$\langle J_V^\mu(x) \rangle [A] = \frac{\int d\mu[\bar{\Psi}, \Psi] J_V^\mu(x) \exp\left(\frac{i}{\hbar} S_D\right)}{\int d\mu[\bar{\Psi}, \Psi] \exp\left(\frac{i}{\hbar} S_D\right)}. \tag{5.508}$$

The result in (5.506) was given originally by Schwinger (1951a), and is valid for any variation. Suppose that we choose a special potential $A_\mu^t(x) = tA_\mu(x)$ which interpolates between $A_\mu = 0$ when $t = 0$ and A_μ when $t = 1$. If A_μ is regarded as fixed, then $\delta A_\mu^t = \delta t A_\mu$, and (5.506) gives

$$\delta W[A_\mu^t] = \int dv_x \langle J_V^\mu(x) \rangle [A_\mu^t] \delta t A_\mu(x). \tag{5.509}$$

Integration from $t = 0$ to $t = 1$ gives

$$W[A] - W[0] = \int dv_x A_\mu(x) \int_0^1 dt \langle J_V^\mu(x) \rangle [tA]. \tag{5.510}$$

This establishes a direct relation between the effective action and the expectation value of the vector current. We may relate $\langle J_V^\mu \rangle$ to the Dirac propagator:

$$\langle J_V^\mu \rangle = -e\,\mathrm{tr}[\gamma^\mu S(x,x)], \tag{5.511}$$

where

$$(i\gamma^\mu \nabla_\mu + e\gamma^\mu A_\mu - m) S(x,x') = \delta(x,x'). \tag{5.512}$$

So far our treatment has been valid for a general background spacetime. For simplicity, we will now restrict attention to flat three-dimensional spacetime. Suppose the background field is such that $A^2 = 0$, with A^0 and A^1 independent of x^2. We will use $i = 0, 1$ in what follows. Consider

$$\langle J_V^2 \rangle = -e\,\mathrm{tr}[\gamma^2 S(x,x)]. \tag{5.513}$$

(Note that the superscript 2 here denotes the second component of the vector, not the square!) From the representations of the γ-matrices given in Section 5.6.2, it should be noted that $\gamma^2 = i\sigma_3 = i\gamma_5$ where γ_5 is the analogue in two dimensions of γ_5 in four dimensions. We will use this notation because we will be making contact with the earlier result for the axial anomaly in two spacetime dimensions.

Since $A_2 = 0$ and A_0, A_1 are independent of x^2, we may write

$$S(x,x') = \int_{-\infty}^{\infty} \frac{dk}{2\pi} \exp\left(ik(x^2 - x'^2)\right) \tilde{S}(k;\bar{x},\bar{x}'), \tag{5.514}$$

where $\tilde{S}(k;\bar{x},\bar{x}')$ satisfies

$$\left[i\gamma^i(\partial_i - ieA_i) - m - ik\gamma_5\right]\tilde{S}(k;\bar{x},\bar{x}') = \delta(\bar{x},\bar{x}'). \tag{5.515}$$

We use $\bar{x} = (x^0, x^1)$ here. Now define

$$\tilde{S}(k;\bar{x},\bar{x}') = \left[-i\gamma^i(\partial_i - ieA_i) - m + ik\gamma_5\right]\tilde{G}(k;\bar{x},\bar{x}'). \tag{5.516}$$

From (5.515) it follows that

$$\left[(\gamma^i D_i)^2 + m^2 + k^2\right]\tilde{G}(k;\bar{x},\bar{x}') = \delta(\bar{x},\bar{x}'), \tag{5.517}$$

where $D_i = \partial_i - ieA_i$. The expression for the current in (5.513) now becomes

$$\langle J_V^2 \rangle = -ie \int_{-\infty}^{\infty} \frac{dk}{2\pi}\,\mathrm{tr}\left[\gamma_5(-i\gamma^i D_i - m + ik\gamma_5)\tilde{G}(k;\bar{x},\bar{x})\right]. \tag{5.518}$$

This expression may be simplified considerably. From (5.517), $\tilde{G}$ is seen to involve only an even number of factors of γ^i. This means that γ_5 must commute with $\tilde{G}$ since $\gamma_5 \gamma^i = -\gamma^i \gamma_5$. The Dirac trace of $\gamma_5 \gamma^i D_i \tilde{G}$ therefore vanishes. It may also be seen from (5.517) that $\tilde{G}$ is an even function of k. This means that $k\tilde{G}$ integrates to zero (at least after regularization). We are left with

5.9 The axial, or chiral, anomaly

$$\langle J_V^2 \rangle = iem \int_{-\infty}^{\infty} \frac{dk}{2\pi} \text{tr}\left[\gamma_5 \tilde{G}(k; \bar{x}, \bar{x})\right]. \tag{5.519}$$

It is now possible to relate $\text{tr}[\gamma_5 \tilde{G}]$ to the axial anomaly in two spacetime dimensions. The axial current $J_A^i = \bar{\Psi}\gamma^i\gamma_5\Psi$ satisfies $\partial_i J_A^i = 2im\bar{\Psi}\gamma_5\Psi$ if the Dirac equation is used. If we use $a(x)$ to denote the two-dimensional axial anomaly given in (5.503), then

$$\langle \partial_i J_A^i \rangle = 2im\langle \bar{\Psi}\gamma_5\Psi \rangle + a(x). \tag{5.520}$$

$\langle \bar{\Psi}\gamma_5\Psi \rangle$ may be expressed in terms of the Dirac propagator as

$$\langle \bar{\Psi}\gamma_5\Psi \rangle = -\text{tr}[\gamma_5 S(\bar{x}, \bar{x})],$$

where

$$(i\gamma^i D_i - m)S(\bar{x}, \bar{x}') = \delta(\bar{x}, \bar{x}').$$

If we let

$$S(\bar{x}, \bar{x}') = (-i\gamma^i D_i - m)G(\bar{x}, \bar{x}'),$$

then

$$\left[(i\gamma^i D_i)^2 + m^2\right]G(\bar{x}, \bar{x}') = \delta(\bar{x}, \bar{x}') \tag{5.521}$$

and

$$\langle \bar{\Psi}\gamma_5\Psi \rangle = -\text{tr}\left[\gamma_5(-i\gamma^i D_i - m)G(\bar{x}, \bar{x})\right]$$
$$= m\text{tr}[\gamma_5 G(\bar{x}, \bar{x})]$$

since $\text{tr}(\gamma_5\gamma^i D_i G) = 0$, as for the case of $\tilde{G}$ above. Thus from (5.520)

$$i\text{tr}\left[\gamma_5 G(\bar{x}, \bar{x})\right] = -\frac{1}{2m^2}a(x) + \frac{1}{2m^2}\langle \partial_i J_A^i \rangle. \tag{5.522}$$

By comparing (5.521) with (5.517), it can be seen that $\tilde{G}$ may be obtained from G by the replacement of m^2 with $m^2 + k^2$. Making this replacement in (5.522), and using the result in (5.519), gives

$$\langle J_V^2 \rangle = -\frac{1}{2}ema(x) \int_{-\infty}^{\infty} \frac{dk}{2\pi}(k^2 + m^2)^{-1}$$
$$+ \frac{1}{2}em \int_{-\infty}^{\infty} \frac{dk}{2\pi} \frac{\langle \partial_i J_A^i \rangle}{k^2 + m^2}.$$

(The replacement $k^2 \to k^2 + m^2$ must also occur in $\langle \partial_i J_A^i \rangle$, but $a(x)$ is unaffected since it is independent of m.) Noting that

$$m \int_{-\infty}^{\infty} \frac{dk}{2\pi}(k^2 + m^2)^{-1} = \frac{m}{2|m|} = \frac{1}{2}\text{sign}(m),$$

we have upon letting $m \to 0$,

$$\langle J_V^2 \rangle = -\frac{e}{4} \text{sign}(m) a(x). \tag{5.523}$$

The sign(m) factor is related to the choice of representation for the Dirac matrices in the original action. (Recall from Section 5.6.2 that in three spacetime dimensions there are two inequivalent representations given by $\{\gamma^\mu\}$ and $\{-\gamma^\mu\}$.) Using the result for the axial anomaly in (5.503), we have

$$\langle J_V^2 \rangle = \frac{\hbar e^2}{8\pi} \text{sign}(m) \epsilon^{ij} F_{ij}.$$

It follows by general covariance that

$$\langle J_V^\mu \rangle = \frac{\hbar e^2}{8\pi} \text{sign}(m) \epsilon^{\mu\nu\lambda} F_{\nu\lambda}. \tag{5.524}$$

The result in (5.524) may also be obtained by assuming a constant background electromagnetic field (Redlich 1984).

This result may now be used in (5.510) to give

$$W[A] - W[0] = -\frac{\hbar e^2}{16\pi} \text{sign}(m) \int dv_x \epsilon^{\mu\nu\lambda} A_\mu F_{\nu\lambda}. \tag{5.525}$$

This was first derived by Redlich (1984) using a different method. Our method follows Dittrich and Reuter (1986). The action on the right-hand side is referred to as the Chern–Simons action, and first arose in a mathematical context (Chern and Simons 1974).

In order to examine the physical consequences of the term just found, suppose that we include it as a term in the action for three-dimensional electromagnetism. Let

$$\mathcal{L} = -\frac{1}{4} F_{\mu\nu} F^{\mu\nu} + \kappa \epsilon^{\mu\nu\lambda} A_\mu F_{\nu\lambda} \tag{5.526}$$

be the Lagrangian density, where κ is a constant. (The value found above for the Chern–Simons term induced from a spinor field was $\kappa = \pm \hbar e^2/16\pi$.) The field equations which follow from (5.526) are

$$0 = \Box A^\mu + 4\kappa \epsilon^{\mu\nu\lambda} \partial_\nu A_\lambda, \tag{5.527}$$

if the gauge condition $\partial_\mu A^\mu = 0$ is imposed. Assume a plane wave solution $A^\mu = E^\mu \exp(ip \cdot x)$ where E^μ is the polarization vector. Equation (5.527) implies

$$0 = \left(p^2 \delta^\mu_\nu - 4i\kappa \epsilon^{\mu\lambda}{}_\nu p_\lambda \right) E^\nu.$$

In order that non-trivial solutions for E^μ exist, it is necessary and sufficient that

$$\det \left(p^2 \delta^\mu_\nu - 4i\kappa \epsilon^{\mu\lambda}{}_\nu p_\lambda \right) = 0.$$

By working out this 3×3 determinant we find

$$0 = (p^2)^2 (p^2 - 16\kappa^2).$$

Therefore, two modes of the field are seen to behave like massless fields, and one mode to behave as if it has a mass given by $4|\kappa| = \hbar e^2/4\pi$. A more complete analysis shows that the massive mode corresponds to the physical transverse degree of freedom, rather than the unphysical longitudinal one (Deser *et al.*, 1982a,b). This is referred to as a topologically massive gauge theory.

6
The effective action: Non-gauge theories

6.1 Introduction

In classical mechanics the main focus of interest is on the equations of motion and their solution. There are two ways of obtaining these. The first way is to write down the Euler–Lagrange equations which, for systems whose Lagrangian contains no more than linear time derivatives of the generalized coordinates, are a set of second-order ordinary differential equations. The Euler–Lagrange equations are the natural generalization of Newton's second law of motion. The second approach is to use a Hamiltonian formulation, and to write down Hamilton's equations of motion which are a set of first-order differential equations. The Hamiltonian equations of motion may be succinctly written by introducing the Poisson bracket. The generalized coordinates q^i and their conjugate momenta p_i satisfy the following fundamental Poisson bracket relations:

$$[q^i, q^j] = 0, \qquad (6.1)$$

$$[p_i, p_j] = 0, \qquad (6.2)$$

$$[q^i, p_j] = \delta^i{}_j. \qquad (6.3)$$

(The analogues of these relations for superclassical systems, which deal with anticommuting coordinates and momenta, are straightforward to write down, but we will not pursue this here.) The usual formulation of classical mechanics may be easily generalized to classical field theory. The basic objects here are Lagrangian and Hamiltonian densities. One crucial feature of the Hamiltonian formulation of classical field theory is that in the Poisson bracket relations the arguments must be given at equal times. (More generally, the two arguments of the bracket must be evaluated on the same spacelike hypersurface.) Thus a time coordinate must be distinguished from the spatial coordinates in the Hamiltonian approach.

Quantum theory is usually presented as an active process: a classical theory is the starting point, and this theory is then quantized. The usual way in

6.1 Introduction

which this is done is by regarding the generalized coordinates and their conjugate momenta as operators, and replacing the Poisson bracket relations with $(i\hbar)^{-1}$ times the commutator. In field theory, these are called the equal time canonical commutation relations, and the approach is called canonical quantization. We have described how it works in Chapters 1 and 2. Again we wish to emphasize that in canonical quantization, the time and spatial coordinates enter in different ways, although we also wish to make it clear that there is nothing wrong with this feature in principle.

In classical mechanics there is an alternate way of obtaining the Euler–Lagrange equations, called the principle of stationary action. For classical field theory this consists, first of all, in writing down the action functional

$$S[\varphi] = \int dv_x \, \mathcal{L}(x), \tag{6.4}$$

where $dv_x = \sqrt{-g(x)} d^n x$ is the invariant spacetime volume element, and $\mathcal{L}(x)$ is the scalar Lagrangian density which involves the fields φ^i and their derivatives $\partial_\mu \varphi^i$ as well as the spacetime metric. (These arguments have all been suppressed in (6.4).) In some cases it may be necessary to add surface terms to (6.4). The Euler–Lagrange equations follow by demanding that $S[\varphi]$ be stationary for an arbitrary variation in the field. Examples of classical actions have been considered in earlier chapters.

The principle of stationary action has some advantages over an approach based on field equations. First of all, by definition, the action functional is a scalar under both general coordinate transformations on spacetime, as well as a scalar under redefinitions of the classical fields. It is always easier to write down scalars than to write down a set of tensor field equations. (This is clearly illustrated in the derivation of the field equations of general relativity from an action principle given by Hilbert (1915, 1917).) We are guaranteed by the stationary action principle that if the action functional is a scalar, the field equations obtained will be tensorial in nature. As we have emphasized, the Hamiltonian approach involves a distinction between the time and spatial coordinates. This means that **manifest** covariance may be lost, and therefore it is necessary to prove that it is not. (The fact that this may not be entirely trivial is evident from Sections 14.6 and 17.9 of Bjorken and Drell (1965), which deals with this question for quantum electrodynamics.) In order to circumvent this cumbersome problem, Dirac (1933) was led to look for a Lagrangian approach to quantum mechanics which could then be generalized to field theory. His paper may be seen as the starting point of covariant quantization, and a motivation for Feynman's pivotal paper (Feynman 1948) on the path integral method, which is undoubtedly the most widely used method of covariant quantization today.

In quantum field theory there are essentially two manifestly covariant methods of quantization. One is the already mentioned Feynman path integral approach. The other is the Schwinger action principle (Schwinger 1951b, 1953). (A concise discussion of the Schwinger action principle is contained in the book of Jauch and Rohrlich (1980). A longer pedagogical exposition may be found in Toms (2007). Some of the early papers on the Schwinger action principle and the Feynman path integral are collected in Schwinger (1958).) Although at first sight these two approaches appear very different, they are in fact closely related. The Schwinger action principle, which contains the quantum version of the classical principle of stationary action, may be regarded as the differential form of the path integral method. This is discussed by DeWitt (1964a), and we will present a version of this connection below. We have already discussed the path integral and the Schwinger action principle at one-loop order in the previous chapter. In the present chapter, we wish to present an analysis that is valid at any order in perturbation theory. Our treatment is not intended to have any pretense of rigor.

Because our concern in this chapter is with the formal properties of field theory, it is convenient to adopt the condensed notation of DeWitt (1964a). The field variables are denoted by φ^i where the index i stands not only for all discrete labels of the field, but also for the spacetime argument. For example, if $\varphi^I(x)$ is a field whose indices are indicated by I in conventional notation (e.g., I may be a vector index), then i stands for the pair (I, x). An object written as g_{ij} in condensed notation really stands for a two-point function $g_{IJ}(x, x')$ in uncondensed notation, where I and J refer only to discrete labels. A summation convention is used, which involves not only a sum over discrete indices, but also an integration over the repeated spacetime labels. For example, $g_{ij}V^iW^j$ in condensed notation is really shorthand for

$$g_{ij}V^iW^j = \int d^n x\, d^n x'\, g_{IJ}(x, x')V^I(x)W^J(x') \tag{6.5}$$

in normal notation. It is possible to deal with spinor fields using a superspace formalism, but this will not be dealt with here. (See DeWitt (1984b) for a systematic development of this.) The utility of a superspace formulation of field theory appears in supersymmetric theories (Salam and Strathdee 1974 or Arnowitt and Nath 1976). A more rigorous mathematical treatment has been given by Rogers (1980).

Let $\delta(x, x')$ be the biscalar (i.e., scalar under coordinate transformations in each argument) Dirac δ-distribution which is defined by

$$\varphi^I(x) = \int dv_{x'}\, \delta(x, x')\, \varphi^I(x') \tag{6.6}$$

for any tensor field $\varphi^I(x)$. In condensed notation the right-hand side may be written as $\delta^i{}_j \varphi^j$, which from our summation convention gives the correspondence

$$\delta^i{}_j = |g(x')|^{1/2} \delta^I{}_J \delta(x, x') \tag{6.7}$$

between condensed and normal notation. ($\delta^I{}_J$ on the right-hand side is just the usual Kronecker delta.) To save a bit of writing, it proves convenient to define

$$\tilde{\delta}(x, x') = |g(x')|^{1/2}\delta(x, x'), \tag{6.8}$$

which transforms as a scalar at x and a scalar density at x'.

One last bit of notation which we wish to mention concerns functional differentiation. Let $F[\varphi]$ be any functional. We want, using condensed notation,

$$F[\varphi + \psi] = F[\varphi] + F_{,i}[\varphi]\psi^i \tag{6.9}$$

to first order in ψ^i, where $F_{,i}[\varphi]$ denotes the functional derivative. But in normal notation, according to our summation convention for condensed indices, we have

$$F_{,i}[\varphi]\psi^i = \int d^n x' \frac{\delta F[\varphi(x)]}{\delta \varphi^I(x')} \psi^I(x'). \tag{6.10}$$

This suggests that the functional derivative be defined by

$$\lim_{\epsilon \to 0} \frac{1}{\epsilon}\{F[\varphi + \epsilon\psi] - F[\varphi]\} = \int d^n x' \frac{\delta F[\varphi(x)]}{\delta \varphi^I(x')} \psi^I(x'). \tag{6.11}$$

The functional derivative is a distribution in general. A particular consequence of (6.11) is

$$\frac{\delta \varphi^I(x)}{\delta \varphi^J(x')} = \delta^I{}_J \tilde{\delta}(x, x'). \tag{6.12}$$

The analogous result in condensed notation is $\varphi^i{}_{,j} = \delta^i{}_j$, which can be seen to be consistent with (6.7).

6.2 The Schwinger action principle

The Schwinger action principle, like the Feynman path integral, concentrates on the transition amplitude between two quantum states. Here we wish to formulate the Schwinger action principle for a local field theory and derive some of its consequences. The next section will examine in a nonrigorous way the relationship between the Schwinger action principle and the Feynman path integral, complementing the earlier discussion in Chapter 1.

Classically, a local field is a function which depends only on a single spacetime point rather than an extended region of spacetime. The theory is quantized by replacing the classical fields $\varphi^i(x)$ with field operators $\phi^i(x)$. Here 'i' represents some type of index (e.g., vector), and is not to be confused with condensed notation. (The context should make it clear whether or not condensed notation is being used.) It is not necessary to use condensed notation in this section.

Let Σ denote a spacelike hypersurface. This means that any two points of Σ have a spacelike separation, and therefore must be causally disconnected. As a consequence, the values of the field at different points on Σ must be independent. If x_1 and x_2 are two points on Σ, then

$$[\phi^i(x_1), \phi^j(x_2)] = 0, \tag{6.13}$$

since a measurement at x_1 cannot influence one at x_2. The fundamental assumption in local field theory is that a complete set of commuting observables can be constructed from the fields and their derivatives on Σ. Let ζ represent such a complete set of commuting observables on Σ, with ζ' denoting the eigenvalues of the observables.[1] We can denote a quantum state by $|\zeta', \Sigma\rangle$. Note that the Heisenberg picture is adopted here, where the states are time-independent with the time dependence in the operators. This is necessary for manifest covariance since we do not wish to treat the time and space arguments of the field differently.

Suppose that Σ_1 and Σ_2 are two spacelike hypersurfaces with all points of Σ_2 to the future of those of Σ_1. Let ζ_1 be a complete set of commuting observables defined on Σ_1, and similarly let ζ_2 be a complete set of commuting observables defined on Σ_2. We will assume that ζ_1 and ζ_2 have the same eigenvalue spectrum. Then ζ_1 and ζ_2 should be related by a unitary transformation:

$$\zeta_2 = U_{12} \zeta_1 U_{12}^{-1}, \tag{6.14}$$

and their respective eigenstates are related by

$$|\zeta_2', \Sigma_2\rangle = U_{12} |\zeta_1', \Sigma_1\rangle. \tag{6.15}$$

Here U_{12} is a unitary matrix giving the evolution of the state between the two spacelike hypersurfaces. The transformation function, or transition amplitude, is defined by

$$\langle \zeta_2', \Sigma_2 | \zeta_1', \Sigma_1 \rangle = \langle \zeta_1', \Sigma_1 | U_{12}^{-1} | \zeta_1', \Sigma_1 \rangle. \tag{6.16}$$

The unitary operator U_{12} depends on the details of the quantum system, the choice made for the commuting observables ζ, as well as the spacelike hypersurfaces Σ_1 and Σ_2. A change in any of these quantities will induce a change in the transformation function:

$$\delta \langle \zeta_2', \Sigma_2 | \zeta_1', \Sigma_1 \rangle = \langle \zeta_1', \Sigma_1 | \delta U_{12}^{-1} | \zeta_1', \Sigma_1 \rangle. \tag{6.17}$$

If we write the unitary operator U_{12} as

$$U_{12} = \exp\left(-\frac{i}{\hbar} S_{12}\right), \tag{6.18}$$

where $S_{12}^\dagger = S_{12}$ is Hermitian, then

$$\delta U_{12}^{-1} = \frac{i}{\hbar} U_{12}^{-1} \delta S_{12}. \tag{6.19}$$

---

[1] We again use the Dirac (1958) notation of using a $'$ to distinguish an eigenvalue from the associated operator.

6.2 The Schwinger action principle

The change in the transformation function expressed in (6.17) may then be written as

$$\delta\langle \zeta_2', \Sigma_2 | \zeta_1', \Sigma_1 \rangle = \frac{i}{\hbar} \langle \zeta_2', \Sigma_2 | \delta S_{12} | \zeta_1', \Sigma_1 \rangle. \tag{6.20}$$

It is possible, as in Schwinger (1953), to regard (6.20) as the definition of δS_{12}. It follows that δS_{12} must be Hermitian in order that (6.20) be consistent with the basic requirement

$$\langle \zeta_2', \Sigma_2 | \zeta_1', \Sigma_1 \rangle^* = \langle \zeta_1', \Sigma_1 | \zeta_2', \Sigma_2 \rangle. \tag{6.21}$$

If Σ_3 is a spacelike hypersurface, all of whose points lie to the future to those of Σ_2, then the basic law for the composition of probability amplitudes is

$$\langle \zeta_3', \Sigma_3 | \zeta_1', \Sigma_1 \rangle = \sum_{\zeta_2'} \langle \zeta_3', \Sigma_3 | \zeta_2', \Sigma_2 \rangle \langle \zeta_2', \Sigma_2 | \zeta_1', \Sigma_1 \rangle. \tag{6.22}$$

Varying both sides of this relation, and using (6.20) gives

$$\langle \zeta_3', \Sigma_3 | \delta S_{13} | \zeta_1', \Sigma_1 \rangle = \sum_{\zeta_2'} \Big\{ \langle \zeta_3', \Sigma_3 | \delta S_{23} | \zeta_2', \Sigma_2 \rangle \langle \zeta_2', \Sigma_2 | \zeta_1', \Sigma_1 \rangle$$
$$+ \langle \zeta_3', \Sigma_3 | \zeta_2', \Sigma_2 \rangle \langle \zeta_2', \Sigma_2 | \delta S_{12} | \zeta_1', \Sigma_1 \rangle \Big\}$$
$$= \langle \zeta_3', \Sigma_3 | (\delta S_{23} + \delta S_{12}) | \zeta_1', \Sigma_1 \rangle.$$

From this result we can deduce that

$$\delta S_{13} = \delta S_{12} + \delta S_{23}. \tag{6.23}$$

If we take the limit $\Sigma_2 \to \Sigma_1$ in (6.20), it follows that

$$\delta S_{12} \to 0. \tag{6.24}$$

If we take the limit $\Sigma_3 \to \Sigma_1$ in (6.23), it follows that

$$\delta S_{21} = -\delta S_{12}. \tag{6.25}$$

If the operators in ζ_1 and ζ_2 undergo infinitesimal unitary transformations on the hypersurfaces Σ_1 and Σ_2 respectively, and only on these two hypersurfaces, then the change in the transformation function must have the form

$$\delta\langle \zeta_2', \Sigma_2 | \zeta_1', \Sigma_1 \rangle = \frac{i}{\hbar} \langle \zeta_2', \Sigma_2 | (F_2 - F_1) | \zeta_1', \Sigma_1 \rangle. \tag{6.26}$$

Here F_1 and F_2 are Hermitian operators constructed from a knowledge of the fields and their derivatives on Σ_1 and Σ_2 respectively. (This follows from the infinitesimal version of (6.15).) This result is seen to be of the form given in (6.20) with the identification

$$\delta S_{12} = F_2 - F_1. \tag{6.27}$$

The generator F on a spacelike hypersurface Σ should be expressible as

$$F = \int_\Sigma d\sigma_x n^\mu F_\mu(x), \tag{6.28}$$

where $d\sigma_x$ is the element of area on Σ, n^μ is the outward unit normal to Σ, and $F_\mu(x)$ is constructed out of a knowledge of the fields on Σ. Equation (6.28) results from the fact that the points making up Σ are all spacelike separated and hence independent; the result follows from adding up all of these independent contributions. Applying (6.28) to Σ_1 and Σ_2, and assuming that $F_\mu(x)$ is defined throughout the spacetime region bounded by Σ_1 and Σ_2, we can write (6.27) as

$$\delta S_{12} = \int_{\Sigma_2} d\sigma_x n^\mu F_\mu(x) - \int_{\Sigma_1} d\sigma_x n^\mu F_\mu(x)$$

$$= \int_{\Omega_{12}} dv_x \nabla^\mu F_\mu(x). \tag{6.29}$$

Here Ω_{12} is the spacetime region bounded by Σ_1 and Σ_2, and dv_x is the invariant volume element. This result may be seen to be consistent with (6.23)–(6.25).

The result in (6.29) has assumed that the operators are changed only on Σ_1 and Σ_2. Suppose that we now consider the case where the operators are changed in the spacetime region between Σ_1 and Σ_2. We will assume that δS_{12} can again be expressed as a volume integral:

$$\delta S_{12} = \int_{\Omega_{12}} dv_x \delta \mathcal{L}(x), \tag{6.30}$$

for some $\delta\mathcal{L}(x)$. This is again consistent with (6.23)–(6.25). Combining the two possible types of variation (6.27) and (6.30) gives

$$\delta S_{12} = F_2 - F_1 + \int_{\Omega_{12}} dv_x \delta\mathcal{L}(x) \tag{6.31}$$

$$= \int_{\Omega_{12}} dv_x [\delta\mathcal{L}(x) + \nabla^\mu F_\mu(x)]. \tag{6.32}$$

The result in the last equation shows that altering $\delta\mathcal{L}$ by the addition of the divergence of a vector field results in a unitary transformation of the states on Σ_1 and Σ_2.

The fundamental assumption of the Schwinger action principle is that δS_{12} may be obtained from a variation of

$$S_{12} = \int_{\Omega_{12}} dv_x \mathcal{L}(x), \tag{6.33}$$

where $\mathcal{L}(x)$ is the Lagrangian density, which depends on the fields and their derivatives at a single spacetime point. Since δS_{12} is required to be Hermitian, S_{12} in (6.33) must be Hermitian, and hence so must the Lagrangian density.

For a given quantum system, the transformation function $\langle \zeta_2', \Sigma_2 | \zeta_1', \Sigma_1 \rangle$ describes the transition amplitude between two states of this given system. It

6.2 The Schwinger action principle

can therefore only alter if the states on Σ_1 and Σ_2 are altered. This means that for a given dynamical system

$$\delta S_{12} = F_2 - F_1, \tag{6.34}$$

as in (6.27). In particular, if the variation is with respect to the fields in the spacetime region Ω_{12} which are fixed on the boundaries Σ_1 and Σ_2, then

$$\delta S_{12} = 0. \tag{6.35}$$

(This follows because the variation described cannot affect the transformation function.) The result in (6.35) is the operator principle of stationary action.

Suppose that only the fields are varied, everything else being held fixed. If $\mathcal{L}$ only depends on the fields and their first derivatives, then using (6.33) we have

$$\delta S_{12} = \int_{\Omega_{12}} dv_x \left\{ \frac{\partial \mathcal{L}}{\partial \phi^i} \delta \phi^i + \frac{\partial \mathcal{L}}{\partial (\partial_\mu \phi^i)} \delta(\partial_\mu \phi^i) \right\}. \tag{6.36}$$

The field variation $\delta \phi^i$ is assumed to be a c-number here (equivalently a multiple of the unit operator); thus, the order of the variation with respect to ϕ^i is not important. If the second term in (6.36) is integrated by parts, it may be seen that

$$\delta S_{12} = \int_{\Omega_{12}} dv_x \delta \phi^i \left\{ \frac{\partial \mathcal{L}}{\partial \phi^i} - \nabla_\mu \left(\frac{\partial \mathcal{L}}{\partial (\partial_\mu \phi^i)} \right) \right\}$$
$$+ \int_{\Omega_{12}} d^n x \, \partial_\mu \left\{ \sqrt{-g} \frac{\partial \mathcal{L}}{\partial (\partial_\mu \phi^i)} \delta \phi^i \right\},$$

using the identity $\partial_\mu(\sqrt{-g}\, v^\mu) = \sqrt{-g}\, \nabla_\mu v^\mu$ for any vector field v^μ. The last term may be reduced to give

$$\delta S_{12} = \int_{\Omega_{12}} dv_x \delta \phi^i \left\{ \frac{\partial \mathcal{L}}{\partial \phi^i} - \nabla_\mu \left(\frac{\partial \mathcal{L}}{\partial (\partial_\mu \phi^i)} \right) \right\} + F_2 - F_1, \tag{6.37}$$

where

$$F_1 = \int_{\Sigma_1} d\sigma_x n_{1\mu} \frac{\partial \mathcal{L}}{\partial (\partial_\mu \phi^i)} \delta \phi^i, \tag{6.38}$$

with a similar expression for F_2.

If $\delta \phi^i$ vanishes on Σ_1 and Σ_2, then F_1 and F_2 must vanish. The transformation function cannot change, as discussed above, and we obtain

$$0 = \frac{\partial \mathcal{L}}{\partial \phi^i} - \nabla_\mu \left(\frac{\partial \mathcal{L}}{\partial (\partial_\mu \phi^i)} \right), \tag{6.39}$$

since the operator principle of stationary action holds. This is the operator equation of motion.

For completeness, we will show how the canonical commutation relations may be obtained from the formalism described above. Consider a variation

$\phi^i \to \phi^i + \delta\phi^i$ of the fields on the spacelike hypersurface Σ. From our earlier discussion, this induces a unitary transformation on the states $|\zeta', \Sigma\rangle$,

$$|\zeta', \Sigma\rangle \to U|\zeta', \Sigma\rangle, \tag{6.40}$$

where

$$U \simeq 1 - \frac{i}{\hbar}F, \tag{6.41}$$

with

$$F = \int_\sigma d\sigma_x n_\mu \pi_i^\mu \delta\phi^i. \tag{6.42}$$

We have defined π_i^μ by

$$\pi_i^\mu = \frac{\partial \mathcal{L}}{\partial(\partial_\mu \phi^i)}. \tag{6.43}$$

(See (6.38).) If the states $|\zeta', \Sigma\rangle$ are chosen to be eigenstates of the field operator ϕ^i on Σ, then the eigenvalues of ϕ^i are not affected by the transformation. The field operator transforms as

$$\phi^i \to U\phi^i U^{-1}. \tag{6.44}$$

Using the infinitesimal form (6.41) for U, we have

$$\delta\phi^i = -\frac{i}{\hbar}[F, \phi^i] \tag{6.45}$$

as the infinitesimal transformation of the field operator. Using (6.42) it may be seen that

$$\delta\phi^i = -\frac{i}{\hbar} \int_\Sigma d\sigma_x n_\mu [\pi_j^\mu, \phi^i] \delta\phi^j. \tag{6.46}$$

If Σ is specified by a constant value of t, as it can be in flat spacetime, or more generally in any spacetime which is globally hyperbolic, then $n^\mu = \delta_0^\mu$, and it may be seen that

$$[\pi_j(t, \vec{x}), \phi^i(t, \vec{x}')] = i\hbar \delta_j^i \delta(\vec{x}, \vec{x}'), \tag{6.47}$$

where $\delta(\vec{x}, \vec{x}')$ is the Dirac δ-distribution on the spacelike hypersurface Σ, and

$$\pi_j(t, \vec{x}) = \frac{\partial \mathcal{L}}{\partial(\partial_0 \phi^j)} \tag{6.48}$$

is the field momentum, which is canonically conjugate to ϕ^i. This may be recognized as one of the canonical commutation relations. The others may be derived as discussed in Schwinger (1951b).

6.3 The Feynman path integral

Suppose that we alter the classical theory described by the action $S[\varphi]$ by coupling the field to an external source J_i, where by external we mean that it has no dependence on the field φ^i. The idea of introducing external sources originates with Schwinger (1951c,d). We will choose[2]

$$S_J[\varphi] = S[\varphi] + J_i\varphi^i. \tag{6.49}$$

The transition amplitude $\langle \zeta_2', \Sigma_2 | \zeta_1', \Sigma_1 \rangle$ may be regarded as a functional of the source J_i, which we will denote explicitly by writing $\langle \zeta_2', \Sigma_2 | \zeta_1', \Sigma_1 \rangle [J]$. If we apply the Schwinger action principle to this modified theory, we have

$$\delta \langle \zeta_2', \Sigma_2 | \zeta_1', \Sigma_1 \rangle [J] = \frac{i}{\hbar} \langle \zeta_2', \Sigma_2 | \delta S_J | \zeta_1', \Sigma_1 \rangle [J]. \tag{6.50}$$

The classical field φ^i has been replaced with the field operator ϕ^i. In addition to its previous meaning, δ now includes a possible change in the source. By considering the variation to be with respect to the dynamical variables which are held fixed on Σ_1 and Σ_2, just as in the previous section, we obtain the operator field equations

$$0 = S_{,i}[\phi] + J_i. \tag{6.51}$$

Assume that the variation in (6.50) is one in which the dynamical variables are held fixed and only the source is altered. Because the source enters in the simple way given in (6.49), we have

$$\delta \langle \zeta_2', \Sigma_2 | \zeta_1', \Sigma_1 \rangle [J] = \frac{i}{\hbar} \delta J_i \langle \zeta_2', \Sigma_2 | \phi^i | \zeta_1', \Sigma_1 \rangle [J]. \tag{6.52}$$

This may be rewritten in the equivalent form

$$\frac{\delta \langle 2|1\rangle [J]}{\delta J_i} = \frac{i}{\hbar} \langle 2|\phi^i|1\rangle [J], \tag{6.53}$$

where we have used an abbreviated notation for the initial and final states to save writing.

We now perform a variation of (6.53) with respect to the source. This gives

$$\delta \frac{\delta \langle 2|1\rangle [J]}{\delta J_i} = \frac{i}{\hbar} \delta \langle 2|\phi^i|1\rangle [J]. \tag{6.54}$$

In order to evaluate this expression, consider a spatial hypersurface Σ' which lies to the future of Σ_1 and the past of Σ_2, and which contains the spacetime point represented by 'i' in (6.54). Any variation of the source can be represented as the sum of a variation which vanishes to the future of Σ' (but is non-zero to the past), and one which vanishes to the past of Σ' (but is non-zero to the future). Consider first the case where δJ_i vanishes to the future of Σ'. In this event, any

---

[2] Condensed notation is used here.

amplitude of the form $\langle 2|\phi^i|\zeta'\rangle[J]$, where $|\zeta'\rangle$ represents a state on Σ', cannot be affected by the variation of the source since δJ_i vanishes to the future of Σ'. If we use

$$\langle 2|\phi^i|1\rangle[J] = \sum_{\zeta'} \langle 2|\phi^i|\zeta'\rangle[J]\langle\zeta'|1\rangle[J], \tag{6.55}$$

which follows upon the use of $\sum_{\zeta'} |\zeta'\rangle\langle\zeta'| = 1$, it follows that the right-hand side of (6.54) may be re-expressed using

$$\delta\langle 2|\phi^i|1\rangle[J] = \sum_{\zeta'} \langle 2|\phi^i|\zeta'\rangle[J]\delta\langle\zeta'|1\rangle[J]. \tag{6.56}$$

The Schwinger action principle gives

$$\delta\langle\zeta'|1\rangle[J] = \frac{i}{\hbar}\delta J_j \langle\zeta'|\phi^j|1\rangle[J], \tag{6.57}$$

which when used in (6.56) results in

$$\delta\langle 2|\phi^i|1\rangle[J] = \frac{i}{\hbar}\delta J_j \sum_{\zeta'} \langle 2|\phi^i|\zeta'\rangle[J]\langle\zeta'|\phi^j|1\rangle[J]$$

$$= \frac{i}{\hbar}\delta J_j \langle 2|\phi^i\phi^j|1\rangle[J]. \tag{6.58}$$

Since δJ_j vanishes to the future of Σ', which contains the spacetime point represented by 'i', the spacetime point represented by 'j' must lie to the past of 'i'.

If we now consider the case where δJ_j vanishes to the past of Σ', a similar argument shows that in this situation

$$\delta\langle 2|\phi^i|1\rangle[J] = \frac{i}{\hbar}\delta J_j \langle 2|\phi^j\phi^i|1\rangle[J]. \tag{6.59}$$

The spacetime point represented by 'j' now lies to the future of that represented by 'i'. Combining the results of (6.58) and (6.59) leads to the conclusion that

$$\frac{\delta\langle 2|\phi^i|1\rangle[J]}{\delta J_j} = \frac{i}{\hbar}\langle 2|T(\phi^i\phi^j)|1\rangle[J], \tag{6.60}$$

where T is the chronological, or time, ordering operator which orders any product of fields in order of increasing time, with those furthest to the past to the right.

Combining (6.54) and (6.60) shows that

$$\frac{\delta^2\langle 2|1\rangle[J]}{\delta J_i \delta J_j} = \left(\frac{i}{\hbar}\right)^2 \langle 2|T(\phi^i\phi^j)|1\rangle[J]. \tag{6.61}$$

The generalization

$$\frac{\delta^n\langle 2|1\rangle[J]}{\delta J_{i_1}\cdots\delta J_{i_n}} = \left(\frac{i}{\hbar}\right)^n \langle 2|T(\phi^{i_1}\cdots\phi^{i_n})|1\rangle[J] \tag{6.62}$$

follows by induction.

We may regard the amplitude $\langle 2|1\rangle[J]$ as defined by its Taylor expansion about $J_i = 0$:

$$\langle 2|1\rangle[J] = \sum_{n=0}^{\infty} \frac{1}{n!} J_{i_1} \cdots J_{i_n} \left.\frac{\delta^n \langle 2|1\rangle[J]}{\delta J_{i_1} \cdots \delta J_{i_n}}\right|_{J=0}. \tag{6.63}$$

Using (6.62) allows us to write

$$\langle 2|1\rangle[J] = \sum_{n=0}^{\infty} \frac{1}{n!} \left(\frac{i}{\hbar}\right)^n J_{i_1} \cdots J_{i_n} \langle 2|T(\phi^{i_1} \cdots \phi^{i_n})|1\rangle[J=0]. \tag{6.64}$$

The series may be formally summed to give

$$\langle 2|1\rangle[J] = \langle 2|T\left(\exp\left(\frac{i}{\hbar}J_i\phi^i\right)\right)|1\rangle[J=0], \tag{6.65}$$

where J_i is set to zero everywhere on the right-hand side except in the exponential.

Now consider the action $S[\phi]$ expanded in a Taylor series about $\phi^i = 0$:

$$S[\phi] = \sum_{n=0}^{\infty} \frac{1}{n!} S_{,i_1 \cdots i_n}[\phi=0] \phi^{i_1} \cdots \phi^{i_n}. \tag{6.66}$$

For many theories of interest this expansion is already given in the form of the action, involving only local polynomials or monomials in the fields. It is usually a finite series, but it is not necessary to assume this in what follows.

Differentiation of (6.66) with respect to the field gives

$$S_{,i}[\phi] = \sum_{n=0}^{\infty} \frac{1}{n!} S_{,ii_1 \cdots i_n}[\phi=0] \phi^{i_1} \cdots \phi^{i_n}. \tag{6.67}$$

If we replace ϕ^i by $\frac{\hbar}{i}\delta/\delta J_i$ in this expression, and operate with $S_{,i}[\frac{\hbar}{i}\delta/\delta J_i]$ on $\langle 2|1\rangle[J]$ using (6.65) or (6.66), it may be seen that

$$S_{,i}\left[\frac{\hbar}{i}\frac{\delta}{\delta J_i}\right]\langle 2|1\rangle[J] = \langle 2|T\left(S_{,i}[\phi]\exp\left(\frac{i}{\hbar}J_i\phi^i\right)\right)|1\rangle[J=0]. \tag{6.68}$$

Using the operator equation of motion (6.51) results in

$$S_{,i}\left[\frac{\hbar}{i}\frac{\delta}{\delta J_i}\right]\langle 2|1\rangle[J] = -J_i\langle 2|1\rangle[J]. \tag{6.69}$$

This provides us with a differential equation for the transition amplitude which we can try to solve.

In order to solve (6.69), we may use the functional analogue of a Fourier transform:

$$\langle 2|1\rangle[J] = \int \left(\prod_i d\varphi^i\right) F[\varphi]\exp\left(\frac{i}{\hbar}J_i\varphi^i\right), \tag{6.70}$$

where the integration extends over all fields which correspond to the choice of states described by $|1\rangle$ and $|2\rangle$. The functional $F[\varphi]$, which is to be thought of as the Fourier transform of the transformation function, is to be determined by requiring that (6.70) satisfies (6.69). It is immediately found that

$$0 = \int \left(\prod_i d\varphi^i\right) \{S_{,i}[\varphi] + J_i\} F[\varphi] \exp\left(\frac{i}{\hbar} J_i \varphi^i\right)$$

$$= \int \left(\prod_i d\varphi^i\right) \left\{S_{,i}[\varphi]F[\varphi] + \frac{\hbar}{i} F[\varphi]\frac{\delta}{\delta\varphi^i}\right\} \exp\left(\frac{i}{\hbar} J_i \varphi^i\right).$$

Upon performing an integration by parts on the second term of the second line, the above result becomes

$$0 = \int \left(\prod_i d\varphi^i\right) \left\{S_{,i}[\varphi]F[\varphi] - \frac{\hbar}{i} F_{,i}[\varphi]\right\} \exp\left(\frac{i}{\hbar} J_i \varphi^i\right)$$

$$+ \frac{\hbar}{i} F[\varphi] \exp\left(\frac{i}{\hbar} J_i \varphi^i\right)\bigg|_{\varphi_1}^{\varphi_2}. \tag{6.71}$$

Assuming that the surface term in (6.71) vanishes, it may be seen that

$$F[\varphi] = f \exp\left(\frac{i}{\hbar} S[\varphi]\right), \tag{6.72}$$

where f is any field-independent constant. The condition for the surface term in (6.71) to vanish is that the action $S[\varphi]$ be the same on Σ_1 and Σ_2. (The source is only non-zero between Σ_1 and Σ_2.) This condition is usually met in field theory by assuming that the fields are in the vacuum state on the initial and final hypersurface.

We have therefore found that the transformation function can be expressed as

$$\langle 2|1\rangle[J] = f \int \left(\prod_i d\varphi^i\right) \exp\frac{i}{\hbar}\{S[\varphi] + J_i \varphi^i\}. \tag{6.73}$$

This is the Feynman path, or functional, integral representation for the transformation function. The multiplicative constant f is not important if we only compare one amplitude with another, and usually we will not indicate it explicitly. Two early papers on the Schwinger action principle and the relation to Feynman's approach are Burton (1955) and Polkinghorne (1955). The derivation of the Feynman functional integral from the Schwinger action principle is the approach taken by DeWitt (1964a, 1965).

6.4 The effective action

The previous chapter contained a result for the one-loop effective action. In this section, we wish to present the general definition of the effective action. We will then show how this general result reduces to the one-loop expression in a special case. Later on, we will develop a consistent perturbative expansion for the effective action which will recover the one-loop expression used earlier in Chapter 5 as the first-order correction to the classical action.

Suppose that we define a functional $W[J]$ in terms of the transformation function by

$$\langle 2|1\rangle[J] = \exp\left(\frac{i}{\hbar}W[J]\right). \tag{6.74}$$

It is clear from this definition that $W[J]$ contains information about the values of the fields on the initial and final hypersurfaces, but this dependence will not be explicitly indicated. If $F[\phi]$ represents any functional of the field operator ϕ, let

$$\langle F[\phi]\rangle[J] = \frac{\langle 2|T(F[\phi])|1\rangle[J]}{\langle 2|1\rangle[J]}. \tag{6.75}$$

If (6.74) is differentiated with respect to the source, it follows that

$$\frac{\delta}{\delta J_i}\langle 2|1\rangle[J] = \frac{i}{\hbar}\langle 2|1\rangle[J]\frac{\delta W[J]}{\delta J_i}. \tag{6.76}$$

Using (6.53) and the definition (6.75), it may be seen that this last result becomes

$$\frac{\delta W[J]}{\delta J_i} = \langle \phi^i\rangle[J]. \tag{6.77}$$

It is customary to define

$$\bar{\varphi}^i = \langle \phi^i\rangle[J], \tag{6.78}$$

so that

$$\frac{\delta W[J]}{\delta J_i} = \bar{\varphi}^i. \tag{6.79}$$

Since the left-hand side of (6.79) is an explicit function of the source J, this result may be interpreted as defining an implicit relationship between J_i and $\bar{\varphi}^i$. It proves convenient to remove the dependence on the source in favor of a dependence on the field $\bar{\varphi}^i$. This is most easily done through a Legendre transformation:

$$\Gamma[\bar{\varphi}] = W[J] - J_i\bar{\varphi}^i. \tag{6.80}$$

$\Gamma[\bar{\varphi}]$ will be called the effective action. Upon differentiation of both sides of (6.80) with respect to $\bar{\varphi}$, it follows that

$$\frac{\delta \Gamma[\bar{\varphi}]}{\delta \bar{\varphi}^i} = \frac{\delta W[J]}{\delta J_j}\frac{\delta J_j}{\delta \bar{\varphi}^i} - \frac{\delta J_j}{\delta \bar{\varphi}^i}\bar{\varphi}^j - J_i$$
$$= -J_i. \tag{6.81}$$

Equation (6.79) has been used to obtain the result in the last line.

It is possible to provide a functional integral representation for the effective action as follows. First of all, note that from (6.80) we have

$$\exp\left(\frac{i}{\hbar}\Gamma[\overline{\varphi}]\right) = \exp\left(\frac{i}{\hbar}W[J] - \frac{i}{\hbar}J_i\overline{\varphi}^i\right)$$

$$= \langle 2|1\rangle[J]\exp\left(-\frac{i}{\hbar}J_i\overline{\varphi}^i\right)$$

$$= \int\left(\prod_i d\varphi^i\right)\exp\left\{\frac{i}{\hbar}S[\varphi] + \frac{i}{\hbar}J_i(\varphi^i - \overline{\varphi}^i)\right\}, \qquad (6.82)$$

if the path integral representation for the transformation function (6.73) is used. Finally, if we note from (6.81) that $J_i = -\Gamma_{,i}[\overline{\varphi}]$, we may eliminate all dependence on the source in (6.82) and write

$$\exp\left(\frac{i}{\hbar}\Gamma[\overline{\varphi}]\right) = \int\left(\prod_i d\varphi^i\right)\exp\left\{\frac{i}{\hbar}S[\varphi] - \frac{i}{\hbar}(\varphi^i - \overline{\varphi}^i)\Gamma_{,i}[\overline{\varphi}]\right\}. \qquad (6.83)$$

Although $\Gamma[\overline{\varphi}]$ appears on both sides of this equation, it may be solved iteratively for $\Gamma[\overline{\varphi}]$ as we will show later.

Although it is not possible to evaluate $W[J]$ or $\Gamma[\overline{\varphi}]$ exactly in general, it is possible to do so in the special case where $S[\varphi]$ is no more than quadratic in φ. The examples dealt with in the previous chapter satisfied this. Suppose that

$$S[\varphi] = \frac{1}{2}\varphi^i S_{,ij}\varphi^j, \qquad (6.84)$$

where $S_{,ij}$ is independent of φ. This is the case if φ is a free field, or is coupled to background fields which are not quantized. $S_{,ij}$ is a differential operator in general, whose inverse, Δ^{ij}, defined by $S_{,ij}\Delta^{jk} = \delta_i^k$, is a Green function. Since

$$\exp\left(\frac{i}{\hbar}W[J]\right) = \int\left(\prod_i d\varphi^i\right)\exp\left\{\frac{i}{\hbar}S[\varphi] + \frac{i}{\hbar}\varphi^i J_i\right\}, \qquad (6.85)$$

it is easy to show that

$$W[J] = -\frac{1}{2}J_i\Delta^{ij}J_j + \frac{i}{2}\hbar\ln\det\left(\ell^2 S_{,ij}\right), \qquad (6.86)$$

using the definition of Gaussian functional integration that was described in Chapter 5. (ℓ is a unit of length introduced to keep the argument of the logarithm dimensionless.) We may compute $\overline{\varphi}^i$ using (6.79), leading to

$$\overline{\varphi}^i = \frac{\delta W[J]}{\delta J_i} = -\Delta^{ij}J_j.$$

This may be inverted to give

$$J_i = -S_{,ij}\overline{\varphi}^j, \qquad (6.87)$$

noting that $S_{,ij}$ is the inverse of Δ^{ij}. It is now easy to show using (6.80) that

$$\Gamma[\overline{\varphi}] = S[\overline{\varphi}] + \frac{i}{2}\hbar \ln \det(\ell^2 S_{,ij}). \qquad (6.88)$$

To lowest order in $\hbar$ the effective action reduces to the classical action. The one-loop correction to the classical action is seen to be

$$\Gamma^{(1)}[\overline{\varphi}] = \frac{i}{2}\hbar \ln \det(\ell^2 S_{,ij}), \qquad (6.89)$$

in agreement with the previous chapter.

6.5 The geometrical effective action

Our definition of the classical action has required it to be a scalar under a general change of spacetime coordinates. The resulting field equations will then be generally covariant under this coordinate change. Since the spacetime coordinates may be arbitrarily specified, covariance of the field equations is regarded as a desirable property. Choosing the action to be a scalar under general coordinate transformations in spacetime is not specific to any type of theory.

A further requirement which we will impose is that the action functional be a scalar under a general redefinition of the fields. Specifically, if we replace the original fields φ with any new fields φ' which are expressible in terms of φ, then the action is required to be a scalar:

$$S'[\varphi'] = S[\varphi]. \qquad (6.90)$$

The naturalness of this requirement is best illustrated with an example. Suppose that we have a complex scalar field $\Phi(x)$ whose action is

$$S[\Phi] = \int dv_x \{\partial^\mu \Phi^\dagger \partial_\mu \Phi - m^2 \Phi^\dagger \Phi\}. \qquad (6.91)$$

Instead of dealing with a complex field Φ, we may wish to deal with real fields. One natural way to do this is to decompose Φ into its real and imaginary parts by defining

$$\Phi(x) = 2^{-1/2}[\phi_1(x) + i\phi_2(x)], \qquad (6.92)$$

where $\phi_1(x)$ and $\phi_2(x)$ are real. With this choice the action (6.91) is

$$S[\phi_1, \phi_2] = \frac{1}{2}\int dv_x \{\partial^\mu \phi_1 \partial_\mu \phi_1 + \partial^\mu \phi_2 \partial_\mu \phi_2 - m^2 \phi_1^2 - m^2 \phi_2^2\} \qquad (6.93)$$

and may be recognized as describing two real fields, each of mass m.

The choice of field parameterization (6.92) is totally arbitrary. For example, in place of (6.91) we might choose to express Φ in polar form

$$\Phi(x) = 2^{-1/2}\rho(x)\exp(i\theta(x)) \qquad (6.94)$$

instead. We still have two real scalar fields, this time $\rho(x)$ and $\theta(x)$, and the action is

$$S'[\rho, \theta] = \frac{1}{2} \int dv_x \{\partial^\mu \rho \partial_\mu \rho + \rho^2 \partial^\mu \theta \partial_\mu \theta - m^2 \rho^2\}. \tag{6.95}$$

It is easily seen that $S'[\rho, \theta] = S[\phi_1, \phi_2]$. At the classical level it does not matter which parameterization is chosen; all physical consequences will be the same. We will therefore require the classical action to be a scalar under a general field redefinition.

In this section we will extend the notion of general covariance under field redefinitions to the quantum theory using the effective action. Because the effective action reduces to the classical action functional in the classical limit, from the discussion above it seems natural to require the effective action be a scalar under a general change of spacetime coordinates, as well as under a general field redefinition. If the effective action is not a scalar under a general field redefinition, then a given classical theory written in two different ways, such as the complex scalar field example discussed above, may give rise to two different quantum theories. (Some explicit examples of this are discussed in Kobes et al. (1988).) One way around this is to state that some particular field parameterization is to be preferred over all others. For example, it might be argued that (6.92) is better than (6.94) because the theory is easier to quantize using canonical methods since the second parameterization leads to a complicated kinetic term. (See (6.95).) However, a polar form like (6.94) has been used in practice to elucidate the physical content of gauge theories. (See the discussion in Abers and Lee (1973) for example. In gauge theories the polar parameterization is related to the unitary gauge, and will be discussed in more detail in Section 7.9.) It would appear difficult to find some fundamental reason to prefer one choice of field parameterization over another.

In flat spacetime, a result known as Borchers' theorem (Borchers 1960) proves that the S-matrix is unaffected by field redefinitions which preserve the asymptotic values of the fields in the in- and out-regions. Because the S-matrix can be obtained from a knowledge of the effective action, if we have a field reparameterization invariant effective action, Borchers' theorem is guaranteed. However, the S-matrix is unaffected by whether or not a reparameterization invariant effective action is used (Rebhan 1988). If everything of interest can ultimately be reduced to a knowledge of some S-matrix elements then a field reparameterization invariant effective action would seem unnecessary. Even if this is the case, in practice the S-matrix is often not used to calculate quantities of interest. For example, the study of spontaneous symmetry breaking is best done using the effective potential part of the effective action (Coleman and Weinberg 1973).

There are theories of interest, such as non-linear sigma models, in which the kinetic terms in the action do not take the simple form in (6.93) but are more akin to (6.95). To begin, consider a set of N real scalar fields $\varphi^I(x)$ where $I = 1, \ldots, N$. The action is chosen to be

$$S[\varphi] = \frac{1}{2} \int dv_x \delta_{IJ} \partial^\mu \varphi^I \partial_\mu \varphi^J. \tag{6.96}$$

The fields φ^I may be regarded as the Cartesian coordinates of a point in $\mathbb{R}^N$. δ_{IJ} is the metric tensor on $\mathbb{R}^N$. This suggests a possible generalization in which $\mathbb{R}^N$ is replaced with an N-dimensional Riemannian manifold M with metric tensor $G_{IJ}(\varphi)$. The action is

$$S[\varphi] = \frac{1}{2} \int dv_x G_{IJ}(\varphi(x)) \partial^\mu \varphi^I \partial_\mu \varphi^J, \tag{6.97}$$

which is seen to reduce to (6.96) in the case where M is a flat manifold and the coordinates φ^I are chosen to be Cartesian. (The manifold M is not to be confused with spacetime.) The theory described by (6.97) is invariant under an arbitrary change of coordinates φ^I on M which, for this theory, is to be interpreted as a general field redefinition. Clearly there is no preferred choice of field parameterization (i.e., no preferred choice of coordinates on the manifold M.) This non-linear sigma model may be kept in mind as the prototype for a generally covariant field theory. In the special case when the manifold M is flat, it describes a normal scalar field theory.

The non-linear sigma model illustrates many of the features present in more general covariant field theories. In the general case, we will regard the fields as local coordinates of points in a field space $\mathcal{F}$. (For the non-linear sigma model, $\mathcal{F}$ was a Riemannian manifold.) The situation is analogous to x^μ being regarded as local coordinates of a point in spacetime. General field redefinitions are then analogous to local changes of coordinates in $\mathcal{F}$, and the requirement that the theory should not depend on any special choice of coordinates elevates the principle of general covariance from just spacetime to include field space as well. With this view, as with general relativity, it makes sense to consider the geometry of the space of fields, and to introduce a metric and related geometrical notions on $\mathcal{F}$.

It is important to emphasize that the choice of field space metric must be regarded as part of the definition of the theory, just as it is for the non-linear sigma model. The usual scalar field theory chooses an action such as that of (6.96) in which the metric is trivial, and it is not necessary to adopt a geometrical approach. The essential difference between the two parameterizations (6.92) and (6.94) is that the first parameterization corresponds to Cartesian coordinates with the usual Euclidean metric, whereas the second parameterization corresponds to polar coordinates with the metric not independent of the coordinates. Thus the origin of the ρ^2 in front of the $\partial^\mu \theta \partial_\mu \theta$ kinetic term in (6.95) may be attributed precisely to the non-trivial field space metric.

From the viewpoint of general covariance, it is clear that the definition of the effective action given in (6.83) cannot be correct. The reason for this is that the definition involves $\varphi^i - \overline{\varphi}^i$, the difference of two field space coordinates, which is not covariant under coordinate changes. The usual definition of the effective action does depend on the choice of field parameterization.

The reason for the occurrence of $\varphi^i - \overline{\varphi}^i$ in the definition of the effective action can be traced to the coupling between the external source term J_i and the field as expressed in (6.49). For a general field space it is clear that $J_i \varphi^i$ is not generally covariant. The only case in which $J_i \varphi^i$ does make good geometrical sense occurs when the field space $\mathcal{F}$ is a vector space. In this case, the usual procedure of coupling the field to a source works just as well if we take the coupling to be $J_i(\varphi^i - \varphi^i_\star)$ where $\varphi^i_\star$ is fixed. J_i cannot depend on the field φ^i if it is to be regarded as external, although a dependence on $\varphi^i_\star$ is allowed. The term $J_i \varphi^i_\star$ does not involve the dynamical fields and therefore does not affect the classical field equations.

For a general field space, we will assume that there is a metric tensor g_{ij} in local coordinates φ^i. (Again the non-linear sigma model can be kept in mind as an example.) Using this metric, it is possible to construct a connection

$$\Gamma^k_{ij} = \frac{1}{2} g^{kl} \left(g_{il,j} + g_{lj,i} - g_{ij,l} \right), \qquad (6.98)$$

and study the geodesics in the usual way. (Here g^{kl} is the inverse of g_{kl}.) Using this connection, the covariant derivative can be formed. For example, the second covariant derivative of the classical action is

$$S_{;ij} = S_{,ij} - \Gamma^k_{ij} S_{,k}. \qquad (6.99)$$

The field space will be flat if $R^k{}_{ilj} = 0$, where $R^k{}_{ilj}$ is the Riemann curvature tensor constructed from Γ^k_{ij} in the normal way.

With this geometrical apparatus, it is now possible to formulate a covariant quantum theory described by the effective action. If we consider the coordinate difference $(\varphi^i - \varphi^i_\star)$, which makes geometrical sense when the field space $\mathcal{F}$ is flat and the coordinates are Cartesian, we can interpret this difference geometrically as a vector which connects the point with coordinates $\varphi^i_\star$ to that with coordinates φ^i. Equivalently, the coordinate difference $(\varphi^i - \varphi^i_\star)$ represents the tangent vector to the geodesic connecting the two points. This suggests that the natural replacement for this difference of coordinates in a general field space is just the tangent vector to the geodesic connecting the two points since the geodesics of a flat space are straight lines. One way of introducing this tangent vector is by means of the geodetic interval $\sigma[\varphi_\star; \varphi]$. By definition,

$$\sigma[\varphi_\star; \varphi] = \frac{1}{2}(length\ of\ geodesic\ from\ \varphi_\star\ to\ \varphi)^2. \qquad (6.100)$$

Thus, the natural replacement for $(\varphi^i - \varphi^i_\star)$ in a general field space is $-\sigma^i[\varphi_\star; \varphi]$, where $\sigma^i[\varphi_\star; \varphi]$ is defined by

6.5 The geometrical effective action

$$\sigma^i[\varphi_*;\varphi] = g^{ij}[\varphi_*]\frac{\delta}{\delta\varphi_*^j}\sigma[\varphi_*;\varphi]$$
$$= g^{ij}[\varphi_*]\sigma_{,j}[\varphi_*;\varphi], \qquad (6.101)$$

with $g_{ij}[\varphi_*]$ the metric on the field space $\mathcal{F}$ evaluated at the point with coordinates φ_*^i. $\Big($Our notation for derivatives of two-point objects such as $\sigma[\varphi_*;\varphi]$ will be that unprimed indices refer to derivatives with respect to the first argument, and primed indices refer to derivatives with respect to the second argument. For example, $\sigma_{,ij'}[\varphi_*;\varphi]$ means $\frac{\delta^2\sigma[\varphi_*;\varphi]}{\delta\varphi^j\delta\varphi_*^i}\Big)$. We are therefore led to define

$$S_J[\varphi] = S[\varphi] - J_i\sigma^i[\varphi_*;\varphi]. \qquad (6.102)$$

This is now completely covariant if J_i transforms like a covariant vector at φ_* and is independent of φ. Note that the introduction of the background field φ_* is necessary if J_i is to have an interpretation as an external source. If we had just added on $J_i v^i[\varphi]$, where $v^i[\varphi]$ is any vector functional, then although the resulting functional $S_J[\varphi]$ would have been completely covariant if J_i transformed like a vector at φ, this would have meant that J_i could not be interpreted as an external source since it would necessarily have to depend on φ.

We may regard $S[\varphi]$ as a functional of φ_* and $\sigma^i[\varphi_*;\varphi]$ that we will call $\hat{S}[\varphi_*;\sigma^i[\varphi_*;\varphi]]$ which is defined by its covariant Taylor series:

$$S[\varphi] = \hat{S}[\varphi_*;\sigma^i[\varphi_*;\varphi]] = \sum_{n=0}^{\infty}\frac{(-1)^n}{n!}S_{;(i_1\cdots i_n)}[\varphi_*]\sigma^{i_1}[\varphi_*;\varphi]\cdots\sigma^{i_n}[\varphi_*;\varphi]. \qquad (6.103)$$

(A derivation of the covariant Taylor series is given in Appendix 6A.1 at the end of this chapter.) The round brackets on $S_{;(i_1\cdots i_n)}[\varphi_*]$ denote a symmetrization over all indices. We may define a similar function $\hat{S}_J[\varphi_*;\sigma^i[\varphi_*;\varphi]]$ by subtracting $J_i\sigma^i[\varphi_*;\varphi]$ from (6.103). (See (6.102).)

It is now possible to repeat the derivation which used the Schwinger action principle and led to the conclusion that the amplitude $\langle 2|1\rangle[J]$ satisfied the equation of motion (6.69). In the present case, this reads

$$E_i\left[\varphi_*;-\frac{\hbar}{i}\frac{\delta}{\delta J_i}\right]\langle 2|1\rangle[J] = J_i\langle 2|1\rangle[J], \qquad (6.104)$$

where

$$E_i[\varphi_*;\sigma^i[\varphi_*;\varphi]] = \frac{\delta \hat{S}[\varphi_*;\sigma^i[\varphi_*;\varphi]]}{\delta\sigma^i[\varphi_*;\varphi]}. \qquad (6.105)$$

The functional E_i may be understood as defined in terms of the Taylor series (6.103). (The steps leading up to (6.104) are left as an exercise for the diligent reader; the less diligent reader can regard (6.104) as a plausible ansatz for the amplitude $\langle 2|1\rangle[J]$.)

As before, we try to solve (6.104) by a Fourier transform. Let

$$\langle 2|1\rangle[J] = \int \left(\prod_i d\sigma^i[\varphi_\star;\varphi]\right) F[\varphi_\star;\sigma^i[\varphi_\star;\varphi]] \exp\left(-\frac{i}{\hbar}J_i\sigma^i[\varphi_\star;\varphi]\right), \quad (6.106)$$

for some functional $F[\varphi_\star;\sigma^i[\varphi_\star;\varphi]]$. The integration extends over all possible values of $\sigma^i[\varphi_\star;\varphi]$ which have $\sigma^i[\varphi_\star;\varphi] = \sigma^i[\varphi_\star;\varphi_1]$ on Σ_1 and $\sigma^i[\varphi_\star;\varphi] = \sigma^i[\varphi_\star;\varphi_2]$ on Σ_2. (An equivalent characterization is that the integration extends over all φ^i, as we will see momentarily.) An analysis similar to that in Section 6.3 leads to the conclusion that

$$F[\varphi_\star;\sigma^i[\varphi_\star;\varphi]] = f[\varphi_\star] \exp\left(\frac{i}{\hbar}\hat{S}[\varphi_\star;\sigma^i[\varphi_\star;\varphi]]\right) \quad (6.107)$$

where $f[\varphi_\star]$ is an arbitrary functional. We therefore have

$$\langle 2|1\rangle[J] = \int \left(\prod_i d\sigma^i[\varphi_\star;\varphi]\right) f[\varphi_\star] \exp\frac{i}{\hbar}\{\hat{S}[\varphi_\star;\sigma^i[\varphi_\star;\varphi]] - J_i\sigma^i[\varphi_\star;\varphi]\}. \quad (6.108)$$

Note that

$$\left(\prod_i d\sigma^i[\varphi_\star;\varphi]\right) = |\det \sigma^i{}_{;j'}[\varphi_\star;\varphi]| \left(\prod_i d\varphi^i\right)$$

$$= \left|\det\left\{g^{ik}[\varphi_\star]\frac{\delta^2\sigma[\varphi_\star;\varphi]}{\delta\varphi_\star^k \delta\varphi^j}\right\}\right| \left(\prod_i d\varphi^i\right)$$

$$= |g[\varphi_\star]|^{-1/2}|g[\varphi]|^{1/2}|\Delta[\varphi_\star;\varphi]| \left(\prod_i d\varphi^i\right) \quad (6.109)$$

where

$$\Delta[\varphi_\star;\varphi] = |g[\varphi_\star]|^{-1/2}|g[\varphi]|^{-1/2}\det(-\sigma_{;ij'}[\varphi_\star;\varphi]) \quad (6.110)$$

is the biscalar Van Vleck–Morette determinant. Then it is seen from (6.109) that

$$\left(\prod_i d\sigma^i[\varphi_\star;\varphi]\right) f[\varphi_\star] = \left(\prod_i d\varphi^i\right) |g[\varphi]|^{1/2}|\Delta[\varphi_\star;\varphi]||g[\varphi_\star]|^{-1/2}f[\varphi_\star]. \quad (6.111)$$

The argument of the exponential in (6.108) is a scalar under redefinitions of both $\varphi_\star$ and φ. The amplitude $\langle 2|1\rangle[J]$ should also be invariant, which means that the measure must be invariant. From (6.111) this is seen to constrain $|g[\varphi_\star]|^{-1/2}f[\varphi_\star]$ to be a scalar under a transformation of $\varphi_\star$. We may therefore take

$$f[\varphi_\star] = |g[\varphi_\star]|^{1/2}f_0[\varphi_\star], \quad (6.112)$$

6.5 The geometrical effective action

where $f_0[\varphi_\star]$ is any scalar functional. Then (6.108) may be written in either of the two following forms:

$$\langle 2|1\rangle[J] = f_0[\varphi_\star] \int \left(\prod_i d\sigma^i[\varphi_\star;\varphi]\right) |g[\varphi_\star]|^{1/2} \exp\frac{i}{\hbar}\{\hat{S}[\varphi_\star;\sigma^i[\varphi_\star;\varphi]]$$
$$- J_i \sigma^i[\varphi_\star;\varphi]\} \qquad (6.113)$$

$$= f_0[\varphi_\star] \int \left(\prod_i d\varphi^i\right) |g[\varphi]|^{1/2} |\Delta[\varphi_\star;\varphi]| \exp\frac{i}{\hbar}\{S[\varphi]$$
$$- J_i \sigma^i[\varphi_\star;\varphi]\}. \qquad (6.114)$$

Normally we only compare one amplitude with another, and so the arbitrary functional $f_0[\varphi_\star]$ is irrelevant. Accordingly, we will set $f_0[\varphi_\star] = 1$ from now on. In addition to the invariant volume element $(\prod_i d\varphi^i)|g[\varphi]|^{1/2}$ in the measure, there also occurs the unconventional factor of $|\Delta[\varphi_\star;\varphi]|$ whose presence has been fixed from the consistency of the functional integral representation for the amplitude with the Schwinger action principle.

In place of (6.74), we will define the functional $W[J;\varphi_\star]$ by

$$\langle 2|1\rangle[J] = \exp\left(\frac{i}{\hbar}W[J;\varphi_\star]\right). \qquad (6.115)$$

We will use the same notation as in (6.75) to denote the average of any functional. Differentiation of both sides of (6.115) with respect to the source leads to

$$\frac{\delta W[J;\varphi_\star]}{\delta J_i} = -\langle \sigma^i[\varphi_\star;\phi]\rangle[J], \qquad (6.116)$$

if it is noted that

$$\frac{\delta}{\delta J_i}\langle 2|1\rangle[J] = -\frac{i}{\hbar}\langle 2|\sigma^i[\varphi_\star;\phi]|1\rangle[J]. \qquad (6.117)$$

(This last result is a consequence of the Schwinger action principle with the variation referring to the external source.)

In order to obtain the effective action, we again wish to eliminate the dependence on the source in $W[J;\varphi_\star]$. This time we can utilize the field space geometry. $\langle \sigma^i[\varphi_\star;\phi]\rangle[J]$ which occurs in (6.116) transforms like a contravariant vector under a change of the coordinates $\varphi^i_\star$. Given the point $\varphi^i_\star$ and a vector at this point, it is always possible to construct, at least in a small neighborhood of $\varphi^i_\star$, a unique geodesic which passes through $\varphi^i_\star$ with $-\langle \sigma^i[\varphi_\star;\phi]\rangle[J]$ as the tangent vector. (The minus sign is because it is conventional to choose the tangent vector to point along the geodesic.) We may define a point $\overline{\varphi}^i$ on this geodesic by

$$\sigma^i[\varphi_\star;\overline{\varphi}] = \langle \sigma^i[\varphi_\star;\phi]\rangle[J]. \qquad (6.118)$$

In a flat field space we recover (6.78). Therefore, (6.116) can be written as

$$\frac{\delta W[J;\varphi_\star]}{\delta J_i} = -\sigma^i[\varphi_\star;\overline{\varphi}]. \qquad (6.119)$$

The result in (6.119) may be seen as defining an implicit relationship between $\overline{\varphi}$ and J. The dependence on the source J may be removed in favor of a dependence on the field $\overline{\varphi}$ through the usual procedure of a Legendre transformation:

$$\Gamma[\overline{\varphi};\varphi_*] = W[J;\varphi_*] + J_i \sigma^i[\varphi_*;\overline{\varphi}]. \quad (6.120)$$

$\Gamma[\overline{\varphi};\varphi_*]$ will be called the effective action. It is seen to depend on two points with local coordinates φ_* and $\overline{\varphi}$.

It is sometimes convenient to regard the effective action as a functional of φ_* and $\sigma^i[\varphi_*;\overline{\varphi}]$. When it is necessary to make this distinction explicit, we will use

$$\hat{\Gamma}[\sigma[\varphi_*;\overline{\varphi}];\varphi_*] = \Gamma[\overline{\varphi};\varphi_*]. \quad (6.121)$$

It follows from (6.120) and (6.121) that

$$\frac{\delta \hat{\Gamma}[\sigma[\varphi_*;\overline{\varphi}];\varphi_*]}{\delta \sigma^i[\varphi_*;\overline{\varphi}]} = J_i. \quad (6.122)$$

It is important to note that φ_* is held fixed in the differentiation in (6.122). The two arguments of $\hat{\Gamma}$ are viewed as independent.

It is possible to obtain a functional integral representation for the effective action using the expression (6.114) for the amplitude. Following the same steps as in Section 6.4, it may be seen easily that

$$\exp\left(\frac{i}{\hbar}\Gamma[\overline{\varphi};\varphi_*]\right) = \int d\mu[\varphi_*;\varphi] \exp\frac{i}{\hbar}\left\{S[\varphi] + J_i\left(\sigma^i[\varphi_*;\overline{\varphi}]\right.\right.$$
$$\left.\left. - \sigma^i[\varphi_*;\varphi]\right)\right\} \quad (6.123)$$
$$= \int d\mu[\varphi_*;\varphi] \exp\frac{i}{\hbar}\left\{S[\varphi] + \frac{\delta\Gamma[\overline{\varphi};\varphi_*]}{\delta\sigma^i[\varphi_*;\overline{\varphi}]}\left(\sigma^i[\varphi_*;\overline{\varphi}]\right.\right.$$
$$\left.\left. - \sigma^i[\varphi_*;\varphi]\right)\right\} \quad (6.124)$$

if (6.122) is used. The effective action may now be regarded as defined through this implicit equation. The functional measure $d\mu[\varphi_*;\varphi]$ is given by

$$d\mu[\varphi_*;\varphi] = \left(\prod_i d\varphi^i\right) |g(\varphi)|^{1/2} |\Delta[\varphi_*;\varphi]|. \quad (6.125)$$

The perturbative solution of (6.124) will be examined in the next section.

The average defined in (6.75) may now be rewritten in terms of the effective action:

$$\langle F[\varphi]\rangle = \exp\left(-\frac{i}{\hbar}\Gamma[\overline{\varphi};\varphi_*]\right)\int d\mu[\varphi_*;\varphi] F[\varphi] \exp\frac{i}{\hbar}\left\{S[\varphi]\right.$$
$$\left. + \frac{\delta\Gamma[\overline{\varphi};\varphi_*]}{\delta\sigma^i[\varphi_*;\overline{\varphi}]}\left(\sigma^i[\varphi_*;\overline{\varphi}] - \sigma^i[\varphi_*;\varphi]\right)\right\}. \quad (6.126)$$

6.5 The geometrical effective action

The expression (6.122) for the source has been used here, and we have dropped the explicit dependence on the source J in the average.

The result in (6.124) gives an effective action $\Gamma[\overline{\varphi}; \varphi_*]$ which depends on two points φ_* and $\overline{\varphi}$. The usual effective action defined in Section 6.4 only depends on a single field $\overline{\varphi}$. We now wish to discuss this following Burgess and Kunstatter (1987). Begin by differentiating (6.126) with respect to φ_*^i holding $\overline{\varphi}$ fixed:

$$\frac{\delta\Gamma[\overline{\varphi}; \varphi_*]}{\delta\varphi_*^i} = \frac{\delta\Gamma[\overline{\varphi}; \varphi_*]}{\delta\sigma^j[\varphi_*; \overline{\varphi}]} \left(\sigma^j{}_{;i}[\varphi_*; \overline{\varphi}] - \langle\sigma^j{}_{;i}[\varphi_*; \varphi]\rangle\right). \tag{6.127}$$

We have used (6.118) here.

In the special case of a flat field space, we have the basic result

$$\sigma^j{}_{;i}[\varphi_*; \varphi] = \delta^j{}_i. \tag{6.128}$$

This is most easily derived by choosing a Cartesian field parameterization (allowed for a flat field space) for which $\sigma^i[\varphi_*; \overline{\varphi}] = \varphi_*^i - \overline{\varphi}^i$, and for which the covariant derivative is the same as the ordinary derivative. The result in (6.128) now follows trivially; but because it is a tensor relation, it must be true regardless of the choice of field parameterization. Using (6.128) it may be seen that the right-hand side of (6.127) vanishes identically. Hence for a flat field space we have the result

$$\frac{\delta\Gamma[\overline{\varphi}; \varphi_*]}{\delta\varphi_*^i} = 0. \tag{6.129}$$

This proves that for a flat field space, $\Gamma[\overline{\varphi}; \varphi_*] = \Gamma[\overline{\varphi}]$, and thus the effective action only depends on the single point $\overline{\varphi}^i$. The above derivation has established this result in a way that is independent of the choice of field parameterization for the flat field space. The definition of the effective action we have adopted is equivalent to that used earlier in Section 6.4.

For a general field space $\Gamma[\overline{\varphi}; \varphi_*]$ does depend on φ_*; however, it only depends on a particular combination of $\overline{\varphi}$ and φ_*, since from (6.127) the effective action satisfies the identity

$$\mathcal{D}_i \Gamma[\overline{\varphi}; \varphi_*] = 0, \tag{6.130}$$

where

$$\mathcal{D}_i = \frac{\delta}{\delta\varphi_*^i} + \left(\langle\sigma^j{}_{;i}[\varphi_*; \varphi]\rangle - \sigma^j{}_{;i}[\varphi_*; \overline{\varphi}]\right) \frac{\delta\overline{\varphi}^k}{\delta\sigma^j[\varphi_*; \overline{\varphi}]} \frac{\delta}{\delta\overline{\varphi}^k}. \tag{6.131}$$

Here $\dfrac{\delta\overline{\varphi}^k}{\delta\sigma^j[\varphi_*; \overline{\varphi}]}$ represents the formal inverse of $\dfrac{\delta\sigma^j[\varphi_*; \overline{\varphi}]}{\delta\overline{\varphi}^k}$.

What the identity in (6.130) shows is that although the effective action does depend on φ_*, this dependence is not important. To see this, consider the change in $\Gamma[\overline{\varphi}; \varphi_*]$ when both φ_* and $\overline{\varphi}$ are varied:

$$\delta\Gamma = \Gamma[\overline{\varphi}+\delta\overline{\varphi};\varphi_*+\delta\varphi_*] - \Gamma[\overline{\varphi};\varphi_*]$$
$$= \left.\frac{\delta\Gamma}{\delta\overline{\varphi}^i}\right|_\varphi \delta\overline{\varphi}^i + \left.\frac{\delta\Gamma}{\delta\varphi_*^i}\right|_{\overline{\varphi}} \delta\varphi_*^i$$
$$= \{\delta\overline{\varphi}^i - E_j{}^i \delta\varphi_*^j\} \left.\frac{\delta\Gamma}{\delta\overline{\varphi}^i}\right|_\varphi, \qquad (6.132)$$

where we have abbreviated

$$E_j{}^i = \left(\langle \sigma^k{}_{;j}[\varphi_*;\varphi]\rangle - \sigma^k{}_{;j}\right) \frac{\delta\overline{\varphi}^i}{\delta\sigma^k} \qquad (6.133)$$

as the expression appearing on the left-hand side of (6.131). We can conclude that Γ is constant on surfaces in field space characterized by

$$0 = \delta\overline{\varphi}^i - E_j{}^i \delta\varphi_*^j. \qquad (6.134)$$

In flat field space, $E_j{}^i = 0$ and these surfaces are simply given by $\delta\overline{\varphi}^i = 0$, or constant values of $\overline{\varphi}$, showing again that $\Gamma[\overline{\varphi};\varphi_*] = \Gamma[\overline{\varphi}]$.

As pointed out by Burgess and Kunstatter (1987), the identity (6.130) is completely analogous to an identity first derived by Nielsen (1975) in the context of gauge theories. In a gauge theory, as we will discuss in the next chapter, the effective potential is dependent on gauge-fixing parameters if it is defined in the usual way. This is a potential problem if we want to use the effective potential to study spontaneous symmetry breaking, since the solution to the effective field equations could then depend on some unphysical parameters. In the standard model of particle physics the masses of the W and Z bosons depend on the solution to the effective field equations, and it is possible that these masses, which are of course measurable quantities, can have a dependence on arbitrary gauge parameters; clearly this is an unacceptable situation. The dependence on gauge parameters is completely analogous to a dependence on the arbitrary point φ_* here. What Nielsen (1975) showed was that physical quantities may still be obtained from a gauge-dependent effective potential, and they will be gauge-independent provided that a change in the gauge parameters is accompanied by a suitable change in the solution for $\overline{\varphi}$ to the effective field equations. In our case if we are at a solution to the effective field equations $\delta\Gamma/\delta\overline{\varphi}^i = 0$, our identity (6.130) ensures that $\delta\Gamma/\delta\varphi_*^i = 0$, showing that the existence of a solution is independent of φ_*. The utility of the generalization of the Nielsen type of identity to gauge theories is demonstrated in Kobes et al. (1990, 1991).

It is possible to go further. Differentiate (6.130) with respect to $\overline{\varphi}^k$. The following results:

$$0 = \frac{\delta^2\Gamma}{\delta\overline{\varphi}^k \delta\varphi_*^i} + \bar{\nabla}_k E_i{}^j \frac{\delta\Gamma}{\delta\overline{\varphi}^j} + E_i{}^j \bar{\nabla}_k \frac{\delta\Gamma}{\delta\overline{\varphi}^j}.$$

Here $\bar{\nabla}_k$ is the covariant derivative computed using the field space connection evaluated at the point $\overline{\varphi}$. This can be re-arranged to read

6.5 The geometrical effective action

$$\left[\frac{\delta}{\delta\varphi_*^i} + E_i{}^j \bar{\nabla}_j\right] \frac{\delta\Gamma}{\delta\bar{\varphi}^k} = -\bar{\nabla}_k E_i{}^j \frac{\delta\Gamma}{\delta\bar{\varphi}^j}. \quad (6.135)$$

The operator which appears in square brackets on the left-hand side of (6.135) can be recognized as $\mathcal{D}_i$ from (6.131). From this new result we can conclude that if $\delta\Gamma/\delta\bar{\varphi}^j = 0$ at some point on the characteristic surface defined in (6.134) then it remains zero at all points on the surface when defined by parallel transport. This means that if we have a solution to $\delta\Gamma/\delta\bar{\varphi}^j = 0$ for some value of φ_*^i, when we change φ_*^i to $\varphi_*^i + \delta\varphi_*^i$ we will still satisfy $\delta\Gamma/\delta\bar{\varphi}^j = 0$ but with a new solution $\bar{\varphi}^i + E_j{}^i \delta\varphi_*^j$. Because Γ is constant on the surface in (6.134) we have $\Gamma[\bar{\varphi} + \delta\bar{\varphi}; \varphi_* + \delta\varphi_*] = \Gamma[\bar{\varphi}; \varphi_*]$. What this means is that we are free to pick any value we like for φ_* to perform calculations and physical quantities will be independent of our choice. In practice the easiest choice to make is to adopt the choice made by DeWitt (1987) of $\varphi_* = \bar{\varphi}$.

The DeWitt (1987) definition for the effective action is obtained by taking the limit $\varphi_* \to \bar{\varphi}$, and defining

$$\Gamma_D[\bar{\varphi}] = \lim_{\varphi \to \bar{\varphi}} \Gamma[\bar{\varphi}; \varphi_*] = \Gamma[\bar{\varphi}; \bar{\varphi}]. \quad (6.136)$$

If we take the limit $\varphi_* \to \bar{\varphi}$ in (6.118), it follows that

$$\langle \sigma^i[\bar{\varphi}; \varphi] \rangle = 0. \quad (6.137)$$

If the limit $\varphi_* \to \bar{\varphi}$ is taken directly in (6.124), the result is

$$\exp\left(\frac{i}{\hbar}\Gamma_D[\bar{\varphi}]\right) = \int d\mu[\bar{\varphi}; \varphi] \exp\frac{i}{\hbar}\Big\{S[\varphi] - \sigma^i[\bar{\varphi}; \varphi]$$
$$\times \lim_{\varphi \to \bar{\varphi}} \frac{\delta\hat{\Gamma}[\sigma[\varphi_*; \bar{\varphi}]; \varphi_*]}{\delta\sigma^i[\varphi_*; \bar{\varphi}]}\Big\}. \quad (6.138)$$

Because φ_* is held fixed in the differentiation of $\hat{\Gamma}[\sigma[\varphi_*; \bar{\varphi}]; \varphi_*]$ in (6.138), we have

$$\frac{\delta\hat{\Gamma}[\sigma[\varphi_*; \bar{\varphi}]; \varphi_*]}{\delta\sigma^i[\varphi_*; \bar{\varphi}]} = \frac{\delta\Gamma[\bar{\varphi}; \varphi_*]}{\delta\bar{\varphi}^j} \frac{\delta\bar{\varphi}^j}{\delta\sigma^i[\varphi_*; \bar{\varphi}]}. \quad (6.139)$$

Now take the limit $\varphi_* \to \bar{\varphi}$, and use the result[3]

$$\lim_{\varphi \to \bar{\varphi}} \frac{\delta\sigma^i[\varphi_*; \bar{\varphi}]}{\delta\bar{\varphi}^j} = -\delta^i{}_j. \quad (6.140)$$

This gives

$$\lim_{\varphi \to \bar{\varphi}} \frac{\delta\hat{\Gamma}[\sigma[\varphi_*; \bar{\varphi}]; \varphi_*]}{\delta\sigma^i[\varphi_*; \bar{\varphi}]} = -\lim_{\varphi \to \bar{\varphi}} \frac{\delta\Gamma[\bar{\varphi}; \varphi_*]}{\delta\bar{\varphi}^i}. \quad (6.141)$$

If $\lim_{\varphi \to \bar{\varphi}} \delta\Gamma[\bar{\varphi}; \varphi_*]/\delta\bar{\varphi}^i$ was simply $\delta\Gamma_D[\bar{\varphi}]/\delta\bar{\varphi}^i$, then the result for $\Gamma_D[\bar{\varphi}]$ would be precisely that postulated by Vilkovisky (1984). But this would only be true

---
[3] This is easily established using normal coordinates.

if the limit $\varphi_* \to \overline{\varphi}$ was taken before the derivative with respect to $\overline{\varphi}^i$, which is not the case here.

In order to compute $\lim_{\varphi \to \overline{\varphi}} \delta\Gamma[\overline{\varphi};\varphi_*]/\delta\overline{\varphi}^i$ properly, take the limit $\varphi_* \to \overline{\varphi}$ in (6.127), where it may be seen using (6.139)–(6.141) that

$$\lim_{\varphi \to \overline{\varphi}} \left\{ \frac{\delta\Gamma[\overline{\varphi};\varphi_*]}{\delta\varphi_*^j} + \frac{\delta\Gamma[\overline{\varphi};\varphi_*]}{\delta\overline{\varphi}^j} \right\} = C^i{}_j[\overline{\varphi}] \lim_{\varphi \to \overline{\varphi}} \frac{\delta\Gamma[\overline{\varphi};\varphi_*]}{\delta\overline{\varphi}^i} \qquad (6.142)$$

with

$$C^i{}_j[\overline{\varphi}] = \langle \sigma^i{}_{;j}[\overline{\varphi};\varphi] \rangle. \qquad (6.143)$$

Because of (6.136), we have

$$\frac{\delta\Gamma_D[\overline{\varphi}]}{\delta\overline{\varphi}^j} = \lim_{\varphi \to \overline{\varphi}} \left\{ \frac{\delta\Gamma[\overline{\varphi};\varphi_*]}{\delta\overline{\varphi}^j} + \frac{\delta\Gamma[\overline{\varphi};\varphi_*]}{\delta\varphi_*^j} \right\}. \qquad (6.144)$$

Hence, combining these last two results, we have proven that

$$\lim_{\varphi \to \overline{\varphi}} \frac{\delta\Gamma[\overline{\varphi};\varphi_*]}{\delta\overline{\varphi}^i} = C^{-1j}{}_i[\overline{\varphi}] \frac{\delta\Gamma_D[\overline{\varphi}]}{\delta\overline{\varphi}^j}. \qquad (6.145)$$

$C^{-1j}{}_i[\overline{\varphi}]$ is the inverse of $C^i{}_j[\overline{\varphi}]$ defined in (6.143). It therefore follows that

$$\lim_{\varphi \to \overline{\varphi}} \frac{\delta\Gamma[\overline{\varphi};\varphi_*]}{\delta\sigma^i[\varphi_*;\overline{\varphi}]} = -C^{-1j}{}_i[\overline{\varphi}] \frac{\delta\Gamma_D[\overline{\varphi}]}{\delta\overline{\varphi}^j}. \qquad (6.146)$$

This differs from the naive expectation mentioned earlier due to the presence of $C^{-1j}{}_i[\overline{\varphi}]$. The significance of this factor will be discussed later. Finally, we have the following functional integral expression for DeWitt's definition of the effective action:

$$\exp\left(\frac{i}{\hbar}\Gamma_D[\overline{\varphi}]\right) = \int d\mu[\overline{\varphi};\varphi] \exp \frac{i}{\hbar} \left\{ S[\varphi] - \sigma^i[\overline{\varphi};\varphi] \right. $$
$$\left. \times C^{-1j}{}_i[\overline{\varphi}] \frac{\delta\Gamma_D[\overline{\varphi}]}{\delta\overline{\varphi}^j} \right\}. \qquad (6.147)$$

This definition of the covariant effective action is a modification of an earlier definition of Vilkovisky (1984), which had $C^i{}_j$ in (6.147) replaced by $\delta^i{}_j$. As was discussed by Ellicott and Toms (1989), it may be seen that this definition, when coupled with the requirement of (6.137), is not consistent except for a flat field space. The case of a flat field space is precisely that where (6.128) does hold, and hence from (6.143) it follows that $C^i{}_j[\overline{\varphi}] = \delta^i{}_j$. Thus Vilkovisky's original definition only makes sense in the case of a flat field space. Finally, it should be emphasized that DeWitt's definition for the effective action really consists of two equations: namely, (6.147) and (6.143). The definition appears rather complicated, since to compute $C^i{}_j[\overline{\varphi}]$ we must know $\Gamma_D[\overline{\varphi}]$, but to know $\Gamma_D[\overline{\varphi}]$ we must know $C^i{}_j[\overline{\varphi}]$. Fortunately, this rather circular appearing situation has a solution within the context of perturbation theory, which is the subject of the next section.

6.6 Perturbative expansion of the effective action

Although the effective action has been defined by (6.126), it is clear that this definition is purely formal since the object of interest, $\Gamma[\overline{\varphi}; \varphi_*]$, occurs on both sides of this equation. The usual method of calculation is to resort to a perturbative expansion in powers of $\hbar$. This is called the loop expansion for reasons which will become apparent when we interpret the results in terms of Feynman diagrams.

In the last section, we showed that the effective action only depended on a single combination of the two points φ_* and $\overline{\varphi}$, and so the DeWitt (1987) definition for the effective action (see (6.147)) could be adopted. The price to be paid for eliminating the dependence on φ_* is that the loop expansion is considerably complicated due to the occurrence of the functional $C^{-1}{}^i{}_j[\overline{\varphi}]$. This results in a rather circular looking definition for the effective action, as was mentioned at the end of the last section. There is a solution to this within the context of perturbation theory; however the details are rather complicated, and have been presented in Toms (1988). We will adopt a different approach here.

Rather than take the limit $\varphi_* \to \overline{\varphi}$ in (6.124), it is in fact much simpler to leave the limit until the end, and to calculate the loop expansion for $\Gamma[\overline{\varphi}; \varphi_*]$ first. This obviates the need for the complicated details described in Toms (1988), and makes the resulting steps look very much like those for the standard flat field space effective action. This possibility was first discussed in Burgess and Kunstatter (1987) and Rebhan (1988).

For brevity, let

$$\sigma^i = \sigma^i[\varphi_*; \varphi], \qquad (6.148)$$
$$v^i = \sigma^i[\varphi_*; \overline{\varphi}]. \qquad (6.149)$$

With this shorthand notation, (6.124) reads

$$\exp\left(\frac{i}{\hbar}\Gamma[\overline{\varphi}; \varphi_*]\right) = \int d\mu[\varphi_*; \varphi] \exp\frac{i}{\hbar}\left\{S[\varphi] + \frac{\delta\Gamma[\overline{\varphi}; \varphi_*]}{\delta v^i}(v^i - \sigma^i)\right\} \qquad (6.150)$$

with

$$d\mu[\varphi_*; \varphi] = \left(\prod_i d\sigma^i\right)|g[\varphi_*]|^{\frac{1}{2}}. \qquad (6.151)$$

We may again regard $S[\varphi]$ as a functional of φ^i_* and σ^i, $\hat{S}[\varphi_*; \sigma^i[\varphi_*; \varphi]]$, defined by its covariant Taylor expansion as in (6.103). By expanding the functional $\hat{S}[\varphi_*; \sigma^i[\varphi_*; \varphi]]$ in a Taylor series about $\sigma^i = v^i$:

$$\hat{S}[\varphi_*; \sigma^i[\varphi_*; \varphi]] = \hat{S}[\varphi_*; v^i] + \sum_{n=1}^{\infty} \frac{1}{n!} \frac{\delta^n \hat{S}[\varphi_*; v^i]}{\delta v^{i_1} \cdots \delta v^{i_n}}(\sigma^{i_1} - v^{i_1}) \cdots (\sigma^{i_n} - v^{i_n}). \qquad (6.152)$$

Note that

$$\hat{S}[\varphi_*; v^i] = \hat{S}[\varphi_*; \sigma^i[\varphi_*; \overline{\varphi}]]$$
$$= S[\overline{\varphi}] \qquad (6.153)$$

by (6.149). The coefficients in the Taylor series (6.152) may be related to $S_{;(i_1 \cdots i_n)}[\varphi_*]$ by replacing σ^i with v^i in (6.103), and then differentiating the result with respect to v^i. This gives

$$\frac{\delta^n \hat{S}[\varphi_*; v^i]}{\delta v^{i_1} \cdots \delta v^{i_n}} = (-1)^n S_{;(i_1 \cdots i_n)}[\varphi_*] + \sum_{m=n+1}^{\infty} \frac{(-1)^m}{(m-n)!} S_{;(i_1 \cdots i_m)}[\varphi_*] v^{i_{n+1}} \cdots v^{i_m}. \qquad (6.154)$$

From (6.149) it may be seen that the limit $\varphi_* \to \overline{\varphi}$ is equivalent to the limit $v^i \to 0$. From (6.154), we have

$$\left. \frac{\delta^n \hat{S}[\varphi_*; v^i]}{\delta v^{i_1} \cdots \delta v^{i_n}} \right|_{v=0} = (-1)^n S_{;(i_1 \cdots i_n)}[\overline{\varphi}]. \qquad (6.155)$$

In order to calculate the $\hbar$ expansion of the effective action, it proves convenient to scale and translate the variable of integration σ^i by[4]

$$\sigma^i \to \hbar^{\frac{1}{2}} \sigma^i + v^i. \qquad (6.156)$$

The key reason for this is that the scaling cancels the $\hbar$ dependence which would otherwise appear in the propagator, making the counting of factors of $\hbar$ easier since internal lines in Feynman diagrams will not carry any powers of $\hbar$. The measure in (6.151) will be invariant under this transformation, because the condition $\sigma^i \sigma_i = 2\sigma$ must be preserved. This entails the transformation of the field space metric $g_{ij} \to \hbar^{-1} g_{ij}$. Note that the covariant derivatives in (6.154) are unaffected by this transformation because the affine connection is invariant under this rescaling of the metric. Equation (6.150) becomes

$$\exp\left(\frac{i}{\hbar} \Gamma[\overline{\varphi}; \varphi_*]\right) = \int d\mu[\varphi_*; \varphi] \exp i A[\overline{\varphi}; \varphi_*; \sigma], \qquad (6.157)$$

where

$$A[\overline{\varphi}; \varphi_*; \sigma] = \frac{1}{\hbar} S[\overline{\varphi}] + \sum_{n=1}^{\infty} \hbar^{n/2-1} \frac{1}{n!} \frac{\delta^n \hat{S}[\varphi_*; v^i]}{\delta v^{i_1} \cdots \delta v^{i_n}} \sigma^{i_1} \cdots \sigma^{i_n} - \hbar^{-1/2} \frac{\delta \Gamma[\overline{\varphi}; \varphi_*]}{\delta v^i} \sigma^i. \qquad (6.158)$$

Assume that the effective action admits an expansion in powers of $\hbar$ which starts at zeroth order and contains only integral powers of $\hbar$:

$$\Gamma[\overline{\varphi}; \varphi_*] = \sum_{n=0}^{\infty} \hbar^n \Gamma^{(n)}[\overline{\varphi}; \varphi_*]. \qquad (6.159)$$

---

[4] The utility of this in the usual effective action approach was illustrated by Jackiw (1974).

6.6 Perturbative expansion of the effective action

Then from (6.158) it is clear that $A[\bar{\varphi}; \varphi_*; \sigma] = \hbar^{-1} \bar{S} + o(\hbar^{-1/2})$, and hence from (6.157) it follows that

$$\Gamma^{(0)}[\bar{\varphi}; \varphi_*] = \bar{S}. \qquad (6.160)$$

Here, $\bar{S}$ denotes $S[\bar{\varphi}]$, and o indicates the order of the term in the $\hbar$ expansion. Thus to lowest order in the $\hbar$ expansion, the effective action coincides with the classical action functional.

In order to obtain the higher-order terms in $\hbar$, use (6.159) and (6.160) in (6.158), and note that the $n = 1$ term in the summation of (6.158) is canceled. It is not necessary to assume that $\bar{\varphi}$ is a solution to the classical equations of motion, or that it is even close to a classical solution, as is sometimes stated. The result for $A[\bar{\varphi}; \varphi_*; \sigma]$ becomes

$$\begin{aligned} A[\bar{\varphi}; \varphi_*; \sigma] = &\frac{1}{\hbar} \bar{S} + \frac{1}{2} \hat{S}_{,ij}[\varphi_*; v^i] \sigma^i \sigma^j \\ &+ \sum_{n=3}^{\infty} \hbar^{n/2-1} \frac{1}{n!} \hat{S}_{,i_1 \cdots i_n}[\varphi_*; v^i] \sigma^{i_1} \cdots \sigma^{i_n} \\ &- \sum_{n=1}^{\infty} \hbar^{n-1/2} \hat{\Gamma}^{(n)}_{,i} \sigma^i, \end{aligned} \qquad (6.161)$$

where the comma now denotes an ordinary derivative with respect to v^i.

Let

$$S_{\text{int}} = \sum_{n=3}^{\infty} \hbar^{n/2-1} \frac{1}{n!} \hat{S}_{,i_1 \cdots i_n}[\varphi_*; v^i] \sigma^{i_1} \cdots \sigma^{i_n} - \sum_{n=1}^{\infty} \hbar^{n-1/2} \hat{\Gamma}^{(n)}_{,i} \sigma^i. \qquad (6.162)$$

S_{int}, as the notation indicates, will be treated as an interaction. Using (6.161) in (6.157) enables us to write

$$\begin{aligned} \Gamma[\bar{\varphi}; \varphi_*] = &\bar{S} - i\hbar \ln \int d\mu[\varphi_*; \varphi] \exp\left(\frac{i}{2} \hat{S}_{,ij} \sigma^i \sigma^j\right) \\ &- i\hbar \ln \left\langle \exp i S_{\text{int}} \right\rangle_G, \end{aligned} \qquad (6.163)$$

where

$$\langle F[\sigma] \rangle_G = \frac{\int d\mu[\varphi_*; \varphi] F[\sigma] \exp\left(\frac{i}{2} \hat{S}_{,ij} \sigma^i \sigma^j\right)}{\int d\mu[\varphi_*; \varphi] \exp\left(\frac{i}{2} \hat{S}_{,ij} \sigma^i \sigma^j\right)} \qquad (6.164)$$

denotes the average of any functional $F[\sigma]$ with the Gaussian weighting factor $\exp\left(\frac{i}{2} \hat{S}_{,ij} \sigma^i \sigma^j\right)$. In particular,

$$\left\langle \sigma^{i_1} \cdots \sigma^{i_{2n+1}} \right\rangle_G = 0; \qquad (6.165)$$

the Gaussian average of an odd number of factors of σ^i vanishes identically. The average of an even number of factors of σ^i may be evaluated as follows. Define

$$Z_0[J] = \int d\mu[\varphi_*;\varphi] \exp\left(\frac{i}{2}\hat{S}_{,ij}\sigma^i\sigma^j + iJ_i\sigma^i\right). \tag{6.166}$$

The variable of integration, σ^i, in (6.166) may be translated by letting $\sigma^i = \tilde{\sigma}^i + \psi^i$, where ψ^i is independent of $\tilde{\sigma}^i$ and is chosen to cancel the term in the argument of the exponential which is linear in $\tilde{\sigma}^i$. It may be easily seen that the required choice is

$$\psi^i = -\hat{\Delta}^{ij} J_j, \tag{6.167}$$

where $\hat{\Delta}^{ij}$ is defined by

$$\hat{S}_{,ij}\hat{\Delta}^{jk} = \delta^k{}_i. \tag{6.168}$$

$\hat{\Delta}^{ij}$ may be interpreted as the inverse of $\hat{S}_{,ij}$, and is the Green function, or propagator, for the field. It therefore follows that

$$Z_0[J] = \exp\left(-\frac{i}{2}\hat{\Delta}^{ij} J_i J_j\right) \int \left(\prod_i d\tilde{\sigma}^i\right) |\det g_{ij}[\varphi_*]|^{\frac{1}{2}} \exp\left(\frac{i}{2}\hat{S}_{,ij}\tilde{\sigma}^i\tilde{\sigma}^j\right)$$

$$= Z_0[0] \exp\left(-\frac{i}{2}\hat{\Delta}^{ij} J_i J_j\right). \tag{6.169}$$

Now note from (6.164) and (6.166) that

$$\langle \sigma^{i_1} \cdots \sigma^{i_n} \rangle_G = \frac{(-i)^n}{Z_0[0]} \frac{\delta^n Z_0[J]}{\delta J_{i_1} \cdots \delta J_{i_n}}\bigg|_{J=0}$$

$$= (-i)^n \frac{\delta^n \exp\left(-\frac{i}{2}\hat{\Delta}^{ij} J_i J_j\right)}{\delta J_{i_1} \cdots \delta J_{i_n}}\bigg|_{J=0}. \tag{6.170}$$

The result given previously in (6.165) follows immediately from this general relation since it is obvious from (6.169) that only an even number of factors of J_i occur. (6.170) is the functional version of Wick's theorem. (See Wick 1950 and Nishijima 1950.) In particular, (6.170) gives

$$\langle \sigma^i \sigma^j \rangle_G = i\hat{\Delta}^{ij} \tag{6.171}$$

and

$$\langle \sigma^i \sigma^j \sigma^k \sigma^l \rangle_G = -\hat{\Delta}^{ij}\hat{\Delta}^{kl} - \hat{\Delta}^{ik}\hat{\Delta}^{jl} - \hat{\Delta}^{il}\hat{\Delta}^{jk}. \tag{6.172}$$

The result for any even number of factors of σ^i follows from a straightforward differentiation of (6.170). It should be clear that the average of any number of factors of σ^i is obtained by taking the sum over all possible pairs of the factors and then replacing each pair with a factor of $i\Delta$. This is the usual statement of Wick's theorem.

It is now easy to obtain the loop expansion of $\Gamma[\bar{\varphi};\varphi_*]$. Because of (6.165) it may be seen that the expansion only contains integral powers of $\hbar$ as assumed in (6.159). The method now consists of expanding $\exp(iS_{\text{int}})$ in powers of the interaction S_{int} and keeping only terms up to some given power of $\hbar$. The Gaussian average of each term may then be evaluated using Wick's theorem in the form of (6.170). For reasons explained, $\langle\exp(iS_{\text{int}})\rangle_G = 1 + o(\hbar)$. $\ln\langle\exp(iS_{\text{int}})\rangle_G$ may then be expanded in powers of $\hbar$ to the desired order using

$$\ln(1+x) = -\sum_{n=1}^{\infty} \frac{(-1)^n}{n} x^n. \tag{6.173}$$

Here x represents a sum of terms of differing orders in $\hbar$. Because

$$\langle\exp(iS_{\text{int}})\rangle_G = 1 + o(\hbar),$$

the last term of (6.163) is seen to start at order $\hbar^2$. This tells us that

$$\Gamma^{(1)}[\overline{\varphi}; \varphi_\star] = -i \ln \int d\mu[\varphi_\star; \varphi] \exp\left(\frac{i}{2}\hat{S}_{,ij}\sigma^i\sigma^j\right) \tag{6.174}$$

gives the one-loop contribution to the effective action. The functional integration in (6.174) may be done by analogy with the finite-dimensional case as discussed in the previous chapter, and gives

$$\Gamma^{(1)}[\overline{\varphi}; \varphi_\star] = -i \ln [\det g_{ij}[\varphi_\star]]^{\frac{1}{2}} - i \ln \left[\det\left(\ell^2 \hat{S}_{,ij}\right)\right]^{-\frac{1}{2}} \tag{6.175}$$

$$= \frac{i}{2} \ln \det\left(\ell^2 \hat{S}^{,i}{}_j\right), \tag{6.176}$$

where

$$\hat{S}^{,i}{}_j = g^{ik}[\varphi_\star]\hat{S}_{,kj}. \tag{6.177}$$

Recall that there was an arbitrariness present in the measure in that it was only defined up to a normalization factor due to the presence of the factor of $f_0(\varphi_\star)$. This factor has been chosen so that (6.176) holds. We will therefore adopt (6.176) as our definition of the functional integral of a Gaussian. This reproduces the result used in the previous chapter. Note that the choice of normalization of the measure does not affect the higher loop terms because an overall multiplicative factor in the measure will cancel between the numerator and denominator of the Gaussian functional average of (6.164).

Define

$$S_{\text{int}} = \sum_{n=1}^{\infty} \hbar^{n/2} A_n, \tag{6.178}$$

where the coefficients A_n may be read off from (6.162):

$$A_{2n-1} = \frac{1}{(2n+1)!} \hat{S}_{,i_1\cdots i_{2n+1}} \sigma^{i_1} \cdots \sigma^{i_{2n+1}} - \hat{\Gamma}^{(n)}_{,i} \sigma^i, \tag{6.179}$$

$$A_{2n} = \frac{1}{(2n+2)!} \hat{S}_{,i_1\cdots i_{2n+2}} \sigma^{i_1} \cdots \sigma^{i_{2n+2}}, \tag{6.180}$$

for $n = 1, 2, \ldots$. Following the steps outlined above, it may be shown that

$$\Gamma[\overline{\varphi}; \varphi_*] = \overline{S} + \frac{i}{2}\hbar \ln \det \left(\ell^2 \hat{S}^{,i}{}_j\right) + \hbar^2 \left\{ \langle A_2 \rangle_G - \frac{i}{2} \langle (A_1)^2 \rangle_G \right\}$$
$$+ \hbar^3 \left\{ \langle [A_4 + iA_1 A_3 + \frac{i}{2}(A_2)^2 - \frac{1}{2}(A_1)^2 A_2 - \frac{i}{24}(A_1)^4] \rangle_G \right.$$
$$\left. - \frac{i}{2} \left(\langle A_2 \rangle_G - \frac{i}{2} \langle (A_1)^2 \rangle_G \right)^2 \right\} + o(\hbar^4). \tag{6.181}$$

Concentrate first on the order $\hbar^2$ contribution given by

$$\Gamma^{(2)}[\overline{\varphi}; \varphi_*] = \langle A_2 \rangle_G - \frac{i}{2} \langle (A_1)^2 \rangle_G. \tag{6.182}$$

From (6.180),

$$\langle A_2 \rangle_G = \frac{1}{4!} \hat{S}_{,ijkl} \langle \sigma^i \sigma^j \sigma^k \sigma^l \rangle_G$$
$$= -\frac{1}{8} \hat{S}_{,ijkl} \hat{\Delta}^{ij} \hat{\Delta}^{kl} \tag{6.183}$$

using (6.172) and relabeling indices. From (6.179),

$$A_1 = \frac{1}{3!} \hat{S}_{,ijk} \sigma^i \sigma^j \sigma^k - \hat{\Gamma}^{(1)}_{,i} \sigma^i. \tag{6.184}$$

Before squaring A_1, it is convenient to evaluate $\hat{\Gamma}^{(1)}_{,i}$. In order to do this, differentiate the following form of (6.176), obtained using the identity $\operatorname{tr} \ln = \ln \det$:

$$\hat{\Gamma}^{(1)} = \frac{i}{2} \operatorname{tr} \ln \hat{S}^{,i}{}_j. \tag{6.185}$$

Then

$$\hat{\Gamma}^{(1)}_{,i} = \frac{i}{2} \hat{\Delta}^{jk} \hat{S}_{,ijk} \tag{6.186}$$

since $\hat{\Delta}^{jk} = (\hat{S}_{,jk})^{-1}$. Thus

$$A_1 = \frac{1}{3!} \hat{S}_{,ijk} \sigma^i \sigma^j \sigma^k - \frac{i}{2} \hat{\Delta}^{jk} \hat{S}_{,ijk} \sigma^i. \tag{6.187}$$

It follows that

$$\langle (A_1)^2 \rangle_G = \frac{1}{36} \hat{S}_{,ijk} \hat{S}_{,lmn} \langle \sigma^i \sigma^j \sigma^k \sigma^l \sigma^m \sigma^n \rangle_G$$
$$- \frac{i}{6} \hat{S}_{,ijk} \hat{S}_{,lmn} \hat{\Delta}^{mn} \langle \sigma^i \sigma^j \sigma^k \sigma^l \rangle_G$$
$$- \frac{1}{4} \hat{S}_{,ijk} \hat{S}_{,lmn} \hat{\Delta}^{jk} \hat{\Delta}^{mn} \langle \sigma^i \sigma^l \rangle_G$$
$$= -\frac{i}{6} \hat{S}_{,ijk} \hat{S}_{,lmn} \hat{\Delta}^{il} \hat{\Delta}^{jm} \hat{\Delta}^{kn} \tag{6.188}$$

if the Gaussian averages are evaluated as described earlier. Using (6.183) and (6.188) in (6.182) leads to

$$\Gamma^{(2)}[\overline{\varphi};\varphi_*] = \frac{1}{12}\hat{S}_{,ijk}\hat{S}_{,lmn}\hat{\Delta}^{il}\hat{\Delta}^{jm}\hat{\Delta}^{kn} - \frac{1}{8}\hat{S}_{,ijkl}\hat{\Delta}^{ij}\hat{\Delta}^{kl}. \quad (6.189)$$

A much lengthier calculation may be used to evaluate the $o(\hbar^3)$ contribution in (6.181). Because we have already indicated the essential features of the calculation, we will not proceed further with this.

The $\hbar$ expansion we have obtained so far is

$$\Gamma[\overline{\varphi};\varphi_*] = \overline{S} + \frac{i}{2}\hbar \ln \det\left(\ell^2 \hat{S}^{,i}{}_j\right)$$
$$+ \hbar^2 \left\{\frac{1}{12}\hat{S}_{,ijk}\hat{S}_{,lmn}\hat{\Delta}^{il}\hat{\Delta}^{jm}\hat{\Delta}^{kn} - \frac{1}{8}\hat{S}_{,ijkl}\hat{\Delta}^{ij}\hat{\Delta}^{kl}\right\} + o(\hbar^3). \quad (6.190)$$

At this stage contact may be made with the traditional Feynman diagram notation. All vertices will correspond to derivatives of $i\hat{S}[\varphi_*; v^i]$ with respect to v^i, the number of lines emanating from a vertex indicating the number of derivatives. Lines 'i' and 'j' in any vertices which are connected together contain a factor of $i\hat{\Delta}^{ij}$. (The symmetry factor for the diagram has followed from using Wick's theorem as described above.) With these rules, the two terms in $\Gamma^{(2)}[\overline{\varphi};\varphi_*]$ given in (6.189) correspond to the graphs shown in Figs 6.1(a) and (b) respectively. There are no terms which correspond to the graphs of Figs 6.1(c) and (d), although they do have the correct form to appear in $\Gamma^{(2)}[\overline{\varphi};\varphi_*]$. These terms are seen to be topologically different from those of Figs 6.1(a) and (b). Fig 6.1(d) consists of two disconnected pieces, whereas Fig 6.1(c) is not disconnected, but can be made so by cutting a single line. A graph which becomes disconnected by cutting a single one of its lines is said to be one-particle reducible. The graphs of Figs 6.1(a) and (b) are connected, and do not become disconnected when any single line is cut. Such graphs are said to be one-particle irreducible. Our calculation shows that with the interpretation of diagrams which has been adopted, the two-loop effective action consists of one-particle irreducible vacuum diagrams

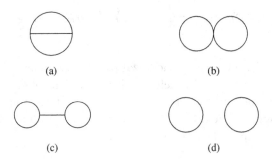

Fig 6.1. Graphical interpretation of the vacuum diagrams of two-loop order.

(i.e., the diagrams have no external lines) only. As we will show in Section 6.11, it is no accident that diagrams such as those in Figs 6.1(c) and (d) are absent.

From the standpoint of conventional Feynman rules, (see Bjorken and Drell (1965) for example) the result we have obtained is horrendously complicated because our vertices defined in (6.154) contain an infinite number of the conventional vertices $S_{;(i_1\cdots i_n)}[\varphi_*]$ in general. However, if we now take the limit $\varphi_* \to \overline{\varphi}$ in (6.190) to obtain the DeWitt effective action, we may use the result of (6.155). This leads to

$$\Gamma_D[\overline{\varphi}] = \overline{S} + \frac{i}{2}\hbar \ln \det\left(\ell^2 \overline{S}^{;i}{}_j\right)$$
$$+ \hbar^2 \left\{\frac{1}{12}\overline{S}_{;(ijk)}\overline{S}_{;(lmn)}\bar{\Delta}^{il}\bar{\Delta}^{jm}\bar{\Delta}^{kn} - \frac{1}{8}\overline{S}_{;(ijkl)}\bar{\Delta}^{ij}\bar{\Delta}^{kl}\right\} + o(\hbar^3),$$
(6.191)

where $\bar{\Delta}^{jk}$ is defined by

$$\overline{S}_{;ij}\bar{\Delta}^{jk} = \delta^k{}_i.$$
(6.192)

We can again give an interpretation in terms of Feynman diagrams, this time where the vertices stand for symmetrized covariant derivatives of the classical action functional evaluated at $\overline{\varphi}$, and where the vertices are connected up using $\bar{\Delta}^{ij}$ defined in (6.192). The result obtained in (6.191) can also be found directly using DeWitt's definition of the effective action given in (6.147), although the calculation is much more complicated than the one just performed. (See Toms (1988) for details.)

6.7 Renormalization of an interacting scalar field theory

Our first application of the results of the last section will be to the renormalization of an interacting scalar field theory in curved four-dimensional spacetime. The matter action S_M describing the scalar field will be chosen to be

$$S_M[\varphi] = \int dv_x \left\{\frac{1}{2}\delta_{ij}\partial^\mu\varphi_B^i\partial_\mu\varphi_B^j - \frac{1}{2}m_{Bij}^2\varphi_B^i\varphi_B^j - \frac{1}{2}\xi_{Bij}\varphi_B^i\varphi_B^j\right.$$
$$\left. - \frac{1}{3!}\eta_{Bijk}\varphi_B^i\varphi_B^j\varphi_B^k - \frac{1}{4!}\lambda_{Bijkl}\varphi_B^i\varphi_B^j\varphi_B^k\varphi_B^l - \tau_{Bi}\varphi_B^i - \gamma_{Bi}R\varphi_B^i\right\},$$
(6.193)

where $i = 1, \ldots, N$. This is a generalization of the scalar field action considered in the previous chapter to include self-interactions of the form φ^3 and φ^4. The subscript 'B' on the various terms denotes that they are bare quantities which must be renormalized. In addition to the matter action, it is necessary for renormalizability to include a gravitational action

$$S_G = \int dv_x \{\Lambda_B + \kappa_B R + \alpha_{1B}R_{\mu\nu\rho\sigma}R^{\mu\nu\rho\sigma} + \alpha_{2B}R_{\mu\nu}R^{\mu\nu} + \alpha_{3B}R^2\}, \quad (6.194)$$

6.7 Renormalization of an interacting scalar field theory

which involves curvature squared terms. The gravitational counterterms are found to be necessary even in the free field theories of the earlier chapters.

The action (6.193) is seen to correspond to a linear sigma model, so that the natural choice of field space metric is $G_{ij} = \delta_{ij}$ with the fields φ^i regarded as Cartesian coordinates in a flat field space. This means that the field space connection $\Gamma^i_{jk} = 0$, resulting in a simplification of the analysis. We will return to the case of general curvilinear coordinates, as well as to the non-linear sigma model later.

First of all, consider the one-loop effective action, given by

$$\Gamma[\bar\varphi] = S[\bar\varphi] + \frac{i}{2}\hbar \ln \det\left(\ell^2 \bar S_{,ij}\right), \qquad (6.195)$$

where $\bar\varphi$ is the background field, and $\bar S_{,ij} = \delta^2 S_M[\bar\varphi]/\delta\bar\varphi^j \delta\bar\varphi^i$. (Because the field space metric is flat, we can replace $\bar S^{i}{}_{j}$ in (6.191) with $\bar S_{,ij}$. Also, recall that only the scalar fields are quantized here, so that only the matter field part of the action contributes to the one-loop part of the effective action.) From (6.193), it is easily found that

$$\bar S_{,ij} = \Big[-\delta_{ij}\Box_x - m^2_{Bij} - \xi_{Bij}R(x) - \eta_{Bijk}\bar\varphi^k_B(x)$$

$$-\frac{1}{2}\lambda_{Bijkl}\bar\varphi^k_B(x)\bar\varphi^l_B(x)\Big]\delta(x,x'). \qquad (6.196)$$

(Note that η_{Bijk} and λ_{Bijkl} can be taken to be totally symmetric in their indices.)

As usual, $\ln \det \bar S_{,ij}$ is divergent and requires regularization. We will adopt dimensional regularization by taking the spacetime to have dimension $4+\epsilon$, with $\epsilon \to 0$ taken at the end. Associated with the ϵ extra dimensions there will be a length scale ℓ defined so that the volume of the extra dimensions is ℓ^ϵ. Instead of using a unit of length, it is more customary to use a unit of mass $\mu = \ell^{-1}$. (See 't Hooft 1973.) The background field $\bar\varphi^i(x)$ will be taken to have no dependence on the coordinates of the extra dimensions. Only the quantum fields which are integrated over in the functional integral have a dependence on the extra dimensions.

The bare quantities occurring in (6.193) and (6.194) are seen to have the following dimensions in units of length. These results all follow by simple dimensional analysis noting that the action (6.193, 6.194) is dimensionless in $\hbar = 1$ units. Following 't Hooft (1973), we will choose renormalized quantities to have the same dimensions in $(4+\epsilon)$ dimensional spacetime as they do in four spacetime dimensions. We will define

$$\Lambda_B = \ell^{-\epsilon}(\Lambda + \delta\Lambda), \qquad (6.197)$$

$$\kappa_B = \ell^{-\epsilon}(\kappa + \delta\kappa), \qquad (6.198)$$

$$\alpha_{iB} = \ell^{-\epsilon}(\alpha_i + \delta\alpha_i)(i=1,2,3), \qquad (6.199)$$

$$\overline{\varphi}_B^i = \ell^{-\epsilon/2} Z_{ij} \overline{\varphi}_B^j \text{(summed } j = 1, \ldots, N\text{)}, \tag{6.200}$$

$$m_{Bij}^2 = m_{ij}^2 + \delta m_{ij}^2, \tag{6.201}$$

$$\xi_{Bij} = \xi_{ij} + \delta \xi_{ij}, \tag{6.202}$$

$$\eta_{Bijk} = \ell^{\epsilon/2}(\eta_{ijk} + \delta \eta_{ijk}), \tag{6.203}$$

$$\lambda_{Bijkl} = \ell^{\epsilon}(\lambda_{ijkl} + \delta \lambda_{ijkl}), \tag{6.204}$$

$$\tau_{Bi} = \ell^{-\epsilon/2}(\tau_i + \delta \tau_i), \tag{6.205}$$

$$\gamma_{Bi} = \ell^{-\epsilon/2}(\gamma_i + \delta \gamma_i). \tag{6.206}$$

Quantities without a subscript 'B' denote renormalized expressions, and the δ denotes a counterterm. Z_{ij} is the field renormalization matrix, which we may write as

$$Z_{ij} = \delta_{ij} + \delta Z_{ij}. \tag{6.207}$$

All of the counterterms appearing in (6.197)–(6.207) may be given an expansion in powers of $\hbar$. If δq represents any of the counterterms above, write

$$\delta q = \hbar \delta q^{(1)} + \hbar^2 \delta q^{(2)} + \cdots . \tag{6.208}$$

The term $\delta q^{(n)}$ is of nth order in the loop expansion and will contain terms which diverge as $\epsilon \to 0$ in order to cancel divergences in the effective action. If we work only to order $\hbar$, it follows that we may take

$$\bar{S}_{,ij} = [-\delta_{ij}\Box_x - m_{ij}^2 - \xi_{ij}R - \eta_{ijk}\overline{\varphi}^k - \frac{1}{2}\lambda_{ijkl}\overline{\varphi}^k\overline{\varphi}^l]\delta(x,x') \tag{6.209}$$

since $\ln \det \bar{S}_{,ij}$ occurs multiplied by a factor of $\hbar$ in (6.195). The analysis of the divergent part of $\ln \det \bar{S}_{,ij}$ is basically the same as in the previous chapter, the only difference being the more complicated differential operator in (6.209). The general result in (5.68) reads

$$\mathbf{divp}\left\{\frac{i}{2}\hbar \ln \det (\ell^2 \bar{S}_{,ij})\right\} = -\frac{\hbar}{16\pi^2 \epsilon} \int dv_x \mathrm{tr} E_2(x), \tag{6.210}$$

where $E_2(x)$ was given in (5.59) and we use **divp** to denote the divergent, or pole part, of any expression as in Chapter 5. The operator in (6.209) is of the general form considered earlier in (5.56), with

$$Q_{ij} = m_{ij}^2 + \xi_{ij}R + \eta_{ijk}\overline{\varphi}^k + \frac{1}{2}\lambda_{ijkl}\overline{\varphi}^k\overline{\varphi}^l, \tag{6.211}$$

and $W_{\mu\nu} = 0$. (For scalar fields, $[\nabla_\mu, \nabla_\nu]\varphi^i = 0$.) A simple calculation gives

6.7 Renormalization of an interacting scalar field theory

$$\mathrm{tr}E_2(x) = \left(\frac{N}{72} + \frac{1}{2}\xi_{ij}\xi_{ij} - \frac{1}{6}\xi_{ii}\right)R^2 + \frac{N}{180}\left(R^{\mu\nu\rho\sigma}R_{\mu\nu\rho\sigma} - R^{\mu\nu}R_{\mu\nu}\right)$$

$$+ \left(\xi_{ij}m_{ij}^2 - \frac{1}{6}m_{ii}^2\right)R + \frac{1}{2}m_{ij}^2 m_{ij}^2 + m_{ij}^2 \eta_{ijk}\overline{\varphi}^k$$

$$+ \left(\xi_{ij}\eta_{ijk} - \frac{1}{6}\eta_{iik}\right)R\overline{\varphi}^k + \frac{1}{2}\left(m_{ij}^2\lambda_{ijkl} + \eta_{ijk}\eta_{ijl}\right)\overline{\varphi}^k\overline{\varphi}^l$$

$$+ \frac{1}{2}\left(\xi_{ij}\lambda_{ijkl} - \frac{1}{6}\lambda_{iikl}\right)R\overline{\varphi}^k\overline{\varphi}^l + \frac{1}{2}\eta_{ijk}\lambda_{ijlm}\overline{\varphi}^k\overline{\varphi}^l\overline{\varphi}^m$$

$$+ \frac{1}{8}\lambda_{ijkl}\lambda_{ijmn}\overline{\varphi}^k\overline{\varphi}^l\overline{\varphi}^m\overline{\varphi}^n + \frac{1}{6}\left(\xi_{ii} - \frac{N}{5}\right)\Box R + \frac{1}{6}\eta_{iij}\Box\overline{\varphi}^j$$

$$+ \frac{1}{12}\lambda_{iijk}\Box(\overline{\varphi}^j\overline{\varphi}^k). \tag{6.212}$$

(Repeated Latin indices $i, j, k, \ldots$ are summed over the number of components of the scalar field.)

Taking the one-loop counterterm contributions in $S[\overline{\varphi}]$, we have

$$\mathbf{divp}\{S[\overline{\varphi}]\} = \hbar \int dv_x \Big\{\delta\Lambda^{(1)} + \delta\kappa^{(1)}R + \delta\alpha_1^{(1)}R_{\mu\nu\rho\sigma}R^{\mu\nu\rho\sigma}$$

$$+ \delta\alpha_2^{(1)}R_{\mu\nu}R^{\mu\nu} + \delta\alpha_3^{(1)}R^2 + \delta Z_{ij}^{(1)}\partial^\mu\overline{\varphi}^i\partial_\mu\overline{\varphi}^j$$

$$- \left(\frac{1}{2}\delta m_{ij}^{2(1)} + m_{ik}^2\delta Z_{kj}^{(1)}\right)\overline{\varphi}^i\overline{\varphi}^j$$

$$- \left(\frac{1}{2}\delta\xi_{ij}^{(1)} + \xi_{ik}\delta Z_{kj}^{(1)}\right)R\overline{\varphi}^i\overline{\varphi}^j - \left(\delta\tau_i^{(1)} + \tau_j\delta Z_{ij}^{(1)}\right)\overline{\varphi}^i$$

$$- \left(\delta\gamma_i^{(1)} + \gamma_j\delta Z_{ij}^{(1)}\right)\overline{\varphi}^j - \left(\frac{1}{3!}\delta\eta_{ijk}^{(1)} + \frac{1}{2}\eta_{ijl}\delta Z_{kl}^{(1)}\right)\overline{\varphi}^i\overline{\varphi}^j\overline{\varphi}^k$$

$$- \left(\frac{1}{4!}\delta\lambda_{ijkl}^{(1)} + \frac{1}{3!}\lambda_{ijkm}\delta Z_{lm}^{(1)}\right)\overline{\varphi}^i\overline{\varphi}^j\overline{\varphi}^k\overline{\varphi}^l\Big\}. \tag{6.213}$$

The terms in (6.213) are seen to be of the same type as those which occur in (6.212), and therefore the theory will be renormalizable.

When (6.212) is substituted into (6.210), terms which are the integral of a total divergence may be discarded. It is then seen that (6.210) does not involve any divergences with derivatives of the background field. We may therefore set

$$\delta Z_{ij}^{(1)} = 0, \tag{6.214}$$

in (6.213). This shows that no renormalization of the background field is necessary at one-loop order. The lack of one-loop field renormalization simplifies (6.213). The remaining counterterms are fixed by demanding that $\Gamma[\overline{\varphi}]$ in (6.195) remain finite as $\epsilon \to 0$. It is readily seen that

$$\delta\Lambda^{(1)} = (32\pi^2\epsilon)^{-1}m_{ij}^2 m_{ij}^2, \tag{6.215}$$

$$\delta\kappa^{(1)} = (16\pi^2\epsilon)^{-1}\left(\xi_{ij}m_{ij}^2 - \frac{1}{6}m_{ii}^2\right), \tag{6.216}$$

$$\delta\alpha_1^{(1)} = (16\pi^2\epsilon)^{-1}\frac{N}{180}, \tag{6.217}$$

$$\delta\alpha_2^{(1)} = -\left(16\pi^2\epsilon\right)^{-1}\frac{N}{180}, \tag{6.218}$$

$$\delta\alpha_3^{(1)} = (16\pi^2\epsilon)^{-1}\left(\frac{N}{72} + \frac{1}{2}\xi_{ij}\xi_{ij} - \frac{1}{6}\xi_{ii}\right), \tag{6.219}$$

$$\delta m_{kl}^{2(1)} = -(16\pi^2\epsilon)^{-1}(m_{ij}^2\lambda_{ijkl} + \eta_{ijk}\eta_{ijl}), \tag{6.220}$$

$$\delta\xi_{kl}^{(1)} = (16\pi^2\epsilon)^{-1}\left(\frac{1}{6}\lambda_{iikl} - \xi_{ij}\lambda_{ijkl}\right), \tag{6.221}$$

$$\delta\tau_i^{(1)} = -(16\pi^2\epsilon)^{-1}m_{jk}^2\eta_{ijk}, \tag{6.222}$$

$$\delta\gamma_i^{(1)} = (16\pi^2\epsilon)^{-1}\left(\frac{1}{6}\eta_{ijj} - \xi_{jk}\eta_{ijk}\right), \tag{6.223}$$

$$\delta\eta_{ijk}^{(1)} = -(16\pi^2\epsilon)^{-1}(\eta_{ilm}\lambda_{jklm} + \eta_{jlm}\lambda_{kilm} + \eta_{klm}\lambda_{ijlm}), \tag{6.224}$$

$$\delta\lambda_{ijkl}^{(1)} = -(16\pi^2\epsilon)^{-1}(\lambda_{ijmn}\lambda_{klmn} + \lambda_{ikmn}\lambda_{jlmn} + \lambda_{ilmn}\lambda_{jkmn}) \tag{6.225}$$

are the required counterterms. Care must be taken to ensure that the counterterms have the same index symmetries as the corresponding coupling constants. With the counterterms chosen as above, the effective action is finite as $\epsilon \to 0$.

It should be apparent that an approach to renormalization theory based on the effective action is much more efficient than the traditional approach based on the analysis of Feynman diagrams. For the theory considered here, we would have to work out all of the diagrams which are of one-loop order and which contribute to the one-, two-, three-, and four-point functions. This involves rather a large number of diagrams. There is no way to calculate the gravitational counterterms for general curvature by consideration of traditional Feynman diagrams since the curvature contribution comes from vacuum diagrams, or zero-point functions, and these are traditionally not considered. In order to use Feynman diagrams to calculate the gravitational counterterms, it is necessary to use a weak-field expansion about flat space for the metric ($g_{\mu\nu} = \eta_{\mu\nu} + h_{\mu\nu}$), and work to order $h_{\mu\nu}^2$. This is also rather involved. Thus, the background-field method coupled with heat kernel techniques is a very powerful way of treating renormalization of interacting scalar field theory in curved spacetime. It was first used in Toms (1982).

The two-loop contribution to the effective action was obtained in (6.191) and for the theory considered here reads

$$\bar{\Gamma}^{(2)} = \hbar^2\left\{\frac{1}{12}\bar{S}_{,ijk}\bar{S}_{,lmn}\bar{\Delta}^{il}\bar{\Delta}^{jm}\bar{\Delta}^{kn} - \frac{1}{8}\bar{S}_{,ijkl}\bar{\Delta}^{ij}\bar{\Delta}^{kl}\right\}. \tag{6.226}$$

In order to avoid inessential complications, we will only consider the case of a single scalar field. The condensed indices in (6.226) now refer only to the spacetime labels. From (6.196) we find that

6.7 Renormalization of an interacting scalar field theory

$$\bar{S}_{,ijk} = -\left[\eta_B + \lambda_B \bar{\varphi}_B(x)\right] \delta(x,x')\delta(x,x''), \tag{6.227}$$
$$\bar{S}_{,ijkl} = -\lambda_B \delta(x,x')\delta(x,x'')\delta(x,x'''), \tag{6.228}$$

where $(i,j,k,l) \Leftrightarrow (x,x',x'',x''')$ defines the relationship between condensed and conventional labels. $\bar{\Delta}^{ij}$ is the inverse of $\bar{S}_{,ij}$ and therefore satisfies (denoting $\bar{\Delta}^{ij}$ by $\Delta_B(x,x')$)

$$\left[-\Box_x - m_B^2 - \xi_B R(x) - \eta_B \bar{\varphi}_B(x) - \frac{1}{2}\lambda_B \bar{\varphi}_B^2(x)\right] \Delta_B(x,x') = \delta(x,x'). \tag{6.229}$$

It is then easy to see that (6.226) becomes

$$\bar{\Gamma}^{(2)} = \hbar^2 \Big\{ \frac{1}{12} \int dv_x \, dv_{x'} \, [\eta_B + \lambda_B \bar{\varphi}_B(x)][\eta_B + \lambda_B \bar{\varphi}_B(x')] \Delta_B^3(x,x')$$
$$+ \frac{1}{8}\lambda_B \int dv_x \, \Delta_B^2(x,x') \Big\}. \tag{6.230}$$

Because we are only working to order $\hbar^2$, we may use (6.200)–(6.204) to write

$$\bar{\Gamma}^{(2)} = \hbar^2 \Big\{ \frac{1}{12} \ell^{-2\epsilon} \int dv_x \, dv_{x'} \, [\eta + \lambda \bar{\varphi}(x)][\eta + \lambda \bar{\varphi}(x')] \Delta^3(x,x')$$
$$+ \frac{1}{8}\ell^{-\epsilon}\lambda \int dv_x \, \Delta^2(x,x') \Big\}, \tag{6.231}$$

where

$$\left[-\Box_x - m^2 - \xi R(x) - \eta \bar{\varphi}(x) - \frac{1}{2}\lambda \bar{\varphi}^2(x)\right] \Delta(x,x') = \ell^{\epsilon}\delta(x,x'). \tag{6.232}$$

The factor of ℓ^ϵ on the right-hand side of (6.232) comes about because $\Delta_B = \ell^{-\epsilon}\Delta + o(\hbar)$ using the field renormalization factor (6.200). It is required to balance the dimensions on both sides of the equation. In addition to the terms in (6.231), there will be terms of order $\hbar^2$ coming from the two-loop counterterms in $\bar{S}$, and the one-loop counterterms in $\bar{\Gamma}^{(1)}$. These will be considered later.

The important feature to notice is that (6.231) involves $\Delta^3(x,x')$ which has non-coincident arguments. This makes two, and higher, loop contributions to the effective action much more difficult to evaluate than single-loop contributions. (Recall that $\Delta(x,x)$ can be related to the trace of the heat kernel which has a known asymptotic expansion.) In order to evaluate $\Delta^3(x,x')$ we will proceed as follows. First make a separation of $\Delta(x,x')$ into two parts:

$$\Delta(x,x') = \Delta_L(x,x') + \Delta_{NL}(x,x'). \tag{6.233}$$

Here $\Delta_L(x,x')$ is local in the sense that it is independent of the choice of quantum state, or the global topology of spacetime. It is only sensitive to what happens for x and x' in the same local neighborhood in spacetime. $\Delta_{NL}(x,x')$ contains all of the effects of non-localities, as well as perhaps some local information. Because $\Delta_{NL}(x,x')$ can contain some local information, the split in (6.233) is not uniquely defined.

We know that the pole part of $\Delta(x,x)$ is purely local, as can be seen from the heat kernel. In fact,

$$\Delta(x,x') = -i\ell^2 \int_0^\infty d\tau\, K(\tau;x,x'), \qquad (6.234)$$

where

$$K(\tau;x,x') = \langle x|\exp(-i\ell^2\tau D)|x'\rangle, \qquad (6.235)$$

with D the negative of the differential operator appearing in (6.232). (This is the reason for the sign change in the definition of K with respect to that in Section 5.3.) D is of the general form given in (5.56), so that the coefficients in the asymptotic expansion of the kernel are as given there.[5] Then

$$\mathbf{divp}\,\Delta(x,x) = -i\ell^2 \mathbf{divp} \int_0^1 d\tau\, i(4\pi i\ell^2\tau)^{-n/2} \sum_{k=0}^\infty (i\ell^2\tau)^k E_k(x),$$

since the divergence comes from $\tau \to 0$ as discussed in Section 5.3. It is then easily seen (if $n = 4+\epsilon$) that

$$\begin{aligned}\mathbf{divp}\,\Delta(x,x) &= \frac{i}{8\pi^2\epsilon} E_1(x) \\ &= -\frac{i}{8\pi^2\epsilon}\left[m^2 + \left(\xi - \frac{1}{6}\right)R + \eta\overline{\varphi}(x) + \frac{\lambda}{2}\overline{\varphi}^2(x)\right]\end{aligned} \qquad (6.236)$$

using the expression for E_1 given in (5.58).

Returning to (6.233), we will define $\Delta_{NL}(x,x')$ so that it is non-singular as $x \to x'$, which is always possible given the fact that (6.236) is local; thus, the divergences in $\Delta(x,x')$ as $x \to x'$ always come from $\Delta_L(x,x')$. We also note that so long as $x \neq x'$, $\Delta_L(x,x')$ is non-singular. This procedure is not specific to any spacetime, and in particular we have not restricted the spacetime topology.

We have defined $\Delta_L(x,x')$ to depend only on x and x' in the same local neighborhood, and so we want to solve

$$-\left[\Box_x + m^2 + \xi R(x) + \eta\overline{\varphi}(x) + \frac{\lambda}{2}\overline{\varphi}^2(x)\right]\Delta_L(x,x') = \ell^\epsilon \delta(x,x'),$$

for x' close to x. (Note that $\Delta_{NL}(x,x')$ satisfies a homogeneous equation.) This may be obtained using an expansion in normal coordinates, along with the local momentum-space method first used by Bunch and Parker (1979). We have considered this method previously in Section 3.8. If we define x' to be the origin of normal coordinates, and write

$$x^\mu = x'^\mu + y^\mu, \qquad (6.237)$$

---

[5] See (5.57)–(5.59).

6.7 Renormalization of an interacting scalar field theory

it is easily shown that if $\tilde{\Delta}_L(x, x') = [-g(x)]^{1/4}[-g(x')]^{1/4}\Delta_L(x, x')$,

$$\tilde{\Delta}_L(x, x') = \ell^\epsilon \int \frac{d^n k}{(2\pi)^n} \exp(ik \cdot y) \Big\{ (k^2 - m^2)^{-1} + (k^2 - m^2)^{-2}$$
$$\times \Big[(\xi - \frac{1}{6})R(x') + \eta\bar{\varphi}(x') + \frac{1}{2}\lambda\bar{\varphi}^2(x')\Big]\Big\} \tag{6.238}$$

are the first two terms in a systematic expansion in powers of $(k^2 - m^2)^{-1}$. Terms not included fall off faster than k^{-4} for large k, and therefore lead to finite contributions when $x \to x'$. These finite terms can be included in $\Delta_{NL}(x, x')$.

As a check on (6.238), we will compute $\mathbf{divp}\, \Delta(x', x)$ and show again that it agrees with (6.236). We have

$$\Delta(x', x') = \Delta_L(x', x') + \Delta_{NL}(x', x'). \tag{6.239}$$

$\Delta_L(x', x')$ can easily be evaluated using

$$\int \frac{d^n k}{(2\pi)^n} (k^2 - m^2)^{-p} = i(-1)^p (4\pi)^{-n/2} \frac{\Gamma(p - n/2)}{\Gamma(p)} (m^2)^{n/2 - p}, \tag{6.240}$$

where $p \geq 1$ is any power. It can then be seen that

$$\Delta_L(x', x') = -i(4\pi)^{-2} m^2 \Gamma(-1 - \epsilon/2) \left(\frac{m^2 \ell^2}{4\pi}\right)^{\epsilon/2}$$
$$+ i(4\pi)^{-2} \Gamma(-\epsilon/2) \left(\frac{m^2 \ell^2}{4\pi}\right)^{\epsilon/2} \Big[\Big(\xi - \frac{1}{6}\Big) R + \eta\bar{\varphi} + \frac{1}{2}\lambda\bar{\varphi}^2\Big],$$

where the curvature and scalar fields are evaluated at x'. (In normal coordinates $g(x') = -1$.) Using $\Gamma(-1 - \epsilon/2) = 2/\epsilon + \gamma - 1 + \cdots$, and $\Gamma(-\epsilon/2) = -2/\epsilon - \gamma + \cdots$, it is easily seen that

$$\Delta_L(x', x') = -\frac{i}{8\pi^2 \epsilon}(m^2 + A) - \frac{i}{16\pi^2}\Big[(m^2 + A)\Big(\gamma + \ln\Big(\frac{m^2 \ell^2}{4\pi}\Big)\Big) - m^2\Big], \tag{6.241}$$

with

$$A = \Big(\xi - \frac{1}{6}\Big) R(x') + \eta\bar{\varphi}(x') + \frac{\lambda}{2}\bar{\varphi}^2(x'). \tag{6.242}$$

Terms which vanish as $\epsilon \to 0$ have been dropped. We can write, from (6.239),

$$\Delta(x, x) = -\frac{i}{8\pi^2 \epsilon}(m^2 + A) + \Delta_R(x), \tag{6.243}$$

where

$$\Delta_R(x) = \Delta_{NL}(x, x) - \frac{i}{16\pi^2}(m^2 + A)\Big(\gamma + \ln\Big(\frac{m^2 \ell^2}{4\pi}\Big)\Big) + \frac{im^2}{16\pi^2} \tag{6.244}$$

is well behaved as $\epsilon \to 0$. The divergent part of (6.243) agrees with that in (6.236) and provides a check on our results.

Suppose that we now consider the divergent part of $\Delta^2(x, x')$ for $x \neq x'$. Using (6.233) we find

$$\Delta^2(x, x') = \Delta_L^2(x, x') + 2\Delta_L(x, x')\Delta_{NL}(x, x') + \Delta_{NL}^2(x, x'). \tag{6.245}$$

The last term on the right-hand side must be finite as $\epsilon \to 0$ from our definition of $\Delta_{NL}(x, x')$. The middle term must also be finite as $\epsilon \to 0$, since any potential divergences can come only from $\Delta_L(x, x')$, which is finite provided $x' \neq x$. Thus if $\Delta^2(x, x')$ is divergent, the divergences can come only from $\Delta_L^2(x, x')$. These divergences may be analyzed using (6.238). In normal coordinates we have

$$\tilde{\Delta}_L^2(x, x') = \ell^{2\epsilon} \int \frac{d^n k}{(2\pi)^n} \frac{d^n p}{(2\pi)^n} \exp\left(i(k+p) \cdot y\right) \Delta(k)\Delta(p),$$

where $\Delta(k)$ represents the momentum-space expansion in (6.238). A change of variable gives

$$\tilde{\Delta}_L^2(x, x') = \ell^{2\epsilon} \int \frac{d^n p}{(2\pi)^n} \exp\left(ip \cdot y\right) \int \frac{d^n k}{(2\pi)^n} \Delta(k)\Delta(p-k). \tag{6.246}$$

Simple power counting shows that if we are only interested in the divergent part of $\tilde{\Delta}_L^2(x, x')$, we only need to take $\Delta(k) = (k^2 - m^2)^{-1}$. Thus

$$\mathbf{divp}\, \tilde{\Delta}_L^2(x, x') = \ell^{2\epsilon} \int \frac{d^n p}{(2\pi)^n} \exp\left(ip \cdot y\right) \mathbf{divp} \int \frac{d^n k}{(2\pi)^n} (k^2 - m^2)^{-1}$$
$$\times [(p-k)^2 - m^2]^{-1}. \tag{6.247}$$

The two denominators may be combined using the identity

$$\frac{1}{a_1 a_2} = \int_0^1 dz [a_1 z + a_2(1-z)]^{-2}, \tag{6.248}$$

leading to

$$\mathbf{divp} \int \frac{d^n k}{(2\pi)^n} \Delta(k)\Delta(p-k)$$
$$= \mathbf{divp} \int_0^1 dz \int \frac{d^n k}{(2\pi)^n} \left[(k - zp)^2 + z(1-z)p^2 - m^2\right]^{-2}$$
$$= \mathbf{divp} \int_0^1 dz\, i(4\pi)^{-n/2} \Gamma(2 - n/2) \left[m^2 - z(1-z)p^2\right]^{n/2-2}$$

after (6.240) has been used. It then follows that

$$\mathbf{divp} \int \frac{d^n k}{(2\pi)^n} \Delta(k)\Delta(p-k) = -\frac{i}{8\pi^2 \epsilon},$$

resulting in

$$\mathbf{divp}\, \tilde{\Delta}_L^2(x, x') = -\frac{i}{8\pi^2 \epsilon}\delta(y).$$

6.7 Renormalization of an interacting scalar field theory

Note that to obtain the divergent part we can let $n \to 4$ everywhere except in the Γ-function, since it is $\Gamma(2 - n/2)$ that determines the pole. Returning from normal coordinates to general coordinates, we find that

$$\operatorname{divp} \Delta_L^2(x, x') = -\frac{i}{8\pi^2 \epsilon} \delta(x, x'), \qquad (6.249)$$

and hence

$$\operatorname{divp} \Delta^2(x, x') = -\frac{i}{8\pi^2 \epsilon} \delta(x, x'). \qquad (6.250)$$

We may now apply a similar (although more complicated) analysis to $\Delta^3(x, x')$, which is needed for $\bar{\Gamma}^{(2)}$. Using (6.233), we have

$$\Delta^3(x, x') = \Delta_L^3(x, x') + 3\Delta_L^2(x, x')\Delta_{NL}(x, x')$$
$$+ 3\Delta_L(x, x')\Delta_{NL}^2(x, x') + \Delta_{NL}^3(x, x'). \qquad (6.251)$$

The last two terms on the right-hand side are finite for $x \neq x'$ using a similar reasoning to that which followed (6.245). Therefore, any divergences present in $\Delta^3(x, x')$ can come only from the first two terms on the right-hand side of (6.251). Because $\Delta_{NL}(x, x')$ is finite as $\epsilon \to 0$, we have from (6.249),

$$\operatorname{divp}\{\Delta_L^2(x, x')\Delta_{NL}(x, x')\} = -\frac{i}{8\pi^2 \epsilon} \delta(x, x')\Delta_{NL}(x, x), \qquad (6.252)$$

where we have set $x' = x$ in $\Delta_{NL}(x, x')$ using the Dirac δ-distribution.

Now consider $\Delta_L^3(x, x')$. From (6.238), we can write

$$\tilde{\Delta}_L(x, x') = I_1(x, x') + A(x')I_2(x, x'), \qquad (6.253)$$

where

$$I_1(x, x') = \ell^\epsilon \int \frac{d^n k}{(2\pi)^n} \exp(ik \cdot y)(k^2 - m^2)^{-1}, \qquad (6.254)$$

$$I_2(x, x') = \frac{\partial}{\partial m^2} I_1(x, x'). \qquad (6.255)$$

$A(x')$ is given in (6.242). We then have

$$\tilde{\Delta}_L^3(x, x') = I_1^3(x, x') + 3A(x')I_1^2(x, x')I_2(x, x')$$
$$+ 3A^2(x')I_1(x, x')I_2^2(x, x') + A^3(x')I_2^3(x, x'). \qquad (6.256)$$

The first term on the right-hand side of (6.256) is

$$I_1^3(x, x') = \ell^\epsilon \int \frac{d^n p}{(2\pi)^n} \exp(ip \cdot y) I(p), \qquad (6.257)$$

where

$$I(p) = \ell^{2\epsilon} \int \frac{d^n k}{(2\pi)^n} \frac{d^n q}{(2\pi)^n} (k^2 - m^2)^{-1}(q^2 - m^2)^{-1}[(p - k - q)^2 - m^2]^{-1}. \qquad (6.258)$$

A way of evaluating this last integral was described by Collins (1974). It begins with the identity

$$\frac{1}{a_1 a_2 a_3} = 2 \int_0^\infty d^3z\, \delta\left(1 - \sum_{i=1}^3 z_i\right) \left(\sum_{i=1}^3 a_i z_i\right)^{-3}. \tag{6.259}$$

After some algebra, and using (6.240), it may be shown that

$$I(p) = \ell^{2\epsilon}(4\pi)^{-n}\Gamma(3-n)(m^2)^\epsilon J(p), \tag{6.260}$$

where

$$J(p) = \int_0^\infty d^3z\, \delta\left(1 - \sum_{i=1}^3 z_i\right) (z_1 z_2 + z_2 z_3 + z_3 z_1)^{-n/2}$$
$$\times \left[1 - \frac{z_1 z_2 z_3}{z_1 z_2 + z_2 z_3 + z_3 z_1} \frac{p^2}{m^2}\right]^\epsilon. \tag{6.261}$$

The integral $J(p)$ may be seen to have divergences as $\epsilon \to 0$ coming from points where $z_1 z_2 + z_2 z_3 + z_3 z_1 = 0$. Because the Dirac delta restricts $z_1 + z_2 + z_3 = 1$, not all of z_1, z_2, z_3 can vanish simultaneously. The only way $z_1 z_2 + z_2 z_3 + z_3 z_1$ can vanish is if two of (z_1, z_2, z_3) vanish, with the remaining parameter fixed to be 1 because of the δ-function constraint. Thus the divergent part of $J(p)$ comes from the three points $(1, 0, 0), (0, 1, 0)$, and $(0, 0, 1)$ in parameter space. The symmetry of the integrand in z_1, z_2, z_3 shows that the divergences from each of the three points must be equal. To obtain the contribution from $z_1 = 1, z_2 = z_3 = 0$, make the change of variables

$$z_1 = 1 - \alpha, z_2 = \alpha\beta, z_3 = \alpha(1-\beta). \tag{6.262}$$

The pole at $z_1 = 1, z_2 = z_3 = 0$ then comes from the $\alpha = 0$ limit of the resulting integral. We have

$$\mathbf{divp}\, J(p) = 3\, \mathbf{divp} \int_0^1 d\alpha \int_0^1 d\beta\, \alpha^{-1-\epsilon/2} \left[1 - \frac{\alpha(1-\alpha)\beta(1-\beta)}{1-\alpha+\alpha\beta(1-\beta)}\frac{p^2}{m^2}\right]^\epsilon. \tag{6.263}$$

(The factor of 3 accounts for the other two divergent parts coming from $z_1 = z_2 = 0, z_3 = 1$ and $z_1 = z_3 = 0, z_2 = 1$.) Because the pole comes from $\alpha = 0$, we may expand the integrand in powers of α and find

$$\mathbf{divp}\, J(p) = 3\, \mathbf{divp} \int_0^1 d\alpha \left\{\alpha^{-1-\epsilon/2} + o(\alpha^{-\epsilon/2})\right\}$$
$$= -\frac{6}{\epsilon}. \tag{6.264}$$

In order to obtain $\mathbf{divp}\, I(p)$, it is clear from (6.260) that we require the finite part of $J(p)$ in addition to the pole. Write

$$J(p) = -\frac{6}{\epsilon} + J_0(p) + o(\epsilon), \tag{6.265}$$

6.7 Renormalization of an interacting scalar field theory

where $J_0(p)$ is the term of order ϵ^0 in the Laurent expansion of $J(p)$ in a neighborhood of $\epsilon = 0$. The pole just calculated comes from $\alpha^{-1-\epsilon/2}$ or $(z_2+z_3)^{-n/2}$ in the integrand of $J(p)$. By symmetry, the other two poles will come from $(z_1+z_2)^{-n/2}$ and $(z_1+z_3)^{-n/2}$. If the three terms responsible for the pole contributions of $J(p)$ are subtracted from the integrand, we may safely let $\epsilon \to 0$ and obtain

$$J_0(p) = \int_0^\infty d^3z\, \delta\left(1 - \sum_{i=1}^{3} z_i\right) \left[(z_1z_2 + z_2z_3 + z_3z_1)^{-2}\right.$$
$$\times \left(1 - \frac{z_1z_2z_3}{z_1z_2 + z_2z_3 + z_3z_1}\frac{p^2}{m^2}\right)$$
$$\left. - (z_1+z_2)^{-2} - (z_2+z_3)^{-2} - (z_3+z_1)^{-2}\right].$$

Making the change of variable (6.262) results in

$$J_0(p) = 3 - \frac{p^2}{2m^2} \qquad (6.266)$$

after some calculation. We therefore have

$$\mathbf{divp}\, I(p) = (4\pi)^{-4}\left\{-\frac{6m^2}{\epsilon^2} + \frac{1}{\epsilon}\left[3(3-2\gamma)m^2 - 6m^2\ln\left(\frac{m^2\ell^2}{4\pi}\right) - \frac{1}{2}p^2\right]\right\}, \qquad (6.267)$$

and hence,

$$\mathbf{divp}\, I_1^3(x,x') = (4\pi)^{-4}\ell^\epsilon\left\{-\frac{6m^2}{\epsilon^2} + \frac{1}{\epsilon}\left[3(3-2\gamma)m^2 - 6m^2\ln\left(\frac{m^2\ell^2}{4\pi}\right) + \frac{1}{2}\Box_y\right]\right\}\delta(y). \qquad (6.268)$$

This determines the divergent part of the first term in (6.256).

The divergent parts of the remaining terms may be evaluated as follows. Suppose that in place of (6.254) we define

$$I_1(m^2) = \ell^\epsilon \int \frac{d^n k}{(4\pi)^n} \exp(ik\cdot y)(k^2 - m^2)^{-1}, \qquad (6.269)$$

and consider $I_1(m_1^2)I_1(m_2^2)I_1(m_3^2)$. This is clearly a symmetric function of m_1^2, m_2^2, m_3^2. Using an analysis similar to that above which had $m_1^2 = m_2^2 = m_3^2$, it may be shown that

$$\mathbf{divp}\{I_1(m_1^2)I_1(m_2^2)I_1(m_3^2)\} = (4\pi)^{-4}\ell^\epsilon\left\{-\frac{2}{\epsilon^2}(m_1^2 + m_2^2 + m_3^2)\right.$$
$$\left. + \frac{1}{\epsilon}\left[(3-2\gamma)(m_1^2 + m_2^2 + m_3^2) - 2\sum_{i=1}^{3} m_i^2\ln\left(\frac{m_i^2\ell^2}{4\pi}\right) + \frac{1}{2}\Box_y\right]\right\}\delta(y). \qquad (6.270)$$

This reduces to (6.268) in the case $m_1^2 = m_2^2 = m_3^2$. Differentiating (6.270) with respect to m_1^2, using (6.255), gives

$$\mathbf{divp}\left\{I_2(m_1^2)I_1(m_2^2)I_1(m_3^2)\right\}$$
$$= (4\pi)^{-4}\ell^\epsilon\left\{-\frac{2}{\epsilon^2} + \frac{1}{\epsilon}\left[1 - 2\gamma - 2\ln\left(\frac{m_1^2\ell^2}{4\pi}\right)\right]\right\}\delta(y). \tag{6.271}$$

Differentiation of (6.271) with respect to m_2^2, using (6.255) again, gives

$$\mathbf{divp}\left\{I_2(m_1^2)I_2(m_2^2)I_1(m_3^2)\right\} = 0. \tag{6.272}$$

A further differentiation of (6.272), with respect to m_3^2, leads to

$$\mathbf{divp}\left\{I_2(m_1^2)I_2(m_2^2)I_2(m_3^2)\right\} = 0. \tag{6.273}$$

If we put $m_1^2 = m_2^2 = m_3^2 = m^2$ in (6.271)–(6.273), it may be seen that

$$\mathbf{divp}\left\{I_1^2(x,x')I_2(x,x')\right\}$$
$$= (4\pi)^{-4}\ell^\epsilon\left\{-\frac{2}{\epsilon^2} + \frac{1}{\epsilon}\left[1 - 2\gamma - 2\ln\left(\frac{m^2\ell^2}{4\pi}\right)\right]\right\}\delta(y), \tag{6.274}$$

$$\mathbf{divp}\left\{I_1(x,x')I_2^2(x,x')\right\} = \mathbf{divp}\left\{I_2^3(x,x')\right\} = 0. \tag{6.275}$$

This proves that only the first two terms on the right-hand side of (6.256) are divergent as $\epsilon \to 0$.

It still remains to return from normal coordinates to general coordinates. We have

$$\Delta_L^3(x,x') = [-g(x)]^{-3/4}\tilde{\Delta}_L^3(x,x')$$

in normal coordinates with origin x'. To return to general coordinates, we must find scalars which reduce to $[-g(x)]^{-3/4}\delta(y)$ and $[-g(x)]^{-3/4}\Box_y\delta(y)$ in normal coordinates. It is clear that

$$[-g(x)]^{-3/4}\delta(y) = [-g(x')]^{-3/4}\delta(y) = \delta(y)$$

in normal coordinates; thus we may simply replace $[-g(x)]^{-3/4}\delta(y)$ with $\delta(x,x')$. The analysis of $[-g(x)]^{-3/4}\Box_y\delta(y)$ is more intricate. Consider

$$K = \int d^n y [-g(x)]^{1/2}\left\{[-g(x)]^{-3/4}\Box_y\delta(y)\right\}f(y)$$

for test function $f(y)$. Integration by parts shows that

$$K = \Box_y\left\{[-g(x)]^{-1/4}f(y)\right\}\bigg|_{y=0}.$$

Using the normal coordinate expansion for $[-g(x)]^{-1/4}$ gives

$$K = \Box_y\left\{f(y) - \frac{1}{12}R_{\mu\nu}(x')y^\mu y^\nu f(y) + \cdots\right\}\bigg|_{y=0}$$
$$= \left(\Box_x - \frac{1}{6}R(x)\right)f(x)\bigg|_{x=x'}.$$

6.7 Renormalization of an interacting scalar field theory

This shows that the generally covariant expression

$$\left(\Box_x - \frac{1}{6}R(x)\right)\delta(x,x') \to [-g(x)]^{-3/4}\Box_y\delta(y)$$

in normal coordinates.

Finally we find that

$$\mathbf{divp}\Delta_L^3(x,x') = (4\pi)^{-4}\ell^\epsilon\bigg\{-\frac{6}{\epsilon^2}(m^2+A) + \frac{1}{\epsilon}\Big[3(3m^2+A)$$

$$- 6\gamma(m^2+A) - 6(m^2+A)\ln\left(\frac{m^2\ell^2}{4\pi}\right)\Big] + \frac{1}{2\epsilon}\left(\Box_x - \frac{1}{6}R\right)\bigg\}\delta(x,x'). \quad (6.276)$$

Using (6.252) and (6.276) in (6.251), it may be seen that

$$\mathbf{divp}\,\Delta^3(x,x') = (4\pi)^{-4}\ell^{-\epsilon}\bigg\{-\frac{6}{\epsilon^2}(m^2+A) + \frac{3}{\epsilon}(m^2+A)$$

$$+ \frac{1}{2\epsilon}\left(\Box_x - \frac{1}{6}R\right)\bigg\}\delta(x,x') - \frac{3i\ell^{-\epsilon}}{8\pi^2\epsilon}\Delta_R(x)\delta(x,x'). \quad (6.277)$$

The relation between $\Delta_R(x)$ and $\Delta_{NL}(x,x)$ in (6.244) has been used here.

The other contribution to $\bar{\Gamma}^{(2)}$ in (6.231) involves $\Delta^2(x,x)$ which is obtained simply by using (6.243). It is now straightforward algebra to obtain

$$\bar{\Gamma}^{(2)} = \int dv_x \bigg\{-\frac{1}{512\pi^4\epsilon^2}\Big[m^2+A+(\eta+\lambda\bar{\varphi})^2\Big](m^2+A)$$

$$+ \frac{1}{1024\pi^4\epsilon}\Big[(m^2+A)(\eta+\lambda\bar{\varphi})^2 + \frac{1}{6}\lambda^2\bar{\varphi}\Box\bar{\varphi} - \frac{1}{356}R(\eta+\lambda\bar{\varphi})^2\Big]$$

$$- \frac{i}{32\pi\epsilon}\Big[\lambda(m^2+A) + (\eta+\lambda\bar{\varphi})^2\Big]\Delta_R(x)\bigg\}. \quad (6.278)$$

Note that because the background gravitational and scalar fields only depend on the coordinates in the physical four-dimensional spacetime, the factor of $\ell^{-\epsilon}$ cancels with the volume of the additional flat dimensions. The remaining integral in (6.278) is four-dimensional. Unlike the situation at one-loop order where the pole part of the effective action involved only local curvature terms and a polynomial in the background scalar field, the last term in (6.278) is seen to involve a very complicated pole which is not local in the sense described. Such a divergent part of the effective action cannot be absorbed into the counterterms in the classical action.

If this was the complete contribution to the effective action at two-loop order we would have to conclude that the theory was not renormalizable. Fortunately there is another order $\hbar^2$ contribution coming from

$$\bar{\Gamma}^{(1)} = \frac{i}{2}\hbar\mathrm{tr}\ln\left(\ell^2\bar{S}_{,ij}\right). \quad (6.279)$$

This is because it is the bare quantities that enter $\bar{S}_{,ij}$, and these have an expansion in powers of $\hbar$ via the counterterms. In fact we may write

$$\bar{S}_{,ij} = \bar{S}_{,ij}^{(0)} + \hbar\bar{S}_{,ij}^{(1)} + \cdots, \quad (6.280)$$

where $\bar{S}_{,ij}^{(0)}$ is the part which contributed to the effective action at one-loop order, and

$$\bar{S}_{,ij}^{(1)} = -\left[\delta m^{(1)2} + \delta \xi^{(1)} R + \delta \eta^{(1)} \overline{\varphi} + \frac{1}{2}\delta \lambda^{(1)} \overline{\varphi}^2\right]\delta(x, x') \quad (6.281)$$

comes from the one-loop counterterms. (We have used the fact that to one-loop order there is no field renormalization.) Expanding $\bar{\Gamma}^{(1)}$ in powers of $\hbar$, it may be easily seen that

$$\bar{\Gamma}^{(1)} = \frac{i}{2}\hbar \operatorname{tr} \ln \left(\ell^2 \bar{S}_{,ij}^{(0)}\right) + \frac{i}{2}\hbar^2 \left(\bar{S}_{,ij}^{(0)}\right)^{-1} \bar{S}_{,ij}^{(1)} + \cdots, \quad (6.282)$$

to order $\hbar^2$. Since $\left(\bar{S}_{,ij}^{(0)}\right)^{-1}$ is $\Delta_B(x, x')$ defined in (6.229), $\bar{\Gamma}^{(1)}$ makes an order $\hbar^2$ contribution to the effective action which is

$$-\frac{i}{2}\hbar^2 \int dv_x \left[\delta m^{(1)2} + \delta \xi^{(1)} R + \delta \eta^{(1)} \overline{\varphi} + \frac{1}{2}\delta \lambda^{(1)} \overline{\varphi}^2\right]\Delta(x, x)$$

$$= \int dv_x \left\{ \frac{1}{256\pi^4 \epsilon^2}\left[\lambda(m^2 + A) + (\eta + \lambda\overline{\varphi})^2\right](m^2 + A) \right.$$

$$\left. + \frac{i}{32\pi^2 \epsilon}\left[\lambda(m^2 + A) + (\eta + \lambda\overline{\varphi})^2\right]\Delta_R(x) \right\}, \quad (6.283)$$

using the explicit results for the one-loop counterterms found earlier, as well as (6.243). Only those terms which diverge as $\epsilon \to 0$ have been retained.

The terms of order $\hbar^2$ arising from the one-loop counterterms are also seen to contain a non-local divergence, which is identical in structure to that found in (6.278), and fortunately which has the opposite sign. Therefore, when (6.278) and (6.283) are combined to obtain the complete two-loop order contribution to the effective action, it may be observed that the potentially troublesome non-local divergence cancels, leaving only a local polynomial in the curvature and the background scalar field. The net term of order $\hbar^2$ in the effective action is obtained by combining (6.278) and (6.283). This results in

$$\mathbf{divp}\,\bar{\Gamma}_{2-loop} = \int dv_x \left\{ \frac{1}{512\pi^4 \epsilon^2}\left[(\lambda m^4 + m^2 \eta^2) + (2\lambda m^2 + \eta^2)\left(\xi - \frac{1}{6}\right)R \right.\right.$$

$$+ \lambda\left(1\xi - \frac{1}{6}\right)^2 R^2 + (\eta^3 + 4\lambda\eta m^2)\overline{\varphi} + \left(2\lambda^2 m^2 + \frac{7}{2}\lambda\eta^2\right)\overline{\varphi}^2$$

$$+ 3\lambda^2 \eta\overline{\varphi}^3 + \frac{3}{4}\lambda^3\overline{\varphi}^4 + 4\lambda\eta\left(\xi - \frac{1}{6}\right)R\overline{\varphi} + 2\lambda^2\left(\xi - \frac{1}{6}\right)R\overline{\varphi}^2\right]$$

$$+ \frac{1}{1024\pi^4 \epsilon}\left[m^2\eta^2 + \eta^2\left(\xi - \frac{7}{36}\right)R + \frac{1}{6}\lambda^2\overline{\varphi}\Box\overline{\varphi}\right.$$

$$+ (\eta^3 + 2\lambda\eta m^2)\overline{\varphi} + 2\lambda^2\eta\overline{\varphi}^3\left(\lambda^2 m^2 + \frac{5}{2}\lambda\eta^2\right)\overline{\varphi}^2 + \frac{1}{2}\lambda^3\overline{\varphi}^4$$

$$\left.\left. + 2\lambda\eta\left(\xi - \frac{7}{36}\right)R\overline{\varphi} + \lambda^2\left(\xi - \frac{7}{36}\right)R\overline{\varphi}^2\right]\right\}. \quad (6.284)$$

6.7 Renormalization of an interacting scalar field theory

These divergences are all of the same form as terms appearing in the classical action, and are therefore removable by the counterterms.

The two-loop counterterms coming from the matter action (6.193) and the gravitational action (6.194) are

$$\begin{aligned}S_M^{(2)} + S_G^{(2)} = \int dv_x \Big\{ &- \delta Z^{(2)} \overline{\varphi} \Box \overline{\varphi} - \frac{1}{2}(2m^2 \delta Z^{(2)} + \delta m^{(2)2}) \overline{\varphi}^2 \\ &- \frac{1}{2}(2\xi \delta Z^{(2)} + \delta \xi^{(2)}) R \overline{\varphi}^2 - (\tau \delta Z^{(2)} + \delta \tau^{(2)}) \overline{\varphi} \\ &- (\gamma \delta Z^{(2)} + \delta \gamma^{(2)}) R \overline{\varphi} - \frac{1}{3!}(3\eta \delta Z^{(2)} + \delta \eta^{(2)}) \overline{\varphi}^3 \\ &- \frac{1}{4!}(4\lambda \delta Z^{(2)} + \delta \lambda^{(2)}) \overline{\varphi}^4 + \delta \Lambda^{(2)} + \delta \kappa^{(2)} R \\ &+ \delta \alpha_1^{(1)} R_{\mu\nu\rho\sigma} R^{\mu\nu\rho\sigma} + \delta \alpha_2^{(2)} R_{\mu\nu} R^{\mu\nu} + \delta \alpha_3^{(2)} R^2 \Big\}. \end{aligned} \quad (6.285)$$

The divergences present in (6.284) may therefore be removed by defining the following counterterms:

$$\delta \Lambda^{(2)} = -\frac{1}{512\pi^4 \epsilon^2}(\lambda m^4 + m^2 \eta^2) - \frac{1}{1024\pi^4 \epsilon} m^2 \eta^2, \quad (6.286)$$

$$\delta \kappa^{(2)} = -\frac{1}{512\pi^4 \epsilon^2}\left(\xi - \frac{1}{6}\right)(2\lambda m^2 + \eta^2) - \frac{1}{1024\pi^4 \epsilon}\left(\xi - \frac{7}{36}\right)\eta^2, \quad (6.287)$$

$$\delta \alpha_1^{(2)} = \delta \alpha_2^{(2)} = 0, \quad (6.288)$$

$$\delta \alpha_3^{(2)} = -\frac{1}{512\pi^4 \epsilon^2}\left(\xi - \frac{1}{6}\right)\lambda, \quad (6.289)$$

$$\delta Z^{(2)} = \frac{1}{6144\pi^4 \epsilon}\lambda^2, \quad (6.290)$$

$$\delta m^{(2)2} = \frac{1}{512\pi^4 \epsilon^2}(4\lambda^2 m^2 + 7\lambda \eta^2) + \frac{5}{3072\pi^4 \epsilon}(\lambda^2 m^2 + 3\lambda \eta^2), \quad (6.291)$$

$$\delta \xi^{(2)} = \frac{1}{128\pi^4 \epsilon^2}\left(\xi - \frac{1}{6}\right)\lambda^2 + \frac{5}{3072\pi^4 \epsilon}\left(\xi - \frac{7}{30}\right)\lambda^2, \quad (6.292)$$

$$\delta \tau^{(2)} = \frac{1}{512\pi^4 \epsilon^2}(4\lambda \eta m^2 + \eta^3) + \frac{1}{6144\pi^4 \epsilon}(6\eta^3 + 12\lambda \eta m^2 - \lambda^2 \tau), \quad (6.293)$$

$$\delta \gamma^{(2)} = \frac{1}{128\pi^4 \epsilon^2}\left(\xi - \frac{1}{6}\right)\lambda \eta + \frac{1}{6144\pi^4 \epsilon}\left[12\left(\xi - \frac{7}{36}\right)\lambda \eta - \lambda^2 \gamma\right], \quad (6.294)$$

$$\delta \eta^{(2)} = \frac{9}{256\pi^4 \epsilon^2}\lambda^2 \eta + \frac{23}{2048\pi^4 \epsilon}\lambda^2 \eta, \quad (6.295)$$

$$\delta \lambda^{(2)} = \frac{9}{256\pi^4 \epsilon^2}\lambda^3 + \frac{17}{1536\pi^4 \epsilon}\lambda^3. \quad (6.296)$$

This completes our discussion of renormalization to two-loop order. For different approaches, see Birrell and Taylor (1980), Lee (1982), or Jack and Osborn (1982, 1984).

Note that we make a distinction between proving renormalizability and calculating counterterms. The proof of renormalizability relies in showing that all of the overlapping divergences cancel.[6] If it is assumed that this is so, then there are some very efficient methods for calculating counterterms to high order. (See Brown and Collins (1980) and Hathrell (1982a) for example.) Note also that we have only concentrated on the divergent parts of the effective action. The calculation of the finite part is more involved, even at one-loop order. For one way to treat this problem see Hu and O'Connor (1984).

6.8 The renormalization group and the effective action

Consider a general theory in which the effective action depends on a number of quantities which we denote by the collective symbol q_i. For example, q_i can stand for any of the coupling constants, masses, or background fields. In the latter case, the index is understood in the sense of condensed notation. Suppose that in n spacetime dimensions the bare quantity q_i^B has dimensions of

$$\dim q_i^B = \alpha_i + \gamma_i \epsilon, \tag{6.297}$$

in units of length, where $\epsilon = n - n_0$, with n_0 the physical spacetime dimension. (For example, we might be interested in $n_0 = 4$.) α_i and γ_i are numbers determined by the actual q_i^B. (Table 6.1 lists the results for a scalar field theory.) In addition to a dependence on q_i^B, the effective action can depend on the background spacetime metric which is not quantized. (It is obviously possible to generalize and allow for a dependence on other external, unquantized fields.) Bare quantities q_i^B are related to renormalized ones q_i by

$$q_i^B = \ell^{\epsilon \gamma_i}(q_i + \delta q_i), \tag{6.298}$$

where δq_i is the relevant counterterm. As in the previous section, we choose the renormalized q_i to have the same dimensions for all n as they do for $n = n_0$. The δq_i are given as a series of pole terms in ϵ, where the coefficients depend on certain of the q_i. (Again, the reader can keep the explicit scalar field example of the previous section in mind here.)

Let $\Gamma_{un}[q_i^B, g_{\mu\nu}]$ denote the unrenormalized effective action expressed in terms of bare quantities, and in n spacetime dimensions. The renormalized effective action is obtained by using (6.298), choosing the counterterms to remove potential divergences as $\epsilon \to 0$, followed by taking the limit $\epsilon \to 0$. We have

$$\Gamma[q_i, g_{\mu\nu}, \ell] = \lim_{\epsilon \to 0} \Gamma_{un}[q_i^B, g_{\mu\nu}]. \tag{6.299}$$

---

[6] For a proof that this occurs, with the exception of the purely gravitational part of the theory, see Bunch (1981b).

6.8 The renormalization group and the effective action

Table 6.1 *Table showing the coupling constants and their dimensions in units of length in $(4+\epsilon)$-dimensional spacetime.*

quantity	dimension in units of length
λ_B	$-4-\epsilon$
κ_B	$-2-\epsilon$
$\alpha_{iB}\,(i=1,2,3)$	$-\epsilon$
φ_B^i	$-1-\epsilon/2$
m^2_{Bij}	-2
ξ_{Bij}	0
η_{Bijk}	$-1+\epsilon/2$
λ_{Bijkl}	ϵ
τ_{Bi}	$-3-\epsilon/2$
γ_{Bi}	$-1-\epsilon/2$

as the renormalized effective action. A dependence on the arbitrary renormalization length scale ℓ is unavoidable, as the example of the preceding section illustrates. It might seem at first sight that a dependence on an arbitrary parameter might be a weakness of the formalism we have discussed; however, as we will now discuss, it is actually a great strength.

All bare quantities q_i^B should be independent of the length scale ℓ, since they are defined in n spacetime dimensions. (ℓ only enters upon reducing to the physical spacetime dimension, and integrating out over the extra dimensions.) Thus

$$\ell \frac{d}{d\ell} q_i^B = 0, \qquad (6.300)$$

which gives

$$0 = \epsilon \gamma_i q_i + \epsilon \gamma_i \delta q_i + \ell \frac{dq_i}{d\ell} + \ell \frac{d}{d\ell} \delta q_i, \qquad (6.301)$$

using (6.298). In dimensional regularization we may express δq_i as a sum of pole terms in ϵ:

$$\delta q_i = \sum_{\alpha=1}^{\infty} \epsilon^{-\alpha} \delta q_i^{(\alpha)}. \qquad (6.302)$$

Since the renormalized q_i must be analytic at $\epsilon = 0$, $\ell dq_i/d\ell$ must be finite as $\epsilon \to 0$. Equation (6.301) must hold for each power of ϵ separately, so it is clear that we must have

$$\ell \frac{dq_i}{d\ell} = \beta_i - \gamma_i q_i \epsilon, \qquad (6.303)$$

for some β_i which does not depend on ϵ. The last term on the right-hand side of (6.303) takes care of the linear dependence in (6.301). The ℓ-dependence in δq_i comes through the dependence of δq_i on q_j:

$$\ell \frac{d}{d\ell} \delta q_i = \sum_j \left(\ell \frac{dq_j}{d\ell} \right) \frac{\partial}{\partial q_j} \delta q_i$$

$$= -\sum_j \gamma_j q_j \frac{\partial}{\partial q_j} \delta q_i^{(1)} + \sum_{\alpha=1}^{\infty} \epsilon^{-\alpha} \sum_j \left(\beta_j \frac{\partial}{\partial q_j} \delta q_i^{(\alpha)} \right.$$

$$\left. - \gamma_j q_j \frac{\partial}{\partial q_j} \delta q_i^{(\alpha+1)} \right). \tag{6.304}$$

This leads to

$$0 = \beta_i - \sum_j \gamma_j q_j \frac{\partial}{\partial q_j} \delta q_i^{(1)} + \gamma_i \delta q_i^{(1)}$$

$$+ \sum_{\alpha=1}^{\infty} \epsilon^{-\alpha} \left\{ \gamma_i \delta q_i^{(\alpha+1)} + \sum_j \left(\beta_j \frac{\partial}{\partial q_j} \delta q_i^{(\alpha)} - \gamma_j q_j \frac{\partial}{\partial q_j} \delta q_i^{(\alpha+1)} \right) \right\}. \tag{6.305}$$

The term of order ϵ^0 gives

$$\beta_i = -\gamma_i \delta q_i^{(1)} + \sum_j \gamma_j q_j \frac{\partial}{\partial q_j} \delta q_i^{(1)}, \tag{6.306}$$

while the term of order $\epsilon^{-\alpha}$ gives

$$0 = \gamma_i \delta q_i^{(\alpha+1)} + \sum_j \left(\beta_j \frac{\partial}{\partial q_j} \delta q_i^{(\alpha)} - \gamma_j q_j \frac{\partial}{\partial q_j} \delta q_i^{(\alpha+1)} \right), \tag{6.307}$$

for $\alpha = 1, 2, \ldots$.

These results were first discussed by 't Hooft (1973). The importance of (6.306) is that it shows the dependence of q_i on the scale ℓ is determined by a knowledge of $\delta q_i^{(1)}$, which is the coefficient of ϵ^{-1} in the counterterm. The existence of higher-order poles in the counterterms is irrelevant in determining how q_i scales with ℓ. Equation (6.307) shows that the higher-order poles are determined by the lower-order ones. Equivalently, it provides a consistency check on any calculation, since, for example, $\delta q_i^{(2)}$ is related to $\delta q_i^{(1)}$. This may be used to verify the order ϵ^{-2} parts of the two-loop counterterms in the last section.

The conventional renormalization group examines the effect of scaling the external momenta in Green functions. This probes the behavior of the theory at short distances. As discussed originally by Nelson and Panangaden (1982), the interpretation of this in curved spacetime may be accomplished by considering a rescaling of the background metric

$$g_{\mu\nu}(x) \to s^{-2} g_{\mu\nu}(x). \tag{6.308}$$

This metric rescaling probes the short distance, or high-curvature limit of the theory, since a curvature invariant such as $R_{\mu\nu\rho\sigma} R^{\mu\nu\rho\sigma}$ scales like

$$R_{\mu\nu\rho\sigma} R^{\mu\nu\rho\sigma} \to s^4 R_{\mu\nu\rho\sigma} R^{\mu\nu\rho\sigma}$$

under (6.308). Our approach follows Toms (1983a).

The effective action must be dimensionless. If we choose $g_{\mu\nu}$ to have dimensions of (length)^2, so that the spacetime coordinates are dimensionless, then on purely dimensional grounds

$$\Gamma[q_i, s^{-2}g_{\mu\nu}, \ell] = F[\ell^{-\alpha_i}q_i, \ell^{-2}s^{-2}g_{\mu\nu}], \tag{6.309}$$

for some functional F. Differentiation of (6.309) with respect to the renormalization length gives

$$\ell\frac{\partial}{\partial \ell}\Gamma[q_i, s^{-2}g_{\mu\nu}, \ell] = -\sum_i \alpha_i q_i \frac{\partial}{\partial q_i}\Gamma[q_i, s^{-2}g_{\mu\nu}, \ell] + s\frac{\partial}{\partial s}\Gamma[q_i, s^{-2}g_{\mu\nu}, \ell]. \tag{6.310}$$

If we change ℓ in $\Gamma_{un}[q_i^B, s^{-2}g_{\mu\nu}]$, then

$$\ell\frac{d}{d\ell}\Gamma_{un}[q_i^B, s^{-2}g_{\mu\nu}] = 0, \tag{6.311}$$

because the q_i^B have no dependence on ℓ (nor does the background gravitational field). Letting $\epsilon \to 0$ in (6.311) and using (6.299), we have

$$0 = \left(\ell\frac{\partial}{\partial \ell} + \sum_i \beta_i \frac{\partial}{\partial q_i}\right)\Gamma[q_i, s^{-2}g_{\mu\nu}, \ell], \tag{6.312}$$

since from (6.303),

$$\beta_i = \lim_{\epsilon \to 0}\left(\ell\frac{d}{d\ell}q_i\right). \tag{6.313}$$

Eliminating $\ell\partial\Gamma/\partial\ell$ between (6.310) and (6.312) results in

$$\left\{s\frac{\partial}{\partial s} + \sum_i(\beta_i - \alpha_i q_i)\frac{\partial}{\partial q_i}\right\}\Gamma[q_i, s^{-2}g_{\mu\nu}, \ell] = 0. \tag{6.314}$$

This equation may be solved by defining running quantities $q_i(s)$ by

$$s\frac{d}{ds}q_i(s) = \beta_i(q(s)) - \alpha_i q_i(s), \tag{6.315}$$

$$q_i(1) = q_i. \tag{6.316}$$

The solution to (6.314) is

$$\Gamma[q_i, s^{-2}g_{\mu\nu}, \ell] = \Gamma[q_i(s), g_{\mu\nu}, \ell]. \tag{6.317}$$

This shows that the short-distance, or high-curvature, behavior of the theory is governed by the behavior of the solutions to (6.315), which are determined by the simple poles of the counterterms in dimensional regularization. We will use the analysis presented here to calculate the effective potential in the next section.

An extension of the renormalization group to consider local scalings (i.e., where the length scale ℓ can depend on x) was suggested by Drummond and Shore (1979a). An early reference on the use of dimensional regularization in the renormalization group analysis is Collins and Macfarlane (1974).

6.9 The effective potential

The computation of the full effective action for a general background field is impossible even in flat spacetime. Instead, it is sometimes useful to restrict the background fields to be constant. In this case, since no derivatives of the background fields can occur, we can write

$$\Gamma[\overline{\varphi}] = \int dv_x \left\{ -V(\overline{\varphi}) \right\}, \tag{6.318}$$

where $V(\overline{\varphi})$ is called the effective potential (Coleman and Weinberg 1973; Weinberg 1973). This approximation is useful if it can be argued that the background field should be constant for reasons of symmetry. For example, in flat spacetime the vacuum expectation value of the field would be expected to be constant. More generally, the same would be expected in any static spacetime of constant curvature, provided that constant solutions are not ruled out by the boundary conditions. In other cases, it might be argued that derivatives of the background field are slowly varying, so that while the background field is not strictly constant, the dominant contribution to the effective action comes from the term of the form (6.318).

The effective potential may be obtained perturbatively using the loop expansion of the effective action considered earlier. We will first of all consider the one-loop effective potential in flat spacetime for the $O(N)$ model whose classical bare action is

$$S[\varphi] = \int dv_x \left\{ \frac{1}{2} \delta_{ij} \partial^\mu \varphi_B^i \partial_\mu \varphi_B^j - \frac{1}{2} m_B^2 \delta_{ij} \varphi_B^i \varphi_B^j \right.$$
$$\left. - \frac{\lambda_B}{4!} (\delta_{ij} \varphi_B^i \varphi_B^j)^2 + \Lambda_B \right\}. \tag{6.319}$$

This is a special case of the theory considered in (6.193) which is obtained by demanding that there be an invariance under $\varphi_B^i \to R^i{}_j \varphi_B^j$ where $R^i{}_j \in O(N)$. (There is no essential reason why the theory of (6.193) cannot be considered; however, we will deal with (6.319) to shorten some of the expressions.) At one-loop order,

$$\Gamma^{(1)}[\overline{\varphi}] = \frac{i}{2} \hbar \operatorname{tr} \ln \left(\ell^2 \bar{S}_{,ij} \right), \tag{6.320}$$

where

$$\bar{S}_{,ij} = \left(-\delta_{ij} \Box_x - M_{ij}^2 \right) \delta(x, x'), \tag{6.321}$$

with

$$M_{ij}^2 = m^2 \delta_{ij} + \frac{\lambda}{3!} \overline{\varphi}^2 \delta_{ij} + \frac{\lambda}{3} \overline{\varphi}^i \overline{\varphi}^j. \tag{6.322}$$

Here we have used $\overline{\varphi}^2 = \delta_{ij} \overline{\varphi}^i \overline{\varphi}^j$. If $\overline{\varphi}^i$ is taken to be constant, the effect of the quartic interaction is seen to simply result in a modification of the mass term. The one-loop effective action is then similar to that calculated for the free scalar

6.9 The effective potential

field in the previous chapter. Any of the techniques described there can be used to calculate $\Gamma^{(1)}$. For example, $\Gamma^{(1)}[\overline{\varphi}]$ may be related to a kernel $K_{ij}(\tau; x, x')$ via

$$\Gamma^{(1)}[\overline{\varphi}] = -\frac{i\hbar}{2} \int dv_x \int_0^\infty \frac{d\tau}{\tau} \operatorname{tr} K_{ij}(\tau; x, x), \qquad (6.323)$$

where

$$K_{ij}(\tau; x, x) = i(4\pi i \ell^2 \tau)^{-n/2} \operatorname{tr}\left(\exp(-i\ell^2 \tau M^2)\right). \qquad (6.324)$$

M^2 is regarded as an $N \times N$ matrix with components given by (6.322). (The kernel for a single scalar field of mass m was contained in (5.109). It is not difficult to deduce the generalized result (6.324).) The one-loop effective potential is then

$$V^{(1)}(\overline{\varphi}) = \frac{i\hbar}{2} \ell^\epsilon \int_0^\infty \frac{d\tau}{\tau} \operatorname{tr} K_{ij}(\tau; x, x), \qquad (6.325)$$

where the factor of ℓ^ϵ arises from the volume of the extra dimensions. (Equation (6.318) is defined in the physical spacetime dimension, whereas (6.323) holds in n dimensions.) After Wick rotation of τ, it may be seen that

$$V^{(1)}(\overline{\varphi}) = -\frac{\hbar \ell^\epsilon}{2(4\pi)^{n/2}} \int_0^\infty d\tau\, \tau^{-1-n/2} \operatorname{tr}\left(\exp(-\tau M^2)\right) \qquad (6.326)$$

$$= -\frac{\hbar \ell^\epsilon}{2(4\pi)^{n/2}} \Gamma(-n/2) \operatorname{tr}(M^2)^{n/2}. \qquad (6.327)$$

This result holds irrespective of the explicit form of M^2. If M^2 has the form (6.322), we can easily diagonalize it in terms of its eigenvalues. It is a simple exercise to show that the eigenvalues are $m^2 + \frac{\lambda}{2}\overline{\varphi}^2$ with multiplicity 1, and $m^2 + \frac{\lambda}{6}\overline{\varphi}^2$ with multiplicity $N - 1$. Thus, (6.327) gives us

$$V^{(1)}(\overline{\varphi}) = -\frac{\hbar \ell^\epsilon}{2(4\pi)^{n/2}} \Gamma(-n/2) \left[(N-1)\left(m^2 + \frac{\lambda}{6}\overline{\varphi}^2\right)^{n/2} + \left(m^2 + \frac{\lambda}{2}\overline{\varphi}^2\right)^{n/2} \right] \qquad (6.328)$$

upon taking the trace. Using $n = 4 + \epsilon$, and expanding about $\epsilon = 0$, we find

$$V^{(1)} = \frac{\hbar}{32\pi^2} \left(\frac{1}{\epsilon} + \frac{1}{2}\gamma - \frac{3}{4}\right) \left[(N-1)\left(m^2 + \frac{\lambda}{6}\overline{\varphi}^2\right)^2 + \left(m^2 + \frac{\lambda}{2}\overline{\varphi}^2\right)^2 \right]$$
$$+ \frac{\hbar}{64\pi^2} \left\{ (N-1)\left(m^2 + \frac{\lambda}{6}\overline{\varphi}^2\right)^2 \ln\left[\frac{\ell^2(m^2 + \frac{\lambda}{6}\overline{\varphi}^2)}{4\pi}\right] \right.$$
$$\left. + \left(m^2 + \frac{\lambda}{2}\overline{\varphi}^2\right)^2 \ln\left[\frac{\ell^2(m^2 + \frac{\lambda}{6}\overline{\varphi}^2)}{4\pi}\right] \right\}. \qquad (6.329)$$

The pole term in (6.329) is removable by the same counterterms as in Section 6.7, to leave a result which is finite as $\epsilon \to 0$. However, it is conventional, following Coleman and Weinberg (1973), to proceed somewhat differently, and impose renormalization conditions on the effective potential instead. We will choose

$$V(\overline{\varphi}=0)=0, \tag{6.330}$$

$$\left.\frac{\partial^2 V(\overline{\varphi})}{\partial \overline{\varphi}^i \partial \overline{\varphi}^j}\right|_{\overline{\varphi}=0} = m^2 \delta_{ij}, \tag{6.331}$$

$$\left.\frac{\partial^4 V(\overline{\varphi})}{\partial \overline{\varphi}^i \partial \overline{\varphi}^j \partial \overline{\varphi}^k \partial \overline{\varphi}^l}\right|_{\overline{\varphi}=\varphi_0} = \frac{\lambda}{3}(\delta_{ij}\delta_{kl} + \delta_{ik}\delta_{jl} + \delta_{il}\delta_{jk}). \tag{6.332}$$

The right-hand side of the last two conditions is chosen to agree with the corresponding term in the classical potential

$$V^{(0)} = \frac{1}{2}\left(m^2 + \hbar \delta m^{2(1)}\right)\overline{\varphi}^2 + \frac{1}{4!}\left(\lambda + \hbar \delta \lambda^{(1)}\right)(\overline{\varphi}^2)^2 - (\Lambda + \hbar \delta \Lambda^{(1)}), \tag{6.333}$$

where we will choose $\Lambda = 0$. $V = V^{(0)} + V^{(1)}$ is the effective potential to one-loop order. φ_0 in (6.332) is arbitrary, and is chosen away from the origin so that, as we will see, the massless limit can be taken.

To simplify calculations, it is convenient to use the $O(N)$ symmetry to take only $\overline{\varphi}^1 = \overline{\varphi}$ to be non-zero. The above conditions (6.330)–(6.332) reduce to

$$V(\overline{\varphi}=0)=0, \tag{6.334}$$

$$\left.\frac{d^2 V(\overline{\varphi})}{d\overline{\varphi}^2}\right|_{\overline{\varphi}=0} = m^2, \tag{6.335}$$

$$\left.\frac{d^4 V(\overline{\varphi})}{d\overline{\varphi}^4}\right|_{\overline{\varphi}=\varphi_0} = \lambda. \tag{6.336}$$

This fixes the renormalization counterterms, and results in the following renormalized effective potential:

$$\begin{aligned}
V(\overline{\varphi}) =& \frac{\lambda}{4!}\overline{\varphi}^4 + \frac{1}{2}m^2\overline{\varphi}^2 + \frac{3\hbar N m^4}{128\pi^2} - \frac{\hbar N m^4}{64\pi^2}\ln m^2 \\
&+ \frac{\hbar m^2}{32\pi^2}\left[(N-1)\frac{\lambda}{6} + \frac{\lambda}{2}\right]\overline{\varphi}^2 - \frac{\hbar m^2}{32\pi^2}\left[(N-1)\frac{\lambda}{6} + \frac{\lambda}{2}\right]\overline{\varphi}^2 \ln m^2 \\
&- \frac{\hbar}{64\pi^2}\left[(N-1)\frac{\lambda^2}{36}\ln\left(m^2 + \frac{\lambda}{6}\varphi_0^2\right) + \frac{\hbar \lambda^2}{4}\ln\left(m^2 + \frac{\lambda}{2}\varphi_0^2\right)\right]\overline{\varphi}^4 \\
&- \frac{\varphi_0^2}{16\pi^2}\left[\frac{(N-1)\lambda^3}{216(m^2 + \frac{\lambda}{6}\varphi_0^2)} + \frac{\lambda^3}{8(m^2 + \frac{\lambda}{2}\varphi_0^2)}\right]\overline{\varphi}^4 \\
&+ \frac{\hbar \varphi_0^4}{48\pi^2}\left[\frac{(N-1)\lambda^4}{1296(m^2 + \frac{\lambda}{6}\varphi_0^2)^2} + \frac{\lambda^4}{16(m^2 + \frac{\lambda}{2}\varphi_0^2)^2}\right]\overline{\varphi}^4 \\
&+ \frac{\hbar}{64\pi^2}(N-1)\left(m^2 + \frac{\lambda}{6}\varphi_0^2\right)^2\left[\ln\left(m^2 + \frac{\lambda}{6}\varphi_0^2\right) - \frac{3}{2}\right] \\
&+ \frac{\hbar}{64\pi^2}\left(m^2 + \frac{\lambda}{2}\varphi_0^2\right)^2\left[\ln\left(m^2 + \frac{\lambda}{2}\varphi_0^2\right) - \frac{3}{2}\right].
\end{aligned} \tag{6.337}$$

6.9 The effective potential

If we choose $\varphi_0 = 0$, then we have, after a bit of simplification,

$$V(\overline{\varphi}) = \frac{\lambda}{4!}\overline{\varphi}^4 + \frac{1}{2}m^2\overline{\varphi}^2 - \frac{\hbar Nm^2}{64\pi^2}\left[(N-1)\frac{\lambda}{6} + \frac{\lambda}{2}\right]\overline{\varphi}^2$$
$$- \frac{3\hbar}{128\pi^2}\left[(N-1)\frac{\lambda^2}{36} + \frac{\lambda^2}{4}\right]\overline{\varphi}^4 + \frac{\hbar}{64\pi^2}\left[(N-1)\left(m^2 + \frac{\lambda}{6}\overline{\varphi}^2\right)^2\right.$$
$$\left.\times \ln\left(1 + \frac{\lambda\overline{\varphi}^2}{6m^2}\right) + \left(m^2 + \frac{\lambda}{2}\overline{\varphi}^2\right)^2 \ln\left(1 + \frac{\lambda\overline{\varphi}^2}{2m^2}\right)\right]. \tag{6.338}$$

It should be clear why it is not possible to obtain the massless limit of this result, which is why we kept $\varphi_0 \neq 0$ in (6.332). If we take $N = 1$, corresponding to a single massive scalar field, we have

$$V(\overline{\varphi}) = \frac{\lambda}{4!}\overline{\varphi}^4 + \frac{1}{2}m^2\overline{\varphi}^2 - \frac{3\hbar\lambda^2}{512\pi^2}\overline{\varphi}^4 + \frac{\hbar}{64\pi^2}\left(m^2 + \frac{\lambda}{2}\overline{\varphi}^2\right)^2 \ln\left(1 + \frac{\lambda\overline{\varphi}^2}{2m^2}\right). \tag{6.339}$$

This result was originally given by Coleman and Weinberg (1973).

If we take the massless limit $m^2 \to 0$ in (6.337), then we have

$$V(\overline{\varphi}) = \frac{\lambda}{4!}\overline{\varphi}^4 + \frac{\hbar}{64\pi^2}\left[(N-1)\frac{\lambda^2}{36} + \frac{\lambda^2}{4}\right]\overline{\varphi}^4\left[\ln(\overline{\varphi}^2/\varphi_0^2) - 25/6\right]. \tag{6.340}$$

This result was originally given by Jackiw (1974). In the case $N = 1$, it reduces to the famous Coleman–Weinberg result of

$$V(\overline{\varphi}) = \frac{\lambda}{4!}\overline{\varphi}^4 + \frac{\hbar\lambda^2}{256\pi^2}\overline{\varphi}^4\left[\ln(\overline{\varphi}^2/\varphi_0^2) - 25/6\right]. \tag{6.341}$$

Again it is clear why we cannot take $\varphi_0 = 0$.

It is also possible to obtain the effective potential from renormalization group considerations, as discussed originally by Coleman and Weinberg (1973). We will show this for the case of the massless $O(N)$ model first in flat spacetime, and then in curved spacetime. The extension of the renormalization group method to calculate effective potentials in curved spacetime was considered by Buchbinder and Odintsov (1985).

The massless $O(N)$ model is described by $q^i = (\overline{\varphi}^i, \lambda)$ in flat spacetime. The effective action satisfies the general equation (6.312), which for the $O(N)$ model reads

$$\left\{\ell\frac{\partial}{\partial \ell} + \int dv_x \beta_{\varphi^i}\frac{\delta}{\delta\overline{\varphi}^i(x)} + \beta_\lambda\frac{\partial}{\partial \lambda}\right\}\Gamma[\overline{\varphi}] = 0, \tag{6.342}$$

where

$$\beta_{\varphi^i} = \lim_{\epsilon \to 0}\ell\frac{d\overline{\varphi}^i(x)}{d\ell}, \tag{6.343}$$

$$\beta_\lambda = \lim_{\epsilon \to 0}\ell\frac{d\lambda}{d\ell}. \tag{6.344}$$

β_{φ^i} can be related to the field renormalization factor Z_{φ^i} defined in (6.200):

$$\overline{\varphi}^i_B(x) = \ell^{-\epsilon/2}Z_\varphi\overline{\varphi}^i(x). \tag{6.345}$$

(By $O(N)$ symmetry, all fields should have the same renormalization factor.) As discussed in Section 6.8,

$$\ell \frac{d}{d\ell} \overline{\varphi}^i_B(x) = 0,$$

which using (6.345) leads to

$$\ell \frac{d}{d\ell} \overline{\varphi}^i(x) = \frac{1}{2} \epsilon \overline{\varphi}^i(x) - \left(\ell \frac{d}{d\ell} \ln Z_\varphi\right) \overline{\varphi}^i(x). \tag{6.346}$$

Taking the limit $\epsilon \to 0$ in this result, it follows from (6.343) that

$$\beta_{\varphi^i} = -\gamma(\lambda) \overline{\varphi}^i(x), \tag{6.347}$$

where

$$\gamma(\lambda) = \lim_{\epsilon \to 0} \left(\ell \frac{d}{d\ell} \ln Z_\varphi\right). \tag{6.348}$$

From (6.342), the effective potential can now be seen to satisfy

$$\left\{\ell \frac{\partial}{\partial \ell} - \gamma(\lambda) \overline{\varphi}^i \frac{\partial}{\partial \overline{\varphi}^i} + \beta_\lambda \frac{\partial}{\partial \lambda}\right\} V(\overline{\varphi}) = 0. \tag{6.349}$$

The background scalar field has been taken to be constant here. As discussed earlier, $V(\overline{\varphi})$ can only depend on $\overline{\varphi}$ through $\overline{\varphi}^2 = \delta_{ij} \overline{\varphi}^i \overline{\varphi}^j$; thus, we may use the $O(N)$ symmetry to take $\overline{\varphi}^i = \overline{\varphi} \delta^{i1}$.

Because the Coleman–Weinberg renormalization condition (6.336) is imposed on the fourth derivative of the effective potential, it proves convenient to obtain a renormalization group equation for

$$V^{(4)}(\overline{\varphi}) = \frac{\partial^4 V(\overline{\varphi})}{\partial \overline{\varphi}^4}. \tag{6.350}$$

This is easily obtained by differentiation of both sides of (6.349), and reads

$$\left\{\ell \frac{\partial}{\partial \ell} - 4\gamma(\lambda) - \gamma(\lambda) \overline{\varphi} \frac{\partial}{\partial \overline{\varphi}} + \beta_\lambda \frac{\partial}{\partial \lambda}\right\} V^{(4)}(\overline{\varphi}) = 0. \tag{6.351}$$

Since the only dimensional quantities occurring are ℓ and $\overline{\varphi}$, and since $V^{(4)}$ must be dimensionless, it can only depend on ℓ and $\overline{\varphi}$ through the dimensionless combination $\ell \overline{\varphi}$. If we define

$$t = \ln(\ell^2 \overline{\varphi}^2), \tag{6.352}$$

then (6.351) becomes

$$\left[-\frac{\partial}{\partial t} + \bar{\beta}_\lambda \frac{\partial}{\partial \lambda} + 4\bar{\gamma}\right] V^{(4)} = 0, \tag{6.353}$$

where
$$\bar{\beta}_\lambda(\lambda) = \frac{\beta_\lambda(\lambda)}{2(\gamma(\lambda) - 1)}, \quad (6.354)$$

$$\bar{\gamma}(\lambda) = \frac{\gamma(\lambda)}{2(1 - \gamma(\lambda))}. \quad (6.355)$$

The Coleman–Weinberg renormalization condition (6.336) now becomes the boundary condition
$$\left. V^{(4)} \right|_{t=0} = \lambda. \quad (6.356)$$

The general solution to (6.353) with boundary condition (6.356) is obtained by defining
$$\frac{d}{dt}\lambda'(t,\lambda) = \bar{\beta}_\lambda(\lambda'), \quad (6.357)$$

with
$$\lambda'(t=0,\lambda) = \lambda. \quad (6.358)$$

The solution is
$$V^{(4)} = \lambda'(t,\lambda) \exp\left[4\int_0^t d\tau \, \bar{\gamma}\left(\lambda'(\tau,\lambda)\right)\right]. \quad (6.359)$$

(A nice discussion of the solution is contained in Coleman (1973).)

This presents an exact result for the effective potential, valid to all orders in perturbation theory. If we knew the exact results for the functions appearing in the renormalization group equation, then the effective potential could be recovered easily. However, we only know the first few terms in the perturbative loop expansion. It was shown in Section 6.8 that the renormalization group functions were determined by a knowledge of the coefficients of the simple poles of the counterterms in dimensional regularization. This knowledge is sufficient to deduce that the effective potential can be obtained from knowing the residues of the simple poles. Thus, if the counterterms are known, the effective potential can be found using the renormalization group.

To demonstrate the utility of this approach, we will recover the one-loop result for the effective potential for the $O(N)$ model given earlier. First of all $Z_\varphi = 1 + o(\hbar^2)$, since there is no field renormalization to one-loop order. It then follows from (6.348) that $\gamma(\lambda) = o(\hbar^2)$, and from (6.355) that $\bar{\gamma} = o(\hbar^2)$. Thus to one-loop order we may take
$$V^{(4)} = \lambda'(t,\lambda). \quad (6.360)$$

Also, since $\gamma = o(\hbar^2)$, we have $\bar{\beta}_\lambda = -\frac{1}{2}\beta_\lambda$ to one-loop order. This means that to compute λ' we only need the one-loop contribution to β_λ. To obtain this, use
$$\lambda_B = \ell^\epsilon \left(\lambda + \hbar\delta\lambda^{(1)} + \cdots\right), \quad (6.361)$$

with

$$\frac{\delta\lambda^{(1)}}{4!} = -\frac{1}{32\pi^2\epsilon}\left[(N-1)\frac{\lambda^2}{36} + \frac{\lambda^2}{4}\right]. \tag{6.362}$$

(See (6.225) specialized to the $O(N)$ model.) From (6.344) it follows that

$$\beta_\lambda = -\frac{3\hbar}{16\pi^2}\left[1 + \frac{1}{9}(N-1)\right] + o(\hbar^2). \tag{6.363}$$

It is now easily seen from (6.357) that

$$\lambda' = \lambda + \frac{3\hbar}{32\pi^2}\left[1 + \frac{1}{9}(N-1)\right]\lambda^2 t + o(\hbar^2). \tag{6.364}$$

t was defined in (6.352). Thus

$$\frac{\partial^4 V}{\partial \overline{\varphi}^4} = \lambda + \frac{3\hbar}{32\pi^2}\left[1 + \frac{1}{9}(N-1)\right]\lambda^2 \ln(\ell^2 \overline{\varphi}^2). \tag{6.365}$$

Integration of both sides recovers the result found earlier in (6.340).

We now wish to extend the renormalization group argument to study the effective potential in curved space. Regard the effective potential as being expanded in powers of the curvature:

$$V = V_0 + V_1 + V_2 + \cdots, \tag{6.366}$$

where the subscript denotes the order of the curvature. V_0 is the effective potential in flat spacetime which has been obtained above. Because the curvature requires the introduction of additional coupling constants for renormalization purposes, we will take

$$\bar{q}^i = (\Lambda, \kappa, \alpha_1, \alpha_2, \alpha_3, \xi, \lambda, \overline{\varphi}^i).$$

From the renormalization group equation satisfied by the effective action (6.312), we have

$$\left\{\ell\frac{\partial}{\partial \ell} - \gamma(\lambda)\overline{\varphi}^i\frac{\partial}{\partial \overline{\varphi}^i} + \beta_\lambda\frac{\partial}{\partial \lambda} + \beta_\xi\frac{\partial}{\partial \xi} + \beta_\Lambda\frac{\partial}{\partial \Lambda} + \beta_\kappa\frac{\partial}{\partial \kappa} + \sum_{i=1}^{3}\beta_{\alpha_i}\frac{\partial}{\partial \alpha_i}\right\}V = 0. \tag{6.367}$$

This is the generalization of (6.349) to curved space. Because the metric is not quantized, the only dependence of V on $\Lambda, \kappa, \alpha_1, \alpha_2,$ and α_3 is through the classical Lagrangian. Furthermore, since the theory is assumed to be massless, we must have $\beta_\Lambda = \beta_\kappa = 0$. We therefore find

$$\left\{\ell\frac{\partial}{\partial \ell} - \gamma(\lambda)\overline{\varphi}^i\frac{\partial}{\partial \overline{\varphi}^i} + \beta_\lambda\frac{\partial}{\partial \lambda} + \beta_\xi\frac{\partial}{\partial \xi}\right\}V = \beta_{\alpha_1}R^{\mu\nu\rho\sigma}R_{\mu\nu\rho\sigma}$$
$$+ \beta_{\alpha_2}R^{\mu\nu}R_{\mu\nu} + \beta_{\alpha_3}R^2. \tag{6.368}$$

(Note that this differs from the expression given in Buchbinder and Odintsov (1985).)

6.9 The effective potential

By equating terms of equal powers in the curvature on both sides, we find that V_0 satisfies (6.349). (Since V_0 has no curvature dependence, it cannot involve ξ.) V_1 satisfies

$$0 = \left[\ell\frac{\partial}{\partial \ell} - \gamma(\lambda)\overline{\varphi}^i\frac{\partial}{\partial \overline{\varphi}^i} + \beta_\lambda\frac{\partial}{\partial \lambda} + \beta_\xi\frac{\partial}{\partial \xi}\right]V_1, \tag{6.369}$$

and V_2 satisfies (6.368). We may write

$$V_1 = R\tilde{V}_1, \tag{6.370}$$

where $\tilde{V}_1$ has dimensions of (length)$^{-2}$. Setting $\overline{\varphi}^i = \overline{\varphi}\delta^{i1}$ as before, and taking

$$\tilde{V}_2^{(2)} = \frac{\partial^2 \tilde{V}_1}{\partial \overline{\varphi}^2}, \tag{6.371}$$

which is dimensionless, it is easily seen that

$$\left[-\frac{\partial}{\partial t} + \bar{\beta}_\lambda\frac{\partial}{\partial \lambda} + 2\bar{\gamma} + \bar{\beta}_\xi\frac{\partial}{\partial \xi}\right]\tilde{V}_1^{(2)} = 0, \tag{6.372}$$

where $\bar{\beta}_\lambda, \bar{\gamma}$ were given in (6.354) and (6.355), t was defined in (6.352), and

$$\bar{\beta}_\xi = \frac{\beta_\xi}{2(\gamma - 1)}. \tag{6.373}$$

Analogously to the Coleman–Weinberg renormalization condition (6.336) on $V^{(4)}$, pick

$$\tilde{V}_1^{(2)}\bigg|_{t=0} = \xi. \tag{6.374}$$

If we define $\xi'(t, \lambda, \xi)$ to satisfy

$$\frac{d}{dt}\xi'(t, \lambda, \xi) = \bar{\beta}_\xi, \tag{6.375}$$

with

$$\xi'(t = 0, \lambda, \xi) = \xi, \tag{6.376}$$

then the exact solution for $\tilde{V}_1^{(2)}$ is

$$\tilde{V}_1^{(2)} = \xi'(t, \lambda, \xi) \exp\left[2\int_0^t d\tau\, \bar{\gamma}(\lambda'(\tau, \lambda))\right]. \tag{6.377}$$

The one-loop counterterm $\delta\xi^{(1)}$ for the $O(N)$ model follows directly from (6.221) as

$$\delta\xi^{(1)} = -\frac{\lambda}{48\pi^2\epsilon}(N+2)\left(\xi - \frac{1}{6}\right). \tag{6.378}$$

It is then easily shown that to order $\hbar$,

$$\beta_\xi = -\frac{\hbar\lambda}{48\pi^2}(N+2)\left(\xi - \frac{1}{6}\right), \tag{6.379}$$

and that

$$\xi' = \frac{1}{6} + \left(\xi - \frac{1}{6}\right)\left[1 - \frac{\hbar\lambda}{96\pi^2}(N+8)t\right]^{-\left(\frac{N+2}{N+8}\right)}$$
$$= \xi + \frac{\hbar\lambda}{96\pi^2}(N+2)\left(\xi - \frac{1}{6}\right)t + \cdots, \quad (6.380)$$

where only terms up to order $\hbar$ have been retained in the last line. We then find

$$\tilde{V}_1 = \frac{1}{2}\xi\overline{\varphi}^2 + \frac{\hbar(N+2)\lambda}{384\pi^2}\left(\xi - \frac{1}{6}\right)\overline{\varphi}^2\left[\ln(\ell\overline{\varphi}^2) - 3\right]. \quad (6.381)$$

Turning next to V_2, we may write

$$V_2 = A_1 R_{\mu\nu\rho\sigma} R^{\mu\nu\rho\sigma} + A_2 R_{\mu\nu} R^{\mu\nu} + A_3 R^2, \quad (6.382)$$

where A_1, A_2, A_3 are dimensionless. Introducing $t = \ln(\ell^2\overline{\varphi}^2)$ as before, we have

$$\left[-\frac{\partial}{\partial t} + \bar{\beta}_\lambda \frac{\partial}{\partial \lambda} + \bar{\beta}_\xi \frac{\partial}{\partial \xi}\right] A_i = -\bar{\beta}_{\alpha_i}, \quad (6.383)$$

where

$$\bar{\beta}_{\alpha_i} = \frac{\beta_{\alpha_i}}{2(\gamma - 1)}. \quad (6.384)$$

Choosing the renormalization conditions $A_i|_{t=0} = \alpha_i$, we have the solution to (6.383) as

$$A_i = \alpha_i - \int_0^t d\tau\, \bar{\beta}_{\alpha_i}(\lambda', \xi'). \quad (6.385)$$

Using the one-loop counterterms found earlier (6.217)–(6.219), it may be seen that

$$A_1 = \alpha_1 + \frac{\hbar N}{5760\pi^2}\ln(\ell^2\overline{\varphi}^2), \quad (6.386)$$

$$A_2 = \alpha_2 - \frac{\hbar N}{5760\pi^2}\ln(\ell^2\overline{\varphi}^2), \quad (6.387)$$

$$A_3 = \alpha_3 + \frac{\hbar N}{32\pi^2}\left(\xi - \frac{1}{6}\right)^2 \ln(\ell^2\overline{\varphi}^2). \quad (6.388)$$

This completes the evaluation of the effective potential in curved spacetime to quadratic order in the curvature.

Only in the case where the curvature is constant could the resulting expression for the effective potential be used to study whether or not symmetry breaking occurs. This is because we have assumed that $\overline{\varphi}$ is constant in both space and time. It would be of great utility if the method could be extended to obtain terms in the effective potential which were of more than quadratic order in the curvature. This is not possible in four dimensions due to the lack of a natural boundary condition when solving the renormalization group equation. The boundary conditions were chosen so that the effective potential agreed with the classical potential when $t = 0$, and the classical potential only contained terms up to quadratic order in the curvature. This means that terms of cubic and higher order are not obtained solely from a knowledge of the renormalization group functions. At one-loop order it is possible to calculate terms of cubic and higher orders in the curvature. (See Balakrishnan and Toms (1992) for example.) It might be possible to use the known one-loop terms as boundary conditions for the renormalization group to evaluate two- and higher-loop contributions to the effective potential.

6.10 The renormalization of the non-linear sigma model

In this section we will be concerned with a field theory whose configuration space is intrinsically curved. The non-linear sigma model, described briefly in Section 6.5, is not only of interest in its own right, but also of interest from its role in string theory. (See, for example, Green et al., 1987.) Two early references on the quantization of non-linear sigma models are Ecker and Honerkamp (1971) and Honerkamp (1972).

Before discussing the specifics of the non-linear sigma model we will consider a generic field theory for which the covariant Taylor expansion of the classical action is

$$\hat{S}[\overline{\varphi}; \sigma^i[\overline{\varphi}; \varphi]] = \sum_{n=0}^{\infty} \frac{(-1)^n}{n!} \bar{S}_{;i_1 \cdots i_n} \sigma^{i_1}[\overline{\varphi}; \varphi] \cdots \sigma^{i_n}[\overline{\varphi}; \varphi]. \tag{6.389}$$

(See Section 6.5.) Given an action functional, we require the coefficients $\bar{S}_{;i_1 \cdots i_n}$ in the covariant Taylor expansion. One way of obtaining these inductively was described in Toms (1988).

Introduce Riemann normal coordinates at $\overline{\varphi}$ in the field space. Then $g_{ij}(\overline{\varphi}) = \delta_{ij}$, and

$$\sigma[\overline{\varphi}; \varphi] = \frac{1}{2}\delta_{ij}(\overline{\varphi}^i - \varphi^i)(\overline{\varphi}^j - \varphi^j) \tag{6.390}$$

gives the geodetic interval. Since $\sigma_i[\overline{\varphi}; \varphi] = \frac{\delta}{\delta \overline{\varphi}^i}\sigma[\overline{\varphi}; \varphi]$, it follows from (6.390) that in Riemann normal coordinates

$$\sigma_i[\overline{\varphi}; \varphi]\sigma^i[\overline{\varphi}; \varphi] = 2\sigma[\overline{\varphi}; \varphi]. \tag{6.391}$$

Since this is a tensor relation, it must be true in any system of coordinates. (A different derivation can be found in DeWitt (1964a).) Differentiation of both sides of (6.391) covariantly with respect to $\overline{\varphi}^j$ leads to

$$\sigma^i \sigma_{i;j} = \sigma_j. \tag{6.392}$$

Because σ is a biscalar, $\sigma_{i;j} = \sigma_{;ij} = \sigma_{;ji} = \sigma_{j;i}$. Thus we have

$$\sigma^i \sigma^j{}_{;i} = \sigma^j. \tag{6.393}$$

Since σ^j is tangent to the geodesic connecting $\bar{\varphi}$ and φ, at the point $\bar{\varphi}$, this last result corresponds to the fact that the tangent vector to a geodesic is transported by parallel propagation along the geodesic.

The result in (6.393) may be used to derive an expression which can be utilized to obtain the coefficients in the covariant Taylor expansion recursively. Let

$$S^{(n)} = \frac{(-1)^n}{n!} \bar{S}_{;i_1\cdots i_n} \sigma^{i_1}\cdots \sigma^{i_n} \tag{6.394}$$

denote the nth term in the Taylor expansion (6.389). We now consider how $S^{(n+1)}$ is related to $S^{(n)}$. From (6.394),

$$S^{(n+1)} = \frac{(-1)^{n+1}}{(n+1)!} \bar{S}_{;i_1\cdots i_{n+1}} \sigma^{i_1}\cdots \sigma^{i_n+1}$$

$$= \frac{(-1)^{n+1}}{(n+1)!} \left[\bar{S}_{;i_1\cdots i_n} \sigma^{i_1}\cdots \sigma^{i_n}\right]_{;i_{n+1}} \sigma^{i_{n+1}}$$

$$- \frac{(-1)^{n+1}}{(n+1)!} \sum_{r=1}^{n} \bar{S}_{;i_1\cdots i_r\cdots i_n} \sigma^{i_1}\cdots \sigma^{i_r}{}_{;i_{n+1}} \cdots \sigma^{i_n}\sigma^{i_{n+1}}.$$

The first term on the right-hand side is seen to be simply related to the derivative of $S^{(n)}$. The second set of terms can also be related to $S^{(n)}$ if we use (6.393). It then follows that

$$S^{(n+1)} = -\frac{1}{(n+1)} S^{(n)}{}_{;i}\sigma^i + \frac{n}{(n+1)} S^{(n)}, \tag{6.395}$$

demonstrating how a knowledge of $S^{(n)}$ enables $S^{(n+1)}$ to be obtained.

We must now transcribe the condensed notation to normal notation. First of all the condensed index i stands for the spacetime label x^μ as well as the discrete label I which runs over the dimension of the Riemannian manifold whose coordinates are φ^I. The field space metric g_{ij} is really the biscalar distribution

$$g_{IJ}(x_I, x_J) = G_{IJ}(\varphi(x_I))\delta(x_I, x_J). \tag{6.396}$$

We append labels I and J to distinguish the two different spacetime coordinates. (We could use x and x' equally well, but our notation makes it easier to trace which condensed index corresponds to which spacetime label.) The inverse metric g^{ij} reads

$$g^{IJ}(x_I, x_J) = G^{IJ}(\varphi(x_I))\delta(x_I, x_J) \tag{6.397}$$

with G^{IJ} the usual inverse to G_{IJ} in the sense of Riemannian differential geometry. The Christoffel connection Γ^i_{jk} in field space becomes the triscalar distribution

$$\Gamma^I_{JK}(x_I, x_J, x_K) = \Gamma^I_{JK}(\varphi(x_I))\delta(x_I, x_J)\delta(x_I, x_K) \tag{6.398}$$

with $\Gamma^I_{JK}(\varphi(x_I))$ the normal Christoffel connection for the Riemannian metric G_{IJ}.

6.10 The renormalization of the non-linear sigma model

To save writing we will set (in condensed notation)
$$\phi^i = -\sigma^i[\bar{\varphi}, \varphi]. \tag{6.399}$$

In standard notation (6.395) reads
$$S^{(n+1)} = \frac{1}{n+1} \int dv_{x'} \phi^I(x') \frac{\delta}{\delta \varphi^I(x')} S^{(n)} + \frac{n}{n+1} S^{(n)}. \tag{6.400}$$

It proves convenient to define the functional covariant derivative operator $\mathcal{D}$ by
$$\mathcal{D} = \int dv_{x'} \phi^I(x') \frac{\delta}{\delta \varphi^I(x')} + \Gamma - \text{terms}, \tag{6.401}$$

where the Γ-terms are those found from the usual tensor analysis. For example, when acting on a scalar functional $F[\varphi(x)]$ no Γ-terms are needed. If $V^I[\varphi(x)]$ is a vector functional, then
$$\mathcal{D}V^I[\varphi(x)] = \int dv_{x'} \phi^J(x') \frac{\delta V^I[\varphi(x)]}{\delta \varphi^J(x')}$$
$$+ \Gamma^I_{JK}(\varphi(x)) V^K[\varphi(x)] \phi^J(x). \tag{6.402}$$

The extension to arbitrary tensor functionals should be obvious. If the functional $V^I[\varphi(x)]$ is local in the sense that
$$\frac{\delta V^I[\varphi(x)]}{\delta \varphi^J(x')} = \frac{\partial V^I[\varphi(x)]}{\partial \varphi^J(x)} \delta(x,x'), \tag{6.403}$$

where $\partial/\partial \varphi^J$ denotes the ordinary derivative, then
$$\mathcal{D}V^I[\varphi(x)] = \phi^J(x) V^I{}_{;J}[\varphi(x)] \tag{6.404}$$

is just the usual covariant derivative along the geodesic connecting φ to $\bar{\varphi}$.

With this notation the following result holds true
$$\mathcal{D}\partial_\mu \varphi^I(x) = D_\mu \phi^I(x), \tag{6.405}$$

where
$$D_\mu \phi^I(x) = \partial_\mu \phi^I(x) + \Gamma^I_{JK}(\varphi(x)) \phi^J(x) \partial_\mu \varphi^K(x). \tag{6.406}$$

This is an immediate consequence of (6.404). Another result we require is that if $V^I[\varphi(x)]$ is local in the sense described above, then
$$[\mathcal{D}, D_\mu] V^I[\varphi(x)] = R^I{}_{JKL}(\varphi(x)) \partial_\mu \varphi^K(x) V^J[\varphi(x)] \phi^L(x) \tag{6.407}$$

with $R^I{}_{JKL}$ the Riemannian curvature. We can transcribe (6.393) to read
$$\mathcal{D}\phi^I(x) = -\phi^I(x). \tag{6.408}$$

If we now apply (6.402) to $\phi^I(x)$ and use (6.407) we find
$$\mathcal{D}D_\mu \phi^I = R^I{}_{JKL} \partial_\mu \varphi^K \phi^J \phi^L - D_\mu \phi^I. \tag{6.409}$$

A further useful result is that

$$\mathcal{D}G_{IJ} = 0, \tag{6.410}$$

which follows simply from the fact that Γ^I_{JK} is the Christoffel connection for G_{IJ}.
We can write the recursive result (6.395) as

$$S^{(n+1)} = \frac{1}{n+1}\mathcal{D}S^{(n)} + \frac{n}{n+1}S^{(n)}. \tag{6.411}$$

Putting $n = 0$ in (6.411) and using (6.405) and (6.410), along with the fact that $S^{(0)}$ is given by (6.97) yields

$$S^{(1)} = \int dv_x G_{IJ}(\overline{\varphi}(x))D_\mu\phi^I \partial^\mu \overline{\varphi}^J. \tag{6.412}$$

Putting $n = 1$ in (6.411) and using (6.412) results in

$$S^{(2)} = \frac{1}{2}\int dv_x \left\{ G_{IJ}(\overline{\varphi})D^\mu\phi^I D_\mu\phi^J + \bar{R}^I{}_{KJL}(\overline{\varphi})\phi^I\phi^J \partial_\mu\overline{\varphi}^K \partial^\mu\overline{\varphi}^L \right\}. \tag{6.413}$$

Higher-order terms follow by taking $n \geq 2$ in (6.411) and building up the expressions iteratively. There is no impediment, other than algebraic complexity, from extending the method to arbitrary order. Because we will limit our attention to one-loop order here we will not pursue this. (Our method is in agreement with the results obtained by Mukhi (1986).)

Before considering the one-loop effective action we will discuss a modification to the non-linear sigma model. It is possible to add on a term to the classical action which has the same classical invariance as the original theory. (This includes invariance under $x^\mu \to x'^\mu(x)$, under $\varphi^I \to \varphi'^I(\varphi)$, and under conformal rescaling of the metric tensor $g_{\mu\nu}(x)$.) This term was introduced in Wess and Zumino (1971), and is called the Wess–Zumino term. Quantization of models with a Wess–Zumino term was considered in Curtright and Zachos (1984) and Braaten et al. (1985). The Wess–Zumino term will be defined to be

$$S_{WZ} = \int dv_x \epsilon^{\mu_1\cdots\mu_n}(x) B_{I_1\cdots I_n}(\varphi(x)) \partial_{\mu_1}\varphi^{I_1} \cdots \partial_{\mu_n}\varphi^{I_n}. \tag{6.414}$$

Here $\epsilon^{\mu_1\cdots\mu_n}$ is the Levi–Civita tensor which we define by

$$\epsilon^{\mu_1\cdots\mu_n} = -[-g(x)]^{-1/2}\mathrm{sgn}(\mu_1,\ldots,\mu_n) \tag{6.415}$$

with $\mathrm{sgn}(\mu_1,\ldots,\mu_n)$ the sign of the permutation of $\mu_1,\ldots,\mu_n$. $B_{I_1\cdots I_n}$ is an antisymmetric tensor. (n is the spacetime dimension.)

The recursive result (6.411) holds just as well if we use S_{WZ}, and it is straightforward to obtain

$$S^{(1)}_{WZ} = \int dv_x \epsilon^{\mu_1\cdots\mu_n} H_{I_1\cdots I_{n+1}}(\overline{\varphi}(x))\phi^{I_{n+1}} \partial_{\mu_1}\overline{\varphi}^{I_1} \cdots \partial_{\mu_n}\overline{\varphi}^{I_n}, \tag{6.416}$$

where

$$H_{I_1\cdots I_{n+1}} = (n+1)B_{[I_1\cdots I_n;I_{n+1}]}. \tag{6.417}$$

(The square brackets denote an antisymmetrization of the sum with a factor of $1/(n+1)!$.) The next term is

$$S_{WZ}^{(2)} = \frac{1}{2} \int dv_x \epsilon^{\mu_1 \cdots \mu_n} \{ \overline{H}_{I_1 \cdots I_{n+1}; J} \phi^J \phi^{I_{n+1}} \partial_{\mu_1} \overline{\varphi}^{I_1} \cdots \partial_{\mu_n} \overline{\varphi}^{I_n}$$
$$+ n \overline{H}_{I_1 \cdots I_{n+1}} \phi^{I_{n+1}} D_{\mu_1} \phi^{I_1} \partial_{\mu_2} \overline{\varphi}^{I_2} \cdots \partial_{\mu_n} \overline{\varphi}^{I_n} \}. \qquad (6.418)$$

The bar on H denotes that it is evaluated at $\varphi = \overline{\varphi}$. Once again higher-order terms may be found by taking $n \geq 3$ in (6.411).

The one-loop effective action involves $\ln \det(\ell^2 \bar{S}_{;i}{}^j)$, and we have used the heat kernel method to reduce the evaluation of the pole part of this expression to a knowledge of the coefficients in the asymptotic expansion of the heat kernel. (See Chapter 5.) For $\bar{S}_{;i}{}^j$ of the general form $D^\mu D_\mu + Q$ the pole term for even values of n involves $Q^{n/2}$. For odd n there is no pole term apart from a possible surface term. For the non-linear sigma model with no Wess–Zumino term our result for $S^{(2)}$ in (6.413) gives

$$Q_I{}^J = \bar{R}_{IK}{}^J{}_L \partial_\mu \overline{\varphi}^K \partial^\mu \overline{\varphi}^L. \qquad (6.419)$$

For $n > 2$ this means that the pole term involves more than two derivatives of the background field. As there are no such terms in the classical action the theory will not be renormalizable. However if $n = 2$ the pole term involves a linear dependence on Q which contains only two derivatives of the field. There is still the possibility of renormalizing the non-linear sigma model in two dimensions.

For $n = 2$, using dimensional regularization, we have

$$\mathbf{divp}\,\{\Gamma^{(1)}\} = -\frac{1}{4\pi\epsilon} \int dv_x \left(\frac{N}{6} r - \mathrm{tr} Q_I{}^J \right). \qquad (6.420)$$

Here r is the scalar curvature of the two-dimensional spacetime[7] with metric $g_{\mu\nu}$. In two dimensions we have the result that $\int dv_x r$ is a topological invariant of the manifold related to the Euler characteristic χ by

$$\int dv_x r = 4\pi \chi.$$

We then find

$$\mathbf{divp}\,\{\Gamma^{(1)}\} = -\frac{N}{6\epsilon} \chi + \frac{1}{4\pi\epsilon} \int dv_x \bar{R}_{IJ} \partial^\mu \overline{\varphi}^I \partial_\mu \overline{\varphi}^J. \qquad (6.421)$$

The first pole term is taken care of easily by adding a counterterm to the classical sigma model action involving the Euler characteristic. Because this term is a topological invariant it cannot affect the dynamics. The second pole term is similar to that appearing in the classical action except that in place of G_{IJ} we have R_{IJ}. The theory is therefore not renormalizable in the usual sense, but is renormalizable in a more general sense as discussed originally by Friedan (1980).

---

[7] We must distinguish the spacetime geometry from that of the non-linear sigma model.

The idea is to regard G_{IJ} in the classical action as a bare quantity G_{IJ}^B and to renormalize it by

$$\bar{G}_{IJ}^B = \bar{G}_{IJ} - \frac{\hbar}{2\pi\epsilon}\bar{R}_{IJ}. \tag{6.422}$$

This cancels the pole term of $\Gamma^{(1)}$.

With the Wess–Zumino term present the situation is a bit more complicated. We will take $n=2$ and find

$$S_{WZ}^{(2)} = \frac{1}{2}\int dv_x \epsilon^{\mu\nu}\Big\{\bar{H}_{KLI;J}\partial_\mu\bar{\varphi}^K\partial_\nu\bar{\varphi}^L\phi^I\phi^J$$
$$+ 2\bar{H}_{IJK}\phi^I D_\mu\phi^J \partial_\nu\bar{\varphi}^K\Big\}. \tag{6.423}$$

Because the last term involves $D_\mu\phi^J$, the operator $\bar{S}_{,i}{}^j$ is not of the standard form $D_\mu D^\mu + Q$ that we have been assuming. To use the results established earlier we must manipulate things so that we obtain an operator of the standard form. For any operator of the form $D_\mu D^\mu + 2P^\mu D_\mu + Q$ this can always be done by defining a new differential operator $\tilde{D}_\mu = D_\mu + P_\mu$. This completes the square removing the linear derivative term in the operator, and results in a differential operator of the standard form. For the specific case of the Wess–Zumino model this procedure can be interpreted in terms of the parallelizing torsion of Braaten et al. (1985). We will define

$$\tilde{D}_\mu\phi^I = D_\mu\phi^I + \bar{H}^I{}_{JK}\epsilon_{\mu\nu}\partial^\nu\bar{\varphi}^K\phi^J. \tag{6.424}$$

It can then be found that

$$S^{(2)} + S_{WZ}^{(2)} = \frac{1}{2}\int dv_x\Big\{\bar{G}_{IJ}\tilde{D}^\mu\phi^I\tilde{D}_\mu\phi^J + [\bar{R}_{IKJL}\partial^\mu\bar{\varphi}^K\partial_\mu\bar{\varphi}^L$$
$$+\epsilon^{\mu\nu}\bar{H}_{KLI;J}\partial_\mu\bar{\varphi}^K\partial_\nu\bar{\varphi}^L + \bar{H}_{MKI}\bar{H}^M{}_{LJ}\partial^\mu\bar{\varphi}^K\partial_\mu\bar{\varphi}^L]\phi^I\phi^J\Big\}. \tag{6.425}$$

Q_{IJ} can be read off to be (noting that it must be symmetrical[8] in I and J)

$$Q_{IJ} = (\bar{R}_{IKJL} + \bar{H}_{MKI}\bar{H}^M{}_{LJ})\partial^\mu\bar{\varphi}^K\partial_\mu\bar{\varphi}^L$$
$$+ \bar{H}_{KL(I;J)}\epsilon^{\mu\nu}\partial_\mu\bar{\varphi}^K\partial_\nu\bar{\varphi}^L. \tag{6.426}$$

After working out $\mathrm{tr}Q_I{}^J$, a straightforward calculation shows that the renormalization is effected by making the choices

$$\bar{G}_{IJ}^B = \bar{G}_{IJ} - \frac{\hbar}{2\pi\epsilon}(\bar{R}_{IJ} + \bar{H}_{IKL}\bar{H}_J{}^{KL}), \tag{6.427}$$

$$\bar{B}_{IJ}^B = \bar{B}_{IJ} - \frac{\hbar}{4\pi\epsilon}\bar{H}_{IJ}{}^K{}_{;K}. \tag{6.428}$$

For a more detailed analysis of the renormalization of the non-linear sigma model, see Howe et al. (1988).

---

[8] We use round brackets to denote that there is a symmetrized sum: $F_{(IJ)} = \frac{1}{2}(F_{IJ} + F_{JI})$.

6.11 Formal properties of the effective action

It was shown in Section 6.6 that when the perturbative expansion of the two-loop effective action was interpreted in terms of Feynman diagrams it contained only vacuum diagrams that were one-particle irreducible. In this section we will prove that this feature of the effective action, of being composed of one-particle irreducible vacuum diagrams only, holds to all orders of perturbation theory. A number of other general results will also be obtained.

First of all, although we have only derived the perturbative expansion for the effective action, we could equally well have obtained the perturbative expansion for the amplitude $\langle \varphi_2|\varphi_1\rangle[J]$, or the functional $W[J;\varphi_*]$ defined in (6.115). The functional integral representation for the amplitude $\langle \varphi_2|\varphi_1\rangle[J]$ was given in Section 6.5. (See (6.113) or (6.114).) Instead of using the notation $\langle \varphi_2|\varphi_1\rangle[J]$ for the amplitude, we will adopt the more conventional notation of $Z[J;\varphi_*]$. Then

$$Z[J;\varphi_*] = \int d\mu[\varphi_*;\varphi] \exp\frac{i}{\hbar}\left\{S[\varphi] - J_i\sigma^i[\varphi_*;\varphi]\right\}, \tag{6.429}$$

and from (6.115)

$$W[J;\varphi_*] = -i\hbar \ln Z[J;\varphi_*]. \tag{6.430}$$

We will now consider the perturbative expansion of each of $Z[J;\varphi_*]$ and $W[J;\varphi_*]$ in turn.

From the covariant Taylor expansion of the classical action in (6.103) we have

$$S[\varphi] = S^\star - S^\star_{;i}\sigma^i + \frac{1}{2}S^\star_{;ij}\sigma^i\sigma^j + S_{\text{int}}, \tag{6.431}$$

where

$$S_{\text{int}} = \sum_{n=3}^{\infty} \frac{(-1)^n}{n!} S^\star_{;(i_1\cdots i_n)}\sigma^{i_1}\cdots\sigma^{i_n}. \tag{6.432}$$

(The $\star$ denotes that the expression is evaluated at $\varphi = \varphi_*$, and σ^i stands for $\sigma^i[\varphi_*;\varphi]$.) S_{int} will be regarded as an interaction and treated perturbatively.

Define

$$S^J_{\text{int}} = S_{\text{int}} - \tilde{J}_i\sigma^i, \tag{6.433}$$

where

$$\tilde{J}_i = J_i + S^\star_{;i}. \tag{6.434}$$

We may write

$$Z[J;\varphi_*] = \exp\left(\frac{i}{\hbar}S^\star\right)\left(\int d\mu[\varphi_*;\varphi]\exp\left\{\frac{i}{2\hbar}S^\star_{;ij}\sigma^i\sigma^j\right\}\right)$$
$$\times \left\langle\exp\left\{\frac{i}{\hbar}S^J_{\text{int}}\right\}\right\rangle_G, \tag{6.435}$$

338   *The effective action: Non-gauge theories*

where $\langle\cdots\rangle_G$ was defined in (6.164). Now define

$$\tilde{Z}[J;\varphi_\star] = \left\langle \exp\left\{\frac{i}{\hbar}S^J_{\text{int}}\right\}\right\rangle_G \tag{6.436}$$

$$= \sum_{n=0}^{\infty}\frac{1}{n!}\left(\frac{i}{\hbar}\right)^n \left\langle (S^J_{\text{int}})^n\right\rangle_G. \tag{6.437}$$

We may evaluate $\langle (S^J_{\text{int}})^n\rangle_G$ using the same procedure as we used in Section 6.6. (Recall that this procedure amounted to the functional integral equivalent of Wick's theorem.) Because the Gaussian average gives the sum of all possible pairings of factors of σ^i, $\tilde{Z}[J;\varphi_\star]$ will be the sum of all vacuum diagrams (because there are no factors of σ^i corresponding to external lines), where the vertices are contained in S^J_{int}, and the propagators are $(S^\star_{;ij})^{-1}$. It should be noted that due to the presence of the term $-\tilde{J}_i\sigma^i$ in (6.433), there is a one-point vertex in addition to the vertices $S^\star_{;(i_1\cdots i_n)}$ with $n\geq 3$.

Examples of the diagrams arising from (6.437) are indicated in Figs 6.2 and 6.3 for a scalar field theory with only a quartic self-interaction term in the classical Lagrangian. For this theory S_{int} contains only cubic and quartic vertices since the Taylor series (6.432) terminates. It may be observed that both connected and disconnected diagrams are present in the expansion of $\tilde{Z}[J;\varphi_\star]$. The diagrams in

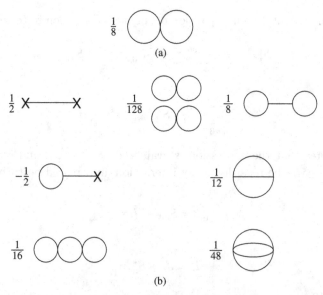

Fig 6.2. The Feynman diagrams of first order (a) and second order (b) arising in the evaluation of $\tilde{Z}$ are indicated for a scalar field theory with a cubic and quartic self-interaction. The number in front of each diagram includes the factor of $1\backslash n!$ in (6.437) as well as the numbers arising from the Wick reduction of $\langle (S^J_{\text{int}})^n\rangle$. The cross denotes the one-point vertex $-\tilde{J}_i\sigma^i$. The factor of $(i/\hbar)^n$ has not been included.

6.11 Formal properties of the effective action

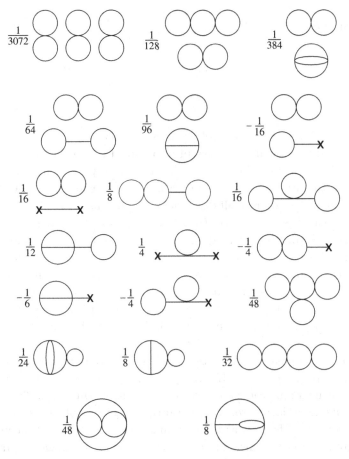

Fig 6.3. The Feynman diagrams of third order arising in the evaluation of $\tilde{Z}$ for a scalar field theory with a quartic self-interaction. All of the symbols are as in the previous figure.

Figs 6.2 and 6.3 correspond to the $n = 1, 2, 3$ terms of (6.437) and represent the sum of all graphs with up to three powers of S_{int}^J.

The diagrammatic expansion of $W[J; \varphi_\star]$ may be obtained from (6.430) and (6.435). We have

$$W[J; \varphi_\star] = S^\star - i\hbar \ln \left(\int d\mu[\varphi_\star; \varphi] \exp\left\{ \frac{i}{2\hbar} S^\star_{;ij} \sigma^i \sigma^j \right\} \right)$$
$$- i\hbar \ln \tilde{Z}[J; \varphi_\star]. \qquad (6.438)$$

The first term in (6.438) is just the classical action evaluated at the background field, and the second term may be recognized as the one-loop part of the effective action. All of the effects of the interaction vertices are contained in the final term of (6.438). If we expand $\tilde{Z}[J; \varphi_\star]$ out to third order in S_{int}^J, it may be seen that

$$\ln \tilde{Z}[J;\varphi_*] = \frac{i}{\hbar}\langle S_{\text{int}}^J\rangle_{\text{G}} + \left(\frac{i}{\hbar}\right)^2 \left\{\frac{1}{2!}\langle (S_{\text{int}}^J)^2\rangle_{\text{G}} - \frac{1}{2}(\langle S_{\text{int}}^J\rangle_{\text{G}})^2\right\}$$
$$+ \left(\frac{i}{\hbar}\right)^3 \left\{\frac{1}{3!}\langle (S_{\text{int}}^J)^3\rangle_{\text{G}} - \frac{1}{2!}\langle (S_{\text{int}}^J)^2\rangle_{\text{G}}\langle S_{\text{int}}^J\rangle_{\text{G}}\right.$$
$$\left. + \frac{1}{3}(\langle S_{\text{int}}^J\rangle_{\text{G}})^3\right\} + \cdots . \tag{6.439}$$

For the simple quartic scalar field theory example considered earlier, it is straightforward to show using the results contained in Figs 6.2 and 6.3 that the diagrammatic expansion of $\ln \tilde{Z}[J;\varphi_*]$ consists of only the connected diagrams contained in the two figures. The logarithm in the definition of $W[J;\varphi_*]$ has resulted in the cancelation of all disconnected diagrams.

We now wish to prove that $W[J;\varphi_*]$ contains all of the connected, and none of the disconnected, vacuum diagrams for an arbitrary theory to all orders of perturbation theory. (Figs 6.2 and 6.3 are specific to a theory with quartic self-interactions, and we have only verified this conclusion to third order in perturbation theory.) This is referred to as the linked cluster theorem in many-body problems. (See Mattuck (1976) for example.) A complete proof was given by Goldstone (1957). The proof given here is based on that in Abrikosov et al. (1975).

Consider a particular diagram that comes from the Wick reduction of the general term appearing in (6.437), $(i/\hbar)^n\langle (S_{\text{int}}^J)^n\rangle_{\text{G}}/n!$. By a single diagram we will mean the diagrammatic representation of a single term in $(i/\hbar)^n\langle (S_{\text{int}}^J)^n\rangle_{\text{G}}/n!$ along with its numerical factor. In general this diagram will consist of a product of various connected factors, as the diagrams of Figs 6.2 and 6.3 show. Suppose that the total number of disjoint connected pieces in the single diagram under consideration is d. In general the same connected piece may appear more than once. Let f_i be the number of times a connected diagram of order $(S_{\text{int}}^J)^n$ occurs in our diagram of order n, where $i \leq k$ for some integer k. Then we have the two relations

$$\sum_{i=1}^{k} f_i = d, \tag{6.440}$$

$$\sum_{i=1}^{k} i f_i = n. \tag{6.441}$$

(The reader should have no difficulty verifying these relations for the diagrams of Figs 6.2 and 6.3.)

We now need to work out the number of times this given diagram occurs in the Wick reduction of $(i/\hbar)^n\langle (S_{\text{int}}^J)^n\rangle_{\text{G}}/n!$. This is simply the number of ways of picking f_i factors of $(S_{\text{int}}^J)^i$ from $(S_{\text{int}}^J)^n$ without replacement. First of all there are $\binom{n}{f_1}$ ways of picking out f_1 factors of S_{int}^J from $(S_{\text{int}}^J)^n$. We are left with $(S_{\text{int}}^J)^{n-f_1}$. There are now $\binom{n-f_1}{2}$ ways of forming the first diagram of order $(S_{\text{int}}^J)^2$, $\binom{n-f_1-2}{2}$

ways of forming the second one, ..., and $\binom{n-f_1-2(f_2-1)}{2}$ ways of forming the f_2^{th}. Because these f_2 connected diagrams are all indistinguishable, there are

$$\frac{1}{f_2!}\binom{n-f_1}{2}\binom{n-f_1-2}{2}\cdots\binom{n-f_1-2(f_2-1)}{2}$$

$$= \frac{(n-f_1)!}{f_2!\,2^{f_2}\,(n-f_1-2f_2)!}$$

ways of obtaining the f_2 diagrams of order $(S_{\text{int}}^J)^2$. Continuing this procedure results in the fact that

$$\frac{(n-f_1-2f_2-\cdots(k-1)f_{k-1})!}{f_k!\,(k!)^{f_k}\,(n-f_1-2f_2-\cdots-kf_k)!}$$

is the number of ways of obtaining the f_k diagrams of order $(S_{\text{int}}^J)^k$. Taking the product of all of these k numbers leads to (noting (6.441))

$$\frac{n!}{\prod_{l=1}^{k}(f_l!)(l!)^{f_l}}$$

as the number of diagrams of the above-mentioned type. The numerical factor for this particular diagram is therefore

$$\frac{\left(\frac{i}{\hbar}\right)^n}{\prod_{i=1}^{k}[f_i!(i!)^{f_i}]}. \tag{6.442}$$

Now let Θ_l denote the sum of all connected vacuum diagrams of order $(S_{\text{int}}^J)^l$. We may write

$$\Theta_l = \left(\frac{i}{\hbar}\right)^l \frac{1}{l!} \left\langle \left(S_{\text{int}}^J\right)^l \right\rangle_C, \tag{6.443}$$

where the subscript 'C' denotes that only connected graphs are retained in the Wick reduction. The sum of all vacuum graphs, apart from the one-loop term which is independent of the interaction, is then

$$\Theta = \sum_{l=1}^{\infty} \Theta_l \tag{6.444}$$

$$= \left\langle \exp \frac{i}{\hbar}\{S_{\text{int}}^J\} \right\rangle_C - 1. \tag{6.445}$$

Recall that the factor in (6.441) refers to that found for a particular diagram in the expansion of Z. If we can now show that the same factor is obtained from $\exp\Theta$, this will complete the proof of the linked cluster expansion and will prove that $\ln Z$ contains only connected vacuum diagrams. (There is an overall factor of $-i\hbar$ between $W[J;\varphi_*]$ and $\ln\tilde{Z}[J;\varphi_*]$, but this is unimportant for this proof.)

The graph that was considered earlier had only terms up to order $(S_{\text{int}}^J)^k$, and hence is contained in

$$\exp\left(\sum_{l=1}^{k}\Theta_l\right) = \sum_{m=0}^{\infty}\frac{1}{m!}\left(\sum_{l=1}^{k}\Theta_l\right)^m.$$

There is a total of d connected vacuum diagrams (some appearing more than once), and therefore we must have a product of d factors of Θ_l's. This means that the diagram of interest is contained in $\frac{1}{d!}(\sum_{l=1}^{k}\Theta_l)^d$. The diagram considered had f_1 terms of order S_{int}^J, ..., f_k terms of order $(S_{\text{int}}^J)^k$ where (6.440) and (6.441) hold. This diagram is therefore contained in the term $\Theta_1^{f_1}\cdots\Theta_k^{f_k}$ arising from the expansion of $(\sum_{l=1}^{k}\Theta_l)^d$ in a sum of products of the Θ_l factors. Denote the coefficient of $\Theta_1^{f_1}\cdots\Theta_k^{f_k}$ by $c(f_1,\ldots,f_k)$. Then the factor for the considered disconnected diagram which arises from $\exp(\sum_{l=1}^{\infty}\Theta_l)$ is

$$\frac{1}{d!}c(f_1,\ldots,f_k)\prod_{l=1}^{k}\left[\left(\frac{i}{\hbar}\right)^l\frac{1}{l!}\right]^{f_1}.$$

(The $\prod_l$ factors come from those in the definition of Θ_l in (6.443).) Using (6.441), this number is seen to be

$$\frac{\left(\frac{i}{\hbar}\right)^n c(f_1,\ldots,f_k)}{d!\left(\prod_{l=1}^{k}(l!)^{f_1}\right)}, \tag{6.446}$$

where d is given by (6.440). The last thing we require is $c(f_1,\ldots,f_k)$, which is just the general coefficient in the multinomial expansion. It is easily shown by repeated use of the binomial expansion that

$$c(f_1,\ldots,f_k) = \frac{d!}{f_1!f_2!\cdots f_k!}. \tag{6.447}$$

Using this in (6.446) leads to a factor identical to that in (6.442). This establishes the linked cluster theorem, which states that $W[J;\varphi_*]$ is the sum of connected vacuum graphs.

Turn next to the effective action which was defined in terms of $W[J;\varphi_*]$ in (6.53) and (6.54). We saw that the vacuum diagrams making up $\Gamma[\bar{\varphi};\varphi_*]$ were all one-particle irreducible to second order in the loop expansion in the last section. This suggests that the effective action is comprised graphically of the subset of all of the connected vacuum diagrams making up $W[J;\varphi_*]$ that are one-particle irreducible. A very elegant proof of this assertion has been given by Burgess and Kunstatter (1987).

Go back to the definition of the effective action given in (6.61) before the relation $J_i = \frac{\delta\hat{\Gamma}}{\delta v^i}$ has been used:

$$\exp\frac{i}{\hbar}\hat{\Gamma}[v;\varphi_*] = \int d\mu[\varphi_*;\varphi]\exp\frac{i}{\hbar}\left\{S[\varphi] + J_i(v^i - \sigma^i)\right\}, \tag{6.448}$$

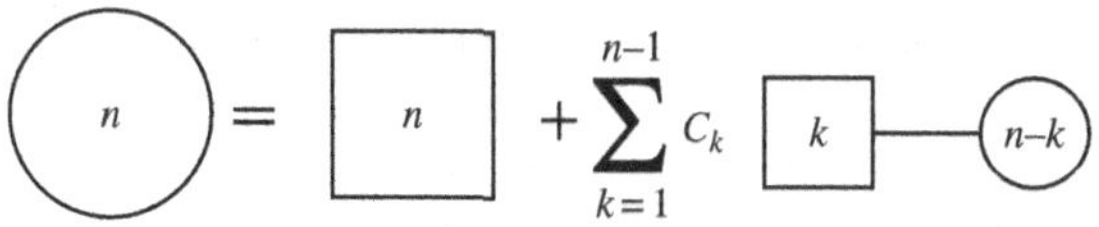

6.11 Formal properties of the effective action

Fig 6.4. The decomposition of vacuum graphs into one-particle reducible and irreducible subsets. The circle denotes the sum of all connected graphs, and the square denotes the sum of all one-particle irreducible graphs. The letter enclosed by the figure denotes the order of that term in the loop expansion. C_k is a numerical factor which counts how many times graphs of a particular type occur in the expansion.

where $v^i = \sigma^i[\varphi_*; \overline{\varphi}]$ and $\sigma^i = \sigma^i[\varphi_*; \varphi]$ as in (6.74) and (6.75). The action functional $S[\varphi]$ may be expanded in a Taylor series about $\sigma^i = v^i$ as before (see (6.78)):

$$S[\varphi] = \overline{S} + \frac{1}{2}\hat{S}_{,ij}\psi^i\psi^j + (\hat{S}_{,i} - J_i)\psi^i + \sum_{n=3}^{\infty} \frac{i}{n!}\hat{S}_{,i_1\cdots i_n}\psi^{i_1}\cdots\psi^{i_n}, \quad (6.449)$$

where $\psi^i = \sigma^i - v^i$. Because J_i is unrestricted, it follows immediately from the linked cluster theorem that $\hat{\Gamma}[v; \varphi_*]$ is the sum of all *connected* vacuum diagrams, where the propagator is given by $(\hat{S}_{,ij})^{-1}$ and the vertices follow from the linear, cubic, and higher-order terms in (6.449). (The only difference between $W[J; \varphi_*]$ and $\hat{\Gamma}[v; \varphi_*]$ with J_i unrestricted is that the meaning of the internal lines and vertices is different. The set of diagrams is the same.)

Consider the term of order $\hbar^n$ in the loop expansion of $\hat{\Gamma}[v; \varphi_*]$ with J_i unrestricted. We can separate all graphs of order $\hbar^n$ into a subset which is one-particle irreducible, and a subset which is not. The graphs which are one-particle reducible can only contain terms which involve one-particle irreducible parts up to order $\hbar^{n-1}$ joined by a single propagator onto connected parts. Graphically, the situation is as shown in Fig 6.4. The combinatorics would come into the calculation of the coefficients C_k, but as we will see, the elegance of the Burgess and Kunstatter proof is that we never need to do any combinatorics.

Now note that because the source term multiplies a term which is linear in ψ^i, it can only be connected by a single propagator to the rest of the diagram, and therefore the source can only occur in the one-particle reducible graphs. The next and last step in the argument is to force the source to obey $J_i = \delta\hat{\Gamma}/\delta v^i$. Differentiation of (6.120) with respect to J_i leads to

$$\frac{\delta\hat{\Gamma}}{\delta J_i} = \frac{\delta W}{\delta J_i} + v^i + J_j\frac{\delta v^j}{\delta J_i}. \quad (6.450)$$

Using (6.119) gives rise to

$$\langle\psi^i\rangle[J] = -\left(\frac{\delta\hat{\Gamma}}{\delta v^j} - J_j\right)\frac{\delta v^j}{\delta J_i}. \quad (6.451)$$

Therefore, forcing the source to obey $J_i = \delta\hat{\Gamma}/\delta v^i$ gives the result that

$$\langle \psi^i \rangle \left[J_i = \frac{\delta\hat{\Gamma}}{\delta v^i} \right] = 0. \qquad (6.452)$$

This must be true order by order in the loop expansion. But (6.452) is just the condition that all of the connected tadpole graphs vanish. (This is proven in Appendix 6A.2 at the end of this chapter.) This means that all of the terms in the summation indicated in Fig 6.4, which involve all of the one-particle reducible graphs, must vanish when the condition $J_i = \frac{\delta\hat{\Gamma}}{\delta v^i}$ is imposed. Only the one-particle irreducible graphs remain, and the proof is complete.

Appendix

6A.1 Covariant Taylor expansion

Because the covariant Taylor expansion plays an important role in the text, we will provide a derivation of it here. The covariant Taylor expansion was first considered by Ruse (1931) who utilized Riemann normal coordinates. The derivation presented here is a simple generalization of one given by Barvinsky and Vilkovisky (1985) and outlined in Toms (1988).

Let $F[\varphi; \varphi']$ be any functional which transforms like a scalar at φ'. Its transformation property at φ is completely arbitrary. Let $\varphi^i(s)$ denote the geodesic which connects φ' to φ with $\varphi'^i = \varphi^i(s=0)$ and $\varphi^i = \varphi^i(s=\tau)$. τ is the length of the geodesic. By definition,

$$\sigma[\varphi; \varphi'] = \frac{1}{2}\tau^2. \qquad (6A.1)$$

(This quantity was first introduced by Ruse (1931).) A basic property is[9]

$$\sigma^i[\varphi; \varphi']\sigma_i[\varphi; \varphi'] = 2\sigma[\varphi; \varphi'], \qquad (6A.2)$$

where

$$\sigma_i[\varphi; \varphi'] = \frac{\delta}{\delta\varphi^i}\sigma[\varphi; \varphi']. \qquad (6A.3)$$

This relation may be easily proven with the aid of Riemann normal coordinates. (A more elegant derivation may be found in DeWitt (1964a).) Because $g_{ij}[\varphi(s)]\dot{\varphi}^i(s)\dot{\varphi}^j(s)$ is conserved along the geodesic, it follows from (6A.2) that

$$\sigma^i[\varphi; \varphi'] = \tau\dot{\varphi}^i(s=\tau). \qquad (6A.4)$$

This gives $\sigma^i[\varphi; \varphi']$ the interpretation of the tangent vector to the geodesic at the point φ.

---

[9] See Section 6.10.

By a straightforward Taylor expansion, we have

$$F[\varphi; \varphi'] = \sum_{n=0}^{\infty} \frac{(-1)^n}{n!} \frac{d^n F[\varphi; \varphi(s)]}{ds^n}\bigg|_{s=\tau}. \tag{6A.5}$$

Now,

$$\frac{d}{ds} F[\varphi; \varphi(s)] = F_{,i'}[\varphi; \varphi(s)]\dot{\varphi}^i(s) \tag{6A.6}$$

(recalling our convention that the prime on the index "i" denotes a derivative with respect to the second argument of the two-point functional). A subsequent differentiation of (6A.6) gives

$$\frac{d^2}{ds^2} F[\varphi; \varphi(s)] = F_{,i'j'}[\varphi; \varphi(s)]\dot{\varphi}^i(s)\dot{\varphi}^j(s) + F_{,i'}[\varphi; \varphi(s)]\ddot{\varphi}^i(s)$$

$$= \left\{ F_{,i'j'}[\varphi; \varphi(s)] - \Gamma^k_{ij}[\varphi(s)] F_{,k'}[\varphi; \varphi(s)] \right\} \dot{\varphi}^i(s)\dot{\varphi}^j(s)$$

if the geodesic equation $\ddot{\varphi}^k(s) = -\Gamma^k_{ij}[\varphi(s)]\dot{\varphi}^i(s)\dot{\varphi}^j(s)$ is used. This shows, by the definition of covariant differentiation, that

$$\frac{d^2}{ds^2} F[\varphi; \varphi(s)] = F_{;i'j'}[\varphi; \varphi(s)]\dot{\varphi}^i(s)\dot{\varphi}^j(s). \tag{6A.7}$$

It is now simple to prove by induction that

$$\frac{d^n}{ds^n} F[\varphi; \varphi(s)] = F_{;i'_1 \cdots i'_n}[\varphi; \varphi(s)]\dot{\varphi}^{i_1}(s) \cdots \dot{\varphi}^{i_n}(s). \tag{6A.8}$$

Note that only the symmetrized covariant derivatives appear. Evaluating (6A.8) at $s = \tau$, and using (6A.4), leads to the final result

$$F[\varphi; \varphi'] = \sum_{n=0}^{\infty} \frac{(-1)^n}{n!} \left(F_{(;i_1' \cdots i_n')}[\varphi; \varphi'] \right)\bigg|_{\varphi'=\varphi} \sigma^{i_1}[\varphi; \varphi'] \cdots \sigma^{i_n}[\varphi; \varphi']. \tag{6A.9}$$

6A.2 $\langle \sigma^i \rangle[J] = 0$ implies the sum of all connected tadpole graphs vanishes

The only J-dependence in $Z[J; \varphi_*]$ comes about through $\tilde{Z}[J; \varphi_*]$ defined in (6.436). It is easy to see from this definition that

$$\frac{\hbar}{i} \frac{\delta}{\delta J_i} \tilde{Z}[J; \varphi_*] = -\sum_{n=1}^{\infty} \left(\frac{i}{\hbar}\right)^n \frac{1}{n!} \left\langle \sigma^i \left(S^J_{\text{int}}\right)^n \right\rangle_G, \tag{6A.10}$$

where the $n = 0$ term has been dropped because $\langle \sigma^i \rangle_G = 0$. The graphical interpretation of the result in (6A.10) is that of the one-point function, where the factor of σ^i gives rise to the external line.

The graphs comprising $\left\langle \sigma^i \left(S^J_{\text{int}}\right)^n \right\rangle_G$ may be divided into two sets, depending upon whether they are connected or not. The connected ones may easily seen to be obtained from the connected vacuum diagrams by attaching the external line, corresponding to the factor σ^i, in all possible ways consistent with the interactions. The same feature does not hold for the disconnected graphs however, because there are graphs which involve the external line connected to the one-point vertex which can never be obtained from attaching the external line onto a vacuum bubble. (The difference for connected graphs arises simply because connected graphs must have the external line paired off with cubic- and higher-order vertices, and not the linear vertex in general.)

Consider the subset of connected graphs in the set of all graphs comprising $\langle \sigma^i (S_{\text{int}}^J)^n \rangle_G$. They can be first, second,...,n^{th} order in S_{int}^J. Let $\langle \sigma^i (S_{\text{int}}^J)^k \rangle_C$ denote the set of all connected graphs of order k. We may then write

$$\langle \sigma^i (S_{\text{int}}^J)^n \rangle_G = \sum_{k=1}^{n} c_k \langle \sigma^i (S_{\text{int}}^J)^k \rangle_C \langle (S_{\text{int}}^J)^{n-k} \rangle_G \quad (6A.11)$$

for some coefficients c_k. The c_k are just the number of ways of picking out k vertices from n, and so $c_k = \binom{n}{k}$. The remaining $(n-k)$ vertices can only pair off with each other in all possible ways, and therefore give rise to vacuum diagrams of order $(n-k)$ (both connected and disconnected). We therefore have

$$-\frac{\hbar}{i} \frac{\delta}{\delta J_i} \tilde{Z}[J; \varphi_*] = \sum_{n=1}^{\infty} \sum_{k=1}^{n} \left(\frac{i}{\hbar}\right)^n \frac{1}{k!(n-k)!} \langle \sigma^i (S_{\text{int}}^J)^k \rangle_C \langle (S_{\text{int}}^J)^{n-k} \rangle_G. \quad (6A.12)$$

The summations may be simplified by using the easily proven identity

$$\sum_{n=1}^{\infty} \sum_{k=1}^{n} c_k d_{n-k} = \left(\sum_{k=1}^{\infty} c_k\right) \left(\sum_{n=0}^{\infty} d_n\right). \quad (6A.13)$$

Using (6A.13) in (6A.12) gives

$$-\frac{\hbar}{i} \frac{\delta}{\delta J_i} \tilde{Z}[J; \varphi_*] = \left(\sum_{k=1}^{\infty} \left(\frac{i}{\hbar}\right)^k \frac{1}{k!} \langle \sigma^i (S_{\text{int}}^J)^k \rangle_C\right)$$
$$\times \left(\sum_{n=0}^{\infty} \left(\frac{i}{\hbar}\right)^n \frac{1}{n!} \langle (S_{\text{int}}^J)^n \rangle_G\right). \quad (6A.14)$$

The second factor on the right-hand side of (6A.14) is seen to be $\tilde{Z}[J; \varphi_*]$; hence,

$$-\frac{\hbar}{i} \frac{\delta}{\delta J_i} \ln \tilde{Z}[J; \varphi_*] = \sum_{k=1}^{\infty} \left(\frac{i}{\hbar}\right)^k \frac{1}{k!} \langle \sigma^i (S_{\text{int}}^J)^k \rangle_C$$
$$= \left\langle \sigma^i \exp\left\{\frac{i}{\hbar} S_{\text{int}}^J\right\} \right\rangle_C. \quad (6A.15)$$

But from (6.430) it then follows that

$$\frac{\delta}{\delta J_i} W[J; \varphi_*] = -\left\langle \sigma^i \exp\left\{\frac{i}{\hbar} S_{\text{int}}^J\right\} \right\rangle_C. \quad (6A.16)$$

is the connected one-point function. From the functional integral representation for $W[J;\varphi_\star]$, it is then seen that

$$\frac{\delta}{\delta J_i} W[J;\varphi_\star] = -\langle \sigma^i \rangle [J]. \qquad (6A.17)$$

Hence, the vanishing of $\langle \sigma^i \rangle_J$ is equivalent to the requirement that the sum of all of the connected graphs contributing to the one-point function vanish. The same proof holds if we replace σ^i by $\sigma^i - v^i = \psi^i$ as in Section 6.11.

7

The effective action: Gauge theories

7.1 Introduction

The first place where the concept of gauge invariance arises is in the study of electromagnetism. In electrostatics, the Coulomb potential Φ is related to the electric field **E** by $\mathbf{E} = -\nabla\Phi$. For a given electric field, this only determines the Coulomb potential up to an additive constant. This is an example of gauge invariance, but is rather uninteresting. A less trivial invariance comes about when magnetism is considered. Here a vector potential **A** may be introduced, and the magnetic field can be recovered from $\mathbf{B} = \nabla \times \mathbf{A}$. From vector calculus we know that if $\theta(x)$ is an arbitrary function whose gradient is $\nabla\theta$, it follows that $\nabla \times (\nabla\theta) = 0$. As a consequence, both **A** and $\mathbf{A} + \nabla\theta$ will give rise to the same magnetic field. The first introduction of the vector potential appears to have been by Neumann. It was used by others as a tool to simplify the solution to magnetic field problems. (Some of the history and early references may be found in Whittaker (1951).)

The fact that $\mathbf{B} = \nabla \times \mathbf{A}$ only determines **A** up to an additive term involving the gradient of an arbitrary scalar appears in the work of Maxwell (1954). In his book, Maxwell also noted that the condition $\nabla \cdot \mathbf{A} = 0$ could be applied. This is an example of a gauge condition, whose role is to reduce the number of degrees of freedom which are present in the potentials. The generalization of this to include the Coulomb potential ϕ was first given by Lorenz (1867) and reads

$$\nabla \cdot \mathbf{A} + \frac{1}{c}\frac{\partial \phi}{\partial t} = 0. \tag{7.1}$$

This condition is now usually referred to as the Lorentz condition, although it should probably be called the Lorenz condition. It has the effect of uncoupling the wave equations for the scalar and vector potential.

Although the basic concept of gauge invariance dates back at least to Maxwell, its use as a basic symmetry of nature is of a more recent vintage. It is now fairly well known that Weyl was the first person to adopt gauge invariance as a foundation for a physical theory in an attempt to give a unified treatment of electromagnetism and gravitation. Although Weyl's original theory proved untenable as a unified field theory, the use of gauge symmetry survived. The adoption of gauge invariance, in the form of what today would be called a local phase transformation, to write down wave equations was due to Fock, London, and Gordon. Weyl noted that there was an intimate connection between local gauge invariance and conservation of charge. (See the book of Pais (1986) for a readable description of Weyl's original theory, and for the original references. Some of the early work is also discussed in Pauli (1980).)

Quantization of electromagnetism in a form which was relativistically covariant was first undertaken by Pauli and Heisenberg (1929). This very important paper also laid the groundwork for the canonical approach to relativistic field theory. The problem with the canonical approach, which is linked to the local gauge invariance, is that A^0 (the time component of the electromagnetic four-potential) is not a dynamical field because there are no terms involving $\dot{A}^0$ in the field equations. (This is evident if we write out the Lagrangian density for electromagnetism given in (7.2) below, directly in terms of the components A^μ.) This means that the momentum π_0, which is canonically conjugate to A^0, vanishes identically, and the usual canonical commutation relations cannot be written down. Pauli and Heisenberg got around this problem by adopting the Coulomb gauge $A^0 = 0$, $\nabla \cdot \mathbf{A} = 0$. This loses manifest Lorentz invariance, and it is necessary to prove that it is there nonetheless, which is what Pauli and Heisenberg did. (This is basically the approach adopted in Bjorken and Drell (1965).)

An alternative procedure, which does not lose manifest Lorentz invariance, and which is similar to the method widely used today, was initiated by Fermi. (A detailed discussion of this approach is contained in Section 16 of the early quantum field theory textbook of Wentzel (1949). A similar procedure was apparently found by Heisenberg. (See Pais 1986.)) This method consists of adding on a term $-\frac{1}{2}(\partial_\mu A^\mu)^2$ to the Lagrangian density for electromagnetism. Such terms are usually called gauge-fixing terms today. Since the choice of gauge-fixing term is arbitrary, although the formalism is manifestly Lorentz invariant, it is now necessary to prove that the results do not depend upon the arbitrary choice of gauge-fixing term. A modern approach to this problem will be dealt with in the present chapter.

The local gauge transformation of electromagnetism may be parametrized by a single, real scalar function $\theta(x)$. Associated with this function is an element $\exp(i\theta(x))$ of the Abelian group U(1). U(1) is said to be the gauge group of electromagnetism. The generalization of the gauge group from U(1) to the non-Abelian group SU(2) was first given by Yang and Mills (1954). The generalization to an arbitrary non-Abelian gauge group was first given by Utiyama (1956). Non-Abelian gauge theories are often referred to as Yang–Mills theories. The application of gauge theories to provide realistic models of particle physics is well described in many review articles (e.g., Abers and Lee 1973; Bernstein

1974; Weinberg 1974) and in many textbooks (e.g., Taylor 1976; Itzykson and Zuber 1980; Quigg 1983; Cheng and Li 1984). An annotated guide to some of the literature is given by Cheng and Li (1988). Our concern in this book is more with the basics of quantum field theory than its application to construct specific models.

The essential difference between non-Abelian gauge theories and electromagnetism is that in the former case the field equations are non-linear, whereas Maxwell's equations of electromagnetism are linear. This leads to tremendous complications when the quantization of non-Abelian gauge theories is considered. The key idea to the successful quantization was supplied by Feynman (1963). He suggested introducing fictitious particles with opposite statistics to the gauge fields to cancel the unphysical degrees of freedom, and showed how this idea could be implemented for one-loop diagrams. The extension to any order was given by DeWitt (1964b).

The prevalent method for the quantization of gauge theories is the functional integral approach. This was first done by DeWitt (1967) and by Faddeev and Popov (1967). The basic problem with straightforward transcription of the results obtained in Chapter 6 for non-gauge theories is that if all gauge fields are integrated over then we will have counted fields which are equivalent under gauge transformations more than once. This overcounting results in an infinite value for the functional integral. What is really required is that one member is picked out from the set of all fields which are related to it under gauge transformations, with only gauge-inequivalent fields then integrated over. A clever way of implementing this was given in Faddeev and Popov (1967), and is sometimes referred to as the Faddeev–Popov ansatz. A more geometrical way of dealing with this problem will be described in this chapter.

7.2 Gauge transformations

The classical action functional which describes electromagnetism is

$$S[A] = -\frac{1}{4} \int dv_x F_{\mu\nu} F^{\mu\nu}, \qquad (7.2)$$

where

$$F_{\mu\nu} = \partial_\mu A_\nu - \partial_\nu A_\mu \qquad (7.3)$$

defines the field strength in terms of the vector four-potential $A_\mu(x)$. (Recall that $dv_x = |g(x)|^{1/2} d^n x$ is the invariant spacetime volume element.) Because the field strength defined in (7.3) is invariant under the gauge transformation

$$A_\mu^\epsilon(x) = A_\mu(x) + \partial_\mu \epsilon(x) \qquad (7.4)$$

7.2 Gauge transformations

for arbitrary function $\epsilon(x)$, the action for electromagnetism defined in (7.2) clearly obeys

$$S[A^\epsilon] = S[A], \qquad (7.5)$$

and therefore is a gauge-invariant functional. By gauge-invariant, we mean that there is no dependence at all on the arbitrary function $\epsilon(x)$ which parameterizes the gauge transformation. The action remains gauge-invariant even with the addition of a source term $\int dv_x J^\mu(x) A_\mu(x)$, provided that the source is conserved (meaning $\nabla_\mu J^\mu = 0$), as can be seen by performing a simple integration by parts.

As we mentioned in Section 7.1, the function $\epsilon(x)$ occurring in the gauge transformation of (7.4) may be thought of as parameterizing the Abelian group $U(1)$, which is called the gauge group of electromagnetism. It is possible to generalize the invariance from one based on the Abelian group $U(1)$ to any compact Lie group G. The restriction to compact Lie groups is necessary in order that the kinetic terms in the action all have the same sign. (If this is not the case, then the action for the Euclidean version of the theory is not bounded from below which leads to convergence problems with the functional integral.) Associated with any compact Lie group G, is a set of Hermitian generators $\{T_a\}$ which forms a basis for the Lie algebra $L(G)$ defined by the commutation relations

$$[T_a, T_b] = i f_{ab}{}^c T_c. \qquad (7.6)$$

$f_{ab}{}^c$ are called the structure constants.[1] The Lie algebra generators must also satisfy the Jacobi relation

$$[[T_a, T_b], T_c] + [[T_b, T_c], T_a] + [[T_c, T_a], T_b] = 0. \qquad (7.7)$$

By using (7.6) in (7.7) along with the linear independence of the generators, the structure constants may be shown to satisfy

$$f_{ab}{}^d f_{dc}{}^e + f_{bc}{}^d f_{da}{}^e + f_{ca}{}^d f_{db}{}^e = 0. \qquad (7.8)$$

The field strength $F^a_{\mu\nu}$ is defined in terms of the non-Abelian gauge field A^a_μ by

$$F^a_{\mu\nu} = \partial_\mu A^a_\nu - \partial_\nu A^a_\mu + g f_{bc}{}^a A^b_\mu A^c_\nu. \qquad (7.9)$$

g is called the gauge coupling constant and is the analogue of the electric charge in electromagnetism. Unlike the $U(1)$ case considered above, $F^a_{\mu\nu}$ is not invariant under the infinitesimal non-Abelian gauge transformation defined by

$$\delta A^a_\mu = f_{bc}{}^a \delta\epsilon^b A^c_\mu - \frac{1}{g} \partial_\mu \delta\epsilon^a, \qquad (7.10)$$

but rather is covariant, transforming as

$$\delta F^a_{\mu\nu} = f_{bc}{}^a \delta\epsilon^b F^c_{\mu\nu}. \qquad (7.11)$$

---

[1] For readers who are unfamiliar with the basic elements of Lie groups and Lie algebras, and their use in gauge field theories, the books of O'Raifeartaigh (1986) and Burgess (2002) are recommended.

Here $\delta\epsilon^a(x)$ is a set of infinitesimal group parameters characterizing the transformation. The Yang–Mills action may be chosen to be

$$S_{YM}[A] = -\frac{1}{4}\int dv_x \delta_{ab} F^a_{\mu\nu} F^{b\,\mu\nu}, \qquad (7.12)$$

and may be seen to be invariant under the gauge transformation (7.10).

As a final example, consider the Einstein–Hilbert action for gravity defined by

$$S_G[g] = (16\pi G)^{-1}\int dv_x\,(R - 2\Lambda). \qquad (7.13)$$

This action is invariant under a general change of coordinates, which in its infinitesimal form may be taken as $x^\mu \to x^\mu_\epsilon$ where

$$x^\mu_\epsilon = x^\mu + \delta\epsilon^\mu(x). \qquad (7.14)$$

If the metric tensor with components $g_{\mu\nu}(x)$ is chosen as the basic field variable, then under the transformation of (7.14), we have $g_{\mu\nu} \to g^\epsilon_{\mu\nu}$ where[2]

$$g^\epsilon_{\mu\nu} = g_{\mu\nu} - \delta\epsilon^\lambda g_{\mu\nu,\lambda} - g_{\mu\lambda}\delta\epsilon^\lambda{}_{,\nu} - g_{\lambda\nu}\delta\epsilon^\lambda{}_{,\mu}. \qquad (7.15)$$

This example, in common with that of non-Abelian gauge theories (7.10), involves a transformation for which the change in the field (in this case $\delta g_{\mu\nu} = g^\epsilon_{\mu\nu} - g_{\mu\nu}$) involves the field itself. This is unlike the simpler case of electromagnetism.

In order to avoid looking at specific examples on a case-by-case basis, it is again advantageous to adopt the condensed notation φ^i for the fields, as we did in Chapter 6. Suppose that we are given a classical action functional $S[\varphi]$ for these fields invariant under infinitesimal transformations of the form

$$\delta\varphi^i = \varphi^i_\epsilon - \varphi^i = K^i_\alpha[\varphi]\delta\epsilon^\alpha. \qquad (7.16)$$

Here $\delta\epsilon^\alpha$ represents a set of parameters characterizing the transformation, and $K^i_\alpha[\varphi]$ may be regarded as the components of a vector $\mathbf{K}_\alpha[\varphi]$ which generates the transformation. In a coordinate basis,

$$\mathbf{K}_\alpha[\varphi] = K^i_\alpha[\varphi]\frac{\delta}{\delta\varphi^i}. \qquad (7.17)$$

The transformation is sufficiently general to include both rigid ($\delta\epsilon^\alpha$ constant) and local ($\delta\epsilon^\alpha$ functions of the spacetime coordinates) gauge transformations. In the latter case, as is evident from the special examples of Yang–Mills theory and gravity considered above, K^i_α may be a differential operator. K^i_α may also depend on the fields, as the Yang–Mills and gravity examples illustrate. We will write out the explicit form for K^i_α for each of these two examples at the end of the present section.

---

[2] This result is obtained from the tensor transformation law for a second-rank covariant tensor $g'_{\mu\nu} = \frac{\partial x}{\partial x'^\mu}\frac{\partial x}{\partial x'^\nu}g_{\alpha\beta}$, using (7.14), and expanding the result to first order in $\delta\epsilon^\mu$.

7.2 Gauge transformations

If the action functional $S[\varphi]$ is to be invariant under the transformation of (7.16), then

$$S[\varphi_\epsilon] = S[\varphi]. \tag{7.18}$$

Using Taylor's theorem on the left-hand side, and expanding to first order in $\delta\epsilon^\alpha$, gives the following relation that the classical action must satisfy:

$$S_{,i}[\varphi]K^i_\alpha[\varphi]\delta\epsilon^\alpha = 0.$$

If this is to hold for arbitrary parameters $\delta\epsilon^\alpha$, then

$$S_{,i}[\varphi]K^i_\alpha[\varphi] = 0. \tag{7.19}$$

Equivalently, using (7.17), this may be written as

$$\mathbf{K}_\alpha[\varphi]S[\varphi] = 0. \tag{7.20}$$

Conversely, given a vector $\mathbf{K}_\alpha[\varphi]$ which obeys (7.20), there always corresponds a transformation (7.16) which is a symmetry of the classical action. Accordingly, in what follows, assume that $\{\mathbf{K}_\alpha\}$ is the set of all vectors that obey (7.20), and that (7.16) represents the complete set of symmetries of the classical action.

Because (7.19) must hold for all φ, we can put $\varphi = \varphi_\epsilon$ and obtain

$$\frac{\delta S[\varphi_\epsilon]}{\delta \varphi_\epsilon^i} K^i_\alpha[\varphi_\epsilon] = 0. \tag{7.21}$$

Because the action is independent of $\delta\epsilon^\alpha$, as expressed by (7.18),

$$\frac{\delta S[\varphi_\epsilon]}{\delta \varphi_\epsilon^i} = \frac{\delta S[\varphi]}{\delta \varphi_\epsilon^i}$$

$$= \frac{\delta S[\varphi]}{\delta \varphi^j} \frac{\delta \varphi^j}{\delta \varphi_\epsilon^i}$$

$$= S_{,j}[\varphi]\left\{\delta^j{}_i - K^j_{\beta,i}[\varphi]\delta\epsilon^\beta\right\}, \tag{7.22}$$

where the last line has resulted from a first-order expansion in $\delta\epsilon$ using (7.16). In a similar way, expanding $K^i_\alpha[\varphi_\epsilon]$ to first order in $\delta\epsilon$, and using the resulting expression along with (7.22) in (7.21), leads to

$$S_{,k}[\varphi]\left(K^j_\beta[\varphi]K^k_{\alpha,j}[\varphi] - K^j_\alpha[\varphi]K^k_{\beta,j}[\varphi]\right) = 0. \tag{7.23}$$

Noting that

$$\left[\mathbf{K}_\alpha[\varphi], \mathbf{K}_\beta[\varphi]\right] = \left(K^j_\alpha[\varphi]K^k_{\beta,j}[\varphi] - K^j_\beta[\varphi]K^k_{\alpha,j}\right)\frac{\delta}{\delta\varphi^k}, \tag{7.24}$$

we may write (7.23) as

$$\left[\mathbf{K}_\alpha[\varphi], \mathbf{K}_\beta[\varphi]\right]S[\varphi] = 0. \tag{7.25}$$

We will assume in the following that[3]

$$\left[\mathbf{K}_\alpha[\varphi], \mathbf{K}_\beta[\varphi]\right] = -f_{\alpha\beta}{}^\gamma[\varphi]\mathbf{K}_\gamma[\varphi] \qquad (7.26)$$

for some functions $f_{\alpha\beta}{}^\gamma[\varphi]$ called the structure functions. This clearly obeys (7.25) by virtue of (7.20). In terms of the components, we may write (7.26) as

$$K_\alpha^j[\varphi]K_{\beta,j}^k[\varphi] - K_\beta^j[\varphi]K_{\alpha,j}^k[\varphi] = -f_{\alpha\beta}{}^\gamma[\varphi]K_\gamma^k[\varphi]. \qquad (7.27)$$

Because the commutator bracket is antisymmetric under an interchange of indices, the structure functions $f_{\alpha\beta}{}^\gamma[\varphi]$ must obey

$$f_{\beta\alpha}{}^\gamma[\varphi] = -f_{\alpha\beta}{}^\gamma[\varphi]. \qquad (7.28)$$

One final requirement that will be imposed is that the bracket must obey the Jacobi identity

$$\left[[\mathbf{K}_\alpha, \mathbf{K}_\beta], \mathbf{K}_\gamma\right] + \left[[\mathbf{K}_\beta, \mathbf{K}_\gamma], \mathbf{K}_\alpha\right] + \left[[\mathbf{K}_\gamma, \mathbf{K}_\alpha], \mathbf{K}_\beta\right] = 0. \qquad (7.29)$$

This means that the bracket has all of the properties required of a Lie bracket.

The Lie bracket written down in (7.26) is not the general solution to (7.25). It is easily seen that if we add on any term of the form $g_{\alpha\beta}^{ij}[\varphi]S_{,j}[\varphi]$, where $g_{\alpha\beta}^{ij}$ is antisymmetric in i and j as well as antisymmetric in α and β, we still satisfy (7.25). The case where $g_{\alpha\beta}^{ij} \neq 0$ is referred to as an open algebra. The three examples described at the beginning of this section (electromagnetism, Yang–Mills theory, and gravity) all result in closed algebras ($g_{\alpha\beta}^{ij} = 0$), as we will verify at the end of this section. It has been shown by Batalin and Vilkovisky (1981) that it is always possible to redefine the generators of the transformation so that a closed algebra is obtained. A related point is that any theory described by a classical action $S[\varphi]$ always has an invariance of the form given in (7.16). The explicit result for K_α^i is

$$K_\alpha^i[\varphi] = A_\alpha^{ij}[\varphi]S_{,j}[\varphi], \qquad (7.30)$$

where $A_\alpha^{ij} = -A_\alpha^{ji}$. The condition for invariance expressed in (7.19) is trivially satisfied in this case. Of course this does not mean that every theory is a gauge theory. One of the features which distinguishes the transformation (7.30) from a gauge transformation is that it will not be a symmetry of the classical field equations in general. Symmetries of the classical field equations result in the differential operator supposed to represent the inverse propagator having zero modes, and therefore being non-invertible. This is why special methods must be used for gauge theories.

A final point that merits some discussion concerns the structure functions $f_{\alpha\beta}{}^\gamma[\varphi]$ in (7.26). If the structure functions are independent of the fields, then

---

[3] The minus sign on the right-hand side is pure convention, and relates to the fact that we are choosing a left action of the symmetry transformation rather than a right action. The structure constants of a Lie algebra are conventionally defined from the right action.

they may be interpreted as the structure constants of a Lie algebra. The set of symmetry transformations then forms a Lie group $\mathfrak{G}$, which may be regarded as acting on the space of fields $\mathcal{F}$ in a way defined by (7.16). In the more general case where the structure functions $f_{\alpha\beta}{}^{\gamma}$ are not constant, the set of symmetry transformations does not form a group. In this case, the standard Faddeev and Popov (1967) prescription for gauge theories fails, and other means must be resorted to. (See de Wit and van Holten (1978); Batalin (1981) or Batalin and Vilkovisky (1981, 1984) for example.)

In what follows, we will assume that the transformation (7.16) is a symmetry of the classical field equations as well as of the action. We will also assume that the algebra (7.26) associated with the transformation is closed, and that the structure functions are constant. This is the case for all examples which will be considered in this book, but the reader should be aware of more general possibilities. If these conditions are met, the theory may be interpreted using the language of fibre bundles. (See Steenrod (1951) or Husemoller (1966) for a mathematical description of fibre bundles. A readable treatment for physicists is contained in Eguchi et al. 1980. Application to Yang–Mills theory has been well described by Daniel and Viallet (1980).) We will not pursue this here, preferring a more pedestrian approach.

Before proceeding with the basic formalism, we will illustrate in the next two subsections what we have discussed so far for the two examples of Yang–Mills theory and gravity. Maxwell's theory of electromagnetism can be obtained as a special case of Yang–Mills theory by taking the gauge group to be $U(1)$.

7.2.1 Yang–Mills theory

For Yang–Mills theory (non-Abelian gauge field) the condensed notation φ^i stands for $A_\mu^a(x)$; thus, the condensed index i is shorthand for the set (a, μ, x). A repeated condensed index i in our general formal expressions stands for a summation over the group index a, a summation over the spacetime vector index μ, and an integration over the spacetime point whose coordinates are x. The condensed notation for the infinitesimal group parameters $\delta\epsilon^\alpha$ represents $\delta\epsilon^a(x)$ in conventional notation, so that the condensed index α refers to (a, x), with a repeated condensed Greek index shorthand for a summation over the group index a and an integration over the spacetime point x.

If we look at (7.16), defining the infinitesimal gauge transformation in condensed notation, then K^i_α involves two group indices (one from i and one from α), one spacetime vector index (from i), and two spacetime points (one from i and one from α). We will associate $i = (a, \mu, x)$ and $\alpha = (b, x')$, so that we have the association of K^i_α in condensed notation with $K^{a\mu}{}_b(x, x')$ in conventional notation. With this association, (7.16) reads

$$\delta A^{a\mu}(x) = \int d^n x' K^{a\mu}{}_b(x, x') \delta\epsilon^b(x'), \qquad (7.31)$$

when uncondensed. Comparison with (7.10) allows us to conclude that

$$K^{a\mu}{}_b(x,x') = f_{bc}{}^a A^{c\mu}(x)\tilde{\delta}(x,x') - \frac{1}{g}g^{\mu\nu}(x)\tilde{\delta}_{,\nu}(x,x')\delta^a_b, \quad (7.32)$$

for the Yang–Mills field. (Here $\tilde{\delta}_{,\nu}(x,x')$ means that the derivative is with respect to the first argument of the Dirac δ-distribution. $\tilde{\delta}(x,x')$ was defined in terms of the biscalar Dirac δ-distribution in (6.8).)

We can now evaluate the structure functions defined in (7.27) by using (7.32) to evaluate the left-hand side, and then reading off $f_{\alpha\beta}{}^\gamma$ by comparison with the right-hand side. We have already mentioned the association of α with a group index and a spacetime point, so if we make the associations $\alpha = (a,x), \beta = (b,x')$, and $\gamma = (c,x'')$ we have $f_{\alpha\beta}{}^\gamma$ in condensed notation standing for the three-point function $f_{ab}{}^c(x,x',x'')$ in conventional notation. We will make the associations $k = (d,\mu,x''')$ and $j = (e,\nu,\bar{x})$ in (7.27), so that $K^k_{\beta,j}$ in condensed notation corresponds to

$$K^k_{\beta,j} = \frac{\delta}{\delta A^{e\nu}(\bar{x})} K^{d\mu}{}_b(x''',x') \quad (7.33)$$

in conventional notation. Using the basic result

$$\frac{\delta A^{f\mu}(x''')}{\delta A^{e\nu}(\bar{x})} = \delta^f_e \delta^\mu_\nu \tilde{\delta}(x''',\bar{x}),$$

it can be seen from (7.32) that

$$K^k_{\beta,j} = f_{be}{}^d \delta^\mu_\nu \tilde{\delta}(x''',\bar{x})\tilde{\delta}(x''',x'), \quad (7.34)$$

gives the correspondence with normal notation. We then find

$$K^j_\alpha K^k_{\beta,j} = \int d^n\bar{x} K^{e\nu}{}_a(\bar{x},x) f_{be}{}^d \delta^\mu_\nu \tilde{\delta}(x''',\bar{x})\tilde{\delta}(x''',x')$$
$$= K^{e\mu}{}_a(x''',x) f_{be}{}^d \tilde{\delta}(x''',x'), \quad (7.35)$$

since the repeated condensed index j involves an integration over its corresponding continuous label $\bar{x}$. By making use of (7.32) along with the Jacobi relation (7.8), it follows that

$$K^j_\alpha K^k_{\beta,j} - K^j_\beta K^k_{\alpha,j} = -f_{ab}{}^e f_{ef}{}^d A^{f\mu}(x''')\tilde{\delta}(x''',x)\tilde{\delta}(x''',x')$$
$$+ \frac{1}{g} f_{ab}{}^d \partial^{\mu'''} \left(\tilde{\delta}(x''',x)\tilde{\delta}(x''',x')\right). \quad (7.36)$$

We now want to compare this with the right-hand side of (7.27). If we write out the right-hand side of (7.27) in conventional notation we find

$$-f_{\alpha\beta}{}^\gamma K^k_\gamma = -\int d^n x'' f_{ab}{}^c(x,x',x'') K^{d\mu}{}_c(x''',x'')$$
$$= -f_{ab}{}^c(x,x',x''') f_{cf}{}^d A^{f\mu}(x''')$$
$$+ \frac{1}{g}\partial^{\mu'''} f_{ab}{}^d(x,x',x'''), \quad (7.37)$$

using (7.32) and performing the integration over x'' using the Dirac δ-distribution. Comparison of this result with (7.36) shows that the structure functions are given by

$$f_{\alpha\beta}{}^\gamma = f_{ab}{}^c(x, x', x'') = f_{ab}{}^c \tilde{\delta}(x'', x) \tilde{\delta}(x'', x'). \tag{7.38}$$

The structure functions are therefore determined by the structure constants of the Yang–Mills gauge group (that multiply the Dirac δ-distributions) and do not have any dependence on the fields themselves. Our requirements for a closed algebra and field-independent structure functions are met in Yang–Mills theory.

7.2.2 Gravity

In the case of gravity, the condensed notation for the field φ^i is shorthand for the spacetime metric $g_{\mu\nu}(x)$. (We could equally well choose $g^{\mu\nu}(x)$, or a metric density if we like.) The condensed index i therefore stands for a pair of spacetime indices as well as a spacetime point. The infinitesimal parameters of the transformation in condensed notation $\delta\epsilon^\alpha$ correspond to $\delta\epsilon^\lambda(x)$ in this example, so we have the correspondence of α with a spacetime index and a spacetime point here. For gravity, (7.16) reads

$$\delta g_{\mu\nu}(x) = \int d^n x' K_{\mu\nu\lambda}(x, x') \delta\epsilon^\lambda(x'). \tag{7.39}$$

Comparison with (7.15) shows that

$$K_{\mu\nu\lambda}(x, x') = -[g_{\mu\nu,\lambda}(x) + g_{\mu\lambda}(x)\partial_\nu + g_{\nu\lambda}(x)\partial_\mu]\tilde{\delta}(x, x'). \tag{7.40}$$

As in the Yang–Mills case, we can evaluate the structure functions by working out the left-hand side of (7.27). We will need the following result for the functional derivative of the metric:

$$\frac{\delta g_{\mu\nu}(x)}{\delta g_{\rho\sigma}(x')} = \delta^\rho_{(\mu}\delta^\sigma_{\nu)}\tilde{\delta}(x, x'), \tag{7.41}$$

where we have abbreviated

$$\delta^\rho_{(\mu}\delta^\sigma_{\nu)} = \frac{1}{2}\left(\delta^\rho_\mu\delta^\sigma_\nu + \delta^\rho_\nu\delta^\sigma_\mu\right). \tag{7.42}$$

The symmetrization over the indices in (7.41) follows from the symmetry of the metric tensor, and (7.41) is a consequence of the obvious identity

$$\delta g_{\mu\nu}(x) = \int d^n x' \delta^\rho_{(\mu}\delta^\sigma_{\nu)}\tilde{\delta}(x, x')\delta g_{\rho\sigma}(x'),$$

and the definition of the functional derivative in (6.11).

Before continuing with our calculation of the structure functions for gravity, we note the following result

$$\frac{\delta}{\delta g_{\mu\nu}(\bar{x})}\tilde{\delta}(x,x') = 0. \tag{7.43}$$

The easiest way to prove this is to take the definition of the Dirac δ-distribution,

$$\int d^n x'\, \tilde{\delta}(x,x') f(x') = f(x),$$

with test function $f(x)$ independent of the metric, and functionally differentiate both sides of this equality with respect to $g_{\mu\nu}(\bar{x})$. As a consequence of (7.43), from (7.41) we find

$$\frac{\delta^2 g_{\mu\nu}(x)}{\delta g_{\lambda\tau}(\bar{x}) \delta g_{\rho\sigma}(x')} = 0,$$

clearly a sensible result. The symmetry of the Dirac δ-distribution means that (7.43) holds equally well with $g(x')$ in $\tilde{\delta}(x,x') = |g(x')|^{1/2}\delta(x,x')$ replaced with $g(x)$.

From (7.40) we then have

$$K^k_{\beta,j} = \frac{\delta}{\delta g_{\rho\sigma}(x')} K_{\mu\nu\lambda}(x,x'')$$

$$= -\delta^\rho_{(\mu}\delta^\sigma_{\nu)}\tilde{\delta}_{,\lambda}(x,x')\tilde{\delta}(x,x'') - \delta^\rho_{(\mu}\delta^\sigma_{\lambda)}\tilde{\delta}(x,x')\tilde{\delta}_{,\nu}(x,x'')$$

$$- \delta^\rho_{(\nu}\delta^\sigma_{\lambda)}\tilde{\delta}(x,x')\tilde{\delta}_{,\mu}(x,x'') \tag{7.44}$$

with the condensed index assignments $k = (\mu\nu, x)$, $j = (\rho,\sigma,x')$ and $\beta = (\lambda, x'')$. Choosing $\alpha = (\tau, \bar{x})$ we then find

$$K^j_\alpha K^k_{\beta,j} = \int d^n x'\, K_{\rho\sigma\tau}(x',\bar{x}) \frac{\delta K_{\mu\nu\lambda}(x,x'')}{\delta g_{\rho\sigma}(x')}$$

$$= -\tilde{\delta}(x,x'') K_{\mu\nu\tau,\lambda}(x,\bar{x}) - K_{\mu\lambda\tau}(x,\bar{x})\tilde{\delta}_{,\nu}(x,x'')$$

$$- K_{\nu\lambda\tau}(x,\bar{x})\tilde{\delta}_{,\mu}(x,x''). \tag{7.45}$$

(All of the derivatives in (7.45) are with respect to the coordinate x.) By using (7.40) in (7.45), and subtracting off the analogous expression with the condensed indices α and β switched (involving not only the interchange of the conventional indices λ and τ, but also the interchange of the spacetime arguments $\bar{x}$ and x''), it can be shown that

$$K^j_\alpha K^k_{\beta,j} - K^j_\beta K^k_{\alpha,j} = g_{\mu\nu,\tau}\tilde{\delta}(x,x'')\tilde{\delta}_{,\lambda}(x,\bar{x}) - g_{\mu\nu,\lambda}\tilde{\delta}_{,\tau}(x,x'')\tilde{\delta}(x,\bar{x})$$

$$+ g_{\mu\tau}\tilde{\delta}_{,\nu}(x,x'')\tilde{\delta}_{,\lambda}(x,\bar{x}) + g_{\mu\tau}\tilde{\delta}(x,x'')\tilde{\delta}_{,\nu\lambda}(x,\bar{x})$$

$$- g_{\mu\lambda}\tilde{\delta}_{,\nu\tau}(x,x'')\tilde{\delta}(x,\bar{x}) - g_{\mu\lambda}\tilde{\delta}_{,\tau}(x,x'')\tilde{\delta}_{,\nu}(x,\bar{x})$$

$$+ g_{\nu\tau}\tilde{\delta}_{,\mu}(x,x'')\tilde{\delta}_{,\lambda}(x,\bar{x}) + g_{\nu\tau}\tilde{\delta}(x,x'')\tilde{\delta}_{,\mu\lambda}(x,\bar{x})$$

$$- g_{\nu\lambda}\tilde{\delta}_{,\mu\tau}(x,x'')\tilde{\delta}(x,\bar{x}) - g_{\nu\lambda}\tilde{\delta}_{,\tau}(x,x'')\tilde{\delta}_{,\mu}(x,\bar{x}). \tag{7.46}$$

We need to compare this with the right-hand side of (7.27). If we make the association of $\gamma = (\sigma, x')$, so that

$$-f_{\alpha\beta}{}^\gamma K^k_\gamma = -\int d^n x' f_{\tau\lambda}{}^\sigma(\bar{x}, x'', x') K_{\mu\nu\sigma}(x, x')$$
$$= g_{\mu\nu,\sigma} f_{\tau\lambda}{}^\sigma(\bar{x}, x'', x) + g_{\mu\sigma}\partial_\nu f_{\tau\lambda}{}^\sigma(\bar{x}, x'', x)$$
$$+ g_{\nu\sigma}\partial_\mu f_{\tau\lambda}{}^\sigma(\bar{x}, x'', x), \qquad (7.47)$$

upon use of (7.40), it can be readily seen that the structure functions are given by

$$f_{\tau\lambda}{}^\sigma(\bar{x}, x'', x) = \delta^\sigma_\tau \tilde{\delta}(x, x'') \tilde{\delta}_{,\lambda}(x, \bar{x}) - \delta^\sigma_\lambda \tilde{\delta}_{,\tau}(x, x'') \tilde{\delta}(x, \bar{x}), \qquad (7.48)$$

for the case of gravity. As for the Yang–Mills example, the algebra is closed with field-independent structure functions. (The right-hand side of (7.48) can be seen to be independent of the spacetime metric by virtue of (7.43).)

7.3 The orbit space and the gauge condition

As explained in the introduction, for purposes of quantization it is necessary to integrate only over fields unrelated by the symmetry or gauge transformations (7.16). Let φ^i be a particular field (i.e., point in the space of all fields $\mathcal{F}$). Define $[\varphi^i]$ to be the set of all other fields related to φ^i by any gauge transformation. This defines the equivalence class of φ^i. (It is straightforward to check that this is a true equivalence relation.[4]) In order to define the quantum theory, we want to pick one and only one member from each equivalence class, otherwise we will have overcounted when we perform the functional integration over the space of fields. The set of all distinct equivalence classes is called the orbit space, and is denoted by $\mathcal{F}/\mathfrak{G}$.

From (7.18), we know that the action functional is invariant under infinitesimal displacements

$$\delta_\parallel \varphi^i = K^i_\alpha[\varphi] \delta\epsilon^\alpha, \qquad (7.49)$$

which lie in the direction of the vector with components $K^i_\alpha[\varphi]$. However, if we consider a general displacement in the space of fields it will not necessarily be of the form given in (7.49), but may have components orthogonal to the K^i_α. (In other words, a general displacement in the space of fields will not be generated by a gauge transformation.) To select out the components orthogonal to K^i_α define a projection operator $P^i{}_j$ obeying

$$P^i{}_j K^j_\alpha = 0, \qquad (7.50)$$
$$K^i_\alpha g_{ij} P^j{}_k = 0, \qquad (7.51)$$
$$P^i{}_j P^j{}_k = P^i{}_k. \qquad (7.52)$$

---

[4] This requires the reflexive, symmetric, and transitive properties, all of which follow as a consequence of the assumption that the set of all gauge transformations forms a group.

Here g_{ij} is the metric on the space of fields $\mathcal{F}$, assumed to exist as in Chapter 6. In order to calculate the projection operator $P^i{}_j$, express it as

$$P^i{}_j = \delta^i{}_j - K^i_\alpha \gamma^{\alpha\beta} K^k_\beta g_{kj} \tag{7.53}$$

for some $\gamma^{\alpha\beta}$ chosen so that (7.53) satisfies all of the properties required in (7.50)–(7.52). Our task now is to determine $\gamma^{\alpha\beta}$.

Begin by operating with the projection operator (7.53) on K^j_γ. This leads to

$$P^i{}_j K^j_\gamma = K^i_\gamma - K^i_\alpha \gamma^{\alpha\beta} K^k_\beta g_{kj} K^j_\gamma. \tag{7.54}$$

Then the property (7.50) holds provided that

$$\gamma^{\alpha\beta} K^k_\beta g_{kj} K^j_\gamma = \delta^\alpha{}_\gamma. \tag{7.55}$$

If we define

$$\gamma_{\alpha\beta} = K^i_\alpha g_{ij} K^j_\beta, \tag{7.56}$$

then we must require

$$\det \gamma_{\alpha\beta} \neq 0, \tag{7.57}$$

and (7.55) shows us that $\gamma^{\alpha\beta}$ may be interpreted as the left inverse of $\gamma_{\alpha\beta}$. The second property (7.51) is then true if

$$\gamma_{\alpha\beta} \gamma^{\beta\gamma} = \delta^\gamma{}_\alpha. \tag{7.58}$$

When this result is combined with (7.55) and (7.56), we can conclude that $\gamma^{\alpha\beta}$ is both a left and a right inverse for $\gamma_{\alpha\beta}$. From (7.56) it is seen that $\gamma_{\alpha\beta} = \gamma_{\beta\alpha}$, and hence it follows that $\gamma^{\alpha\beta} = \gamma^{\beta\alpha}$. The final property (7.52) holds without any further conditions.

If $\delta\varphi^i$ is an arbitrary displacement, then with the aid of the projection operator we can define the components orthogonal to K^i_α by

$$\delta_\perp \varphi^i = P^i{}_j \delta\varphi^j. \tag{7.59}$$

An arbitrary displacement can always be resolved as

$$\delta\varphi^i = \delta_\parallel \varphi^i + \delta_\perp \varphi^i, \tag{7.60}$$

where $\delta_\parallel \varphi^i$ is given in (7.49) and $\delta_\perp \varphi^i$ is given in (7.59). This result extends immediately to the coordinate differential one-form $d\varphi^i$. From (7.53) it is clear that

$$d\varphi^i = P^i{}_j d\varphi^j + K^i_\alpha d\epsilon^\alpha, \tag{7.61}$$

where

$$d\epsilon^\alpha = \gamma^{\alpha\beta} K^i_\beta g_{ij} d\varphi^j. \tag{7.62}$$

Hence we have the coordinate basis one-form expressed as

$$d\varphi^i = \omega^i_\perp + \omega^i_\parallel \tag{7.63}$$

with
$$w^i_\perp = P^i{}_j d\varphi^j, \qquad (7.64)$$
and
$$w^i_\parallel = K^i_\alpha d\epsilon^\alpha. \qquad (7.65)$$

By virtue of properties (7.50) and (7.51), the one-forms (7.64) and (7.65) are orthogonal with respect to the field space metric g_{ij}:
$$w^i_\parallel g_{ij} w^j_\perp = 0. \qquad (7.66)$$

As a consequence of this orthogonality, the line element for $\mathcal{F}$, which is $ds^2 = g_{ij} d\varphi^i d\varphi^j$ in the coordinate basis, may be written as
$$ds^2 = g_{ij} w^i_\perp w^j_\perp + g_{ij} w^i_\parallel w^j_\parallel. \qquad (7.67)$$

Using (7.64) and (7.65), along with the basic property (7.50) of the projection operator, allows us to express the line element for $\mathcal{F}$ as
$$ds^2 = g^\perp_{ij} w^i_\perp w^j_\perp + \gamma_{\alpha\beta} d\epsilon^\alpha d\epsilon^\beta, \qquad (7.68)$$
where
$$g^\perp_{ij} = P^n{}_i P^m{}_j g_{nm}. \qquad (7.69)$$

This has written the line element in such a way as to exhibit the local product structure $\mathcal{F} = (\mathcal{F}/\mathfrak{G}) \times \mathfrak{G}$ of the space of fields. The first term on the right-hand side of (7.68) may be interpreted as the line element on $\mathcal{F}/\mathfrak{G}$ with a metric (using (7.53) in (7.69))
$$g^\perp_{ij} = g_{ij} - g_{in} K^n_\alpha \gamma^{\alpha\beta} K^m_\beta g_{mj} \qquad (7.70)$$
that is induced from g_{ij}. The second term on the right-hand side of (7.68) may be interpreted as the line element on $\mathfrak{G}$, with $\gamma_{\alpha\beta}$ as the metric on $\mathfrak{G}$.

Unfortunately the results we have obtained are still not directly useful as they stand, because normally only the action functional is specified, and there is no preferred way to select out one member from each equivalence class of fields. In order to accomplish this last task, gauge conditions (sometimes called gauge-fixing conditions) are introduced. Let
$$\chi^\alpha[\varphi] = 0 \qquad (7.71)$$
denote a set of gauge conditions. They may depend on other variables such as φ_* or $\bar{\varphi}$ as defined in Chapter 6, but as these are regarded as fixed from the standpoint of the functional integration, we will not indicate their presence explicitly in (7.71). (This will be illustrated in the case of Yang–Mills theory and gravity in the two examples at the end of this section.) In order for the gauge conditions to pick out one member from each equivalence class, we must require
$$\chi^\alpha[\varphi_\epsilon] = \chi^\alpha[\varphi], \qquad (7.72)$$

where φ_ϵ^i is defined in (7.16), have only the solution $\delta\epsilon^\alpha = 0$ for a given φ^i. (If there was a solution to (7.72) other than $\delta\epsilon^\alpha = 0$, then the gauge conditions would pick out more than one member from each equivalence class, and hence not be fit for purpose.) Expanding the left-hand side of (7.72) to first order in $\delta\epsilon$ leads to

$$Q^\alpha{}_\beta[\varphi]\delta\epsilon^\beta = 0, \tag{7.73}$$

where we have defined

$$Q^\alpha{}_\beta[\varphi] = \chi^\alpha{}_{,i}[\varphi]K^i_\beta[\varphi]. \tag{7.74}$$

We then conclude that (7.73) has the trivial solution $\delta\epsilon^\beta = 0$ if and only if

$$\det Q^\alpha{}_\beta \neq 0. \tag{7.75}$$

We will assume that the gauge conditions chosen in (7.71) do satisfy this restriction.

The introduction of the gauge conditions can be given a geometrical interpretation as follows. The action of the gauge group $\mathfrak{G}$ on $\mathcal{F}$ may be regarded as foliating $\mathcal{F}$ into orbits as already discussed. (The situation is illustrated pictorially in Fig 7.1.) The gauge conditions (7.71) may then be viewed as defining a surface $\mathcal{S}$ in $\mathcal{F}$ which intersects each orbit in one and only one place. Therefore, the surface $\mathcal{S}$ may be regarded as homeomorphic to the space of orbits: $\mathcal{S} \sim \mathcal{F}/\mathfrak{G}$. It is in this way that one member from each equivalence class is to be chosen.

One warning should be made here. Although there are no problems locally with the procedure just described, we can say nothing about global features with this approach. (One advantage of the fibre bundle picture is that it can

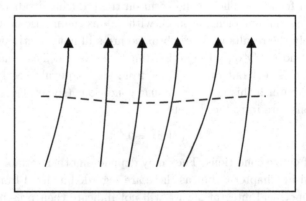

Fig 7.1. This illustrates the field space divided up into distinct orbits under the action of the gauge group. The dotted line indicates the surface specified by the gauge condition, which is assumed to intersect each orbit in one, and only one, place. The arrows represent the action of the gauge group on a given point in field space. All points along a line with an arrow on it are equivalent under a gauge transformation.

7.3 The orbit space and the gauge condition

address these global issues.) It is known from the work of Gribov (1978) that the gauge conditions may pick out two (or more) members from a given equivalence class, where these members are related by a finite gauge transformation. The fact that this was a generic feature of Yang–Mills theory was demonstrated by Singer (1978). Because this Gribov ambiguity, as it is now known, does not arise from local considerations, it is not an issue for a perturbative approach to gauge theories, which is our present interest. However, the reader should be aware that a more complete treatment would have to come to terms with this problem.

Once again it proves helpful to put flesh on the bare formalism we have been presenting by giving the constructs in the familiar examples of Yang–Mills theory and gravity.

7.3.1 Yang–Mills theory, continued

We continue the example started in Section 7.2.1. Recall that we had the associations $\varphi^i \leftrightarrow A_\mu^a(x)$ and $\delta\epsilon^\alpha \leftrightarrow \delta\epsilon^a(x)$ with a the Yang–Mills gauge group index. Our first task is to decide on the metric g_{ij} on the space of Yang–Mills gauge fields. Because of our associations of $i \leftrightarrow (a, \mu, x), j \leftrightarrow (b, \nu, x')$, we will have $g_{ij} \leftrightarrow g_{ab}^{\mu\nu}(x, x')$, and

$$ds^2 = \int d^n x\, d^n x'\, g_{ab}^{\mu\nu}(x, x') dA_\mu^a(x) dA_\nu^b(x'), \tag{7.76}$$

as the line element on the space of Yang–Mills fields. The simplest choice we can make is

$$g_{ab}^{\mu\nu}(x, x') = |g(x)|^{1/2} g^{\mu\nu}(x) \delta_{ab} \tilde{\delta}(x, x'), \tag{7.77}$$

where $g^{\mu\nu}(x)$ on the right-hand side is the inverse of the spacetime metric tensor. (It should be clear from the context and notation whether we are talking about the spacetime metric or the field space metric.) This is the minimal choice we can make for the field space metric; any other choice would entail the introduction of extra dimensional parameters into the theory to balance the dimensions on both sides of (7.76). (We choose ds^2 to have the conventional dimensions of length-squared.) Vilkovisky (1984) has suggested that the field space metric be read off from the highest derivative terms in the classical action functional; for the Yang–Mills action, this can be seen to correspond precisely to (7.77). A feature of (7.77) which will prove useful later is that the field space metric has no dependence on the Yang–Mills gauge fields; in condensed notation, this reads $g_{ij,k} = 0$, and as a consequence the Christoffel connection for the field space metric vanishes.

To save some writing, it is helpful to write (7.10) in the form

$$\delta_\parallel A_\mu^a = -\frac{1}{g} D_\mu{}^a{}_b \delta\epsilon^b \tag{7.78}$$

by introducing the gauge-covariant derivative operator

$$D_\mu{}^a{}_b = \delta^a_b \nabla_\mu - g f_{bc}{}^a A^c_\mu. \tag{7.79}$$

(The subscript $\|$ appended to δ in (7.78) is to remind us that the displacement is generated by a gauge transformation and is not a general displacement in the space of fields.) Since we adopted the group metric δ_{ab} in (7.12), we will define

$$f_{bca} = \delta_{ad} f_{bc}{}^d, \tag{7.80}$$

and use the result that f_{bca} is totally antisymmetric under a permutation of any of its indices. We can therefore write, from (7.79),

$$D_{\mu ab} = \delta_{ab} \nabla_\mu - g f_{abc} A^c_\mu. \tag{7.81}$$

In order to find $\gamma_{\alpha\beta}$, we can either use the direct result (7.58), or else use the fact that for displacements solely in directions specified by the gauge transformations (i.e., $\delta A^a_\mu = \delta_\| A^a_\mu$) we have from (7.68),

$$ds_\|^2 = g_{ij} d_\| \varphi^i d_\| \varphi^j$$
$$= \int d^n x\, d^n x'\, g^{\mu\nu}_{ab}(x,x') d_\| A^a_\mu(x) d_\| A^b_\nu(x')$$
$$= \frac{1}{g^2} \int dv_x g^{\mu\nu}(x) \delta_{ab} \left(D_\mu{}^a{}_c d\epsilon^c\right)\left(D_\nu{}^b{}_d d\epsilon^d\right) \tag{7.82}$$

using (7.77). With $D_\mu{}^a{}_b$ given by (7.81), after an integration by parts, use of the definition (7.80) along with the antisymmetry of the gauge group structure constants, it can be seen that

$$ds_\|^2 = -\frac{1}{g^2} \int dv_x\, d\epsilon^a (D^2)_{ab} d\epsilon^b, \tag{7.83}$$

where

$$(D^2)_{ab} = D^\mu{}_{ac} D_\mu{}^c{}_b. \tag{7.84}$$

This allows us to deduce that

$$\gamma_{\alpha\beta} = \gamma_{ab}(x,x') = -\frac{1}{g^2} |g(x)|^{1/2} (D^2)_{ab} \tilde\delta(x,x'), \tag{7.85}$$

where the derivative acts on the x (rather than the x') coordinate.

It now follows from (7.58) that $\gamma^{\alpha\beta}$, which is the inverse of $\gamma_{\alpha\beta}$, is a Green function for the differential operator (7.85). If we write $\gamma^{\alpha\beta} = \gamma^{ab}(x,x')$, then $\gamma^{ab}(x,x')$ must satisfy

$$-\frac{1}{g^2} (D^2)_{ab} \gamma^{bc}(x,x') = \delta^c_a \delta(x,x'). \tag{7.86}$$

Finally, we turn to the gauge conditions. We have the correspondence of $\chi^\alpha[\varphi]$ with $\chi^a(x)$ in conventional notation, where $\chi^a(x)$ can depend on the background field $A^a_{*\mu}(x)$ as well as the field variable $A^a_\mu(x)$. For reasons which will become apparent later, it proves convenient to adopt the gauge conditions $\chi^a(x) = 0$ with

$$\chi^a(x) = D_{*\mu}{}^a{}_b(A^{b\mu} - A^{b\mu}_*). \tag{7.87}$$

Here $D_{*\mu}{}^a{}_b$ is defined as in (7.79), but with the gauge field taken as the background field $A^c_\mu = A^c_{*\mu}$. These gauge conditions go under various names, such as the background-field gauge, or Landau–DeWitt gauge. With the choice (7.87) we find

$$\chi^\alpha{}_{,j} = \frac{\delta\chi^a(x)}{\delta A^{c\mu}(x'')} = D_{*\mu}{}^a{}_c\tilde\delta(x, x''), \tag{7.88}$$

and using this result in (7.74) it follows that

$$Q^\alpha{}_\beta = Q^a{}_b(x, x') = \int d^n x'' D_{*\mu}{}^a{}_c \tilde\delta(x, x'') K^{c\mu}{}_b(x'', x')$$
$$= D_{*\mu}{}^a{}_c K^{c\mu}{}_b(x, x')$$
$$= -\frac{1}{g} D_{*\mu}{}^a{}_c D^{\mu c}{}_b \tilde\delta(x, x'). \tag{7.89}$$

We will examine the motivation for the choice (7.87) when we continue the Yang–Mills example in Section 7.4.1.

7.3.2 Gravity, continued

In the case of gravity we will write the field space metric as

$$ds^2 = \int d^n x\, d^n x'\, G^{\mu\nu\rho\sigma}(x, x') dg_{\mu\nu}(x) dg_{\rho\sigma}(x'). \tag{7.90}$$

(We will use a capital letter G for the field space metric to avoid any possible confusion with the spacetime metric.) There is an unique one-parameter family of metrics $G^{\mu\nu\rho\sigma}(x, x')$ on the space of all spacetime metrics, referred to as the DeWitt metric, involving the introduction of no extra dimensional parameters into the theory (DeWitt 1967):

$$G^{\mu\nu\rho\sigma}(x, x') = |g(x)|^{1/2}\left\{g^{\mu(\rho}(x)g^{\sigma)\nu}(x) + \frac{c}{2}g^{\mu\nu}(x)g^{\rho\sigma}(x)\right\}\delta(x, x'). \tag{7.91}$$

Here c is an (at this stage) arbitrary dimensionless constant, and to save writing we adopt the standard notation

$$g^{\mu(\rho}g^{\sigma)\nu} = \frac{1}{2}\left(g^{\mu\rho}g^{\sigma\nu} + g^{\mu\sigma}g^{\rho\nu}\right). \tag{7.92}$$

The inverse DeWitt metric $G_{\mu\nu\rho\sigma}(x, x')$ must satisfy

$$\int d^n x\, G^{\mu\nu\rho\sigma}(x, x') G_{\rho\sigma\lambda\tau}(x', x'') = \delta^\mu_{(\lambda}\delta^\nu_{\tau)}\tilde\delta(x, x''), \tag{7.93}$$

and it is easy to show that

$$G_{\rho\sigma\lambda\tau}(x,x') = \left\{g_{\rho(\lambda}(x)g_{\tau)\sigma}(x) - \frac{c}{(2+nc)}g_{\lambda\tau}(x)g_{\rho\sigma}(x)\right\}\delta(x,x'), \qquad (7.94)$$

provided that $c \neq -2/n$ with n the spacetime dimension. If we require, as proposed by Vilkovisky (1984), that the metric be determined from the highest derivative term in the classical action, it can be shown that $c = -1$. This means that a spacetime of dimension $n = 2$ is a special case.

In contrast with Yang–Mills theory, the metric on the space of fields for gravity does have a dependence on the fields themselves, in this case the spacetime metric $g_{\mu\nu}$. As a consequence of this dependence, the Christoffel connection computed from the field space metric (7.91) will not vanish here. Given the identifications $i \leftrightarrow (\mu\nu, x), j \leftrightarrow (\rho\sigma, x')$, and $k \leftrightarrow (\lambda\tau, x'')$ of condensed and normal notation, we have

$$\Gamma^k_{ij} = \Gamma^{\mu\nu\rho\sigma}_{\lambda\tau}(x,x',x'') = \frac{1}{2}\int d^n\bar{x}\, G_{\lambda\tau\alpha\beta}(x,\bar{x}) \left\{ \frac{\delta G^{\mu\nu\alpha\beta}(x,\bar{x})}{\delta g_{\rho\sigma}(x')} \right.$$
$$\left. + \frac{\delta G^{\alpha\beta\rho\sigma}(\bar{x},x')}{\delta g_{\mu\nu}(x)} - \frac{\delta G^{\mu\nu\rho\sigma}(x,x')}{\delta g_{\alpha\beta}(\bar{x})} \right\}, \qquad (7.95)$$

where we have simply used the definition $\Gamma^k_{ij} = \frac{1}{2}g^{kl}(g_{il,j} + g_{lj,i} - g_{ij,l})$ to obtain the second equality with the association $l \leftrightarrow (\alpha\beta, \bar{x})$.

The functional derivative of the DeWitt metric with respect to the spacetime metric is obtained in a straightforward manner to be

$$\frac{\delta}{\delta g_{\lambda\tau}(\bar{x})}G^{\mu\nu\rho\sigma}(x,x') = \frac{1}{2}|g(x)|^{1/2}\tilde{\delta}(x,\bar{x})\tilde{\delta}(x,x')\left\{g^{\mu(\rho}g^{\sigma)\nu}g^{\lambda\tau}\right.$$
$$+ \frac{c}{2}g^{\mu\nu}g^{\rho\sigma}g^{\lambda\tau} - g^{\mu(\lambda}g^{\tau)\rho}g^{\sigma\nu} - g^{\mu\rho}g^{\sigma(\lambda}g^{\tau)\nu}$$
$$- cg^{\mu(\lambda}g^{\tau)\nu}g^{\rho\sigma} - cg^{\mu\nu}g^{\rho(\lambda}g^{\tau)\sigma}$$
$$\left. - g^{\mu(\lambda}g^{\tau)\sigma}g^{\rho\nu} - g^{\mu\sigma}g^{\rho(\lambda}g^{\tau)\nu}\right\}. \qquad (7.96)$$

We have used the result

$$\frac{\delta g^{\mu\nu}(x)}{\delta g_{\lambda\tau}(\bar{x})} = -g^{\mu(\lambda}g^{\tau)\nu}\tilde{\delta}(x,\bar{x}), \qquad (7.97)$$

here, and also noted (7.43). After some calculation, the following result for the field space Christoffel connection is obtained:

$$\Gamma^{\mu\nu\rho\sigma}_{\lambda\tau}(x,x',x'') = \tilde{\delta}(x'',x')\tilde{\delta}(x'',x)\left[-\delta^{(\mu}_{(\lambda}g^{\nu)(\rho}\delta^{\sigma)}_{\tau)} + \frac{1}{4}g^{\mu\nu}\delta^{\rho}_{(\lambda}\delta^{\sigma}_{\tau)}\right.$$
$$+ \frac{1}{4}g^{\rho\sigma}\delta^{\mu}_{(\lambda}\delta^{\nu}_{\tau)} - \frac{1}{2(2+nc)}g_{\lambda\tau}g^{\mu(\rho}g^{\sigma)\nu}$$
$$\left. - \frac{c}{4(2+nc)}g_{\lambda\tau}g^{\mu\nu}g^{\rho\sigma}\right] \qquad (7.98)$$

This expression agrees with the ones given in Fradkin and Tseytlin (1984) and Huggins et al. (1987b) where the role of the Christoffel connection (7.98) in perturbative quantum gravity calculations is explored.

The easiest way to evaluate $\gamma_{\alpha\beta}$ is to consider the displacement along directions in field space generated by gauge transformations, as we did in the Yang–Mills case. Here we have

$$ds_{\parallel}^2 = g_{ij}d_{\parallel}\varphi^i d_{\parallel}\varphi^j$$
$$= \int d^n x \, d^n x' \, G^{\mu\nu\rho\sigma}(x,x') d_{\parallel}g_{\mu\nu}(x) d_{\parallel}g_{\rho\sigma}(x'), \qquad (7.99)$$

where the subscript $\parallel$ is again used to distinguish that the displacement is generated by a gauge transformation. From (7.15) we have

$$d_{\parallel}g_{\mu\nu}(x) = -\nabla_{\mu}d\epsilon_{\nu}(x) - \nabla_{\nu}d\epsilon_{\mu}(x), \qquad (7.100)$$

if we rewrite the right-hand side of (7.15) in terms of the spacetime covariant derivative. (This makes it much easier to perform integration by parts.) Using (7.91) and performing an integration by parts results in

$$ds_{\parallel}^2 = -2 \int dv_x d\epsilon^{\mu} \left[g_{\mu\nu} \Box + \nabla_{\nu}\nabla_{\mu} + c\nabla_{\mu}\nabla_{\nu} \right] d\epsilon^{\nu}$$
$$= -2 \int dv_x d\epsilon^{\mu} \left[g_{\mu\nu} \Box - R_{\mu\nu} + (1+c)\nabla_{\mu}\nabla_{\nu} \right] d\epsilon^{\nu}. \qquad (7.101)$$

To obtain the second equality we have used $[\nabla_{\mu}, \nabla_{\nu}] d\epsilon^{\lambda} = -R^{\lambda}{}_{\sigma\mu\nu} d\epsilon^{\sigma}$ and $R_{\mu\nu} = R^{\lambda}{}_{\mu\lambda\nu}$ to change the order of the covariant derivatives in the middle term of the first line. Comparison of (7.101) with the condensed index version $ds_{\parallel}^2 = \gamma_{\alpha\beta} d\epsilon^{\alpha} d\epsilon^{\beta}$, allows us to conclude that

$$\gamma_{\alpha\beta} = \gamma_{\mu\nu}(x,x') = -2|g(x)|^{1/2} \left[g_{\mu\nu} \Box - R_{\mu\nu} + (1+c)\nabla_{\mu}\nabla_{\nu} \right] \tilde{\delta}(x,x'), \qquad (7.102)$$

where the derivatives act on the x coordinate of the Dirac δ-distribution. If we require the DeWitt metric to coincide with the metric contained in the highest derivative of the classical action, then we have $c = -1$ and the differential operator in (7.102) becomes a minimal operator[5] as defined by Barvinsky and Vilkovisky (1985). The inverse of $\gamma_{\alpha\beta}$ is given by $\gamma^{\alpha\beta} = \gamma^{\mu\nu}(x,x')$ where $\gamma^{\mu\nu}(x,x')$ is a Green function, defined as the solution to

$$-2 \left[g_{\mu\nu} \Box - R_{\mu\nu} + (1+c)\nabla_{\mu}\nabla_{\nu} \right] \gamma^{\nu\lambda}(x,x') = \delta^{\lambda}_{\mu} \delta(x,x'). \qquad (7.103)$$

The analogue of the Landau–DeWitt gauge condition in Yang–Mills theory (7.87) in general is

$$K^i_{\alpha}[\varphi_*] g_{ij}[\varphi_*](\varphi^j - \varphi^j_*) = 0, \qquad (7.104)$$

---

[5] This means that the highest derivative term appearing in the operator involves $\Box$.

where $\varphi_\star^i$ is the background field. If we apply this to gravity we have $\chi_\alpha = K_\alpha^i[\varphi_\star]g_{ij}[\varphi_\star](\varphi^j - \varphi_\star^j)$ corresponding to $\chi_\lambda(x)$ with the association $\alpha \leftrightarrow (\lambda, x)$. Using (7.40) and (7.91) results in

$$\chi_\lambda(x) = 2|g(x)|^{1/2}\left(\nabla^{\star\mu}g_{\mu\lambda} + \frac{c}{2}g^{\star\mu\nu}\nabla_\lambda^\star g_{\mu\nu}\right), \qquad (7.105)$$

after a short calculation. Here $g_{\mu\nu}^\star$ is the background metric and $\nabla_\mu^\star$ is the covariant derivative computed using the Christoffel connection formed from the background metric.[6]

If we form $Q_{\alpha\beta} = \chi_{\alpha,j}K_\beta^j$, we find

$$\begin{aligned}Q_{\mu\nu}(x,x') &= \int d^n x'' \frac{\delta\chi_\mu(x)}{\delta g_{\rho\sigma}(x'')}K_{\rho\sigma\nu}(x'',x')\\ &= -2|g(x)|^{1/2}\Big\{\nabla^{\star\lambda}\left(g_{\mu\nu}\nabla_\lambda\tilde\delta(x,x') + g_{\lambda\nu}\nabla_\mu\tilde\delta(x,x')\right)\\ &\quad + c\nabla_\mu^\star\left(g^{\star\rho\sigma}g_{\rho\nu}\nabla_\sigma\tilde\delta(x,x')\right)\Big\}.\end{aligned} \qquad (7.106)$$

The case of gravity will be continued in Section 7.4.2.

7.4 Field space reparameterization and the Killing equation

Once a set of gauge conditions has been adopted that specifies a surface $\mathcal{S} \sim \mathcal{F}/\mathfrak{G}$ as discussed in the previous section, it is more convenient to choose a set of coordinates that are adapted to the surface $\mathcal{S}$. Let ξ^A be a set of coordinates that parameterize $\mathcal{S}$. Because we assume that $\mathcal{S} \sim \mathcal{F}/\mathfrak{G}$, the coordinates ξ^A must be gauge-invariant functionals of the original coordinates φ^i: $\xi^A = \xi^A[\varphi^i]$ with $\xi^A[\varphi^i + \delta_\parallel \varphi^i] = \xi^A[\varphi^i]$. A consequence of this invariance is

$$K_\alpha^i[\varphi]\xi^A_{,i}[\varphi] = 0, \qquad (7.107)$$

which comes from using (7.16) and expanding $\xi^A[\varphi^i + \delta_\parallel \varphi^i]$ to first order in $\delta\epsilon^a$. If we now consider the coordinate differentials $d\xi^A$, we have

$$\begin{aligned}d\xi^A &= \xi^A_{,i}d\varphi^i\\ &= \xi^A_{,i}\omega_\perp^i,\end{aligned} \qquad (7.108)$$

if we use $d\varphi^i = \omega_\parallel^i + \omega_\perp^i$, and the gauge invariance (7.107). Because the coordinates ξ^A only parameterize part of the full field space, namely the surface $\mathcal{S}$, we will now introduce a set of coordinates $\theta^\alpha = \theta^\alpha[\varphi]$, which label points in a given orbit. A particularly convenient choice is

$$\theta^\alpha = \chi^\alpha[\varphi], \qquad (7.109)$$

---

[6] Note that $\nabla_\lambda^\star g_{\mu\nu}^\star = 0$, but $\nabla_\lambda^\star g_{\mu\nu} = \partial_\lambda g_{\mu\nu} - \Gamma_{\lambda\mu}^{\star\sigma}g_{\sigma\nu} - \Gamma_{\lambda\nu}^{\star\sigma}g_{\mu\sigma} \neq 0$.

7.4 Field space reparameterization and the Killing equation

since different constant values of θ^α will simply pick out different surfaces, all of which select one member from each orbit.

We may now work either with the coordinates $\varphi^i[\xi,\theta]$ or else with the new coordinates $(\xi^A[\varphi],\theta^\alpha[\varphi])$ that have been introduced. Because ξ^A and θ^α are assumed to be independent coordinates, it must be the case that

$$0 = \frac{\delta \xi^A}{\delta \theta^\alpha} = \xi^A{}_{,i}\varphi^i{}_{,\alpha}. \tag{7.110}$$

From (7.74), we have $Q^\alpha{}_\beta = \theta^\alpha{}_{,i} K^i_\beta$, since $\theta^\alpha = \chi^\alpha$ can be taken there, and hence

$$\delta^\alpha{}_\beta = \theta^\alpha{}_{,i} K^i_\gamma (Q^{-1})^\gamma{}_\beta. \tag{7.111}$$

This equality simply expresses the condition that $Q^\alpha{}_\gamma (Q^{-1})^\gamma{}_\beta = \delta^\alpha{}_\beta$. (Recall that we assume $\det Q^\alpha{}_\beta \neq 0$, so that $(Q^{-1})^\gamma{}_\beta$ exists.) We can also write $\delta^\alpha{}_\beta = \theta^\alpha{}_{,i} \varphi^i{}_{,\beta}$, so can conclude that

$$\varphi^i{}_{,\beta} = K^i_\gamma (Q^{-1})^\gamma{}_\beta, \tag{7.112}$$

by comparison with (7.111). (Note that using (7.112) in (7.110) implies our previous result in (7.107).) The right-hand side of (7.112) may be regarded as a particular linear combination of generators of the algebra defined as

$$\mathfrak{K}^i_\beta = K^i_\gamma (Q^{-1})^\gamma{}_\beta. \tag{7.113}$$

It is possible to prove (Batalin and Vilkovisky 1984) that

$$[\mathfrak{K}_\alpha, \mathfrak{K}_\beta] = 0 \tag{7.114}$$

where $\mathfrak{K}_\alpha = \mathfrak{K}^i_\alpha \frac{\delta}{\delta \varphi^i}$. The linear combination of generators in (7.114) turns the original non-Abelian algebra into an Abelian one.[7]

The transformation from φ^i to (ξ^A, θ^α) gives rise to

$$\delta^i{}_j = \varphi^i{}_{,j} = \varphi^i{}_{,A} \xi^A{}_{,j} + \varphi^i{}_{,\alpha} \theta^\alpha{}_{,j}$$
$$= \varphi^i{}_{,A} \xi^A{}_{,j} + K^i_\gamma (Q^{-1})^\gamma{}_\alpha \theta^\alpha{}_{,j}, \tag{7.115}$$

where (7.112) has been used. Because $K^i_\gamma g^\perp_{ij} = 0$, we have using (7.115), the result

$$g^\perp_{jk} = \varphi^i{}_{,A} \xi^A{}_{,j} g^\perp_{ik}. \tag{7.116}$$

Using (7.115) again gives the identity

$$g^\perp_{ij} = \varphi^n{}_{,A} \xi^A{}_{,i} g^\perp_{nm} \varphi^m{}_{,B} \xi^B{}_{,j}. \tag{7.117}$$

From (7.108) it then follows that

$$g^\perp_{ij} \omega^i_\perp \omega^j_\perp = \varphi^n{}_{,A} \varphi^m{}_{,B} g^\perp_{nm} d\xi^A d\xi^B. \tag{7.118}$$

---

[7] The proof of this assertion is given in Appendix (7A.1).

We may therefore define

$$h_{AB} = \varphi^n{}_{,A}\varphi^m{}_{,B}g^\perp_{nm}, \tag{7.119}$$

and interpret h_{AB} as the metric on $\mathcal{S}$. Note that h_{AB} is not simply the metric induced on the surface from g_{ij}, but rather is induced from $g^\perp_{ij}$ which is the projected form of g_{ij}. Equation (7.119) may be inverted (noting (7.116)) to give

$$g^\perp_{ij} = \xi^A{}_{,i}\xi^B{}_{,j}h_{AB}. \tag{7.120}$$

It is expected that any gauge-invariant functional of φ^i should be able to be expressed solely in terms of the coordinates ξ^A that parameterize $\mathcal{S}$. Consider, in particular, a gauge-invariant scalar functional $F[\varphi]$. It must satisfy

$$K^i_\alpha F_{,i} = 0, \tag{7.121}$$

by definition of what we mean by gauge-invariant.[8] Under the change of coordinates $\varphi^i \to (\xi^A, \theta^\alpha)$, we have

$$F[\varphi] = \tilde{F}[\xi, \theta] \tag{7.122}$$

in general, since this is the definition of what is meant by a scalar functional. Using (7.121) and the chain rule, leads to

$$K^i_\alpha\left(\tilde{F}_{,A}\xi^A{}_{,i} + \tilde{F}_{,\beta}\theta^\beta{}_{,i}\right) = 0. \tag{7.123}$$

The first term vanishes by virtue of (7.107), and the second term that remains may be written as

$$\tilde{F}_{,\beta}Q^\beta{}_\alpha = 0 \tag{7.124}$$

using (7.74). Because $(Q^{-1})^\alpha{}_\beta$ is assumed to exist, this last result tells us that

$$\tilde{F}_{,\beta} = 0, \tag{7.125}$$

meaning that when $F[\varphi]$ is expressed in terms of the coordinates (ξ^A, θ^α) as defined in (7.122), it has no dependence on θ^α. Thus $\tilde{F}[\xi, \theta] = \tilde{F}[\xi]$ only. This proves that any gauge-invariant scalar functional can be expressed solely in terms of the coordinates ξ^A that parameterize $\mathcal{S}$. The result holds, in particular, for the action functional, as well as for the effective action for gauge theories, provided that it can be shown to be gauge-invariant.

If the intrinsic geometry of $\mathcal{S}$ is to be independent of which point in the orbit is selected, then the metric h_{AB} on $\mathcal{S}$ should also be independent of the coordinates θ^α. We will now work out the necessary and sufficient conditions for this to be true.

---

[8] That is, $F[\varphi + \delta_\| \varphi] = F[\varphi]$. Equation (7.121) follows by expanding both sides of this result to first order in $\delta\epsilon^a$.

7.4 Field space reparameterization and the Killing equation

It follows directly from (7.119) that

$$\frac{\delta}{\delta\theta^\alpha}h_{AB} = \varphi^i{}_{,A\alpha}\varphi^j{}_{,B}g^\perp_{ij} + \varphi^i{}_{,A}\varphi^j{}_{,B\alpha}g^\perp_{ij} + \varphi^i{}_{,A}\varphi^j{}_{,B}\frac{\delta}{\delta\theta^\alpha}g^\perp_{ij}. \qquad (7.126)$$

Now by interchanging the order of differentiation, we can obtain

$$\begin{aligned}\varphi^i{}_{,A\alpha} &= \left(\varphi^i{}_{,\alpha}\right)_{,A} \\ &= \left(K^i_\beta(Q^{-1})^\beta{}_\alpha\right)_{,A} \qquad \text{(by (7.112))} \\ &= \left(K^i_\beta(Q^{-1})^\beta{}_\alpha\right)_{,j}\varphi^j{}_{,A} \qquad \text{(by the chain rule)}.\end{aligned}$$

Because $K^i_\beta g^\perp_{ij} = 0$, use of this last result for $\varphi^i{}_{,A\alpha}$ leads to

$$\varphi^i{}_{,A\alpha}g^\perp_{ij} = K^i_{\beta,k}(Q^{-1})^\beta{}_\alpha\varphi^k{}_{,A}g^\perp_{ij}. \qquad (7.127)$$

From the chain rule, since $\theta^\alpha = \theta^\alpha[\varphi]$, we have

$$\begin{aligned}\frac{\delta}{\delta\theta^\alpha}g^\perp_{ij} &= \varphi^k{}_{,\alpha}g^\perp_{ij,k} \\ &= K^k_\beta(Q^{-1})^\beta{}_\alpha g^\perp_{ij,k} \qquad \text{(by (7.121))}.\end{aligned} \qquad (7.128)$$

Using (7.127) and (7.128) in (7.126) gives

$$\frac{\delta}{\delta\theta^\alpha}h_{AB} = (Q^{-1})^\beta{}_\alpha\varphi^i{}_{,A}\varphi^j{}_{,B}\{K^k_{\beta,i}g^\perp_{kj} + K^k_{\beta,j}g^\perp_{ik} + K^k_\beta g^\perp_{ij,k}\}. \qquad (7.129)$$

Because all of the factors multiplying the bracket on the right-hand side are non-singular, it follows from this result that h_{AB} is independent of θ^α if and only if

$$K^k_{\alpha,i}g^\perp_{kj} + K^k_{\alpha,j}g^\perp_{ik} + K^k_\alpha g^\perp_{ij,k} = 0. \qquad (7.130)$$

This result may be recognized as Killing's equation for the metric $g^\perp_{ij}$. Therefore, it follows that the metric on the surface $\mathcal{S}$ is independent of the coordinates θ^α, and hence is a gauge-invariant functional, if and only if the vectors $\{\mathbf{K}_\alpha\}$ which generate the gauge transformations are Killing vectors for the metric $g^\perp_{ij}$.

The consequences of (7.130) for the original metric g_{ij} may also be obtained. From (7.69),

$$\begin{aligned}g^\perp_{ij,k} &= P^n{}_i P^m{}_j g_{nm,k} + P^n{}_{i,k} P^m{}_j g_{nm} + P^n{}_i P^m{}_{j,k} g_{nm} \\ &= P^n{}_i P^m{}_j g_{nm,k} + P^n{}_{i,k} g^\perp_{nj} + P^m{}_{j,k} g^\perp_{im},\end{aligned} \qquad (7.131)$$

(since $P^n{}_i g_{nj} = g^\perp_{ij}$). Differentiation of the result for the projection operator defined in (7.53), and using $K^i_\alpha g^\perp_{ij} = 0$, leads to

$$P^n{}_{i,k}g^\perp_{nj} = -K^n_{\alpha,k}\gamma^{\alpha\beta}K^l_\beta g_{li}g^\perp_{nj}. \qquad (7.132)$$

Substitution of this expression into (7.131) and using the result in (7.130) yields

$$0 = P^n{}_i P^m{}_j K^k_\lambda g_{nm,k} - K^k_\lambda K^n_{\delta,k} \gamma^{\delta\beta} K^l_\beta g_{li} g^\perp_{nj}$$
$$- K^k_\lambda K^n_{\delta,k} \gamma^{\delta\beta} K^l_\beta g_{lj} g^\perp_{ni} + K^k_{\lambda,i} g^\perp_{kj} + K^k_{\lambda,j} g^\perp_{ik}. \qquad (7.133)$$

Assuming a closed algebra, we have from (7.15),

$$K^k_\lambda K^n_{\delta,k} g^\perp_{nj} = (K^k_\delta K^n_{\lambda,k} - f_{\lambda\delta}{}^\epsilon K^n_\epsilon) g^\perp_{nj}$$
$$= K^k_\delta K^n_{\lambda,k} g^\perp_{nj}, \qquad (7.134)$$

again using $K^n_\epsilon g^\perp_{nj} = 0$. Thus, we obtain

$$K^k_\lambda K^n_{\delta,k} \gamma^{\delta\beta} K^l_\beta g_{li} g^\perp_{nj} = K^k_\delta K^n_{\lambda,k} \gamma^{\delta\beta} K^l_\beta g_{li} g^\perp_{nj}$$
$$= K^n_{\lambda,k} (\delta^k{}_i - P^k{}_i) g^\perp_{nj} \qquad (\text{see } (7.53))$$
$$= K^n_{\lambda,i} g^\perp_{nj} - P^k{}_i K^n_{\lambda,k} g^\perp_{nj}$$
$$= K^n_{\lambda,i} g^\perp_{nj} - P^k{}_i P^m{}_j K^n_{\lambda,k} g_{nm}. \qquad (7.135)$$

Using this expression in (7.133) gives

$$0 = P^n{}_i P^m{}_j \{ K^k_\lambda g_{nm,k} + K^k_{\lambda,n} g_{km} + K^k_{\lambda,m} g_{kn} \}. \qquad (7.136)$$

The quantity in brackets is observed to be Killing's equation for the metric g_{ij}. However, this time the factors out in front of the bracket may not be pulled off because the projection operator is not invertible on the full field space. We can conclude that if the vectors $\{\mathbf{K}_\alpha\}$ are Killing vectors for the metric g_{ij}, then the metric induced on the gauge surface $\mathcal{S}$ is a function of the invariants ξ^A only. The converse is not true in general.

It is only possible to conclude from (7.136) that the metric g_{ij} must satisfy Killing's equation up to terms proportional to the Killing vectors. (This is because the projection operators in (7.136) will project out anything proportional to the Killing vectors.) We may therefore write

$$K^k_\lambda g_{nm,k} + K^k_{\lambda,n} g_{km} + K^k_{\lambda,m} g_{kn} = K^k_\epsilon g_{kn} B^\epsilon{}_{\lambda m} + K^k_\epsilon g_{km} B^\epsilon{}_{\lambda n} \qquad (7.137)$$

for some $B^\epsilon{}_{\lambda m}$. This equation was first given and discussed by DeWitt (1987). In many examples, as we will see in Sections 7.4.1 and 7.4.2 below, $B^\epsilon{}_{\lambda m}$ will vanish; however, it is not necessary to assume that this is the case in what follows.

7.4.1 Yang–Mills theory, continued

We have defined the field space metric in (7.77) and the gauge transformation generators in (7.32). It is now a straightforward matter to evaluate the left-hand side of (7.137) with these expressions used. The first term on the left-hand side of (7.137) vanishes because the field space metric has no dependence on the Yang–Mills fields. The remaining terms are easily evaluated using (7.34). If we make the index associations $m \leftrightarrow (a,\mu,x), n \leftrightarrow (b,\nu,x'), k \leftrightarrow (c,\rho,\bar{x})$, and $\lambda \leftrightarrow (d,x'')$, we have

$$K^k_{\lambda,n} = f_{db}{}^c \delta^\nu_\rho \tilde{\delta}(\bar{x},x')\tilde{\delta}(\bar{x},x''). \tag{7.138}$$

Using (7.77) we find

$$K^k_{\lambda,n}g_{km} = \int d^n\bar{x} f_{db}{}^c \delta^\nu_\rho \tilde{\delta}(\bar{x},x')\tilde{\delta}(\bar{x},x'') g^{\rho\mu}(\bar{x})\delta_{ca}\tilde{\delta}(\bar{x},x)$$
$$= |g(x)|^{1/2} f_{dba} g^{\mu\nu}(x)\tilde{\delta}(x,x')\tilde{\delta}(x,x''). \tag{7.139}$$

To obtain $K^k_{\lambda,m}g_{kn}$ we merely interchange the two condensed indices m and n in (7.139). With the index associations chosen above, this entails the interchanges of μ with ν, a with b, and x with x'. Making use of the properties of the Dirac δ-distribution[9] results in

$$K^k_{\lambda,n}g_{km} = |g(x)|^{1/2} f_{dab} g^{\mu\nu}(x)\tilde{\delta}(x,x')\tilde{\delta}(x,x''). \tag{7.140}$$

It is now obvious that the Yang–Mills field space metric obeys Killing's equation

$$K^k_\lambda g_{mn,k} + K^k_{\lambda,n}g_{km} + K^k_{\lambda,m}g_{kn} = 0, \tag{7.141}$$

because the first term vanishes, and the last two terms cancel due to the antisymmetry of the structure constants for the Lie algebra. We can conclude that $B^\epsilon{}_{\lambda n} = 0$ for Yang–Mills theory.

7.4.2 Gravity, continued

The field space metric for gravity is the DeWitt metric defined in (7.91). The generator of gauge transformations was given in (7.40), or the equivalent form expressed in terms of the spacetime covariant derivative found on the right-hand side of (7.100). We will work out each of the three terms in Killing's equation (7.141) separately using the associations $m \leftrightarrow (\mu\nu,x), n \leftrightarrow (\rho\sigma,x'), k \leftrightarrow (\lambda\tau,\bar{x})$, and $\lambda \leftrightarrow (\alpha,x'')$ between condensed indices and conventional ones. It is sometimes easiest to deal with the products of Dirac δ-distributions that occur by smearing them with test functions. Let $F^\alpha(x'')$ be an arbitrary test function, corresponding to F^λ in condensed notation, that has no dependence on the spacetime metric. We will define $F^k = K^k_\lambda F^\lambda$ in condensed notation, corresponding to

$$F_{\lambda\tau}(\bar{x}) = \int d^n x'' K_{\lambda\tau\alpha}(\bar{x},x'')F^\alpha(x'') \tag{7.142}$$

---

[9] Specifically, we use $\delta(x',x)\delta(x',x'') = \delta(x,x')\delta(x',x'') = \delta(x,x')\delta(x,x'')$ here. This can be proven by integration of both sides of the equalities after multiplication by an arbitrary test function.

in normal notation. Using the expression (7.40), with the right-hand side expressed in terms of spacetime covariant derivatives, we find

$$F_{\lambda\tau}(\bar{x}) = -\bar{\nabla}_\lambda F_\tau(\bar{x}) - \bar{\nabla}_\tau F_\lambda(\bar{x}). \tag{7.143}$$

Here the bar over ∇ denotes that the covariant derivative is taken at the spacetime point $\bar{x}$. It is easy to show using the derivative of the DeWitt metric given in (7.96) that

$$F^\lambda K^k_{\lambda} g_{mn,k} = \int d^n \bar{x} F_{\lambda\tau}(\bar{x}) \frac{\delta}{\delta g_{\lambda\tau}(\bar{x})} G^{\mu\nu\rho\sigma}(x,x')$$
$$= |g(x)|^{1/2} \tilde{\delta}(x,x') \Big\{ -g^{\mu(\rho} g^{\sigma)\nu} F^\alpha{}_{;\alpha} - c g^{\mu\nu} g^{\rho\sigma} F^\alpha{}_{;\alpha}$$
$$+ g^{\mu\rho} F^{(\sigma;\nu)} + g^{\mu\sigma} F^{(\rho;\nu)} + g^{\nu\rho} F^{(\sigma;\mu)}$$
$$+ g^{\nu\sigma} F^{(\rho;\mu)} + c g^{\mu\nu} F^{(\rho;\sigma)} + c g^{\rho\sigma} F^{(\mu;\nu)} \Big\}. \tag{7.144}$$

Because of our assumption that F^λ is independent of the metric, we have $F^\lambda K^k_{\lambda,n} = (F^\lambda K^k_\lambda)_{,n}$. In conventional notation this reads

$$F^\lambda K^k_{\lambda,n} = \frac{\delta}{\delta g_{\rho\sigma}(x')} F_{\lambda\tau}(\bar{x}).$$

By using (7.143) rewritten as

$$F_{\lambda\tau}(\bar{x}) = -2 g_{\alpha(\tau}(\bar{x}) \partial_{\lambda)} F^\alpha(\bar{x}) - 2\Gamma^\alpha_{\beta(\lambda}(\bar{x}) g_{\tau)\alpha}(\bar{x}) F^\beta(\bar{x}), \tag{7.145}$$

and the DeWitt metric (7.91), it can be shown that

$$F^\lambda K^k_{\lambda,n} g_{km} = |g(x)|^{1/2} \Big\{ -g^{\nu(\rho} F^{\sigma)};\mu - g^{\mu(\rho} F^{\sigma)};\nu - c g^{\mu\nu} F^{(\rho;\sigma)}$$
$$- g^{\rho(\mu} g^{\nu)\sigma} F^\alpha \nabla_\alpha - \frac{c}{2} g^{\rho\sigma} g^{\mu\nu} F^\alpha \nabla_\alpha \Big\} \tilde{\delta}(x,x'). \tag{7.146}$$

The third term in Killing's equation (7.141) is $F^\lambda K^k_{\lambda,m} g_{kn}$, and can be obtained from (7.146) by the interchange of condensed indices n and m; in conventional notation, this corresponds to $(\mu\nu) \leftrightarrow (\rho\sigma)$ and $x \leftrightarrow x'$, giving

$$F^\lambda K^k_{\lambda,m} g_{kn} = |g(x')|^{1/2} \Big\{ -g'^{\sigma(\mu} F'^{\nu)};\rho - g'^{\rho(\mu} F'^{\nu)};\sigma - c g'^{\rho\sigma} F'^{(\mu;\nu)}$$
$$- g'^{\mu(\rho} g'^{\sigma)\nu} F'^\alpha \nabla'_\alpha - \frac{c}{2} g'^{\rho\sigma} g'^{\mu\nu} F'^\alpha \nabla'_\alpha \Big\} \tilde{\delta}(x',x). \tag{7.147}$$

In order to combine this result with that in (7.146), it is necessary to rewrite the δ-distributions in terms of $\tilde{\delta}(x,x')$ and its derivatives rather than $\tilde{\delta}(x',x)$. The biscalar Dirac δ-distribution is symmetric, so we find

$$f(x')|g(x')|^{1/2} \tilde{\delta}(x',x) = f(x')|g(x')|^{1/2} |g(x)|^{1/2} \delta(x',x)$$
$$= f(x)|g(x)|^{1/2} \tilde{\delta}(x,x') \tag{7.148}$$

for any function $f(x)$. All of the functions multiplying the undifferentiated δ-distribution in (7.147) may be evaluated at x rather than x'. It can also be shown that

$$|g(x')|^{1/2} f^\alpha(x') \nabla'_\alpha \tilde{\delta}(x', x) = -|g(x)|^{1/2} f^\alpha(x) \nabla_\alpha \tilde{\delta}(x, x') \\ - |g(x)|^{1/2} (\nabla_\alpha f^\alpha(x)) \tilde{\delta}(x, x'). \quad (7.149)$$

(A formal proof of (7.148) as well as (7.149) may be given by multiplying both sides with an arbitrary scalar test function, integrating over x', and showing that both sides lead to the same result.)

Armed with the two distributional identities (7.148) and (7.149), it is a straightforward matter to show using (7.144), (7.146), and (7.147) that

$$F^\lambda \left(K^k_\lambda g_{mn,k} + K^k_{\lambda,n} g_{km} + K^k_{\lambda,m} g_{kn} \right) = 0.$$

Because this holds for arbitrary F^λ, we conclude that the DeWitt metric satisfies Killing's equation. We can set $B^\epsilon{}_{\lambda m}$ in (7.137) to zero in the case of gravity.

7.5 The connection and its consequences

The key idea for the Vilkovisky (1984) formulation of the effective action resides in the construction of the appropriate connection. Some information can be gleaned about the connection from the following considerations. It was shown in the previous section that any gauge-invariant scalar functional can be expressed solely in terms of the invariants ξ^A that parameterize the surface $\mathcal{S}$ defined by the gauge conditions. In particular, the action functional must look like

$$S[\varphi] = \tilde{S}[\xi] \quad (7.150)$$

in the (ξ^A, θ^α) coordinate system. Just as for the case of non-gauge theories in the previous chapter, the perturbation theory will be obtained from a covariant Taylor expansion of the action. If the result is to be gauge invariant, then this expansion should only involve $\tilde{\sigma}^A[\xi; \xi']$ which contains only the invariants, and not the θ^α or θ'^α coordinates. This means that one of the geodesic equations must read

$$\frac{d^2 \xi^A}{ds^2} + \Gamma^A_{BC}[\xi] \frac{d\xi^B}{ds} \frac{d\xi^C}{ds} = 0 \quad (7.151)$$

if this is to be the case. We will require $\Gamma^A_{BC}[\xi]$ to be the Christoffel connection of the induced metric h_{AB}:

$$\Gamma^A_{BC}[\xi] = \frac{1}{2} h^{AD} \left(h_{DC,B} + h_{BD,C} - h_{BC,D} \right). \quad (7.152)$$

The geodesic equation (7.151) could have involved the components $\Gamma^A_{B\alpha}$ or $\Gamma^A_{\alpha\beta}$. Therefore we must demand that these components vanish:

$$\Gamma^A_{B\alpha} = 0, \tag{7.153}$$

$$\Gamma^A_{\alpha\beta} = 0, \tag{7.154}$$

in addition to (7.152). Because the Christoffel connection does not transform as a tensor, the requirements of (7.152)–(7.154) will not be coordinate-independent in general. However, the only allowed coordinate transformations are those with $\xi^A \to \xi'^A[\xi]$, because the feature of the induced metric being independent of θ discussed in the last section must be maintained. Under these restricted transformations, the conditions (7.152)–(7.154) are preserved.

So far, all that we know about the connection is that when it is written in (ξ^A, θ^α) coordinates, its components satisfy (7.152)–(7.154). (Note that because of the requirements we have imposed on the field space connection, we have no reason to assume that it is simply given by the Christoffel connection of the full field space metric g_{ij}.) The components in the original coordinates are

$$\tilde{\Gamma}^i_{jk}[\varphi] = \varphi^i{}_{,A}\xi^B{}_{,j}\xi^C{}_{,k}\Gamma^A_{BC} + \varphi^i{}_{,A}\xi^A{}_{,jk} + K^i_\alpha A^\alpha_{jk} \tag{7.155}$$

for some A^α_{jk} which involves the Γ^α_{ij} components. This result follows directly from the general properties of the connection under a coordinate transformation, along with equations (7.112) and (7.152)–(7.154). It will be shown later that the term proportional to K^i_α in (7.155) is irrelevant to the effective action, and therefore it follows that equations (7.152)–(7.154) are sufficient, as well as necessary, conditions for obtaining the desired connection.

The result in (7.155) is not in a very useful form. It would be better if we could relate $\tilde{\Gamma}^i_{jk}$ directly to the metric g_{ij} rather than dealing with (7.155). One way to do this is to define

$$\tilde{\nabla}_i g^\perp_{jk} = g^\perp_{jk,i} - g^\perp_{jl}\tilde{\Gamma}^l_{ik} - g^\perp_{kl}\tilde{\Gamma}^l_{ij} \tag{7.156}$$

and then to use (7.155) to prove that the right-hand side vanishes.[10] Using (7.120), we have, upon differentiation of both sides with respect to φ^i,

$$g^\perp_{jk,i} = \xi^A{}_{,ji}\xi^B{}_{,k}h_{AB} + \xi^A{}_{,j}\xi^B{}_{,ki}h_{AB} + \xi^A{}_{,j}\xi^B{}_{,k}\xi^C{}_{,i}h_{AB,C}. \tag{7.157}$$

Also using (7.120) and (7.155), as well as the fact that $g^\perp_{jl}K^l_\alpha = 0$ gives

$$g^\perp_{jl}\tilde{\Gamma}^l_{ik} = \xi^D{}_{,j}\xi^E{}_{,l}h_{DE}\varphi^l{}_{,A}\xi^B{}_{,i}\xi^C{}_{,k}\Gamma^A_{BC} + \xi^D{}_{,j}\xi^E{}_{,l}h_{DE}\varphi^l{}_{,A}\xi^A{}_{,ik}, \tag{7.158}$$

which is independent of A^α_{jk} in (7.155). Because $\xi^E{}_{,l}\varphi^l{}_{,A} = \delta^E{}_A$, this result may be simplified to

$$g^\perp_{jl}\tilde{\Gamma}^l_{ik} = \xi^D{}_{,j}h_{DA}\xi^B{}_{,i}\xi^C{}_{,k}\Gamma^A_{BC} + \xi^D{}_{,j}\xi^A{}_{,ik}h_{DA}. \tag{7.159}$$

---

[10] The fact that $\tilde{\nabla}_i g^\perp_{jk} = 0$ may be used to compute the connection was suggested by S. Carlip (private communication).

7.5 The connection and its consequences

Substitution of (7.157) and (7.159) into (7.156) results in

$$\tilde{\nabla}_i g_{\overline{jk}}^\perp = \xi^A{}_{,j}\xi^B{}_{,k}\xi^C{}_{,i}\{h_{AB,C} - h_{AD}\Gamma^D_{BC} - h_{BD}\Gamma^D_{AC}\}$$
$$= 0 \quad \text{(by (7.152))}. \tag{7.160}$$

We may now use the result of (7.156), and the fact that $\tilde{\nabla}_i g_{\overline{jk}}^\perp = 0$, to try to solve for $\tilde{\Gamma}^i_{jk}$ in the usual way. It is easily seen that

$$\tilde{\Gamma}^l_{ij} g_{\overline{lk}}^\perp = \frac{1}{2}\left(g_{\overline{jk},i}^\perp + g_{\overline{ki},j}^\perp - g_{\overline{ij},k}^\perp\right). \tag{7.161}$$

The usual next step involves inversion of the metric tensor appearing on the left-hand side, and cannot be performed here because $g_{\overline{lk}}^\perp$ is not invertible on the full field space. In fact, it is easy to see that (7.161) only determines the connection $\tilde{\Gamma}^l_{ij}$ up to an arbitrary term proportional to K^i_α, which we have already remarked as being irrelevant.

Contraction of both sides of (7.161) with g^{km}, and use of (7.70) leads to:

$$\tilde{\Gamma}^l_{ij} g_{\overline{lk}}^\perp g^{km} = \Gamma^m_{ij} - \frac{1}{2}g^{mk}\left[\left(\gamma^{\alpha\beta}K_{\alpha j}K_{\beta k}\right)_{,i}\right.$$
$$\left. + \left(\gamma^{\alpha\beta}K_{\alpha k}K_{\beta i}\right)_{,j} - \left(\gamma^{\alpha\beta}K_{\alpha i}K_{\beta j}\right)_{,k}\right], \tag{7.162}$$

where

$$\Gamma^m_{ij} = \frac{1}{2}g^{mk}\left(g_{kj,i} + g_{ik,j} - g_{ij,k}\right) \tag{7.163}$$

is the Christoffel connection of the original metric, and $K_{\alpha i}$ has been defined by

$$K_{\alpha i} = g_{ij} K^j_\alpha. \tag{7.164}$$

From (7.70) it follows that

$$\tilde{\Gamma}^l_{ij} g_{\overline{lk}}^\perp g^{km} = \tilde{\Gamma}^m_{ij} - K^m_\alpha \tilde{\Gamma}^l_{ij} \gamma^{\alpha\beta} K_{\beta l}. \tag{7.165}$$

Because we have argued that terms proportional to K^m_α in $\tilde{\Gamma}^m_{ij}$ are not important, we can write

$$\tilde{\Gamma}^m_{ij} \stackrel{K}{=} \Gamma^m_{ij} - \frac{1}{2}g^{mk}\left[\gamma^{\alpha\beta}K_{\alpha j}K_{\beta k,i} + \gamma^{\alpha\beta}K_{\alpha i}K_{\beta k,j} - \left(\gamma^{\alpha\beta}K_{\alpha i}K_{\beta j}\right)_{,k}\right], \tag{7.166}$$

where $\stackrel{K}{=}$ means equality up to the addition of a term proportional to $K^m_\alpha A^\alpha_{ij}$ for some A^α_{ij}.

Because $\gamma^{\alpha\beta}$ is a scalar under changes of φ^i, $\gamma^{\alpha\beta}{}_{,k} = \gamma^{\alpha\beta}{}_{;k}$ where the covariant derivative denoted by a semicolon is with respect to the Christoffel connection Γ^i_{jk}. It then follows from differentiation of both sides of (7.58) that

$$\gamma^{\alpha\beta}{}_{;k} = -\gamma^{\alpha\epsilon}\gamma^{\beta\sigma}\gamma_{\epsilon\sigma;k}. \tag{7.167}$$

Use of (7.56) leads to

$$\gamma^{\alpha\beta}{}_{;k} = -\gamma^{\alpha\epsilon}\gamma^{\beta\sigma}g_{ij}\left(K^i_\epsilon K^j_{\sigma;k} + K^i_{\epsilon;k}K^j_\sigma\right). \tag{7.168}$$

The result in (7.166) may then be written as

$$\tilde{\Gamma}^m_{ij} \stackrel{K}{=} \Gamma^m_{ij} - \frac{1}{2}g^{mk}\{\gamma^{\alpha\beta}K_{\alpha i}(K_{\beta k,j} - K_{\beta j,k}) + \gamma^{\alpha\beta}K_{\alpha j}(K_{\beta k,i} - K_{\beta i,k})$$
$$+ \gamma^{\alpha\epsilon}\gamma^{\beta\sigma}K_{\alpha i}K_{\beta j}g_{np}(K^n_\epsilon K^p_{\sigma;k} + K^p_\sigma K^n_{\epsilon;k})\}. \tag{7.169}$$

Further simplification on the right-hand side of (7.169) occurs as a result of the symmetry of the metric expressed in (7.137). The left-hand side of (7.137) may be re-expressed using the covariant derivative with respect to the Christoffel symbol to give

$$K_{\lambda m;n} + K_{\lambda n;m} = K_{\epsilon m}B^\epsilon{}_{\lambda n} + K_{\epsilon n}B^\epsilon{}_{\lambda m}. \tag{7.170}$$

Use of this result in (7.169) leads to

$$\tilde{\Gamma}^m_{ij} \stackrel{K}{=} \Gamma^m_{ij} - \frac{1}{2}g^{mk}\{2\gamma^{\alpha\beta}K_{\alpha i}K_{\beta k;j} - \gamma^{\alpha\beta}K_{\alpha i}K_{\epsilon j}B^\epsilon{}_{\beta k}$$
$$- \gamma^{\alpha\beta}K_{\alpha i}K_{\epsilon k}B^\epsilon{}_{\beta j} + 2\gamma^{\alpha\beta}K_{\alpha j}K_{\beta k;i} - \gamma^{\alpha\beta}K_{\alpha j}K_{\epsilon i}B^\epsilon{}_{\beta k}$$
$$- \gamma^{\alpha\beta}K_{\alpha j}K_{\epsilon k}B^\epsilon{}_{\beta i} - \gamma^{\alpha\epsilon}\gamma^{\beta\sigma}K_{\alpha i}K_{\beta j}(K^n_\epsilon K_{\sigma k;n} + K^n_\sigma K_{\epsilon k;n})$$
$$+ \gamma^{\alpha\epsilon}\gamma^{\beta\sigma}K_{\alpha i}K_{\beta j}(K^n_\epsilon K_{\delta n}B^\delta{}_{\sigma k} + K^n_\epsilon K_{\delta k}B^\delta{}_{\sigma n})$$
$$+ K_{\sigma n}K^n_\delta B^\delta{}_{\epsilon k} + K^n_\sigma K_{\delta k}B^\delta{}_{en}\}.$$

Terms proportional to K^m_α may be dropped, and (7.56) and (7.58) may be used, leading to

$$\tilde{\Gamma}^m_{ij} = \Gamma^m_{ij} - \gamma^{\alpha\beta}K_{\alpha i}K^m_{\beta;j} - \gamma^{\alpha\beta}K_{\alpha j}K^m_{\beta;i}$$
$$+ \frac{1}{2}\gamma^{\alpha\epsilon}\gamma^{\beta\sigma}K_{\alpha i}K_{\beta j}(K^n_\epsilon K^m_{\sigma;n} + K^n_\sigma K^m_{\epsilon;n}) + K^m_\alpha A^\alpha_{ij}, \tag{7.171}$$

where we have added on the term proportional to K^m_α in order to replace the $\stackrel{K}{=}$ sign with an $=$ sign. (See (7.155).) Note that the expressions involving $B^\epsilon{}_{\beta i}$ which appear in (7.170) have canceled in the final expression, so it does not matter whether or not they vanish.

The connection (7.171) may now be used to derive three important results to be used later.

First, consider $\tilde{\nabla}_j K^i_\alpha$ where $\tilde{\nabla}_j$ denotes the covariant derivative formed using the connection $\tilde{\Gamma}^k_{ij}$. We have

$$\tilde{\nabla}_j K^i_\alpha = K^i_{\alpha,j} + \tilde{\Gamma}^i_{jk}K^k_\alpha \tag{7.172}$$

by definition. Straightforward use of (7.171) along with (7.56) immediately leads to

$$\tilde{\nabla}_j K^i_\alpha = \frac{1}{2}\gamma^{\delta\beta}K_{\delta j}(K^k_\beta K^i_{\alpha;k} - K^k_\alpha K^i_{\beta;k}) + K^i_\beta A^\beta_{jk}K^k_\alpha. \tag{7.173}$$

The first term may be simplified by using the algebra (7.15), resulting in

$$\tilde{\nabla}_j K^i_\alpha = \frac{1}{2}f_{\alpha\beta}{}^\gamma K^i_\gamma \gamma^{\delta\beta}K_{\delta j} + K^i_\delta A^\delta_{jk}K^k_\alpha. \tag{7.174}$$

7.5 The connection and its consequences

This proves that $\tilde{\nabla}_j K_\alpha^i \propto K_\beta^i$. The more general result

$$\tilde{\nabla}_{j_1} \cdots \tilde{\nabla}_{j_n} K_\alpha^i = K_\beta^i T^\beta{}_{j_1\cdots j_n} \tag{7.175}$$

for some $T^\beta{}_{j_1\cdots j_n}$, now follows easily by induction.

The other two results needed later involve the geodetic interval. One of the requirements which led to the form found above for the connection, was that $\sigma^i[\varphi;\varphi']$ took the form $(\tilde{\sigma}^A[\xi;\xi'], \tilde{\sigma}^\alpha[\xi,\theta;\xi',\theta'])$ when written in the (ξ^A, θ^α) coordinate system. It then follows by performing a coordinate transformation that

$$\sigma^i[\varphi;\varphi'] = \varphi^i{}_{,A}\tilde{\sigma}^A[\xi;\xi'] + \varphi^i{}_{,\alpha}\tilde{\sigma}^\alpha[\xi,\theta;\xi',\theta']. \tag{7.176}$$

Consider $\mathfrak{K}_\alpha^j[\varphi']\frac{\delta}{\delta\varphi'^j}\sigma^i[\varphi;\varphi'] = \mathfrak{K}_\alpha^j[\varphi']\sigma^i{}_{;j'}[\varphi;\varphi']$ where $\mathfrak{K}_\alpha^j$ was defined in (7.113). Using (7.176), we have

$$\mathfrak{K}_\alpha^j[\varphi']\sigma^i{}_{;j'}[\varphi;\varphi'] = \mathfrak{K}_\alpha^j[\varphi']\varphi^i{}_{,A}\left\{\frac{\delta\tilde{\sigma}^A}{\delta\xi'^B}\frac{\delta\xi'^B}{\delta\varphi'^j}\right\}$$
$$+ \mathfrak{K}_\alpha^j[\varphi']\varphi^i{}_{,\beta}\left\{\frac{\delta\tilde{\sigma}^\beta}{\delta\xi'^A}\frac{\delta\xi'^A}{\delta\varphi'^j} + \frac{\delta\tilde{\sigma}^\beta}{\delta\theta'^\gamma}\frac{\delta\theta'^\gamma}{\delta\varphi'^j}\right\} \tag{7.177}$$

upon use of the chain rule. From (7.112) and (7.113), it may be seen that

$$\mathfrak{K}_\alpha^j[\varphi']\frac{\delta\xi'^A}{\delta\varphi'^j} = \frac{\delta\varphi'^j}{\delta\theta'^\alpha}\frac{\delta\xi'^A}{\delta\varphi'^j} = \frac{\delta\xi'^A}{\delta\theta'^\alpha} = 0 \tag{7.178}$$

$$\mathfrak{K}_\alpha^j[\varphi']\frac{\delta\theta'^\gamma}{\delta\varphi'^j} = \delta^\gamma{}_\alpha. \tag{7.179}$$

We then have from (7.177) the result

$$\mathfrak{K}_\alpha^j[\varphi']\sigma^i{}_{;j'}[\varphi;\varphi'] = \mathfrak{K}_\beta^i[\varphi]\frac{\delta\tilde{\sigma}^\beta}{\delta\theta'^\alpha}. \tag{7.180}$$

This proves the important result,

$$K_\alpha^j[\varphi']\sigma^i{}_{;j'}[\varphi;\varphi'] \propto K_\beta^i[\varphi]. \tag{7.181}$$

Finally, consider $\mathfrak{K}_\alpha^j[\varphi]\frac{\delta}{\delta\varphi^j}\sigma^i[\varphi;\varphi']$. Again using (7.176), the chain rule, (7.178) and (7.179) (which hold equally well when the coordinates are unprimed), it may be seen that

$$\mathfrak{K}_\alpha^j[\varphi]\sigma^i{}_{,j}[\varphi;\varphi'] = \varphi^i{}_{,A\alpha}\tilde{\sigma}^A + \mathfrak{K}_\alpha^j K^i_{\beta,j}\tilde{\sigma}^\beta + \mathfrak{K}_\beta^i\tilde{\sigma}^\beta{}_{,\alpha}. \tag{7.182}$$

Interchanging the order of differentiation (noting that $\varphi^i{}_{,\alpha} = \mathfrak{K}_\alpha^i$) gives

$$\varphi^i{}_{,A\alpha} = \frac{\delta}{\delta\xi^A}\mathfrak{K}_\alpha^i[\varphi]$$
$$= \mathfrak{K}^i_{\alpha,j}\varphi^j{}_{,A}. \tag{7.183}$$

Also, (7.114) may be used to rewrite the second term on the right-hand side of (7.182) in the form $\mathfrak{K}^j_\beta \mathfrak{K}^i_{\alpha,j} \tilde{\sigma}^\beta$. We therefore have

$$\mathfrak{K}^j_\alpha[\varphi]\sigma^i{}_{,j}[\varphi;\varphi'] = \mathfrak{K}^i_{\alpha,j}\varphi^j{}_{,A}\tilde{\sigma}^A + \mathfrak{K}^j_\beta \mathfrak{K}^i_{\alpha,j}\tilde{\sigma}^\beta + \mathfrak{K}^i_\beta \tilde{\sigma}^\beta{}_{,\alpha}$$
$$= \mathfrak{K}^i_{\alpha,j}\sigma^i + \mathfrak{K}^i_\beta \tilde{\sigma}^\beta{}_{,\alpha} \qquad \text{(by (7.176))}. \qquad (7.184)$$

Using this result it follows that

$$\mathfrak{K}^j_\alpha[\varphi]\tilde{\nabla}_j \sigma^i[\varphi;\varphi'] = (\tilde{\nabla}_j \mathfrak{K}^i_\alpha)\sigma^j + \mathfrak{K}^i_\beta \tilde{\sigma}^\beta{}_{,\alpha}. \qquad (7.185)$$

But then (7.174) may be used to show $\tilde{\nabla}_j \mathfrak{K}^i_\alpha \propto \mathfrak{K}^i_\beta$. Hence,

$$K^j_\alpha[\varphi]\tilde{\nabla}_j \sigma^i[\varphi;\varphi'] \propto K^i_\beta[\varphi]. \qquad (7.186)$$

The results established in (7.181) and (7.186) were first given by Vilkovisky (1984) by introducing a new metric of a specific type on $\mathcal{F}$. The derivations and interpretation of the theory presented here are independent of the precise nature of his metric. They were first presented in Ellicott et al. (1991) for the more general class of theories considered here.

7.5.1 The Landau–DeWitt gauge conditions

Having obtained the connection $\tilde{\Gamma}^m_{ij}$, it is possible to now explain why the Landau–DeWitt gauge conditions defined in (7.87) for Yang–Mills theory, and in (7.104) for gravity provide a convenient choice. The perturbative expansion of the effective action can be obtained much as described in Chapter 6 (with some additional complications due to gauge symmetry as we will discuss later). This expansion involves the covariant Taylor expansion of the classical action in powers of $\sigma^i[\varphi_\star;\varphi]$, and we must now use the connection $\tilde{\Gamma}^m_{ij}$ to calculate the various covariant derivatives that appear in the expansion. Because the classical action functional is most naturally expressed in terms of φ^i rather than $\sigma^i[\varphi_\star;\varphi]$, we must look at how $\sigma^i[\varphi_\star;\varphi]$ depends on $\varphi^i_\star$ and φ^i. In general this dependence is very complicated, but we will see how in some cases a much simpler result can be obtained in the Landau–DeWitt gauge.

Suppose that we define η^i by

$$\eta^i = \varphi^i - \varphi^i_\star. \qquad (7.187)$$

$\sigma^i[\varphi_\star;\varphi]$ is the tangent vector, at the point specified by $\varphi^i_\star$, to the geodesic connecting $\varphi_\star$ to φ, and satisfies the geodesic equation

$$\sigma^j[\varphi_\star;\varphi]\left(\sigma^i{}_{,j}[\varphi_\star;\varphi] + \tilde{\Gamma}^i_{jk}[\varphi_\star]\sigma^k[\varphi_\star;\varphi]\right) = \sigma^i[\varphi_\star;\varphi]. \qquad (7.188)$$

(Note that $\sigma^i{}_{,j}[\varphi_\star;\varphi] = \frac{\delta}{\delta\varphi^j}\sigma^i[\varphi_\star;\varphi]$ here.) If the field space was flat with $\tilde{\Gamma}^i_{jk}$ the Christoffel connection, then we would simply have $\sigma^i[\varphi_\star;\varphi] = -\eta^i$ if we chose to work in Cartesian coordinates. Obviously the case of a general field

7.5 The connection and its consequences

space, and the introduction of gauge symmetry (so that $\tilde{\Gamma}^i_{jk}$ differs from the Christoffel connection) complicates things; however we can still expand $\sigma^i[\varphi_*; \varphi]$ about $\varphi = \varphi_*$ in a Taylor series whose first term is $-\eta^i$. Define this expansion to be

$$\sigma^i[\varphi_*; \varphi] = -\eta^i + \sum_{n=2}^{\infty} \sigma^i_{(n)}, \qquad (7.189)$$

where

$$\sigma^i_{(n)} = \frac{1}{n!} \sigma^i{}_{j_1 \cdots j_n}[\varphi_*] \eta^{j_1} \cdots \eta^{j_n}. \qquad (7.190)$$

The subscript (n) serves to keep track of the order of the terms in powers of η^i. $\sigma^i{}_{j_1 \cdots j_n}[\varphi_*]$ is a coefficient that depends only on φ_*.

The object now is to substitute the expansion (7.189) into (7.188) and equate terms of equal order in η^i on both sides. From (7.187) we have

$$\eta^i{}_{,j} = -\delta^i{}_j, \qquad (7.191)$$

where $,j$ is short for $\frac{\delta}{\delta \varphi^j}$ as in the parenthetical remark below (7.188). Differentiation of (7.189) with respect to φ^j_* noting (7.191), shows that

$$\sigma^i{}_{,j}[\varphi_*; \varphi] = \delta^i{}_j + \sum_{n=1}^{\infty} a^i_{(n)j}, \qquad (7.192)$$

where

$$a^i_{(1)j} = -\sigma^i{}_{jk}[\varphi_*] \eta^k, \qquad (7.193)$$

$$a^i_{(n)j} = \frac{1}{n!} \left(\sigma^i{}_{j_1 \cdots j_n, j}[\varphi_*] - \sigma^i{}_{jj_1 \cdots j_n}[\varphi_*] \right) \eta^{j_1} \cdots \eta^{j_n}, \qquad (7.194)$$

with $n \geq 2$ in (7.194). We will stop writing the argument φ_* of the expansion coefficients to save ink; unless indicated otherwise, all expansion coefficients will be understood to be evaluated at φ_*.

Substitution of (7.189) and (7.192) into (7.188) leads to

$$0 = -a^i_{(1)j}\eta^j - a^i_{(2)j}\eta^j + a^i_{(1)j}\sigma^j_{(2)} + \tilde{\Gamma}^i_{jk}\eta^j\eta^k - 2\tilde{\Gamma}^i_{jk}\eta^j\sigma^k_{(2)}$$

$$+ \sum_{n=4}^{\infty} \left\{ a^i_{(1)j}\sigma^j_{(n-1)} - a^i_{(n-1)j}\eta^j - 2\tilde{\Gamma}^i_{jk}\eta^j\sigma^k_{(n-1)} \right.$$

$$\left. + \sum_{m=2}^{n-2} \left[a^i_{(n-m)j} + \tilde{\Gamma}^i_{jk}\sigma^k_{(n-m)} \right] \sigma^j_{(m)} \right\}. \qquad (7.195)$$

We may solve this equation order by order in η. The lowest order term is of order η^2 and gives

$$a^i_{(1)j}\eta^j = \tilde{\Gamma}^i_{jk}\eta^j\eta^k. \qquad (7.196)$$

Using (7.193) it is easily seen that

$$\sigma^i{}_{jk} = -\tilde{\Gamma}^i_{jk}, \tag{7.197}$$

and hence from (7.190),

$$\sigma^i_{(2)} = -\frac{1}{2}\tilde{\Gamma}^i_{jk}\eta^j\eta^k. \tag{7.198}$$

From (7.193) and (7.197) we find

$$a^i_{(1)j} = \tilde{\Gamma}^i_{jk}\eta^k, \tag{7.199}$$

which can be used to simplify the remaining terms in (7.195). The terms of order η^3 in (7.195) give (using (7.199))

$$a^i_{(2)j}\eta^j = -\tilde{\Gamma}^i_{jk}\eta^j\sigma^k_{(2)}, \tag{7.200}$$

and for $n \geq 4$ we have

$$a^i_{(n-1)j}\eta^j = -\tilde{\Gamma}^i_{jk}\eta^j\sigma^k_{(n-1)} + \sum_{m=2}^{n-2}\left[a^i_{(n-m)j} + \tilde{\Gamma}^i_{jk}\sigma^k_{(n-m)}\right]\sigma^j_{(m)}. \tag{7.201}$$

These last two equations may be used to solve for the coefficients appearing in (7.190) iteratively, starting from (7.197). From (7.200) it can be shown that

$$\sigma^i_{(3)} = -\frac{1}{6}\left(\tilde{\Gamma}^i_{j_1j_2,j_3} + \tilde{\Gamma}^i_{j_1k}\tilde{\Gamma}^k_{j_2j_3}\right)\eta^{j_1}\eta^{j_2}\eta^{j_3}. \tag{7.202}$$

With $n = 4$ in (7.201) it can be shown that

$$\sigma^i_{(4)} = -\frac{1}{24}\Big(\tilde{\Gamma}^i_{j_1j_2,j_3j_4} + 2\tilde{\Gamma}^i_{kj_1,j_2}\tilde{\Gamma}^k_{j_3j_4} + 2\tilde{\Gamma}^i_{kj_1}\tilde{\Gamma}^k_{j_2j_3,j_4}$$
$$- \tilde{\Gamma}^i_{j_1j_2,k}\tilde{\Gamma}^k_{j_3j_4} + \frac{1}{6}\tilde{\Gamma}^i_{jk}\tilde{\Gamma}^j_{j_1j_2}\tilde{\Gamma}^k_{j_3j_4}\Big)\eta^{j_1}\eta^{j_2}\eta^{j_3}\eta^{j_4}. \tag{7.203}$$

Obviously an explicit evaluation of $\sigma^i_{(n)}$ for large n becomes increasingly difficult, and anything that can be done to simplify matters is desirable. This is where the Landau–DeWitt gauge comes in. Suppose that we adopt

$$K_{\alpha i}[\varphi_*]\eta^i = 0 \tag{7.204}$$

as our gauge conditions. (Recall that $K_{\alpha i} = g_{ij}K^j_\alpha$.) Then if we write the connection $\tilde{\Gamma}^i_{jk}$ in (7.171) as

$$\tilde{\Gamma}^m_{ij} = \Gamma^m_{ij} + T^m_{ij} + K^m_\alpha A^\alpha_{ij}, \tag{7.205}$$

where

$$T^m_{ij} = \frac{1}{2}\gamma^{\alpha\epsilon}\gamma^{\beta\sigma}K_{\alpha i}K_{\beta j}\left(K^n_\epsilon K^m_{\sigma;n} + K^n_\sigma K^m_{\epsilon;n}\right)$$
$$- \gamma^{\alpha\beta}\left(K_{\alpha i}K^m_{\beta;j} + K_{\alpha j}K^m_{\beta;i}\right), \tag{7.206}$$

7.5 The connection and its consequences

it can be seen that when T^m_{ij} is contracted with η^i or η^j some of the terms will vanish. In fact, using (7.204) it is obvious that

$$T^m_{ij}\eta^j = -\gamma^{\alpha\beta} K_{\alpha i} K^m_{\beta;j}\eta^j, \tag{7.207}$$

and

$$T^m_{ij}\eta^i\eta^j = 0. \tag{7.208}$$

Thus it is not necessary to know the full complicated form of T^m_{ij} if we adopt the Landau–DeWitt gauge.

In the Landau–DeWitt gauge (7.204), we find

$$\sigma^i_{(2)} = -\frac{1}{2}\Gamma^i_{jk}\eta^j\eta^k - \frac{1}{2}K^i_\alpha A^\alpha_{jk}\eta^j\eta^k, \tag{7.209}$$

if we use (7.208). The last term is proportional to K^i_α, so that when $\sigma^i[\varphi_*;\varphi]$ multiplies $F_{,i}[\varphi_*]$, where $F[\varphi_*]$ is any gauge-invariant functional, the last term of (7.209) makes no contribution. All that we require in this case for $\sigma^i_{(2)}$ is the Christoffel connection for the field space metric.

At the next order of the expansion, involving $\sigma^i_{(3)}$, the situation is more complicated in general. In the first place, the first term on the right-hand side of (7.202) involves $\tilde{\Gamma}^i_{j_1j_2,j_3}$, that from (7.205) is seen to involve $\eta^j K^i_{\alpha,j}$. This will not be proportional to K^i_α in general, and we cannot argue that it may be ignored on the same basis as a term proportional to the undifferentiated K^i_α. However, we are free to choose $A^\alpha_{ij} = 0$ if we wish, as Vilkovisky (1984) did. In this case it is easy to show that

$$\tilde{\Gamma}^i_{j_1j_2,j_3}\eta^{j_1}\eta^{j_2}\eta^{j_3} = \Gamma^i_{j_1j_2,j_3}\eta^{j_1}\eta^{j_2}\eta^{j_3}. \tag{7.210}$$

(This follows directly from (7.208).) However, the second term in (7.202) reads

$$\tilde{\Gamma}^i_{j_1k}\tilde{\Gamma}^k_{j_2j_3}\eta^{j_1}\eta^{j_2}\eta^{j_3} = \left(\Gamma^i_{j_1k} + T^i_{j_1k}\right)\Gamma^k_{j_2j_3}\eta^{j_1}\eta^{j_2}\eta^{j_3} \tag{7.211}$$

and is not determined solely by the Christoffel connection. This is the situation for gravity, and it is necessary to utilize (7.207) when working beyond lowest order in perturbation theory. Fortunately, as we will see, at one-loop order we do not have to worry about this because we only require $\sigma^i_{(2)}$.

In the special case where $g_{ij,k} = 0$, so that the Christoffel connection vanishes, it is possible to proceed much further. This includes the important example of Yang–Mills theory as first noted by Fradkin and Tseytlin (1984), and proven by Rebhan (1987) for the Vilkovisky connection (obtained with $A^\alpha_{ij} = 0$ in (7.171)). For the Vilkovisky connection, we have

$$\tilde{\Gamma}^m_{ij} = T^m_{ij}, \tag{7.212}$$

if $\Gamma^m_{ij} = 0$. We can also replace the semicolons in (7.206) with commas representing ordinary derivatives. From (7.209) we immediately find

$$\sigma^i_{(2)} = 0.$$

From (7.202) we also find

$$\sigma^i_{(3)} = 0,$$

noting (7.210) and (7.211) with $\Gamma^k_{ij} = 0$. We will now prove by induction that $\sigma^i_{(n)} = 0$ for all $n \geq 2$.

Suppose that we have established that $\sigma^i_{(2)}, \ldots, \sigma^i_{(n-1)}$ all vanish. We need to show that this implies that $\sigma^i_{(n)} = 0$. The result in (7.201) implies that

$$a^i_{(n-1)j}\eta^j = 0, \tag{7.213}$$

since the right-hand side of this result is determined solely by $\sigma^i_{(m)}$ for $m \leq n-1$ all of which are assumed to vanish. If we use (7.194) in (7.213) and relabel $j \to j_n$, we find

$$\left(\sigma^i{}_{j_1 \cdots j_{n-1}, j_n} - \sigma^i{}_{j_1 \cdots j_n}\right) \eta^{j_1} \cdots \eta^{j_n} = 0. \tag{7.214}$$

Using this result in the definition (7.190) shows us that $\sigma^i_{(n)}$ can be written as

$$\sigma^i_{(n)} = \frac{1}{n!} \sigma^i{}_{j_1 \cdots j_{n-1}, j_n} \eta^{j_1} \cdots \eta^{j_n}. \tag{7.215}$$

By assumption, we have $\sigma^i_{(n-1)} = 0$, so that

$$\sigma^i{}_{j_1 \cdots j_{n-1}} \eta^{j_1} \cdots \eta^{j_{n-1}} = 0. \tag{7.216}$$

Differentiation of (7.216) with respect to $\varphi^{j_n}_*$, followed by multiplication with η^{j_n} leads to

$$\sigma^i{}_{j_1 \cdots j_{n-1}, j_n} \eta^{j_1} \cdots \eta^{j_n} - (n-1) \sigma^i{}_{j_1 \cdots j_{n-1}} \eta^{j_1} \cdots \eta^{j_{n-1}} = 0,$$

using the symmetry of $\sigma^i{}_{j_1 \cdots j_{n-1}}$ in its lower indices, along with (7.191). Because of (7.216) we find

$$\sigma^i{}_{j_1 \cdots j_{n-1}, j_n} \eta^{j_1} \cdots \eta^{j_n} = 0,$$

and hence from (7.215) follows our desired result that $\sigma^i_{(n)} = 0$.

To summarize, we have shown that for any theory whose field space metric is constant, in the Landau–DeWitt gauge

$$\sigma^i[\varphi_*; \varphi] = -\eta^i, \tag{7.217}$$

exactly as if the field space was flat. (Note that the proof given here does not require the metric to satisfy Killing's equation, in contrast to the treatment given by Rebhan (1987).)

7.6 The functional measure for gauge theories

The line element for $\mathcal{F}$ was written in the form given in (7.68). From this expression it is apparent that the natural volume element for $\mathcal{F}$ is

$$d\mu[\varphi] = \left(\prod_i \omega^i_\perp\right)\left(\prod_\alpha d\epsilon^\alpha\right)(\det_\perp g^\perp_{ij})^{1/2}(\det \gamma_{\alpha\beta})^{1/2}. \tag{7.218}$$

Here the subscript "$\perp$" on $\det_\perp g^\perp_{ij}$ denotes the fact that the determinant is only taken with respect to the space of eigenvectors orthogonal to K^i_α.

The first thing that we will do is study the dependence of $(\det_\perp g^\perp_{ij})^{1/2}$ and $(\det \gamma_{\alpha\beta})^{1/2}$ on the gauge parameters ϵ^α. We may use the following identity to do this:

$$\delta(\det M)^{1/2} = \frac{1}{2}(\det M)^{1/2}\operatorname{tr}(M^{-1}\delta M), \tag{7.219}$$

where M is any operator with an inverse M^{-1} and "δ" denotes a generic variation.

When we apply this identity to $(\det_\perp g^\perp_{ij})^{1/2}$, it is seen to require a knowledge of the inverse to $g^\perp_{ij}$. Although $g^\perp_{ij}$ is not invertible on the full field space, it is invertible on the subspace orthogonal to K^i_α. Define $(g^\perp)^{ij}$ to be the inverse of $g^\perp_{ij}$ on this subspace. Just as in (7.120),

$$(g^\perp)^{ij} = \varphi^i{}_{,A}\varphi^j{}_{,B}h^{AB}. \tag{7.220}$$

Using (7.120), it is easily shown that

$$(g^\perp)^{ij}g^\perp_{jk} = \varphi^i{}_{,A}\xi^A{}_{,k}. \tag{7.221}$$

We are regarding φ^i as $\varphi^i[\xi,\epsilon]$ here, so the chain rule gives

$$\varphi^i{}_{,A}\xi^A{}_{,k} = \delta^i{}_k - \frac{\delta\varphi^i}{\delta\epsilon^\alpha}\frac{\delta\epsilon^\alpha}{\delta\varphi^k}$$

$$= \delta^i{}_k - K^i_\alpha \frac{\delta\epsilon^\alpha}{\delta\varphi^k} \quad \text{(by (7.2))}.$$

From (7.62), it follows that $\frac{\delta\epsilon}{\delta\varphi^k} = \gamma^{\alpha\beta}K^i_\beta g_{ik}$, and so

$$\varphi^i{}_{,A}\xi^A{}_{,k} = \delta^i{}_k - K^i_\alpha \gamma^{\alpha\beta} K^j_\beta g_{jk}$$

$$= P^i{}_k \quad \text{(by (7.53))}. \tag{7.222}$$

Therefore, using (7.222) in (7.221), we see that

$$(g^\perp)^{ij}g^\perp_{jk} = P^i{}_k. \tag{7.223}$$

Note that when operating on transverse vectors, $P^i{}_k$ acts like a Kronecker delta, so that (7.223) is what might have been expected.

Applying (7.219) to $\left(\det_\perp g_{ij}^\perp\right)^{1/2}$ gives

$$\frac{\delta}{\delta\epsilon^\alpha}\left(\det_\perp g_{ij}^\perp\right)^{1/2} = \frac{1}{2}\left(\det_\perp g_{ij}^\perp\right)^{1/2}(g^\perp)^{ij}\frac{\delta}{\delta\epsilon^\alpha}g_{ij}^\perp$$

$$= \frac{1}{2}\left(\det_\perp g_{ij}^\perp\right)^{1/2}(g^\perp)^{ij}g_{ij;k}^\perp K_\alpha^k, \qquad (7.224)$$

using the chain rule and the gauge transformation (7.2). Killing's equation (7.130) may be used to obtain the result

$$K_\alpha^k g_{ij;k}^\perp = -\left(K_{\alpha;i}^l g_{lj}^\perp + K_{\alpha;j}^l g_{il}^\perp\right). \qquad (7.225)$$

Substitution of this into the right-hand side of (7.224) followed by use of (7.223) leads to

$$\frac{\delta}{\delta\epsilon^\alpha}\left(\det_\perp g_{ij}^\perp\right)^{1/2} = -\left(\det_\perp g_{ij}^\perp\right)^{1/2} K_{\alpha;i}^l P^i{}_l. \qquad (7.226)$$

From the definition of $P^i{}_l$, (see (7.53)), we have

$$K_{\alpha;i}^l P^i{}_l = K_{\alpha;i}^i - K_{\alpha;i}^l K_\epsilon^i \gamma^{\epsilon\tau} K_\tau^m g_{ml}. \qquad (7.227)$$

The algebra (7.15) may be used in the second term to give

$$K_{\alpha;i}^l P^i{}_l = K_{\alpha;i}^i + f_{\epsilon\alpha}{}^\epsilon - K_{\epsilon;i}^l K_\alpha^i \gamma^{\epsilon\tau} K_\tau^m g_{ml}. \qquad (7.228)$$

(Equation (7.55) has also been used.) Now note from (7.56) that

$$\gamma_{\epsilon\tau;i} = K_{\epsilon;i}^l K_\tau^m g_{ml} + K_{\tau;i}^m K_\epsilon^l g_{ml}, \qquad (7.229)$$

and so

$$K_{\alpha;i}^l P^i{}_l = K_{\alpha;i}^i + f_{\epsilon\alpha}{}^\epsilon - \frac{1}{2}K_\alpha^i \gamma^{\epsilon\tau}\gamma_{\epsilon\tau;i}. \qquad (7.230)$$

With this result, (7.226) becomes

$$\frac{\delta}{\delta\epsilon^\alpha}\left(\det_\perp g_{ij}^\perp\right)^{1/2} = \left(\det_\perp g_{ij}^\perp\right)^{1/2}\left\{-K_{\alpha;i}^i - f_{\epsilon\alpha}{}^\epsilon + \frac{1}{2}K_\alpha^i \gamma^{\epsilon\tau}\gamma_{\epsilon\tau;i}\right\}. \qquad (7.231)$$

Now,

$$K_\alpha^i \gamma^{\epsilon\tau}\gamma_{\epsilon\tau;i} = K_\alpha^i \gamma^{\epsilon\tau}\gamma_{\epsilon\tau,i}$$
$$= K_\alpha^i \gamma^{\epsilon\tau}\left(2K_{\epsilon,i}^j K_\tau^k g_{jk} + K_\epsilon^j K_\tau^k g_{jk,i}\right) \quad \text{(by (7.56))}$$
$$= 2K_\alpha^i \gamma^{\epsilon\tau} K_{\epsilon,i}^j K_\tau^k g_{jk} + \gamma^{\epsilon\tau} K_\epsilon^j K_\tau^k \left(-2K_{\alpha,j}^l g_{lk}\right.$$
$$\left. + 2K_\beta^i g_{ij} B^\beta{}_{\alpha k}\right) \quad \text{(by (7.137))}$$
$$= 2f_{\epsilon\alpha}{}^\epsilon + 2K_\epsilon^i B^\epsilon{}_{\alpha i} \qquad (7.232)$$

using (7.15) and (7.55), as well as the antisymmetry property of the structure constants. This expression may be written in another form by contraction of the Killing equation. First rewrite (7.137) as

$$K_{\lambda n;m} + K_{\lambda m;n} = K_{\epsilon n} B^\epsilon{}_{\lambda m} + K_{\epsilon m} B^\epsilon{}_{\lambda n} \qquad (7.233)$$

7.6 The functional measure for gauge theories

using the definition (7.164). Contraction of both sides with g^{nm} leads to the result

$$K_\epsilon^i B^\epsilon{}_{\alpha i} = K^i_{\alpha;i}. \tag{7.234}$$

We therefore have from (7.232) that

$$K^i_\alpha \gamma^{\epsilon\tau} \gamma_{\epsilon\tau;i} = 2\left(f_{\epsilon\alpha}{}^\epsilon + K^i_{\alpha;i}\right). \tag{7.235}$$

Using this in (7.231) shows that

$$\frac{\delta}{\delta\epsilon^\alpha}\left(\det\nolimits_\perp g^\perp_{ij}\right)^{1/2} = 0, \tag{7.236}$$

proving that $\left(\det\nolimits_\perp g^\perp_{ij}\right)^{1/2}$ is independent of the gauge parameters.

Applying (7.219) to $\left(\det \gamma_{\alpha\beta}\right)^{1/2}$ leads to

$$\begin{aligned}\frac{\delta}{\delta\epsilon^\alpha}\left(\det \gamma_{\alpha\beta}\right)^{1/2} &= \frac{1}{2}\left(\det \gamma_{\alpha\beta}\right)^{1/2} \gamma^{\epsilon\tau} \frac{\delta}{\delta\epsilon^\alpha} \gamma_{\epsilon\tau} \\ &= \frac{1}{2}\left(\det \gamma_{\alpha\beta}\right)^{1/2} \gamma^{\epsilon\tau} \gamma_{\epsilon\tau;i} K^i_\alpha \\ &= \left(\det \gamma_{\alpha\beta}\right)^{1/2}\left\{f_{\epsilon\alpha}{}^\epsilon + K^i_{\alpha;i}\right\}, \end{aligned} \tag{7.237}$$

using (7.235).

The result obtained in (7.237) involves the following two expressions with contracted indices: $f_{\epsilon\alpha}{}^\epsilon$ and $K^i_{\alpha;i}$. For both Yang–Mills theory and gravity we showed that $B^\epsilon{}_{\alpha i} = 0$; thus, from (7.234) we can conclude that $K^i_{\alpha;i} = 0$. If we lower the index on the structure functions $f_{\epsilon\alpha}{}^\epsilon$ and use the anti-symmetry property, we can also conclude that $f_{\epsilon\alpha}{}^\epsilon = 0$. However, for more general theories, we might not be able to apply similar reasoning to conclude that $K^i_{\alpha;i} = 0$. Nevertheless, the distributional nature of both $K^i_{\alpha;i}$ and $f_{\epsilon\alpha}{}^\epsilon$ means that they will involve Dirac δ-distributions with coincident arguments, $\delta(x,x)$, when the indices are uncondensed. It is customary to define $\delta(x,x)$ to be zero, and we will follow this practice. As a consequence of this regularization, $\left(\det \gamma_{\alpha\beta}\right)^{1/2}$ is independent of ϵ^α, and can therefore be expressed in terms of invariants ξ^A only. It could be that a more rigorous treatment of the functional measure would lead to this result in a more satisfactory way.

The importance of showing that $\det \gamma_{\alpha\beta}$ and $\det\nolimits_\perp g^\perp_{ij}$ are gauge invariant resides in the fact that they can each be expressed solely in terms of the invariants ξ^A if we use the (ξ^A, θ^α) coordinate system. This is one feature that allows us to obtain a gauge-invariant functional integral for the effective action. Let

$$I = \int d\mu[\varphi] F[\varphi], \tag{7.238}$$

where $F[\varphi]$ is any gauge-invariant scalar functional and $d\mu[\varphi]$ was given in (7.218). Because $F[\varphi]$ is gauge invariant, this means that $\delta F[\varphi]/\delta\epsilon^\alpha = 0$. Nothing in the integrand or measure depends on the group parameters ϵ^α. We may therefore ignore the integration over ϵ^α in the integral and take

$$I = \int \left(\prod_i \omega_\perp^i\right) \left(\det{}_\perp g_{ij}^\perp\right)^{1/2} \left(\det \gamma_{\alpha\beta}\right)^{1/2} F[\varphi]. \tag{7.239}$$

Note that even though there is no longer an integration over ϵ^α, the factor $\left(\det \gamma_{\alpha\beta}\right)^{\frac{1}{2}}$ remains. This term still involves the invariants ξ^A which parameterize $\mathcal{F}/\mathfrak{G}$, and may be interpreted geometrically as associated with the volume of the orbit over each point in $\mathcal{F}/\mathfrak{G}$.

Because $F[\varphi]$ is gauge invariant, it is possible to express it solely in terms of the invariants ξ^A:

$$F[\varphi] = \tilde{F}[\xi]. \tag{7.240}$$

(See (7.121)–(7.125).) From (7.108) and (7.120), we have

$$\left(\prod_i \omega_\perp^i\right) \left(\det{}_\perp g_{ij}^\perp\right)^{1/2} = \left(\prod_A d\xi^A\right) \left(\det h_{AB}[\xi]\right)^{1/2}. \tag{7.241}$$

We may therefore write

$$I = \int \left(\prod_A d\xi^A\right) \left(\det h_{AB}[\xi]\right)^{1/2} \left(\det \tilde{\gamma}_{\alpha\beta}[\xi]\right)^{1/2} \tilde{F}[\xi], \tag{7.242}$$

where $\tilde{\gamma}_{\alpha\beta}[\xi] = \gamma_{\alpha\beta}[\varphi[\xi,\theta]]$, and the integration is now over the surface $\mathcal{S}$ specified by the gauge conditions.

In order to obtain a more useful form for the integral, it is helpful to write it back in a form involving an integration over all fields φ^i. To this end, consider the space spanned by (ξ^A, θ^α) where θ^α is chosen to be the gauge condition χ^α as in (7.109). We will introduce an integration over χ^α in (7.242), and then make a change of variables from (ξ^A, χ^α) to φ^i. The integration over χ is easily obtained by use of the Dirac δ-distribution $\tilde{\delta}[\chi^\alpha; 0]$ defined by

$$\int \left(\prod_\alpha d\chi^\alpha\right) \tilde{\delta}[\chi^\alpha; 0] = 1. \tag{7.243}$$

$\tilde{\delta}[\chi^\alpha; 0]$ transforms like a scalar density under a change of coordinates in the first argument, and like a scalar under a change of coordinates in its second argument. (This is the functional analogue of the normal Dirac δ-distribution introduced in (6.8).) Use of (7.243) enables us to write (7.242) as

$$I = \int \left(\prod_A d\xi^A\right) \left(\prod_\alpha d\chi^\alpha\right) \left(\det h_{AB}[\xi]\right)^{1/2} \left(\det \tilde{\gamma}_{\alpha\beta}[\xi]\right)^{1/2} \tilde{\delta}[\chi^\alpha; 0] \tilde{F}[\xi]. \tag{7.244}$$

All that remains now is to work out the Jacobian of the coordinate change from (ξ^A, χ^α) to φ^i.

7.6 The functional measure for gauge theories

The simplest way to do this is to begin by taking the differential of χ^α and expressing it in terms of $\omega_\perp^i$ and $\omega_\parallel^\alpha$ defined in (7.63)–(7.65). It can be seen that

$$dx^\alpha = \chi^\alpha{}_{,i} d\varphi^i$$
$$= \chi^\alpha{}_{,i}(\omega_\perp^i + K_\beta^i d\epsilon^\beta)$$
$$= \chi^\alpha{}_{,i}\omega_\perp^i + Q^\alpha{}_\beta d\epsilon^\beta.$$

We have used the definition (7.74) in the last line. Solving for $d\epsilon^\alpha$ results in

$$d\epsilon^\alpha = (Q^{-1})^\alpha{}_\beta \left(d\chi^\beta - \chi^\beta{}_{,i}\omega_\perp^i\right). \tag{7.245}$$

We may use this result in the field space line element (7.68) to obtain

$$ds^2 = g_{ij}^\perp \omega_\perp^i \omega_\perp^j + \gamma_{\alpha\beta}(Q^{-1})^\alpha{}_\rho(Q^{-1})^\beta{}_\sigma \left(d\chi^\rho - \chi^\rho{}_{,i}\omega_\perp^i\right)\left(d\chi^\sigma - \chi^\sigma{}_{,j}\omega_\perp^j\right). \tag{7.246}$$

In this form the line element resembles that for a Kaluza–Klein theory. By choosing the obvious non-coordinate basis indicated by the form of (7.246) we obtain the key result,

$$\left(\prod_i d\varphi^i\right)(\det g_{ij})^{1/2} = \left(\prod_i \omega_\perp^i\right)\left(\prod_\alpha d\chi^\alpha\right)(\det_\perp g_{ij}^\perp)^{1/2}$$
$$\times \left[\det\left(\gamma_{\alpha\beta}(Q^{-1})^\alpha{}_\rho(Q^{-1})^\beta{}_\sigma\right)\right]^{1/2}. \tag{7.247}$$

By changing basis from $\omega_\perp^i$ to $d\xi^A$, using (7.108) and (7.120), it can be seen that

$$\left(\prod_i d\varphi^i\right)(\det g_{ij})^{1/2} = \left(\prod_A d\xi^A\right)\left(\prod_\alpha d\chi^\alpha\right)(\det h_{AB})^{1/2}$$
$$\times (\det \gamma_{\alpha\beta})^{1/2}(\det Q^\alpha{}_\beta)^{-1}. \tag{7.248}$$

This is sufficient to prove that the Jacobian of the transformation from (ξ^A, χ^α) to φ^i is

$$(\det g_{ij})^{1/2}(\det h_{AB})^{-1/2}(\det \gamma_{\alpha\beta})^{-1/2}(\det Q^\alpha{}_\beta).$$

Using this result in (7.244) shows that

$$I = \int \left(\prod_i d\varphi^i\right)\left(\det g_{ij}\right)^{1/2}(\det Q^\alpha{}_\beta)\tilde\delta[\chi^\alpha;0]F[\varphi]. \tag{7.249}$$

This presents a geometrical derivation of the Faddeev and Popov (1967) ansatz. Such an interpretation was first given for pure Yang–Mills theory in a special gauge by Babelon and Viallet (1979). (See also Babelon and Viallet (1981) and Viallet (1987).) The more general treatment described here follows Ellicott et al. (1989, 1991).

7.7 Gauge-invariant effective action

The effective action $\hat{\Gamma}[v; \varphi_*]$ for non-gauge theories was expressed in functional integral form in the previous chapter. Here we will generalize to gauge theories. The only significant new feature that arises in the formal definition of the effective action is that the measure does not simply involve $(\det g_{ij})^{1/2}$ as we found in the last section. We will define

$$\exp\left(\frac{i}{\hbar}\hat{\Gamma}[v;\varphi_*]\right) = \int \left(\prod_i \omega_\perp^i\right)(\det_\perp g_{ij}^\perp)^{1/2}(\det\gamma_{\alpha\beta})^{1/2}$$
$$\times \exp\left\{\frac{i}{\hbar}S[\varphi] + \frac{i}{\hbar}(v^i - \sigma^i)\frac{\delta}{\delta v^i}\hat{\Gamma}[v;\varphi_*]\right\}, \qquad (7.250)$$

where v^i and σ^i are defined as before by

$$v^i = \sigma^i[\varphi_*; \overline{\varphi}], \qquad (7.251)$$
$$\sigma^i = \sigma^i[\varphi_*; \varphi]. \qquad (7.252)$$

We can equally well regard the effective action as a functional of φ_* and $\overline{\varphi}$, instead of φ_* and v^i if we choose. We will define

$$\Gamma[\overline{\varphi}; \varphi_*] = \hat{\Gamma}\left[v^i = \sigma^i[\varphi_*; \overline{\varphi}], \varphi_*\right]. \qquad (7.253)$$

7.7.1 Gauge invariance

Our intention now is to use the result in Section 7.6 to write the functional integral as an integral over the basic field variables φ^i. To obtain (7.249) we made the assumption that the integrand of the functional integral was a gauge-invariant functional; thus, we must show that the argument of the exponential in (7.250) is gauge invariant. The classical action functional is gauge invariant, so we really only need to check the second term in the argument of the exponential. The proof will amount to showing that the effective action, as defined in (7.250), is gauge invariant. We will do this using the loop expansion that was discussed in Chapter 6 for non-gauge theories.

Let $\Gamma^{(n)}[v; \varphi_*]$ be the term of nth order in the loop counting parameter $\hbar$. Because (7.250) is really a perturbative definition of $\hat{\Gamma}[v; \varphi_*]$, to obtain $\Gamma^{(n)}[v; \varphi_*]$ we only need to use $\Gamma^{(0)}[v; \varphi_*], \ldots, \Gamma^{(n-1)}[v; \varphi_*]$ on the right-hand side of the equation. (This was shown in the previous chapter for a non-gauge theory.) To lowest order in the loop expansion we have

$$\Gamma^{(0)}[v; \varphi_*] = S[\overline{\varphi}], \qquad (7.254)$$
$$= \hat{S}[v; \varphi_*], \qquad (7.255)$$

where $\hat{S}[v;\varphi_*]$ is obtained from $S[\overline{\varphi}]$ by solving (7.251) for $\overline{\varphi}$ in terms of v^i and φ_*. This follows since to evaluate the functional integral we expand the right-hand side about $\varphi = \overline{\varphi}$. (The lowest order term is obtained simply by taking $\varphi = \overline{\varphi}$ in the exponential.) We are guaranteed that $\Gamma^{(0)}[v;\varphi_*]$ is gauge invariant.

In order to generate $\Gamma^{(1)}[v;\varphi_*]$ using (7.250), it is only necessary to use $\hat{\Gamma}[v;\varphi_*] = \hat{S}[v;\varphi_*]$ on the right-hand side. If we can prove that $(v^i - \sigma^i)\delta\hat{S}[v;\varphi_*]/\delta v^i$ depends only on gauge-invariant quantities at φ_* and $\overline{\varphi}$, this will be sufficient for us to deduce that $\Gamma^{(1)}[v;\varphi_*]$ is gauge invariant. We will begin with our knowledge that the classical action functional is gauge invariant. This can be expressed in differential form as

$$K^i_\alpha[\overline{\varphi}]\frac{\delta}{\delta\overline{\varphi}^i}S[\overline{\varphi}] = 0. \tag{7.256}$$

Now regard $S[\overline{\varphi}] = \hat{S}[v;\varphi_*]$, and use the chain rule to obtain

$$K^i_\alpha[\overline{\varphi}]\frac{\delta v^j}{\delta\overline{\varphi}^i}\frac{\delta\hat{S}[v;\varphi_*]}{\delta v^j} = 0 \tag{7.257}$$

from (7.256). If we use (7.251) we find $\delta v^j/\delta\overline{\varphi}^i = \sigma^j{}_{;i'}[\varphi_*;\overline{\varphi}]$ with our convention that the prime on the index i denotes differentiation with respect to the second argument of the two-point functional. In (7.181) we proved that $K^i_\alpha[\overline{\varphi}]\sigma^j{}_{;i'}[\varphi_*;\overline{\varphi}] \propto K^j_\beta[\varphi_*]$. If we use this in (7.257) we find

$$K^i_\alpha[\varphi_*]\frac{\delta}{\delta v^i}\hat{S}[v;\varphi_*] = 0. \tag{7.258}$$

(Note that this result holds for any functional that obeys (7.256), not just the classical action.)

The properties of the connection discussed in Section 7.5 resulted in (7.176) for the geodetic interval when we transformed to the (ξ^A,θ^α) coordinate system. Furthermore, from (7.112) we had $\varphi^i_{,\alpha} \propto K^i_\beta$. This means that we can write σ^i defined in (7.252) as

$$\sigma^i = \sigma^i[\varphi_*;\varphi] = \varphi^i_{*,A}\tilde{\sigma}^A[\xi_*;\xi] + K^i_\alpha[\varphi_*]T^\alpha[\varphi_*;\varphi], \tag{7.259}$$

for some $T^\alpha[\varphi_*;\varphi]$. Similarly, if we replace φ in (7.259) with $\overline{\varphi}$, then v^i defined in (7.251) may be written as

$$v^i = \sigma^i[\varphi_*;\overline{\varphi}] = \varphi^i_{*,A}\tilde{\sigma}^A[\xi_*;\bar{\xi}] + K^i_\alpha[\varphi_*]T^\alpha[\varphi_*;\overline{\varphi}]. \tag{7.260}$$

We may now use (7.258) to see that

$$(v^i - \sigma^i)\frac{\delta}{\delta v^i}\hat{S}[v;\varphi_*] = \varphi^i_{*,A}\left(\tilde{\sigma}^A[\xi_*;\bar{\xi}] - \tilde{\sigma}^A[\xi_*;\xi]\right)\frac{\delta}{\delta v^i}\hat{S}[v;\varphi_*]. \tag{7.261}$$

We know that $\hat{S}[v;\varphi_*] = S[\overline{\varphi}]$ is gauge invariant. This means that in the (ξ^A,θ^α) coordinate system $\hat{S}[v;\varphi_*]$ can only depend on the coordinates $\bar{\xi}^A$, with no dependence at all on $\bar{\theta}^\alpha$. We can therefore regard $\hat{S}[v;\varphi_*]$ as a functional of $\tilde{v}^A[\xi_*;\bar{\xi}]$ and ξ_*:

$$\hat{S}[v;\varphi_*] = \tilde{S}[\tilde{v};\xi_*]. \tag{7.262}$$

This means that the right-hand side of (7.261) can be rewritten to give

$$(v^i - \sigma^i)\frac{\delta}{\delta v^i}\hat{S}[v;\varphi_*] = (\tilde{v}^A - \tilde{\sigma}^A)\frac{\delta}{\delta \tilde{v}^A}\tilde{S}[\tilde{v};\xi_*]. \tag{7.263}$$

(Note from (7.260) that $\delta v^i/\delta \tilde{v}^A = \varphi^i_{*,A}$.) $\tilde{v}^A$ and $\tilde{\sigma}^A$ are defined by an obvious extension of (7.251) and (7.252) to stand for $\tilde{\sigma}^A[\xi_*;\bar{\xi}]$ and $\tilde{\sigma}^A[\xi_*;\xi]$ respectively.

The result in (7.263) is sufficient to deduce what we are trying to show; namely, that the expression on the left-hand side that enters $\Gamma^{(1)}[v;\varphi_*]$ is gauge invariant. This is true because the right-hand side of (7.263) has been expressed only in terms of the coordinates ξ_*^A and $\bar{\xi}^A$, which are invariant under gauge transformations. It therefore follows that the one-loop approximation for the effective action defined through (7.250) is gauge invariant, meaning that it depends only on the invariants ξ_* and $\bar{\xi}$. This invariance may be expressed in differential form as

$$K^i_\alpha[\varphi_*]\frac{\delta}{\delta \varphi_*^i}\Gamma^{(1)}[v;\varphi_*] = 0, \tag{7.264}$$

$$K^i_\alpha[\bar{\varphi}]\frac{\delta}{\delta \bar{\varphi}^i}\Gamma^{(1)}[v;\varphi_*] = 0. \tag{7.265}$$

The next stage is to consider the two-loop expression for the effective action, $\Gamma^{(2)}[v;\varphi_*]$. We must use $\Gamma^{(0)}[v;\varphi_*]$ and $\Gamma^{(1)}[v;\varphi_*]$ on the right-hand side of (7.250). The steps which led from (7.256) to (7.258) may be repeated with $\hat{S} = \Gamma^{(0)}$ replaced with $\Gamma^{(1)}$ since we have now established (7.264) and (7.265). It is easily deduced that $(v^i - \sigma^i)\delta\Gamma^{(1)}[v;\varphi_*]/\delta v^i$ can be expressed in terms of the invariants ξ_* and $\bar{\xi}$ only, exactly as we did in (7.263) for $\hat{S}$. We conclude that $\Gamma^{(2)}[v;\varphi_*]$ can be expressed solely in terms of the invariants ξ_* and $\bar{\xi}$. This argument is sufficient to establish by induction that the effective action defined by (7.250) is gauge invariant order by order in the loop expansion. We are therefore justified in applying the results of Section 7.6 for the functional measure to write (7.250) in the equivalent form

$$\exp\left(\frac{i}{\hbar}\hat{\Gamma}[v;\varphi_*]\right) = \int [d\varphi]\, \mathcal{M}[\chi;v;\varphi_*;\varphi]$$
$$\times \exp\left\{\frac{i}{\hbar}S[\varphi] + \frac{i}{\hbar}(v^i - \sigma^i)\frac{\delta}{\delta v^i}\hat{\Gamma}[v;\varphi_*]\right\}, \tag{7.266}$$

where we have abbreviated

$$[d\varphi] = \left(\prod_i d\varphi^i\right)(\det g_{ij})^{1/2}, \tag{7.267}$$

$$\mathcal{M}[\chi;v;\varphi_*;\varphi] = \left(\det Q^\alpha_{\ \beta}\right)\tilde{\delta}[\chi^\alpha;0]. \tag{7.268}$$

The loop expansion of this definition for the effective action $\hat{\Gamma}[v;\varphi_*]$ generates a functional $\Gamma[\overline{\varphi};\varphi_*]$ which is gauge invariant in both $\overline{\varphi}$ and φ_*.

7.7.2 Gauge condition independence

We will now give a formal proof that (7.266) gives an effective action that does not depend on the choice of gauge conditions χ^α.

Suppose that we alter the gauge conditions from χ^α to χ'^α with φ_*^i and $\overline{\varphi}^i$ held fixed. (From (7.251) this will keep v^i fixed.) Let the change be expressed as

$$\chi'^\alpha[\varphi] = \chi^\alpha[\varphi] + \Delta\chi^\alpha[\varphi], \qquad (7.269)$$

where we only indicate the functional dependence on the variable of integration φ^i which appears in the functional integration. Call $\hat{\Gamma}'$ the effective action obtained from (7.266) with the replacement of χ^α with χ'^α. We have

$$\exp\left(\frac{i}{\hbar}\hat{\Gamma}'\right) = \int [d\varphi]\, \mathcal{M}[\chi';v;\varphi_*;\varphi]$$
$$\times \exp\left\{\frac{i}{\hbar}S[\varphi] + \frac{i}{\hbar}(v^i - \sigma^i)\frac{\delta}{\delta v^i}\hat{\Gamma}'\right\}, \qquad (7.270)$$

with

$$\mathcal{M}[\chi';v;\varphi_*;\varphi] = \left(\det Q'^\alpha{}_\beta\right)\tilde{\delta}[\chi'^\alpha;0]. \qquad (7.271)$$

Now we will make a change of integration variable in the functional integral from φ^i to φ'^i, where

$$\chi'^\alpha[\varphi] = \chi^\alpha[\varphi']. \qquad (7.272)$$

Let $\varphi'^i = \varphi^i + \delta\varphi^i$. Then

$$\chi'^\alpha[\varphi] = \chi^\alpha[\varphi^i + \delta\varphi^i]$$
$$= \chi^\alpha[\varphi^i] + \delta\varphi^i \chi^\alpha{}_{,i}[\varphi]$$

if we expand to first order in $\delta\varphi^i$. Treating $\Delta\chi^\alpha[\varphi]$ in (7.269) as infinitesimal results in

$$\delta\varphi^i \chi^\alpha{}_{,i}[\varphi] = \Delta\chi^\alpha[\varphi].$$

If we use the definition of $Q^\alpha{}_\beta = \chi^\alpha{}_{,i} K^i_\beta$ given in (7.74) it is easily seen that

$$\delta\varphi^i = K^i_\alpha[\varphi](Q^{-1})^\alpha{}_\beta[\varphi]\Delta\chi^\beta[\varphi]. \qquad (7.273)$$

This takes the form of an infinitesimal gauge transformation $K^i_\alpha \delta\epsilon^\alpha$.

The measure factor (7.271) involves $\det Q'^\alpha{}_\beta$. Using the definition (7.74), along with (7.272), shows that

$$\begin{aligned} Q'^\alpha{}_\beta[\varphi] &= \chi'^\alpha{}_{,i}[\varphi] K^i_\beta[\varphi] \\ &= \frac{\delta \chi^\alpha[\varphi']}{\delta \varphi^i} K^i_\beta[\varphi] \\ &= \frac{\delta \chi^\alpha[\varphi']}{\delta \varphi'^j} \frac{\delta \varphi'^j}{\delta \varphi^i} K^i_\beta[\varphi] \\ &= \frac{\delta \chi^\alpha[\varphi']}{\delta \varphi'^j} K'^j_\beta[\varphi'] \\ &= Q'^\alpha{}_\beta[\varphi']. \end{aligned} \qquad (7.274)$$

(In the third line we have used the chain rule for functional differentiation, and in the fourth line the definition of a vector field transformation was used.) Using (7.272) and (7.274) in (7.271) shows that

$$\mathcal{M}[\chi'; v; \varphi_*; \varphi] = \mathcal{M}[\chi; v; \varphi_*; \varphi']. \qquad (7.275)$$

Clearly, $[d\varphi] = [d\varphi']$ from (7.267). We therefore find, using (7.270), that

$$\exp\left(\frac{i}{\hbar}\hat{\Gamma}\right) = \int [d\varphi']\, \mathcal{M}[\chi; v; \varphi_*; \varphi'] \\ \times \exp\left\{\frac{i}{\hbar} S[\varphi'] + \frac{i}{\hbar}(v^i - \sigma'^i)\frac{\delta}{\delta v^i}\hat{\Gamma}\right\}, \qquad (7.276)$$

where

$$\sigma'^i = \sigma^i[\varphi_*; \varphi' - \delta\varphi]. \qquad (7.277)$$

(Note that the classical action functional is invariant under the gauge transformation (7.273).)

Now let

$$\hat{\Gamma} = \hat{\Gamma}[v; \varphi_*] + \Delta\hat{\Gamma}[v; \varphi_*], \qquad (7.278)$$

with $\Delta\hat{\Gamma}[v; \varphi_*]$ representing the first-order change in the effective action under the change of gauge condition. We will define

$$\langle F[\varphi] \rangle = \exp\left(-\frac{i}{\hbar}\hat{\Gamma}[v; \varphi_*]\right) \int [d\varphi]\, \mathcal{M}[\chi; v; \varphi_*; \varphi]\, F[\varphi] \\ \times \exp\left\{\frac{i}{\hbar} S[\varphi] + \frac{i}{\hbar}(v^i - \sigma^i)\frac{\delta}{\delta v^i}\hat{\Gamma}[v; \varphi_*]\right\} \qquad (7.279)$$

7.7 Gauge-invariant effective action

to be the average of any functional $F[\varphi]$. We now substitute (7.278) into (7.276), and expand both sides to first order in $\Delta\hat{\Gamma}[v; \varphi_*]$ using (7.273) to expand σ'^i defined in (7.277). The result of this is

$$\Delta\hat{\Gamma}[v; \varphi_*] = \langle \sigma^i{}_{;j'}[\varphi_*; \varphi] K^j_\alpha[\varphi] (Q^{-1})^\alpha{}_\beta[\varphi] \Delta\chi^\beta[\varphi] \rangle \frac{\delta\hat{\Gamma}[v; \varphi_*]}{\delta v^i}$$

$$+ \langle (v^i - \sigma^i) \rangle \frac{\delta}{\delta v^i} \Delta\hat{\Gamma}[v; \varphi_*], \qquad (7.280)$$

with $\sigma^i{}_{;j'}[\varphi_*; \varphi] = \delta\sigma^i[\varphi_*; \varphi]/\delta\varphi^j$. The identity established in (7.181) proved that $\sigma^i{}_{;j'}[\varphi_*; \varphi] K^j_\alpha[\varphi] \propto K^i_\alpha[\varphi_*]$. If we use this identity in (7.280), along with the gauge invariance of the effective action established in Section 7.7.1, the first term on the right-hand side of (7.280) can be seen to vanish. (i.e., we use $K^i_\alpha[\varphi_*]\frac{\delta}{\delta v^i}\hat{\Gamma}[v; \varphi_*] = 0$.) For a non-gauge theory, we had $\langle v^i - \sigma^i \rangle = 0$, which if true here would result in the last term on the right-hand side of (7.280) also vanishing; however, we have no reason to assume that $\langle v^i - \sigma^i \rangle$ vanishes. If we regard $\langle v^i - \sigma^i \rangle$ as being defined perturbatively in terms of the loop expansion, then to zeroth order (i.e., at the classical level) $\langle v^i - \sigma^i \rangle$ will vanish provided that χ^α vanishes when we set the quantum fluctuation η^i defined in (7.187) to zero. (This requirement ensures that there is no contribution from the non-trivial measure (7.266) to the classical limit of the theory.) It must then be the case that $\langle v^i - \sigma^i \rangle$ is at least of order $\hbar$ in the loop expansion parameter. Using this knowledge in (7.280) is sufficient to show that $\Delta\hat{\Gamma}[v; \varphi_*]$ must vanish order by order in perturbation theory.

There is another way to argue that the second term on the right-hand side of (7.280) must vanish. Because we have set up the whole formalism of the effective action to be covariant under an arbitrary change in the field space coordinates, whatever the result for $\langle v^i - \sigma^i \rangle$ is when calculated, it must transform like a vector under a change of coordinates at φ_*. (This follows immediately from (7.251) and (7.252).) The only natural object that is available to form a vector at φ_* is the generator of gauge transformations $K^i_\alpha[\varphi_*]$. Thus, we must have

$$\langle v^i - \sigma^i \rangle = K^i_\alpha[\varphi_*] T^\alpha[\varphi_*; \overline{\varphi}], \qquad (7.281)$$

for some expression $T^\alpha[\varphi_*; \overline{\varphi}]$. Once again we can use the established gauge invariance of the effective action to deduce that the second term on the right-hand side of (7.280) must vanish. Of course $T^\alpha[\varphi_*; \overline{\varphi}]$ can only be calculated perturbatively, and from the argument given in the previous paragraph we know that it will begin at first order, at least, in $\hbar$. Similar conclusions were reached in Kobes et al. (1988) in a slightly different way. Rebhan (1987) gave a proof of gauge invariance and independence on gauge conditions in the special case $v^i = 0$.

7.7.3 The loop expansion

We now have a result for the effective action that is both gauge invariant and independent of the choice of gauge conditions. The problem of performing the functional integral over gauge fields has been dealt with by a proper treatment of the measure, and we have only to discuss the evaluation of the effective action. Were it not for the non-trivial factors appearing in the measure entering the definition in (7.266)–(7.268), we could simply refer back to the perturbative evaluation discussed in the previous chapter. This suggests that the simplest approach is to try to "remove" the factors which enter $\mathcal{M}$ and write the integrand in a way that involves an exponential only.

The factor of $\det Q^\alpha{}_\beta$ is easily dealt with by introducing an integration over a new set of fields. Recall from our discussion of integration over Grassmann variables in Section 5.6.5 that if we let $\bar{c}_\alpha$ and c^β be independent Grassmann variables, up to an unimportant numerical factor, we have

$$\det Q^\alpha{}_\beta = \int \left(\prod_\alpha dc^\alpha \right) \left(\prod_\beta d\bar{c}_\beta \right) \exp\left(\frac{i}{\hbar} \bar{c}_\alpha Q^\alpha{}_\beta c^\beta \right). \tag{7.282}$$

We can interpret $\bar{c}_\alpha Q^\alpha{}_\beta c^\beta$ as the action functional for a set of anticommuting Grassmann fields $\bar{c}_\alpha$ and c^α. These anticommuting fields are usually referred to as Faddeev–Popov ghosts.

The δ-distribution in $\mathcal{M}$ can also be written as an exponential if we make use of a particular representation for it. There are several ways in which this can be done, by direct analogy with the many representations possible for the normal Dirac δ-function. The first way we will discuss makes use of the one-dimensional representation

$$\delta(x) = \lim_{\alpha \to 0} (4\pi i\hbar\alpha)^{-1/2} \exp\left(\frac{i}{4\hbar\alpha} x^2 \right).$$

This definition can be extended to n dimensions in a straightforward way; hence, we can write

$$\tilde{\delta}[\chi^\alpha; 0] = \lim_{\alpha \to 0} \left[\det\left(\frac{f_{\alpha\beta}}{4\pi i\hbar\alpha} \right) \right]^{1/2} \exp\left(\frac{i}{4\hbar\alpha} f_{\alpha\beta} \chi^\alpha \chi^\beta \right) \tag{7.283}$$

in the functional case, where $f_{\alpha\beta}$ is any symmetric, positive definite operator. If we use this expression in the effective action, the factor involving the determinant of $f_{\alpha\beta}$, being field-independent, makes no non-trivial contribution to $\hat{\Gamma}[v; \varphi_*]$, and may be ignored in most calculations.

We can use

$$\hat{\Gamma}[v; \varphi_*] = \lim_{\alpha \to 0} \left[\det\left(\frac{f_{\alpha\beta}}{4\pi i\hbar\alpha} \right) \right]^{1/2} \hat{\Gamma}_\alpha[v; \varphi_*], \tag{7.284}$$

where

$$\exp\left(\frac{i}{\hbar} \hat{\Gamma}_\alpha[v; \varphi_*] \right) = \int [d\varphi][dcd\bar{c}] \exp \frac{i}{\hbar} \left\{ S_{eff}[\varphi, c, \bar{c}] + (v^i - \sigma^i) \frac{\delta \hat{\Gamma}_\alpha}{\delta v^i} \right\}, \tag{7.285}$$

7.7 Gauge-invariant effective action

with
$$S_{eff}[\varphi, c, \bar{c}] = S_{GC}[\varphi] + S_{GH}[c, \bar{c}], \tag{7.286}$$

an effective classical action,
$$S_{GC}[\varphi] = S[\varphi] + \frac{1}{4\alpha} f_{\alpha\beta} \chi^\alpha \chi^\beta, \tag{7.287}$$

the classical action for the original theory with the addition of the term arising from the gauge conditions, and
$$S_{GH}[c, \bar{c}] = \bar{c}_\alpha Q^\alpha{}_\beta c^\beta, \tag{7.288}$$

the action for the ghost fields. (We have suppressed writing any possible dependence on the background field $\varphi_\star$ in these expressions.) The DeWitt effective action is obtained by letting $\varphi_\star \to \bar{\varphi}$ to remove all dependence on the arbitrary origin $\varphi_\star$ in field space. We have
$$\Gamma_D[\bar{\varphi}] = \hat{\Gamma}[v = 0; \varphi_\star = \bar{\varphi}] \tag{7.289}$$
$$= \Gamma[\bar{\varphi}; \varphi_\star = \bar{\varphi}], \tag{7.290}$$

as in Chapter 6. (Remember that the $\varphi_\star \to \bar{\varphi}$ limit must only be taken after the effective action has been calculated.)

The loop expansion is expressed as described in Section 6.6. (See (6.159).) The counting is facilitated, as before, by scaling the integration variables $(\eta^i, \bar{c}_\alpha, c^\beta) \to \hbar^{1/2}(\eta^i, \bar{c}_\alpha, c^\beta)$ in the functional integral. Assuming that χ^α is at least linear in η^i, as it is in the Landau–DeWitt gauge, it is easily seen that
$$\Gamma_\alpha^{(0)}[\bar{\varphi}; \varphi_\star] = S[\bar{\varphi}], \tag{7.291}$$

exactly as we found for a non-gauge theory. (The subscript α denotes that the limit appearing in (7.284) has not yet been taken.) It is easy to show that the one-loop part of the effective action is
$$\Gamma_{D\alpha}^{(1)}[\bar{\varphi}] = \frac{i}{2} \ln \det \bar{S}_{GC}{}^i{}_j - i \ln \det \bar{Q}^\alpha{}_\beta, \tag{7.292}$$

where
$$\bar{Q}^\alpha{}_\beta = Q^\alpha{}_\beta\big|_{\varphi=\varphi_\star=\bar{\varphi}}. \tag{7.293}$$

The two-loop approximation can be found if desired using a similar analysis to that of the previous chapter. The only new feature present here is the integration over the anticommuting ghost fields. We will not pursue this, contenting ourselves with the one-loop approximation. Rebhan (1987) has investigated the two-loop self-energy for Yang–Mills theory.

Another often used representation for the δ-functional, in place of (7.283), makes use of the gauge condition independence of the effective action to replace $\chi^\alpha = 0$ with $\chi^\alpha = f^\alpha$ where f^α is an arbitrary function. This has the effect of replacing $\tilde{\delta}[\chi^\alpha; 0]$ in the functional integral with $\tilde{\delta}[\chi^\alpha; f^\alpha]$, and gauge condition independence of $\hat{\Gamma}[v; \varphi_\star]$ means that it is unchanged under this replacement. If $G[f]$ is any functional of f^α, we may multiply both sides of the definition (7.266) (with $\tilde{\delta}[\chi^\alpha; f^\alpha]$ now occurring in $\mathcal{M}$) and perform a functional integral over f^α to give

$$\exp\left(\frac{i}{\hbar}\hat{\Gamma}[v; \varphi_\star]\right) \int [df] G[f] = \int [d\varphi] (\det Q^\alpha{}_\beta) G[\chi]$$
$$\times \exp\left\{\frac{i}{\hbar} S[\varphi] + \frac{i}{\hbar}(v^i - \sigma^i)\frac{\delta}{\delta v^i}\hat{\Gamma}[v; \varphi_\star]\right\}, \quad (7.294)$$

(The presence of $\tilde{\delta}[\chi^\alpha; f^\alpha]$ on the right-hand side removes the integral over f^α there, and the gauge condition independence of $\hat{\Gamma}[v; \varphi_\star]$ means that the only place where f^α appears is in $G[f]$.) A typical choice for $G[f]$ is the Gaussian weight factor

$$G[f] = \exp\left(\frac{i}{2\hbar}\eta_{\alpha\beta} f^\alpha f^\beta\right), \quad (7.295)$$

with $\eta_{\alpha\beta}$ any constant symmetric matrix. This again allows us to write the definition for $\hat{\Gamma}[v; \varphi_\star]$ in a way that resembles the effective action for a non-gauge theory. (The $\det Q^\alpha{}_\beta$ factor can be dealt with as in (7.282).) An important point here is that if we adopt this method for dealing with the δ-distribution that occurs in the functional measure, we cannot use the simplifying features of the Landau–DeWitt gauge discussed in Section 7.5.1. (This is because the Landau–DeWitt gauge is defined by $K_{\alpha i}\eta^i = 0$, and not $K_{\alpha i}\eta^i = f_\alpha$.) In order to maintain gauge condition independence of the effective action, it is necessary to use the full field space connection described in Section 7.5. If this is not done (as is the case with the standard definition of the effective action) a result may be obtained that depends in general on the choice of gauge conditions.

Finally, we discuss a third way to deal with the δ-distribution. This time we base our discussion around the analogous result for the one-dimensional Dirac δ-distribution which is useful in Fourier integral expansions. An often used representation for $\delta(x)$ is

$$\delta(x) = \int_{-\infty}^{\infty} \frac{dp}{2\pi} e^{ipx}.$$

In our case, to define the functional δ-distribution, we must introduce a functional integral over another function λ_α, often referred to as an auxiliary field. We will define

$$\tilde{\delta}[\chi^\alpha; 0] = \int [d\lambda_\alpha] \exp\left(\frac{i}{\hbar}\lambda_\alpha \chi^\alpha\right), \quad (7.296)$$

with the functional integral extending over all values of λ_α. The factor of $\hbar$ is inserted for convenience, since we can then combine $\lambda_\alpha \chi^\alpha$ with the original action $S[\varphi]$ to form an effective classical action in much the same way as we did in (7.287). In this case this effective classical action will involve both λ_α and φ^i. Once again the loop expansion proceeds as discussed in the previous chapter.

The choice of representation for the δ-distribution is largely a matter of convenience and ease of computation. From the formal point of view we have proved that the effective action is independent of gauge conditions and, as a consequence, of how the δ-distribution is handled.

7.8 Yang–Mills theory, concluded

We have already established that the result found for $\hat{\Gamma}[v; \varphi_*]$ is independent of the choice of gauge condition, as well as gauge invariant. Furthermore, we have discussed how the choice of the Landau–DeWitt gauge condition leads to a great simplification in certain situations like Yang–Mills theory. In this final section on Yang-Mills theory we will consider the effective action at one-loop order and demonstrate some of the features that have been obtained using the effective action formalism described earlier in this chapter.

The functional integral representation for $\hat{\Gamma}[v; \varphi_*]$ is (abbreviating expressions by not explicitly indicating the arguments)

$$\exp\left(\frac{i}{\hbar}\hat{\Gamma}\right) = \int [d\varphi]\, (\det Q^\alpha{}_\beta)\, \tilde{\delta}[\chi^\alpha; 0]\, \exp\left(\frac{i}{\hbar}S + \frac{i}{\hbar}(v^i - \sigma^i)\frac{\delta\hat{\Gamma}}{\delta v^i}\right). \quad (7.297)$$

We adopt the Landau–DeWitt gauge condition

$$\chi_\alpha = K_{\alpha i}[\varphi_*](\varphi^i - \varphi^i_*), \quad (7.298)$$

and we have

$$v^i - \sigma^i = \varphi^i - \overline{\varphi}^i \quad (7.299)$$

in this gauge as discussed in Section 7.5.1. Here $\overline{\varphi}^i$ is defined by

$$\hat{\Gamma}[v; \varphi_*] = \Gamma[\overline{\varphi}; \varphi_*], \quad (7.300)$$

and we find

$$\left.\frac{\delta\hat{\Gamma}}{\delta v^i}\right|_\varphi = -\left.\frac{\delta\Gamma[\overline{\varphi}; \varphi_*]}{\delta\overline{\varphi}^i}\right|_\varphi \quad (7.301)$$

from (7.299) in the Landau–DeWitt gauge. We also have

$$Q_{\alpha\beta} = \chi_{\alpha,i} K^i_\beta = K_{\alpha i}[\varphi_*] K^i_\beta[\varphi] \quad (7.302)$$

with the gauge choice (7.298). If we take the limit $\varphi^i_* \to \overline{\varphi}^i$ to obtain the DeWitt effective action

$$\Gamma[\overline{\varphi}] = \Gamma[\overline{\varphi}; \overline{\varphi}] = \lim_{\varphi \to \overline{\varphi}} \hat{\Gamma}[v; \varphi_*], \quad (7.303)$$

we find from (7.297)

$$\exp\left(\frac{i}{\hbar}\Gamma[\overline{\varphi}]\right) = \int [d\varphi]\det\left(K_{\alpha i}[\overline{\varphi}]K_\beta^i[\varphi]\right)\delta\left[K_{\alpha i}[\overline{\varphi}](\varphi^i - \overline{\varphi}^i); 0\right]$$
$$\times \exp\left(\frac{i}{\hbar}S[\varphi] - \frac{i}{\hbar}(\varphi^i - \overline{\varphi}^i)\left.\frac{\delta\Gamma[\overline{\varphi};\varphi_*]}{\delta\overline{\varphi}^i}\right|_\varphi\right). \quad (7.304)$$

Recall that we can only take $\varphi_* \to \overline{\varphi}$ in the last term of the exponential after performing the indicated differentiation. By expanding $S[\varphi]$ about $\varphi = \overline{\varphi}$ and using the fact that to lowest order in the loop expansion (powers of $\hbar$) we have $\Gamma[\overline{\varphi}] = S[\overline{\varphi}]$, it is straightforward to conclude that the one-loop part of $\Gamma[\overline{\varphi}]$ is given by $\Gamma^{(1)}[\overline{\varphi}]$ where

$$\exp\left(\frac{i}{\hbar}\Gamma^{(1)}[\overline{\varphi}]\right) = \det(\bar{K}_{\alpha i}\bar{K}_\beta^i)\int[d\varphi]\delta[\bar{K}_{\alpha i}\varphi^i;0]\exp\left(\frac{i}{2}\bar{S}_{,ij}^{LD}\varphi^i\varphi^j\right). \quad (7.305)$$

Here $\bar{S}_{,ij}^{LD}\varphi^i\varphi^j$ denotes that the Taylor series expansion of $S[\varphi]$ about $\varphi^i = \overline{\varphi}^i$ that is quadratic in φ^i and responsible for the one-loop contribution to the effective action ($\bar{S}_{,ij}\varphi^i\varphi^j$) has been simplified using the Landau–DeWitt gauge condition $\bar{K}_{\alpha i}\varphi^i = 0$ because of the presence of the Dirac δ-distribution in the functional integration. Here $\bar{K}_{\alpha i} = K_{\alpha i}[\overline{\varphi}]$. We then find

$$\Gamma^{(1)}[\overline{\varphi}] = -i\hbar\ln\det(\bar{K}_{\alpha i}\bar{K}_\beta^i) - i\hbar\ln\int[d\varphi]\delta[\bar{K}_{\alpha i}\varphi^i;0]\,e^{\frac{i}{2}\bar{S}_{,ij}^{LD}\varphi^i\varphi^j}. \quad (7.306)$$

All that remains now is to perform the functional integration over φ^i, a task that is complicated by the presence of the δ-distribution in the integrand.

We will deal with the δ-distribution by using an auxiliary field λ_α as discussed in Section 7.7.3. We write

$$\delta[\bar{K}_{\alpha i}\varphi^i;0] = \int[d\lambda]\,e^{i\lambda\cdot\bar{K}_{\alpha i}\varphi^i}. \quad (7.307)$$

The functional integral in (7.306) can now be written as one over the two fields φ^i and λ^α. We can integrate over φ^i first by completing the square. To do this define the propagator in the Landau–DeWitt gauge to be the inverse of $\bar{S}_{,ij}^{LD}$,

$$\bar{S}_{,ij}^{LD}\Delta^{jk} = \delta_i^k. \quad (7.308)$$

Note that there are no problems inverting $\bar{S}_{,ij}^{LD}$ because the gauge has been fixed. (This is not the case for $\bar{S}_{,ij}$ before gauge fixing.) The result for $\Gamma^{(1)}$ after performing the Gaussian functional integrations is

$$\Gamma^{(1)} = -i\hbar\ln\det(\bar{K}_{\alpha i}\bar{K}_\beta^i) + \frac{i\hbar}{2}\ln\det(\bar{S}_{,ij}^{LD}) + \frac{i\hbar}{2}\ln\det(-\bar{K}_{\alpha i}\Delta^{ij}\bar{K}_{\beta j}). \quad (7.309)$$

(We have ignored the renormalization length scale that is introduced to keep the arguments of the logarithms dimensionless here as it is not important to our subsequent analysis.) Apart from the final term of (7.309) the result that we have obtained for $\Gamma^{(1)}$ is exactly what would have been found using the conventional background-field method and adopting the Feynman gauge.

We now specialize to Yang–Mills theory with action (7.12) where the field strength $F^a_{\mu\nu}$ is defined in terms of the gauge field A^a_μ by (7.9). K^i_α was given by (7.32). If we adopt standard notation

$$D_\mu{}^a{}_b = \delta^a_b \nabla_\mu - g f_{bc}{}^a A^c_\mu \qquad (7.310)$$

for the gauge-covariant derivative, with $\bar{D}_\mu{}^a{}_b$ obtained from $D_\mu{}^a{}_b$ by replacing A^c_μ with $\bar{A}^c_\mu$, then the Landau–DeWitt gauge condition ($\bar{K}_{\alpha i}\varphi^i = 0$) reads

$$\bar{D}_\mu{}^a{}_b A^{b\mu} = 0. \qquad (7.311)$$

We will assume for simplicity that the Yang–Mills gauge group is such that group indices are raised and lowered with the Kronecker delta. Any repeated group indices will be summed. The structure constants can then be expressed as f_{abc} and we take them to be totally antisymmetric in their indices.

A straightforward Taylor expansion of the Yang–Mills action to quadratic order in A^a_μ (obtained by replacing A^a_μ in (7.12) with $\bar{A}^a_\mu + A^a_\mu$) gives

$$S_2 = -\frac{1}{2}\int dv_x \left\{ -A^a_\mu (\bar{D}^2)_{ab} A^{b\mu} + A^a_\nu (\bar{D}_\mu \bar{D}^\nu)_{ab} A^{b\mu} + g f_{abc} \bar{F}^a_{\mu\nu} A^{b\mu} A^{c\nu} \right\}. \qquad (7.312)$$

(Here $(\bar{D}^2)_{ab} = \bar{D}_{\mu ac}\bar{D}^\mu{}_{cb}$ is the gauge-covariant generalization of the Laplacian.) We want to commute the order of differentiation in the middle term of S_2 so that the result can be simplified using the Landau–DeWitt gauge condition (7.311). It is easy to show that

$$[\bar{D}_\mu, \bar{D}_\nu]_{ab} A^{b\mu} = -\left(R_{\mu\nu}\delta_{ab} + g f_{abc}\bar{F}^c_{\mu\nu}\right) A^{b\mu}. \qquad (7.313)$$

We then find, after using the gauge condition (7.311) to simplify the result,

$$S_2^{LD} = \frac{1}{2}\int dv_x \left\{ A^a_\mu (\bar{D}^2)_{ab} A^{b\mu} + R_{\mu\nu} A^{a\mu} A^\nu_a - 2g f_{abc}\bar{F}^a_{\mu\nu} A^{b\mu} A^{c\nu} \right\}. \qquad (7.314)$$

This result enables us to read off $\bar{S}^{LD}_{,ij}$ that we need to compute $\Gamma^{(1)}$ in (7.309). ($\bar{S}^{LD}_{,ij}$ is the differential operator that appears on the left-hand side of (7.315) below.) The propagator Δ^{ij} becomes $\Delta^{a\mu\,b\nu}(x,x')$ in standard notation and obeys the defining equation

$$\left[\delta^\mu_\nu (\bar{D}^2)_{ab} + R^\mu{}_\nu \delta_{ab} - 2g f_{eab} \bar{F}^{e\mu}{}_\nu\right] \Delta^{b\nu\,c\lambda}(x,x') = g^{\mu\lambda}(x)\delta^c_a \delta(x,x'). \qquad (7.315)$$

The last term of $\Gamma^{(1)}$ involves $\bar{K}_{\alpha i}\Delta^{ij}\bar{K}_{\beta j}$ in condensed notation that when uncondensed reads

$$\bar{K}_{\alpha i}\Delta^{ij}\bar{K}_{\beta j} = -\frac{1}{g^2}\bar{D}_{\nu ab}\bar{D}'_{\lambda cd}\Delta^{b\nu\,d\lambda}(x,x'). \qquad (7.316)$$

Dealing with this expression involves a bit of work.

We can easily obtain a result for $\bar{D}'_{\lambda cd}\Delta^{b\nu\, d\lambda}(x,x')$ by operating on both sides of (7.315) with $\bar{D}'_{\lambda cd}$. There is no problem with interchanging the order of differentiation in the resulting expression because $\bar{D}'_{\lambda cd}$ involves a different spacetime coordinate (x'^λ in place of x^λ) than that occurring in the differential operator appearing in (7.315). In addition, because $\nabla'_\mu \delta(x,x') = -\nabla_\mu \delta(x,x')$ we have

$$\bar{D}'_{\lambda cd}\delta(x,x') = -\bar{D}_{\lambda dc}\delta(x,x'). \tag{7.317}$$

If we operate on both sides of (7.315) with $\bar{D}_\mu{}_f{}^a \bar{D}'_{\lambda dc}$ we find

$$\bar{D}_\mu{}_f{}^a \left[\delta^\mu_\nu(\bar{D}^2)_{ab} + R^\mu{}_\nu \delta_{ab} - 2g f_{eab}\bar{F}^{e\mu}{}_\nu\right] \bar{D}'_{\lambda cd}\Delta^{b\nu\,\lambda c}(x,x')$$
$$= -(\bar{D}^2)_{fd}\delta(x,x'). \tag{7.318}$$

We now need to know how to commute $\bar{D}_\mu$ through $\bar{D}^2$ in order to obtain a result for the expression in (7.316). This is most readily accomplished if we suppress the group indices by using a vector and matrix notation. Let Ψ have any transformation property under spacetime transformations, but assume that it transforms like

$$\Psi \to U\Psi U^{-1} \tag{7.319}$$

under a gauge transformation. The gauge-covariant derivative of Ψ, which by definition also transforms like (7.319) under a gauge transformation, is given by

$$\bar{D}_\mu \Psi = \nabla_\mu \Psi - ig[\bar{A}_\mu, \Psi] \tag{7.320}$$

in matrix form. Here we view Ψ as a matrix $\Psi = \Psi^a T_a$ and $\bar{A}_\mu$ as a matrix $\bar{A}^a_\mu T_a$, with T_a the Lie algebra generators defined by $[T_a, T_b] = if_{ab}{}^c T_c$. It is easily verified that

$$(\bar{D}_\mu \Psi)^a = \nabla_\mu \Psi^a - g f_{bc}{}^a \bar{A}^c_\mu \Psi^b$$

in agreement with the component form of $\bar{D}_{\mu ab}$ given in (7.310). The field strength matrix is defined by

$$\bar{F}_{\mu\nu} = \nabla_\mu \bar{A}_\nu - \nabla_\nu \bar{A}_\mu - ig[\bar{A}_\mu, \bar{A}_\nu], \tag{7.321}$$

and we have

$$[\bar{D}_\mu, \bar{D}_\nu]\Psi = [\nabla_\mu, \nabla_\nu]\Psi - ig[\bar{F}_{\mu\nu}, \Psi]. \tag{7.322}$$

If we take Ψ in (7.322) to be $\bar{D}_\sigma \Psi^\lambda$ where Ψ^λ transforms like a vector under a spacetime transformation, then

$$[\bar{D}_\mu, \bar{D}_\nu]\bar{D}_\sigma \Psi^\lambda = -R^\tau{}_{\sigma\nu\mu}\bar{D}_\tau \Psi^\lambda + R^\lambda{}_{\tau\nu\mu}\bar{D}_\sigma \Psi^\tau - ig[\bar{F}_{\mu\nu}, \bar{D}_\sigma \Psi^\lambda]. \tag{7.323}$$

Contraction of both sides of (7.323) with $g^{\nu\sigma}$ and setting $\mu = \lambda$ result in

$$[\bar{D}_\mu, \bar{D}^\nu]\bar{D}_\nu \Psi^\mu = -ig[\bar{F}_{\mu\nu}, \bar{D}^\nu \Psi^\mu], \tag{7.324}$$

with the spacetime curvature terms canceling out. We can now write

$$\bar{D}_\lambda \bar{D}^2 \Psi^\lambda = [\bar{D}_\lambda, \bar{D}^\mu]\bar{D}_\mu \Psi^\lambda + \bar{D}^\mu[\bar{D}_\lambda, \bar{D}_\mu]\Psi^\lambda + \bar{D}^2 \bar{D}_\lambda \Psi^\lambda \tag{7.325}$$

and use (7.323) and (7.324) to find

$$\bar{D}_\lambda \{\bar{D}^2 \Psi^\lambda + R^\lambda{}_\sigma \Psi^\sigma - 2ig[\bar{F}^\lambda{}_\sigma, \Psi^\sigma]\} = \bar{D}^2 \bar{D}_\lambda \Psi^\lambda - ig[\bar{D}_\mu \bar{F}^\mu{}_\lambda, \Psi^\lambda]. \quad (7.326)$$

In the special case where the background gauge field is covariantly constant, $\bar{D}_\mu \bar{F}^\mu{}_\lambda = 0$, the second term on the right-hand side of (7.326) is absent and the result is very simple. A covariantly constant gauge field is the natural generalization to non-Abelian gauge fields of the classic Schwinger calculation for quantum electrodynamics described in Section 5.4.1. However we do not assume that $\bar{D}_\mu \bar{F}^\mu{}_\lambda = 0$ in what follows, so that the background gauge field is kept general.

The identity (7.326) is exactly what is required to commute the derivative $\bar{D}_\mu$ through the differential operator in (7.318) to find the desired expression for the gauge-covariant derivatives of the propagator. In the special case where $\bar{D}_\mu \bar{F}^\mu{}_\lambda = 0$ it follows from our discussion that

$$\bar{D}_{\nu ac} \bar{D}'_{\lambda bd} \Delta^{c\nu\, d\lambda}(x, x') = -\delta_{ab}\delta(x, x') \quad (7.327)$$

and therefore that the last term in the one-loop effective action (7.309) makes no contribution at all. Even in the general case that we are considering, where $\bar{D}_\mu \bar{F}^\mu{}_\lambda \neq 0$, we can argue that the third term of (7.309) makes no non-trivial contribution to the divergent part of the effective action $\Gamma^{(1)}$. The simplest way to deal with the last term of $\Gamma^{(1)}$ in (7.309) is to first define a Green function $G(x, x')$ by the requirement

$$\bar{D}^2 G(x, x') = \delta(x, x'). \quad (7.328)$$

(Thus $G(x, x')$ is viewed as the inverse of the operator $\bar{D}^2$.) We can operate on both sides of (7.318) with $(\bar{D}^2)^{-1}$ after using the identity (7.326) to obtain (suppressing group indices)

$$\bar{D}_\nu \bar{D}'_\lambda \Delta^{\nu\lambda}(x, x') = -\delta(x, x') + \int dv_{x''} G(x, x'') T(x'', x'), \quad (7.329)$$

where $T(x'', x')$ stands for the term that arises from the second term in (7.326), which involves the commutator of $\bar{D}_\mu \bar{F}^\mu{}_\lambda$ and $\bar{D}'_\lambda \Delta^{\nu\lambda}(x, x')$. The one-loop effective action involves the logarithm of (7.329) that we can expand in powers of $T(x'', x')$ thereby generating an expansion in powers of $\bar{D}_\mu \bar{F}^\mu{}_\lambda$ whose coefficients involve products of Green functions $G(x, x')$ and $\bar{D}'_\lambda \Delta^{\nu\lambda}(x, x')$. We can then use the normal coordinate expansion method to deal with the Green functions as described in Chapter 6. The leading term in the normal coordinate expansion (i.e., always the most divergent) is just the flat spacetime expression and we can analyze the degree of divergence by simple power counting. (Higher-order terms in the normal coordinate expansion that involve the spacetime curvature can be analyzed by dimensional analysis and power counting.) It is easy to argue that there can be no pole terms present in dimensional regularization. The integrand of $\Gamma^{(1)}$ must be of dimension -4 using units of length. $\bar{D}_\mu \bar{F}^\mu{}_\lambda$ has dimension -3

so that it is not possible to have any poles present at second and higher order in the expansion of the logarithm (because they must be formed from the spacetime curvature or $\bar{F}_{\mu\nu}$ that are both of dimension -2, and gauge-covariant derivatives decrease the dimension further). The linear term in the expansion involves $\bar{D}_\mu \bar{F}^\mu{}_\lambda$ and its coefficient is of dimension -1. The coefficient of $\bar{D}_\mu \bar{F}^\mu{}_\lambda$ must be a vector (to obtain a spacetime-covariant expression) and the only gauge-covariant vector is $\bar{D}^\lambda$; this will give rise to $\bar{D}^\lambda \bar{D}_\mu \bar{F}^\mu{}_\lambda = 0$. Thus there can be no divergence coming from the linear term in the expansion of the logarithm either. We can conclude that the third term of $\Gamma^{(1)}$ can have no poles in dimensional regularization. Note that we are not claiming that the third term of $\Gamma^{(1)}$ can be ignored altogether; merely, that it cannot make any contribution to the divergent part of $\Gamma^{(1)}$. The finite part of $\Gamma^{(1)}$ is a different story.

It now follows that the divergent part of $\Gamma^{(1)}$ can come only from the first two terms of (7.309). Since $\bar{K}_{\alpha i} \bar{K}^i_\beta = \bar{Q}_{\alpha\beta} = -\frac{1}{g^2} \bar{D}^2 \tilde{\delta}(x, x')$ (see (7.89)), we have

$$\mathbf{divp}(\Gamma^{(1)}) = -i\hbar \, \mathbf{divp} \left\{ \ln\det(-\bar{D}^2) - \frac{1}{2} \ln\det \bar{S}^{LD}_{,ij} \right\}, \qquad (7.330)$$

with $\mathbf{divp}$ denoting the divergent pole part as before. $\bar{S}^{LD}_{,ij}$ is the operator that occurs on the right-hand side of (7.315). Both expressions in (7.330) are in the general form of the operator that was assumed in the discussion of Section 5.3 culminating in

$$\mathbf{divp} \left\{ \ln\det(D^2 + Q) \right\} = \frac{i}{8\pi^2 \epsilon} \int dv_x \, \mathrm{tr} E_2 \qquad (7.331)$$

with E_2 given by (5.59) and the remaining trace on the right-hand side over any group and spacetime indices. We will use the same conventions as in Parker and Toms (1984). If N is the dimension of the gauge group (equal to the number of linearly independent generators for the Lie algebra associated with the gauge group), the quadratic Casimir invariant is defined by (I being the group identity matrix)

$$T_a T_a = C_2(G_R) I \qquad (7.332)$$

in any representation G_R of the gauge group G that has dimension d_R. (So T_a can be viewed as a $d_R \times d_R$ Hermitian matrix.) We also have

$$\mathrm{tr}\,(T_a T_b) = \frac{d_R}{N} C_2(G_R) \delta_{ab}. \qquad (7.333)$$

The Lie algebra generators in the adjoint representation ($d_{adj} = N$) have matrix elements given by

$$(T_a)_{bc} = -i f_{abc}, \qquad (7.334)$$

and (7.333) gives us

$$f_{acd}f_{bcd} = C_2(G_{adj})\delta_{ab}, \tag{7.335}$$

$$f_{abc}f_{abc} = NC_2(G_{adj}). \tag{7.336}$$

For the special case of $SU(n)$ we have $N = n^2 - 1$ and $C_2(G_{adj}) = n$.

In the adjoint representation we have the matrix elements of $\bar{F}_{\mu\nu}$ given by

$$(\bar{F}_{\mu\nu})_{ab} = \bar{F}^c_{\mu\nu}(T_c)_{ab} = -if_{cab}\bar{F}^c_{\mu\nu}, \tag{7.337}$$

if we note (7.334). With group indices suppressed we have $\bar{S}^{LD}_{,ij}$ given by the differential operator $\mathcal{D}^\mu{}_\nu$ where

$$\mathcal{D}^\mu{}_\nu = \delta^\mu_\nu \bar{D}^2 + Q^\mu{}_\nu, \tag{7.338}$$

with

$$Q^\mu{}_\nu = R^\mu{}_\nu I - 2ig\bar{F}^\mu{}_\nu. \tag{7.339}$$

In addition,

$$(W_{\mu\nu})^\lambda{}_\sigma = [\bar{D}_\mu, \bar{D}_\nu]^\lambda{}_\sigma$$
$$= IR^\lambda{}_{\sigma\nu\mu} - ig\delta^\lambda_\sigma \bar{F}_{\mu\nu} \tag{7.340}$$

if we note (7.313) and (7.337). It is a straightforward matter to substitute (7.339) and (7.340) into (5.59) to find

$$\operatorname{tr}(E_2) = -\frac{N}{9}R^2 + \frac{43}{90}NR^{\mu\nu}R_{\mu\nu} - \frac{11}{180}NR_{\mu\nu\rho\sigma}R^{\mu\nu\rho\sigma}$$
$$+ \frac{5}{3}g^2 C_2(G_{adj})\bar{F}^a_{\mu\nu}\bar{F}^{a\mu\nu} \tag{7.341}$$

up to total derivative terms that we are not interested in.

The operator $\bar{D}^2$ in the first term of (7.330) acts on spacetime scalars. We take $Q = 0$ and

$$W_{\mu\nu} = [\bar{D}_\mu, \bar{D}_\nu] = -ig\bar{F}_{\mu\nu} \tag{7.342}$$

in this case. A straightforward calculation (again ignoring total derivatives) results in

$$\operatorname{tr}(E_2) = \frac{N}{72}R^2 - \frac{N}{180}R^{\mu\nu}R_{\mu\nu} + \frac{N}{180}R_{\mu\nu\rho\sigma}R^{\mu\nu\rho\sigma}$$
$$- \frac{1}{12}g^2 C_2(G_{adj})\bar{F}^a_{\mu\nu}\bar{F}^{a\mu\nu}. \tag{7.343}$$

Using (7.331) for each of the two terms in (7.330) along with the heat kernel coefficients (7.341) and (7.343) leads to

$$\mathbf{divp}(\Gamma^{(1)}) = -\frac{\hbar}{8\pi^2\epsilon}\int dv_x \Big\{ -\frac{5}{72}NR^2 + \frac{11}{45}NR_{\mu\nu}R^{\mu\nu}$$
$$- \frac{13}{360}NR_{\mu\nu\rho\sigma}R^{\mu\nu\rho\sigma} + \frac{11}{12}g^2 C_2(G_{adj})\bar{F}^a_{\mu\nu}\bar{F}^{a\mu\nu} \Big\}. \tag{7.344}$$

The pole terms that involve the spacetime curvature are identical in structure to those found for the scalar field in Chapter 6 and can be dealt with in the same way as we did there. The pole term that involves the Yang–Mills field strength has the same form as the classical Yang–Mills action; thus, it follows that Yang–Mills theory is renormalizable (at least to one-loop order) in curved spacetime.

The pole term in (7.344) that involves the Yang–Mills field strength can be dealt with by a renormalization of the background gauge field. Introduce gauge coupling constant and background field renormalizations

$$g_B = \ell^{n/2-2} Z_g g, \qquad (7.345)$$

$$\bar{A}^a_{\mu B} = \ell^{2-n/2} Z_A^{1/2} \bar{A}^a_\mu. \qquad (7.346)$$

The subscript B denotes a bare quantity as before. The bare field strength tensor is

$$\bar{F}^a_{\mu\nu B} = \partial_\mu \bar{A}^a_{\nu B} - \partial_\nu \bar{A}^a_{\mu B} + g_B f_{abc} \bar{A}^b_{\mu B} \bar{A}^c_{\nu B}$$
$$= \ell^{2-n/2} Z_A^{1/2} \left\{ \partial_\mu \bar{A}^a_\nu - \partial_\nu \bar{A}^a_\mu + Z_g Z_A^{1/2} g f_{abc} \bar{A}^b_\mu \bar{A}^c_\nu \right\}. \qquad (7.347)$$

We know that the effective action must be gauge invariant. The only way that $\bar{F}^a_{\mu\nu B}$ can be related to $\bar{F}^a_{\mu\nu}$ in a gauge-invariant way is if

$$Z_g Z_A^{1/2} = 1. \qquad (7.348)$$

Thus the field renormalization factor and the gauge coupling constant renormalization factor are related to one another. The utility of the background field method was first illustrated by Abbott (1981) to two-loop order in flat spacetime. The application to curved spacetime was first given by Toms (1983b).

The bare Yang–Mills action must be added to $\Gamma^{(1)}$ to obtain the full effective action to one-loop order. We have

$$S_{YM} = -\frac{1}{4} \int d^n v_x \hat{F}^a_{\mu\nu B} \hat{F}^{a\mu\nu B}$$
$$= -\frac{1}{4} \int d^4 v_x \, Z_A \hat{F}^a_{\mu\nu} \hat{F}^{a\mu\nu}. \qquad (7.349)$$

(Note that the factor of ℓ^{4-n} that arises from expressing the bare field strength in terms of the renormalized one is canceled by the volume associated with the $(n-4)$ extra dimensions.) If we now expand the renormalization factors in the loop expansion,

$$Z_g = 1 + \hbar \delta Z_g^{(1)} + \cdots, \qquad (7.350)$$

$$Z_A = 1 + \hbar \delta Z_A^{(1)} + \cdots, \qquad (7.351)$$

it follows from (7.348) that

$$\delta Z_g^{(1)} = -\frac{1}{2} \delta Z_A^{(1)}. \qquad (7.352)$$

The pole term of $\Gamma^{(1)}$ in (7.344) that involves the Yang–Mills field strength can be canceled with the choice

$$\delta Z_A = -\frac{11}{24\pi^2 \epsilon} C_2(G_{adj}) g^2. \tag{7.353}$$

Using the method described in Section 6.8 we can compute the β-function associated with g (to one-loop order) to be

$$\beta_g(g) = \frac{11}{48\pi^2} C_2(G_{adj}) g^2. \tag{7.354}$$

The associated renormalization group equation is

$$\ell \frac{dg}{d\ell} = \beta_g(g), \tag{7.355}$$

showing that $g \to 0$ as $\ell \to 0$. This means in the short distance limit the behavior of the theory is like a free field theory. This important feature of Yang–Mills theory was first discussed by Gross and Wilczek (1973) and Politzer (1973) and is called asymptotic freedom. Renormalization of Yang–Mills theory in flat spacetime was considered by 't Hooft (1971a,b).

Other approaches to the renormalization of Yang–Mills theory in curved spacetime are found in Leen (1983); Jack (1984, 1985); Jack and Osborn (1985). The case of non-Abelian quantum electrodynamics treated by Toms (1983b) is given in an elegant way by Lee and Rim (1985). (For Abelian quantum electrodynamics, see Panangaden (1981).) Powerful methods for calculating counterterms are contained in Hathrell (1982b) and Freeman (1984). An efficient method for one-loop counterterms is Omote and Ichinose (1983).

7.9 Scalar quantum electrodynamics

In this section we will describe how to use the gauge-invariant effective action method to study the problem of scalar electrodynamics in curved spacetime. We will also illustrate explicitly how the reduction of a gauge theory to a non-gauge theory works with this example. We will first evaluate the divergences present in the one-loop effective action in curved spacetime. Next we will examine the effective potential in flat spacetime, illustrating the importance of the geometrical approach in obtaining a result that is consistent.

7.9.1 One-loop divergences

We begin with a charged scalar field with a general coupling to the background curvature. If we initially describe the theory using a single complex scalar field

$\Phi(x)$ along with its Hermitian adjoint $\Phi^\dagger(x)$, then the action functional for the classical theory is

$$S = \int dv_x \left\{ (D^\mu \Phi)^\dagger (D_\mu \Phi) - m^2 |\Phi|^2 - \xi R |\Phi|^2 - \frac{\lambda}{6} |\Phi|^4 - \frac{1}{4} F_{\mu\nu} F^{\mu\nu} \right\}. \quad (7.356)$$

Here

$$D_\mu \Phi = \partial_\mu \Phi - ie A_\mu \Phi \quad (7.357)$$

denotes the gauge-covariant derivative. The theory is invariant under the $U(1)$ local gauge transformation

$$\delta \Phi(x) = ie \delta \epsilon(x) \Phi(x), \quad (7.358)$$
$$\delta \Phi^\dagger(x) = -ie \delta \epsilon(x) \Phi^\dagger(x), \quad (7.359)$$
$$\delta A_\mu(x) = \partial_\mu \delta \epsilon(x), \quad (7.360)$$

with $\delta \epsilon(x)$ the infinitesimal parameter of the transformation.[11]

The formalism described in the earlier parts of the present chapter was set up to deal with real, rather than complex, fields. It is therefore expedient to rewrite the theory by expressing the complex scalar field in terms of two real fields. (The alternative is to rework the general formalism for complex fields.) There are two obvious ways that we can do this. We could express Φ in terms of its real and imaginary parts as

$$\Phi = \frac{1}{\sqrt{2}} (\varphi_1 + i \varphi_2), \quad (7.361)$$

or in polar form as

$$\Phi = \frac{1}{\sqrt{2}} \rho \exp(i\theta). \quad (7.362)$$

The first choice will lead to a field space analogue of Cartesian coordinates, whereas the second choice will lead to the field space analogue of polar coordinates. The field space line element, from which the field space metric is obtained, will be chosen as

$$ds^2 = \int dv_x \left\{ d\varphi_1^2(x) + d\varphi_2^2(x) - g^{\mu\nu}(x) dA_\mu(x) dA_\nu(x) \right\} \quad (7.363)$$

$$= \int dv_x \left\{ d\rho^2(x) + \rho^2(x) d\theta^2(x) - g^{\mu\nu}(x) dA_\mu(x) dA_\nu(x) \right\} \quad (7.364)$$

in these two different field parameterizations. The minus sign in front of the spacetime metric in the third term of the line element is chosen so that ds^2 will be positive definite if we adopt a Riemannian metric by performing a Wick rotation.

---

[11] The finite transformations corresponding to (7.358) and (7.359) are $\Phi \to \exp(ie\epsilon(x))\Phi$, $\Phi^\dagger \to \exp(-ie\epsilon(x))\Phi^\dagger$, with (7.360) unchanged.

7.9 Scalar quantum electrodynamics

Because we are interested in the reduction to a non-gauge theory we will examine only the polar form of the theory. From (7.358), (7.359), and (7.362) we find

$$\delta\rho(x) = 0, \quad (7.365)$$
$$\delta\theta(x) = e\delta\epsilon(x). \quad (7.366)$$

This shows that $\rho(x)$ is gauge invariant, but that there is still a gauge dependence in the theory through $\theta(x)$ as well as $A_\mu(x)$. Before we eliminate this gauge dependence we will work out the field space metric, generators of gauge transformations, and connection components.

Generally, we have the correspondence between condensed notation and normal notation

$$ds^2 = g_{ij}d\varphi^i d\varphi^j$$
$$= \int d^n x d^n x' g_{IJ}(x,x')d\varphi^I(x)d\varphi^J(x'),$$

where we use $\varphi^I(x)$ to denote the complete set of fields. Here we have $\varphi^I = (\rho(x), \theta(x), A_\mu(x))$. Comparison with (7.364) shows that the non-zero field space metric components are (in an obvious notation chosen to avoid confusion with the spacetime metric)

$$g_{\rho(x)\rho(x')} = |g(x)|^{1/2}\tilde{\delta}(x,x') = |g(x)|^{1/2}|g(x')|^{1/2}\delta(x,x'), \quad (7.367)$$
$$g_{\theta(x)\theta(x')} = |g(x)|^{1/2}\rho^2(x)\tilde{\delta}(x,x'), \quad (7.368)$$
$$g_{A_\mu(x)A_\nu(x')} = -|g(x)|^{1/2}g^{\mu\nu}(x)\tilde{\delta}(x,x'). \quad (7.369)$$

The only dependence on the fields occurs through $g_{\theta(x)\theta(x')}$. The inverse metric components are easily evaluated, and the Christoffel symbols that are non-zero are

$$\Gamma^{\theta(x)}_{\rho(x')\theta(x'')} = \Gamma^{\theta(x)}_{\theta(x')\rho(x'')} = \frac{1}{\rho(x)}\tilde{\delta}(x,x')\tilde{\delta}(x',x''), \quad (7.370)$$
$$\Gamma^{\rho(x)}_{\theta(x')\theta(x'')} = -\rho(x)\tilde{\delta}(x,x')\tilde{\delta}(x',x''). \quad (7.371)$$

The infinitesimal gauge transformation was expressed in condensed notation as $\delta\varphi^i = K^i_\alpha \delta\epsilon^\alpha$. In the present case of a local $U(1)$ transformation, the condensed index α stands for just a spacetime coordinate. The uncondensed version of $\delta\varphi^i = K^i_\alpha \delta\epsilon^\alpha$ is

$$\delta\varphi^I(x) = \int d^n x' K^{\varphi^I(x)}_{x'} \delta\epsilon(x'). \quad (7.372)$$

From (7.365), (7.366), and (7.360) we find

$$K^{\rho(x)}_{x'} = 0, \quad (7.373)$$
$$K^{\theta(x)}_{x'} = e\tilde{\delta}(x,x'), \quad (7.374)$$
$$K^{A_\mu(x)}_{x'} = \nabla_\mu \tilde{\delta}(x,x'), \quad (7.375)$$

all of which are independent of the fields. For later purposes we lower the index on K_α^i using $K_{\alpha i} = g_{ij} K_\alpha^j$ with the field space metric given in (7.367)–(7.369):

$$K_{x'\rho(x)} = 0, \qquad (7.376)$$
$$K_{x'\theta(x)} = e\rho^2(x)|g(x)|^{1/2}\tilde{\delta}(x,x'), \qquad (7.377)$$
$$K_{x'A_\mu(x)} = -|g(x)|^{1/2}\nabla^\mu\tilde{\delta}(x,x'). \qquad (7.378)$$

The condensed expression $\gamma_{\alpha\beta} = K_\alpha^i g_{ij} K_\beta^j$ becomes

$$\gamma_{xx'} = |g(x)|^{1/2}\left(\Box_x + e^2\rho^2(x)\right)\tilde{\delta}(x,x') \qquad (7.379)$$

and its inverse (defined through $\gamma_{\alpha\beta}\gamma^{\beta\gamma} = \delta_\alpha^\gamma$ in condensed notation) satisfies the differential equation

$$\left(\Box_x + e^2\rho^2(x)\right)\gamma^{xx'} = \delta(x,x'). \qquad (7.380)$$

We can now evaluate the full connection components using (7.171). For reasons explained earlier we do not require any terms in the connection that are proportional to the Killing vectors K_α^i. After some calculation the non-zero components turn out to be (apart from those that follow from the symmetry $\tilde{\Gamma}_{ji}^k = \tilde{\Gamma}_{ij}^k$)

$$\tilde{\Gamma}^{\rho(x)}_{\theta(x')\theta(x'')} = -|g(x')|^{1/2}|g(x'')|^{1/2}\rho(x)(\Box_{x''}\gamma^{x''x})(\Box_{x'}\gamma^{x'x}), \qquad (7.381)$$

$$\tilde{\Gamma}^{\rho(x)}_{\theta(x')A\ (x'')} = -e|g(x')|^{1/2}|g(x'')|^{1/2}\rho(x)(\nabla''^\nu\gamma^{x''x})(\Box_{x'}\gamma^{x'x}), \qquad (7.382)$$

$$\tilde{\Gamma}^{\rho(x)}_{A_\mu(x')A\ (x'')} = -e^2|g(x')|^{1/2}|g(x'')|^{1/2}\rho(x)(\nabla'^\mu\gamma^{x'x})(\nabla''^\nu\gamma^{x''x}), \qquad (7.383)$$

$$\tilde{\Gamma}^{\theta(x)}_{\rho(x')\theta(x'')} = \frac{1}{\rho(x)}|g(x'')|^{1/2}(\Box_{x''}\gamma^{x''x})\tilde{\delta}(x,x'), \qquad (7.384)$$

$$\tilde{\Gamma}^{\theta(x)}_{\rho(x')A\ (x'')} = \frac{e}{\rho(x)}|g(x'')|^{1/2}(\nabla''^\nu\gamma^{x''x})\tilde{\delta}(x,x'). \qquad (7.385)$$

At this stage we could proceed to evaluate the effective action by choosing a gauge condition and continuing on as described earlier in the chapter. Instead of this, we will reduce the theory to a non-gauge theory by performing a further field transformation. Let

$$\bar{\rho}(x) = \rho(x), \qquad (7.386)$$
$$\bar{\theta}(x) = \theta(x), \qquad (7.387)$$
$$\bar{A}_\mu(x) = A_\mu(x) - \frac{1}{e}\nabla_\mu\theta(x) \qquad (7.388)$$

be the transformation from (ρ, θ, A_μ) to $(\bar{\rho}, \bar{\theta}, \bar{A}_\mu)$. Clearly only the gauge field is altered by this transformation; this is sometimes referred to (Abers and Lee

1973) as the unitary gauge. Under the infinitesimal gauge transformation (7.366, 7.360) it can be seen that

$$\delta \bar{A}_\mu(x) = 0, \qquad (7.389)$$

meaning that $\bar{A}_\mu(x)$ is a gauge-invariant field. The action functional (7.356) becomes

$$S = \int dv_x \left\{ \frac{1}{2} \partial^\mu \bar{\rho} \partial_\mu \bar{\rho} - \frac{1}{2} m^2 \bar{\rho}^2 - \frac{1}{2} \xi R \bar{\rho}^2 - \frac{\lambda}{4!} \bar{\rho}^4 \right. $$
$$\left. + \frac{1}{2} e^2 \bar{\rho}^2 \bar{A}^\mu \bar{A}_\mu - \frac{1}{4} \bar{F}^{\mu\nu} \bar{F}_{\mu\nu} \right\} \qquad (7.390)$$

when expressed in terms of new fields. The important observation is that $\bar{\theta}(x)$ does not appear in the action functional. Only $\bar{\rho}(x)$ and $\bar{A}_\mu(x)$ appear, and from (7.365) and (7.389), these new fields are gauge invariant. These fields correspond to the gauge-invariant quantities ξ^A of Section 7.4.

Although we have reduced the original gauge theory to a non-gauge theory there is still a residue of the original gauge invariance present in the field space connection. The most straightforward way to obtain the connection components in the new coordinates is simply to perform the field space transformation to the new coordinates on the old components of the field space connection given in (7.381)–(7.385). Because the transformation in (7.386)–(7.388) is linear, we can simply use

$$\bar{\Gamma}^k_{ij} = \frac{\partial \bar{\varphi}^k}{\partial \varphi^l} \frac{\partial \varphi^m}{\partial \bar{\varphi}^i} \frac{\partial \varphi^n}{\partial \bar{\varphi}^j} \tilde{\Gamma}^l_{mn}. \qquad (7.391)$$

Also, because θ has disappeared from the action functional (7.390), we only need the components of the connection where the condensed indices i, j, k refer to $\bar{\rho}$ or $\bar{A}_\mu$. It can be shown that

$$\bar{\Gamma}^{\bar{\rho}(x)}_{\bar{A}_\mu(x') \bar{A}(x'')} = -e^2 \bar{\rho}(x) |g(x')|^{1/2} |g(x'')|^{1/2} \left(\nabla'^\mu \gamma^{x'x} \right) \left(\nabla''^\nu \gamma^{x''x} \right), \qquad (7.392)$$

$$\bar{\Gamma}^{\bar{A}_\mu(x)}_{\bar{\rho}(x') \bar{A}(x'')} = -|g(x')|^{1/2} |g(x'')|^{1/2} \nabla''^\nu \gamma^{x''x} \nabla_\mu \left(\frac{1}{\bar{\rho}(x)} \delta(x, x') \right). \qquad (7.393)$$

We will now evaluate the one-loop effective action by expanding about a constant value of $\bar{\rho} = \rho$ and a vanishing value for $\bar{A}_\mu = 0$. From (7.390) we find

$$S_{,\bar{\rho}(x)} = \left[-m^2 \rho - \xi R \rho - \frac{\lambda}{6} \rho^3 \right] |g(x)|^{1/2}, \qquad (7.394)$$

$$S_{,\bar{A}_\mu(x)} = 0, \qquad (7.395)$$

when evaluated at the background fields described. We then obtain

$$S_{;\bar{\rho}(x)\bar{\rho}(x')} = S_{,\bar{\rho}(x)\bar{\rho}(x')}$$
$$= |g(x)|^{1/2}\left[-\Box_x - m^2 - \xi R - \frac{\lambda}{2}\rho^2\right]\tilde{\delta}(x,x'), \tag{7.396}$$

$$S_{;\bar{A}_\mu(x)\bar{A}(x')} = |g(x)|^{1/2}\left[g^{\mu\nu}\Box_x - \nabla^\nu\nabla_\mu + e^2\rho^2 g^{\mu\nu}\right]\tilde{\delta}(x,x')$$
$$- e^2\rho^2 |g(x)|^{1/2}|g(x')|^{1/2}\int dv_z\left[m^2 + \xi R(z) + \frac{\lambda}{6}\rho^2\right]$$
$$\times (\nabla^\mu\gamma^{xz})(\nabla'^\nu\gamma^{x'z}). \tag{7.397}$$

The last term of (7.397) presents the correction due to the non-trivial field space connection. Because the components of the field space metric responsible for raising and lowering indices do not involve the fields $\bar{\rho}$ or $\bar{A}_\mu$, it does not matter if we concentrate on $S_{;ij}$ rather than $S_{;i}{}^j$.

We can now evaluate the one-loop effective action. To do this it is instructive to begin with that for the original theory as found from our general expression (7.266)–(7.268) and follow through the field redefinitions that led to $(\bar{\rho}, \bar{A}_\mu)$ as the dependent variables in the classical action. Because we have established that the effective action is independent of the gauge choice, we can pick what we like. We will choose the standard Landau gauge condition

$$\chi(x) = \nabla^\mu A_\mu. \tag{7.398}$$

By varying χ under the gauge transformation (7.360) and making use of the definition $\delta\chi_\alpha = Q_{\alpha\beta}\delta\epsilon^\beta$ in condensed notation, we can read off

$$Q(x,x') = \Box_x\tilde{\delta}(x,x'). \tag{7.399}$$

The formal functional integral measure for (7.356) reads

$$d\mu = \left(\prod_x d\Phi^\dagger(x)d\Phi(x)\prod_\mu dA_\mu(x)\right)\delta[\nabla^\mu A_\mu; 0]\det(\Box_x) \tag{7.400}$$

according to (7.268). Making the transformation to a polar coordinate representation (7.362) leads to

$$d\mu = \left(\prod_x d\rho(x)d\theta(x)\prod_\mu dA_\mu(x)\right)\det\rho(x)\,\delta[\nabla^\mu A_\mu; 0]\det(\Box_x), \tag{7.401}$$

with $\det\rho(x)$ the Jacobian of the transformation. We next make the transformation (7.386)–(7.388) to find

$$d\mu = \left(\prod_x d\bar{\rho}(x)d\bar{\theta}(x)\prod_\mu d\bar{A}_\mu(x)\right)\det\bar{\rho}(x)\,\delta\left[\nabla^\mu \bar{A}_\mu + \frac{1}{e}\Box_x\bar{\theta}; 0\right]\det(\Box_x). \tag{7.402}$$

7.9 Scalar quantum electrodynamics

At this stage, the classical action (7.390) is independent of $\bar{\theta}$. It is easy to perform the functional integral over $\bar{\theta}(x)$ using the δ-function identity

$$\int \left(\prod_x d\bar{\theta}(x)\right) \delta\left[\nabla^\mu \bar{A}_\mu + \frac{1}{e}\Box_x \bar{\theta}; 0\right] = \left[\det\left(\frac{1}{e}\Box_x\right)\right]^{-1}. \quad (7.403)$$

The factor of $\det\Box_x$ in (7.402) therefore cancels with the similar factor in (7.403) leaving us with

$$d\mu = \left(\prod_x d\bar{\rho}(x) \prod_\mu d\bar{A}_\mu(x)\right) \det(e\bar{\rho}(x)). \quad (7.404)$$

Apart from the factor of $\det(e\bar{\rho}(x))$ this is just what would have been expected for a non-gauge theory based on the action (7.390).

We can now perform the functional integrals over $\bar{\rho}(x)$ and $\bar{A}_\mu(x)$ in terms of the usual Gaussian results to obtain the one-loop effective action:

$$\Gamma = \bar{S} - i\hbar \ln\det(e\rho) + \frac{i}{2}\hbar \ln\det S_{;\bar{\rho}(x)\bar{\rho}(x')} + \frac{i}{2}\hbar \ln\det S_{;\bar{A}_\mu(x)\bar{A}(x')}. \quad (7.405)$$

The factor of $\det(e\rho)$ arises from expanding the analogous factor in the measure (7.404) about the constant background field ρ. We have not bothered with the inclusion of the unit of length inside of the logarithms of (7.405) since its role is not essential for what we wish to study.

The last term of (7.405) is complicated due to the nature of (7.397). The term in $S_{;\bar{\rho}(x)\bar{\rho}(x')}$ involves an operator of the standard form considered in Chapter 5. To deal with the last term of (7.405) we will begin by defining a Green function $G^\lambda{}_\nu(x,x')$ by

$$\left(\Box_x \delta^\mu_\lambda - \nabla_\lambda \nabla^\mu + e^2\rho^2 \delta^\mu_\lambda\right) G^\lambda{}_\nu(x,x') = \delta^\mu_\nu \delta(x,x'). \quad (7.406)$$

It then follows from (7.397) that

$$\ln\det S_{;\bar{A}_\mu(x)\bar{A}(x')} = \ln\det\left(\Box\delta^\mu_\nu - \nabla_\nu\nabla^\mu + e^2\rho^2\delta^\mu_\nu\right)$$
$$+ \ln\det\left[\delta^\mu_\nu \delta(x,x') - P^\mu{}_\nu(x,x')\right], \quad (7.407)$$

where

$$P^\mu{}_\nu(x,x') = e^2\rho^2 \int dv_y dv_z G^\mu{}_\lambda(x,y)\left[m^2 + \xi R(y) + \frac{\lambda}{6}\rho^2\right]$$
$$\times \left(\nabla^\lambda_y \gamma^{yz}\right)\left(\nabla'_\nu \gamma^{x'z}\right). \quad (7.408)$$

(We append the y on ∇^λ_y to denote that it operates on the y coordinate.) In order to simplify the calculation we will assume that ξR is constant. This includes either minimal coupling ($\xi = 0$) but arbitrary curvature, or non-minimal coupling ($\xi \neq 0$) but constant curvature (e.g., de Sitter spacetime). We can integrate by parts over y in this case and put the covariant derivative on the Green function:

$$P^\mu{}_\nu(x,x') = -e^2\rho^2 M^2 \int dv_y dv_z \nabla^\lambda_y G^\mu{}_\lambda(x,y)\gamma^{yz}\left(\nabla'_\nu \gamma^{x'z}\right), \quad (7.409)$$

with
$$M^2 = m^2 + \xi R + \frac{\lambda}{6}\rho^2. \tag{7.410}$$

If we operate on both sides of (7.406) with ∇_μ it is possible to show that
$$e^2\rho^2 \nabla_\mu G^\mu{}_\nu(x,x') = \nabla_\nu \delta(x,x'). \tag{7.411}$$

This is essentially the identity first given by Endo (1984). Another way to obtain the result is described by Barvinsky and Vilkovisky (1985). If we assume that the vector Green function is symmetric, $G^\mu{}_\lambda(x,y) = G_\lambda{}^\mu(y,x)$, then it follows that
$$e^2\rho^2 \nabla^\lambda_y G^\mu{}_\lambda(x,y) = -\nabla^\mu_x \delta(x,y). \tag{7.412}$$

(This uses the symmetry of the δ-distribution $\delta(x,y) = \delta(y,x)$ along with the identity $\nabla^\mu_y \delta(x,y) = -\nabla^\mu_x \delta(x,y)$.) Using (7.412) to simplify (7.409) results in
$$P^\mu{}_\nu(x,x') = M^2 \int dv_z \left(\nabla^\mu_x \gamma^{xz}\right)\left(\nabla'_\nu \gamma^{x'z}\right). \tag{7.413}$$

To evaluate the last term of (7.407) note that the expression is formally expressed using $\ln\det(I-P) = \operatorname{tr}\ln(I-P)$. By expanding the logarithm in its Taylor series we have
$$\ln\det(I-P) = -\sum_{n=1}^\infty \frac{1}{n}\operatorname{tr}P^n. \tag{7.414}$$

Here the trace involves a sum over the vector indices as well as an integration over the spacetime labels. From (7.413) we have
$$\operatorname{tr}P = \int dv_x\, P^\mu{}_\mu(x,x)$$
$$= M^2 \int dv_x dv_z \left(\nabla^\mu_x \gamma^{xz}\right)\left(\nabla_{x\mu}\gamma^{xz}\right)$$
$$= -M^2 \int dv_x dv_z \left(\Box_x \gamma^{xz}\right)\gamma^{xz}, \tag{7.415}$$

where in the last line we integrate by parts over x. If we define
$$E(x,x') = \int dv_z \left(\Box_x \gamma^{xz}\right)\gamma^{x'z}, \tag{7.416}$$
then
$$\operatorname{tr}P = -M^2 \operatorname{tr}E. \tag{7.417}$$

By performing similar steps, it is possible to show that
$$\operatorname{tr}P^n = \int dv_x dv_{y_1}\cdots dv_{y_{n-1}} P^\mu{}_{\lambda_1}(x,y_1) P^{\lambda_1}{}_{\lambda_2}(y_1,y_2)\cdots$$
$$\times P^{\lambda_{n-2}}{}_{\lambda_{n-1}}(y_{n-2},y_{n-1}) P^{\lambda_{n-1}}{}_\mu(y_{n-1},x)$$
$$= (-M^2)^n \operatorname{tr}E^n. \tag{7.418}$$

7.9 Scalar quantum electrodynamics

Thus,
$$\ln \det(I - P) = \ln \det(I + M^2 E). \tag{7.419}$$

We can use (7.380) to show that

$$\begin{aligned}
E(x, x') &= \int dv_z \Box_x \left[(\Box_x + e^2 \rho^2)^{-1} \delta(x, z) \right] \gamma^{x'z} \\
&= \Box_x \left(\Box_x + e^2 \rho^2 \right)^{-1} \gamma^{x'x} \\
&= \Box_x \left(\Box_x + e^2 \rho^2 \right)^{-2} \delta(x, x').
\end{aligned} \tag{7.420}$$

We then have

$$\begin{aligned}
\ln \det(I - P) &= \ln \det \left[I + M^2 \Box_x \left(\Box_x + e^2 \rho^2 \right)^{-2} \right] \\
&= -2 \ln \det(\Box_x + e^2 \rho^2) + \ln \det \left(\Box_x + e^2 \rho^2 + \frac{1}{2} M^2 + b \right) \\
&\quad + \ln \det \left(\Box_x + e^2 \rho^2 + \frac{1}{2} M^2 - b \right),
\end{aligned} \tag{7.421}$$

where

$$b^2 = M^2 \left(e^2 \rho^2 + \frac{1}{4} M^2 \right). \tag{7.422}$$

All of the operators appearing in (7.421) are of the general form $\Box + Q$ and the method of Chapter 5 can be used to evaluate the poles in dimensional regularization.

The first term on the right-hand side of (7.407) also must be manipulated into the form $\Box + Q$ to enable the results of Chapter 5 to be used. Formally the determinant of any operator can be viewed as the product of its eigenvalues. We will consider the eigenvalue equation

$$(\Box \delta_\nu^\mu - \nabla_\nu \nabla^\mu + e^2 \rho^2 \delta_\nu^\mu) \psi_n^\nu = \lambda_n \psi_n^\mu. \tag{7.423}$$

Set
$$\psi_n^\mu = \psi_{n\perp}^\mu + \nabla^\mu \psi_n, \tag{7.424}$$

where
$$\nabla_\mu \psi_{n\perp}^\mu = 0. \tag{7.425}$$

It now follows that

$$\Box \psi_{n\perp}^\mu + R^\mu{}_\nu \psi_{n\perp}^\nu + e^2 \rho^2 \psi_{n\perp}^\mu + e^2 \rho^2 \nabla^\mu \psi_n = \lambda_n \psi_{n\perp}^\mu + \lambda_n \nabla^\mu \psi_n. \tag{7.426}$$

We now find

$$\begin{aligned}
\ln \det \left(\Box \delta_\nu^\mu - \nabla_\nu \nabla^\mu + e^2 \rho^2 \delta_\nu^\mu \right) &= \ln \det(e^2 \rho^2) \\
&\quad + \ln \det_\perp \left(\Box \delta_\nu^\mu + R^\mu{}_\nu + e^2 \rho^2 \delta_\nu^\mu \right),
\end{aligned} \tag{7.427}$$

where the subscript $\perp$ on the determinant denotes that there is a restriction to transverse vector fields that satisfy (7.425). It is necessary to remove this restriction if we are to utilize the results of Chapter 5. To this end consider the eigenvalue problem

$$\left(\Box\delta^\mu_\nu + R^\mu{}_\nu + e^2\rho^2\delta^\mu_\nu\right)\phi^\nu_n = \lambda_n\phi^\mu_n, \tag{7.428}$$

where we do not assume that $\nabla_\mu\phi^\mu_n = 0$. Using the eigenvalues of (7.428) will clearly overcount when compared with the operator in (7.427), and we now determine this overcounting. Operate on both sides of (7.428) with ∇_μ to find

$$(\Box + e^2\rho^2)\nabla_\mu\phi^\mu_n = \lambda_n\nabla_\mu\phi^\mu_n. \tag{7.429}$$

The eigenvalues of the operator in (7.428) that overcount those in relation to those restricted to transverse vectors are seen to be the same as those of the operator $(\Box + e^2\rho^2)$. We therefore have

$$\ln\det{}_\perp\left(\Box\delta^\mu_\nu + R^\mu{}_\nu + e^2\rho^2\delta^\mu_\nu\right) = \ln\det\left(\Box\delta^\mu_\nu + R^\mu{}_\nu + e^2\rho^2\delta^\mu_\nu\right)$$
$$- \ln\det(\Box + e^2\rho^2). \tag{7.430}$$

Combining (7.407), (7.421), (7.427), and (7.430) shows that

$$\ln\det S_{;\bar{A}_\mu(x)\bar{A}(x')} = \ln\det\left(\Box\delta^\mu_\nu + R^\mu{}_\nu + e^2\rho^2\delta^\mu_\nu\right)$$
$$+ \ln\det\left(\Box + e^2\rho^2 + \frac{1}{2}M^2 + b\right) + \ln\det\left(\Box + e^2\rho^2 + \frac{1}{2}M^2 - b\right)$$
$$- 3\ln\det(\Box + e^2\rho^2) + 2\ln\det(e\rho). \tag{7.431}$$

When this result is used back in (7.405), the last term of (7.431) cancels with a similar term that came from the measure. The effective action to one-loop order becomes

$$\Gamma = \bar{S} + \frac{i}{2}\hbar\ln\det\left(\Box + m^2 + \xi R + \frac{\lambda}{2}\rho^2\right)$$
$$+ \frac{i}{2}\hbar\ln\det\left(\Box\delta^\mu_\nu + R^\mu{}_\nu + e^2\rho^2\delta^\mu_\nu\right) - \frac{3i}{2}\hbar\ln\det(\Box + e^2\rho^2)$$
$$+ \frac{i}{2}\hbar\ln\det\left(\Box + e^2\rho^2 + \frac{1}{2}M^2 + b\right) + \frac{i}{2}\hbar\ln\det\left(\Box + e^2\rho^2 + \frac{1}{2}M^2 - b\right). \tag{7.432}$$

(M^2 is given by (7.410) and b by (7.422).)

We can evaluate the pole terms of the one-loop part of Γ by using the general results of Chapter 5 since all operators are now of the general form $D^2 + Q$. For the operator with vector indices we must use $(W_{\mu\nu})^\lambda{}_\sigma = R^\lambda{}_{\sigma\nu\mu}$ with our

7.9 Scalar quantum electrodynamics

curvature conventions. After some straightforward calculation it can be shown that the divergent pole part of the one-loop effective action is given by

$$\mathbf{divp}(\Gamma^{(1)}) = -\frac{\hbar}{16\pi^2 \epsilon} \int dv_x \left\{ \frac{1}{6} \left(\xi + \frac{1}{5} \right) \Box R + \left(\xi^2 - \frac{1}{3}\xi - \frac{1}{9} \right) R^2 \right.$$
$$+ \frac{43}{90} R_{\mu\nu} R^{\mu\nu} - \frac{11}{180} R_{\mu\nu\rho\sigma} R^{\mu\nu\rho\sigma} + m^4$$
$$+ 2 \left(\xi - \frac{1}{6} \right) m^2 R + \left[\left(\xi - \frac{1}{6} \right) \left(2e^2 + \frac{2\lambda}{3} \right) + \frac{5}{6} e^2 \right] \rho^2 R$$
$$\left. + 2 \left(e^2 + \frac{\lambda}{3} \right) m^2 \rho^2 + \left(\frac{5}{36} \lambda^2 + \frac{1}{3} \lambda e^2 + \frac{3}{2} e^4 \right) \rho^4 \right\}. \quad (7.433)$$

It is important to remember that this result is restricted to situations where ξR and ρ are constant. This restriction was for simplicity when performing various integrations by parts. Without this restriction extra terms involving derivatives of R and ρ will arise that require a more complicated analysis. The poles that occur in (7.433) are all of a type that can be dealt with by local counterterms of the form considered in Chapter 6 for the interacting scalar field. We do not pursue this analysis any further.

This section presents a generalization of some of the results given in Balakrishnan and Toms (1991, 1992) where the concern was effective potentials in de Sitter spacetime. Earlier work on the gauge-independent effective action applied to de Sitter space was given by Odintsov (1990a,b).

7.9.2 Effective potential

We now evaluate the effective potential in the special case of flat spacetime. There are various places we could start, including the final expression (7.432); however, as a check on our manipulations we will go back to (7.405). In flat spacetime we can make use of momentum space to Fourier transform Green functions and operators. Feynman boundary conditions are assumed as usual. From (7.380) we have

$$\gamma^{xx'} = \int \frac{d^n k}{(2\pi)^n} \frac{e^{ik \cdot (x-x')}}{-k^2 + e^2 \rho^2}. \quad (7.434)$$

It is then easy to see that

$$\int dv_z (\nabla^\mu \gamma^{xz})(\nabla'_\nu \gamma^{x'z}) = \int \frac{d^n k}{(2\pi)^n} \frac{k^\mu k_\nu}{(-k^2 + e^2 \rho^2)^2} e^{ik \cdot (x-x')}. \quad (7.435)$$

From (7.397) we find

$$S_{;\bar{A}_\mu(x)\bar{A}(x')} = \int \frac{d^n k}{(2\pi)^n} e^{ik \cdot (x-x')} \left\{ (-k^2 \delta^\mu_\nu + k^\mu k_\nu + e^2 \rho^2 \delta^\mu_\nu) \right.$$
$$\left. - v e^2 \rho^2 \left(m^2 + \frac{\lambda}{6} \rho^2 \right) \frac{k^\mu k_\nu}{(-k^2 + e^2 \rho^2)^2} \right\}, \quad (7.436)$$

where we have included a book-keeping parameter v to keep track of how the Vilkovisky–DeWitt connection alters the result from the naive effective action where the connection is absent. By taking the determinant over the vector indices in n spacetime dimensions it can be shown that

$$\det(A\delta^\mu_\nu + Bk^\mu k_\nu) = A^{n-1}(A + Bk^2).$$

For the operator appearing in (7.436) we take

$$A = -k^2 + e^2\rho^2,$$
$$B = 1 - \frac{ve^2\rho^2(m^2 + \frac{\lambda}{6}\rho^2)}{(-k^2 + e^2\rho^2)^2},$$

to find

$$A + Bk^2 = e^2\rho^2 \left[1 - \frac{v(m^2 + \frac{\lambda}{6}\rho^2)k^2}{(-k^2 + e^2\rho^2)^2}\right].$$

Performing a Wick rotation, allowed due to Feynman boundary conditions, the one-loop part of the effective potential becomes[12]

$$V^{(1)} = -\frac{\hbar}{2}\int \frac{d^n k}{(2\pi)^n} \ln\left(k^2 + m^2 + \frac{\lambda}{2}\rho^2\right)$$
$$- \frac{\hbar}{2}(n-1)\int \frac{d^n k}{(2\pi)^n} \ln(k^2 + e^2\rho^2)$$
$$- \frac{\hbar}{2}\int \frac{d^n k}{(2\pi)^n} \ln\left[1 + \frac{v(m^2 + \frac{\lambda}{6}\rho^2)k^2}{(k^2 + e^2\rho^2)^2}\right]. \quad (7.437)$$

The first term on the right-hand side follows from using

$$S_{;\bar{\rho}(x)\bar{\rho}(x')} = \int \frac{d^n k}{(2\pi)^n} e^{ik\cdot(x-x')}\left(-k^2 + m^2 + \frac{\lambda}{2}\rho^2\right) \quad (7.438)$$

that arises from Fourier transforming (7.396). We must add in the classical potential to (7.437) to obtain the complete effective potential.

The last term of (7.437) can be evaluated by noting that

$$\ln\left[1 + \frac{v(m^2 + \frac{\lambda}{6}\rho^2)k^2}{(k^2 + e^2\rho^2)^2}\right] = -2\ln(k^2 + e^2\rho^2) + \ln(k^2 + M_+^2) + \ln(k^2 + M_-^2), \quad (7.439)$$

where

$$M_\pm^2 = e^2\rho^2 + \frac{v}{2}\left(m^2 + \frac{\lambda}{6}\rho^2\right) \pm v\left(m^2 + \frac{\lambda}{6}\rho^2\right)^{1/2}\left(m^2 + e^2\rho^2 + \frac{\lambda}{6}\rho^2\right)^{1/2}. \quad (7.440)$$

---

[12] Compare the result to that in Section 5.1.

7.9 Scalar quantum electrodynamics

(We are only interested in the case of $v = 0$ (no connection term present), or $v = 1$ here, so we can use $v^2 = v$ whenever appropriate.) We then find

$$V^{(1)} = \frac{\hbar}{2}(4\pi)^{-n/2}\Gamma(-n/2)\left\{\left(m^2 + \frac{\lambda}{2}\rho^2\right)^{n/2}\right.$$
$$\left. + (n-3)(e^2\rho^2)^{n/2} + (M_+^2)^{n/2} + (M_-^2)^{n/2}\right\}, \quad (7.441)$$

as the unrenormalized one-loop contribution to the effective potential. The result can be expanded about the pole term using $n = 4 + \epsilon$ and

$$(4\pi)^{-n/2}\Gamma(-n/2)(M^2)^{n/2} = -\frac{M^4}{16\pi^2\epsilon} - \frac{M^4}{32\pi^2}\left[\ln\left(\frac{M^2}{4\pi}\right) + \gamma - \frac{3}{2}\right] + \cdots . \quad (7.442)$$

We have presented sufficient information to enable the calculation of the effective potential for arbitrary values of m^2. We can fix a set of renormalization conditions and remove the pole terms with local counterterms as usual. Because the result is rather lengthy we will not pursue this rather straightforward calculation.

As a check on our results we can compute the pole part of $V^{(1)}$ from (7.441) and (7.442) and compare it with what we found in (7.433). It can be seen that

$$\mathbf{divp}(V^{(1)}) = -\frac{\hbar}{32\pi^2\epsilon}\left\{m^4(v+1) + m^2\rho^2\left[\left(1 + \frac{v}{3}\right)\lambda + 4ve^2\right]\right.$$
$$\left. + \rho^4\left[3e^4 + \frac{2}{3}v\lambda e^2 + \frac{\lambda^2}{4}\left(1 + \frac{v}{9}\right)\right]\right\}. \quad (7.443)$$

If we set $v = 1$ we find a result that agrees with (7.433). The role of the connection in obtaining the correct pole terms is evident in the presence of v in (7.443).

If we set $m^2 = 0$ and impose the renormalization condition

$$\left.\frac{\partial^4 V(\rho)}{\partial\rho^4}\right|_{\rho=M} = \lambda, \quad (7.444)$$

we obtain

$$V(\rho) = \frac{\lambda}{4!}\rho^4 + \frac{\hbar}{64\pi^2}\left[3e^4 + \frac{1}{4}\left(1 + \frac{v}{9}\right)\lambda^2 + \frac{2}{3}v\lambda e^2\right]\rho^4\left[\ln\left(\frac{\rho^2}{M^2}\right) - \frac{25}{6}\right] \quad (7.445)$$

as the effective potential.

If we take $v = 1$, so that the non-trivial effect of the Vilkovisky–DeWitt field space connection is included, and the result is gauge condition independent, the result agrees with the calculation of Fradkin and Tseytlin (1984) who used a Cartesian parameterization for the complex scalar field and adopted a general background field gauge. The gauge condition independence (in the case $v = 1$) can be verified by comparing the result found here with that of Kunstatter (1987), who used the same polar coordinate representation for the scalar field as we used here, but calculated using the Lorentz gauge. If on the other hand we choose $v = 0$, so that the field space connection is not included, the result is gauge condition dependent, and the result of Dolan and Jackiw (1974) is obtained. The calculation described in this section was first given by Russell and Toms (1989). A more complete analysis of scalar electrodynamics can be found in Epp et al. (1993).

Appendix

7A.1 Proof that $[\mathfrak{K}_\alpha, \mathfrak{K}_\beta] = 0$.

By definition of the Lie bracket,

$$[\mathfrak{K}_\alpha, \mathfrak{K}_\beta] = \left(\mathfrak{K}_\alpha^j \mathfrak{K}_{\beta,j}^i - \mathfrak{K}_\beta^j \mathfrak{K}_{\alpha,j}^i\right) \frac{\delta}{\delta\varphi^i}, \tag{7A.1}$$

where $\mathfrak{K}_\alpha^i$ was defined in (7.113). It follows from this definition that

$$\mathfrak{K}_\alpha^j \mathfrak{K}_{\beta,j}^i = \mathfrak{K}_\alpha^j K_{\gamma,j}^i (Q^{-1})^\gamma{}_\beta + \mathfrak{K}_\alpha^j K_\gamma^i (Q^{-1})^\gamma{}_{\beta,j}. \tag{7A.2}$$

Upon differentiation of the relation $Q^\alpha{}_\gamma (Q^{-1})^\gamma{}_\beta = \delta^\alpha{}_\beta$, we find

$$(Q^{-1})^\alpha{}_{\beta,j} = -(Q^{-1})^\alpha{}_\gamma (Q^{-1})^\delta{}_\beta Q^\gamma{}_{\delta,j}. \tag{7A.3}$$

This leads to the expression

$$[\mathfrak{K}_\alpha, \mathfrak{K}_\beta] = \left(K_\delta^j K_{\gamma,j}^i - K_\gamma^j K_{\delta,j}^i\right)(Q^{-1})^\delta{}_\alpha (Q^{-1})^\gamma{}_\beta \frac{\delta}{\delta\varphi^i}$$
$$+ \left[\mathfrak{K}_\beta^j (Q^{-1})^\rho{}_\alpha - \mathfrak{K}_\alpha^j (Q^{-1})^\rho{}_\beta\right] \mathfrak{K}_\tau^i Q^\tau{}_{\rho,j} \frac{\delta}{\delta\varphi^i}. \tag{7A.4}$$

The algebra (7.15) may be used in the first term on the right-hand side, and the definition of $Q^\tau{}_\rho = \chi^\tau{}_{,k} K_\rho^k$ may be used in the second term, resulting in

$$[\mathfrak{K}_\alpha, \mathfrak{K}_\beta] = -f_{\delta\gamma}{}^\epsilon K_\epsilon^i (Q^{-1})^\delta{}_\alpha (Q^{-1})^\gamma{}_\beta \frac{\delta}{\delta\varphi^i}$$
$$+ \left[\mathfrak{K}_\beta^j (Q^{-1})^\rho{}_\alpha - \mathfrak{K}_\alpha^j (Q^{-1})^\rho{}_\beta\right] K_{\rho,j}^k \mathfrak{K}_\tau^k \chi^\tau{}_{,k} \frac{\delta}{\delta\varphi^i}. \tag{7A.5}$$

Using the definition $\mathfrak{K}_\beta^j = K_\gamma^j (Q^{-1})^\gamma{}_\beta$, it is seen easily that

$$\mathfrak{K}_\beta^j (Q^{-1})^\rho{}_\alpha K_{\rho,j}^k - \mathfrak{K}_\alpha^j (Q^{-1})^\rho{}_\beta K_{\rho,j}^k$$
$$= \left(K_\delta^j K_{\rho,j}^k - K_\gamma^j K_{\delta,j}^k\right)(Q^{-1})^\rho{}_\alpha (Q^{-1})^\delta{}_\beta$$
$$= f_{\rho\delta}{}^\epsilon K_\epsilon^k (Q^{-1})^\rho{}_\alpha (Q^{-1})^\delta{}_\beta. \tag{7A.6}$$

The algebra (7.15) has been used to obtain the last line. It then follows that

$$\left[\mathfrak{R}^j_\beta(Q^{-1})^\rho{}_\alpha - \mathfrak{R}^j_\alpha(Q^{-1})^\rho{}_\beta\right] K^k_{\rho,j}\mathfrak{R}^k_\tau \chi^\tau{}_{,k} \tag{7A.7}$$

$$= f_{\rho\delta}{}^\epsilon K^k_\epsilon (Q^{-1})^\rho{}_\alpha (Q^{-1})^\delta{}_\beta \mathfrak{R}^i_\tau \chi^\tau{}_{,k}$$

$$= f_{\rho\delta}{}^\epsilon (Q^{-1})^\rho{}_\alpha (Q^{-1})^\delta{}_\beta \mathfrak{R}^i_\tau Q^\tau{}_\epsilon$$

$$= f_{\rho\delta}{}^\epsilon K^i_\epsilon (Q^{-1})^\rho{}_\alpha (Q^{-1})^\delta{}_\beta, \tag{7A.8}$$

where we have used the definition $Q^\tau{}_\epsilon = \chi^\tau{}_{,k} K^k_\epsilon$ in the second last line, and the definition $\mathfrak{R}^i_\tau = K^i_\epsilon (Q^{-1})^\epsilon{}_\tau$ to obtain the last line. The result in (7A.8) leads to a cancelation of the two terms on the right-hand side of (7A.5), leaving

$$[\mathfrak{R}_\alpha, \mathfrak{R}_\beta] = 0. \tag{7A.9}$$

This result holds even if the structure constants of the algebra are replaced with structure functions, but does require that the algebra be closed. Use of the $\mathfrak{R}_\alpha$ as generators was first noted by Batalin and Vilkovisky (1984), and was referred to as "abelianization".

Appendix
Quantized Inflaton Perturbations

In this Appendix, we discuss properties of the quantized inflaton perturbation $\delta\phi(\vec{x},t)$ in both the discrete and continuous momentum-space representations. We define the Fourier components $\delta\phi_{\vec{k}}(t)$ of $\delta\phi(\vec{x},t)$ in both representations. We also show how the power spectrum $\mathcal{P}(k,t)$ is related to the coincident and separated two-point functions formed from $\delta\phi(\vec{x},t)$.

Recall that the inflaton perturbation field can be written as in (2.173) and (2.174); namely,

$$\delta\phi(\vec{x},t) = \sum_{\vec{k}} \left\{ A_{\vec{k}} f_{\vec{k}}(\vec{x},t) + A_{\vec{k}}^\dagger f_{\vec{k}}^*(\vec{x},t) \right\}, \qquad (A.1)$$

where, for consistency with (2.129), we let

$$f_{\vec{k}} = (2Va^3(t))^{-1/2} e^{i\vec{k}\cdot\vec{x}} h_k(t) . \qquad (A.2)$$

Here $k^i = 2\pi n^i/L$ with n^i an integer, $k = |\vec{k}|$, and we have imposed periodic boundary conditions on a cube of coordinate volume $V = L^3$ and physical volume $Va^3(t)$. The function $h_k(t)$ depends on the magnitude, $k = |\vec{k}|$, but not the direction of $\vec{k}$.

Define the operator $\delta\phi_{\vec{k}}(t)$ as

$$\delta\phi_{\vec{k}}(t) = (2Va^3(t))^{-1/2} \left\{ A_{\vec{k}} h_k(t) + A_{-\vec{k}}^\dagger h_k^*(t) \right\} . \qquad (A.3)$$

This satisfies $\delta\phi_{-\vec{k}}(t)^\dagger = \delta\phi_{\vec{k}}(t)$. It is straightforward to show that $\delta\phi(\vec{x},t)$ in (A.1) can be written as

$$\delta\phi(\vec{x},t) = \sum_{\vec{k}} \delta\phi_{\vec{k}}(t) e^{i\vec{k}\cdot\vec{x}} . \qquad (A.4)$$

In the discrete case, the dimension[1] of $\delta\phi_{\vec{k}}(t)$ is l^{-1}, the same as that of $\delta\phi(\vec{x},t)$. It follows from the commutation relations of the creation and annihilation operators that

$$\langle 0|\delta\phi_{\vec{k}}(t)^\dagger \delta\phi_{\vec{k}'}(t)|0\rangle = \langle 0|\delta\phi_{\vec{k}}(t)^\dagger \delta\phi_{\vec{k}}(t)|0\rangle \delta_{\vec{k}\ \vec{k}'} . \tag{A.5}$$

This resembles the ensemble average taken over an ensemble of systems in which the perturbation components have random phases.

The perturbation spectrum $\mathcal{P}_\phi(k,t)$ of $\delta\phi(\vec{x},t)$ in the limit of large L is independent of the direction of $\vec{k}$, and can be defined as (Liddle and Lyth 2000)

$$\mathcal{P}_\phi(k,t) = \left(\frac{L}{2\pi}\right)^3 4\pi k^3 \langle 0|\delta\phi_{\vec{k}}(t)^\dagger \delta\phi_{\vec{k}}(t)|0\rangle . \tag{A.6}$$

This is equivalent to

$$\langle 0|\delta\phi(\vec{x},t)^\dagger \delta\phi(\vec{x},t)|0\rangle = \int_0^\infty \mathcal{P}_\phi(k,t) k^{-1} dk , \tag{A.7}$$

where we have used (A.4)–(A.6) and $(2\pi/L)^3 \sum_{\vec{k}} \to \int d^3k$ in the continuum limit. The dimension of the power spectrum $\mathcal{P}_\phi(k,t)$ is l^{-2}.

A similar calculation shows that the non-coincident two-point correlation function can be written in terms of the spectrum as

$$\langle 0|\delta\phi(\vec{x}_1,t)^\dagger \delta\phi(\vec{x}_2,t)|0\rangle = \sum_{\vec{k}} \langle 0|\delta\phi_{\vec{k}}(t)^\dagger \delta\phi_{\vec{k}}(t)|0\rangle e^{i\vec{k}\cdot(\vec{x}_1-\vec{x}_2)}$$

$$= \int_0^\infty \mathcal{P}_\phi(k,t) \frac{\sin(k|\vec{x}_1-\vec{x}_2|)}{k|\vec{x}_1-\vec{x}_2|} \frac{dk}{k} . \tag{A.8}$$

In the coincidence limit, $\vec{x}_1 \to \vec{x}_2$, this reduces to the previous equation. Therefore, we expect that point-splitting regularization and renormalization should give the same result as a regularization and renormalization scheme such as adiabatic regularization applied to the coincident two-point function. We will use adiabatic regularization because it is simpler in the present context and directly yields an expression for the spectrum.

Finally, for completeness, we explain how to go to the continuum expression for the field, from which we can obtain the various results given here directly in the continuum limit. We define the continuum annihilation operators $A(\vec{k})$ for large L by the replacement

$$A_{\vec{k}} \to (2\pi/L)^{3/2} A(\vec{k}) . \tag{A.9}$$

The operator for the number of particles with momenta near $\vec{k}$ in the range d^3k is

$$A(\vec{k})^\dagger A(\vec{k}) d^3k. \tag{A.10}$$

---

[1] Here l is an arbitrary unit of length.

Heuristically, these equations can be understood because the number of particles of momentum $\vec{k}$ in the box of side L is $A_{\vec{k}}^\dagger A_{\vec{k}}$, and the separation between adjacent values of each component of $\vec{k}$ is $2\pi/L$. Therefore, (A.9) is the logical expression of the fact that the number of particles with a given discrete value of $\vec{k}$ is held constant as L is taken to infinity, while the volume in $\vec{k}$-space that corresponds to a discrete value of $\vec{k}$ is $(2\pi/L)^3 \to d^3k$.

The continuum operators satisfy

$$[A(\vec{k}), A(\vec{k}')^\dagger] = \delta(\vec{k}-\vec{k}') , \qquad (A.11)$$

because $V/(2\pi)^3 \delta_{\vec{k}\,\vec{k}'}$ is a representation of the Dirac δ-function in the limit that $L \to \infty$.

It follows from the discrete expression for $\delta\phi(\vec{x},t)$ in (A.1) that

$$\delta\phi(\vec{x},t) = \frac{1}{\sqrt{2(2\pi)^3 a^3(t)}} \int d^3k \left\{ A(\vec{k}) h_k(t) e^{i\vec{k}\cdot\vec{x}} + A(\vec{k})^\dagger h_k(t)^* e^{-i\vec{k}\cdot\vec{x}} \right\}, \qquad (A.12)$$

where we have used (A.2). The dimension of $A(\vec{k})$ is $l^{3/2}$. The dimension of $h_k(t)$ is $l^{1/2}$. We are taking $a(t)$ as dimensionless. Therefore, the dimension of $\delta\phi(\vec{x},t)$ is $l^{-3} l^{3/2} l^{1/2} = l^{-1}$. This serves as a check, since the dimension of $\delta\phi(\vec{x},t)$ cannot depend on whether we use the discrete or continuum method of representing its Fourier expansion.

We can obtain the expressions corresponding to the previous results by working directly in the continuum limit.

Using (A.9) in (A.3), we obtain the continuum expression for $\delta\phi_{\vec{k}}(t)$:

$$\delta\phi_{\vec{k}}(t) = \left(2(2\pi)^3 a^3(t)\right)^{-1/2} \left\{ A(\vec{k}) h_k(t) + A(-\vec{k})^\dagger h_k^*(t) \right\}, \qquad (A.13)$$

where we have absorbed a factor of L^3 into the continuum definition of $\delta\phi_{\vec{k}}(t)$. Thus, the dimension of $\delta\phi_{\vec{k}}(t)$ in the continuum case is l^2. In the continuum limit,

$$\delta\phi(\vec{x},t) = \int d^3k\, \delta\phi_{\vec{k}}(t) e^{i\vec{k}\cdot\vec{x}} . \qquad (A.14)$$

We find that

$$\langle 0 | \delta\phi_{\vec{k}}(t)^\dagger \delta\phi_{\vec{k}'}(t) | 0 \rangle = \left(2(2\pi)^3 a^3(t)\right)^{-1} |h_k(t)|^2 \delta(\vec{k}-\vec{k}'). \qquad (A.15)$$

The power spectrum is defined by

$$\langle 0 | \delta\phi(\vec{x},t)^\dagger \delta\phi(\vec{x},t) | 0 \rangle = \int_0^\infty \mathcal{P}_\phi(k,t) k^{-1} dk . \qquad (A.16)$$

This form of the definition does not depend on whether we work in the discrete or continuum representation.

We can also write $\langle 0| \delta\phi(\vec{x},t)^\dagger \delta\phi(\vec{x},t) |0\rangle$ as

$$\langle 0| \delta\phi(\vec{x},t)^\dagger \delta\phi(\vec{x},t) |0\rangle = \int d^3k \int d^3k' \langle 0|\delta\phi_{\vec{k}}(t)^\dagger \delta\phi_{\vec{k}'}(t)|0\rangle$$

$$= (2(2\pi)^3 a^3(t))^{-1} \int_0^\infty d^3k |h_k(t)|^2$$

$$= ((2\pi)^2 a^3(t))^{-1} \int_0^\infty dk\, k^2 |h_k(t)|^2 \,. \quad (A.17)$$

Thus, in the continuum case, we can write

$$\mathcal{P}_\phi(k,t) = 4\pi k^3 \int d^3k' \langle 0|\delta\phi_{\vec{k}}(t)^\dagger \delta\phi_{\vec{k}'}(t)|0\rangle \,, \quad (A.18)$$

which is the continuum analogue of the corresponding discrete expression given in (A.6).

Comparing (A.16) and (A.17), we obtain

$$\mathcal{P}_\phi(k,t) = \left((2\pi)^2 a^3(t)\right)^{-1} k^3 |h_k(t)|^2 \,. \quad (A.19)$$

We also obtain the same result from (A.15) and (A.18). This result agrees with (2.185) if the adiabatic subtraction terms are dropped, as would be expected in the present unregularized expression. Regularization is discussed in connection with (2.185).

References

Abbott, L. (1981). The background field method beyond one loop, *Nucl. Phys. B* **185**, 189–203.

Abers, E. S. and Lee, B. W. (1973). Gauge theories, *Phys. Rep. C* **9**, 1–144.

Abramowitz, M. and Stegun, I. A. (1972). *Handbook of Mathematical Functions* (National Bureau of Standards, Washington, DC, USA).

Abrikosov, A. A., Gorkov, L. P., and Dzyaloshinski, I. E. (1975). *Methods of Quantum Field Theory in Statistical Physics* (Dover Publications, New York).

Adams, F. C., Bond, J. R., Freese, K., Frieman, J. A., and Olinto, A. V. (1993). Natural inflation: Particle physics models, power-law spectra for large-scale structure, and constraints from the cosmic background explorer, *Phys. Rev. D* **47**, 426–455.

Adams, F. C., Freese, K., and Guth, A. H. (1991). Constraints on the scalar-field potential in inflationary models, *Phys. Rev. D* **43**, 965–976.

Adler, S. L. (1973). Massless electrodynamics on the 5-dimensional unit hypersphere: An amplitude-integral formulation, *Phys. Rev. D* **8**, 2400–2418.

Adler, S. L. (1982). Einstein gravity as a symmetry-breaking effect in quantum field theory, *Rev. Mod. Phys.* **54**, 729–766; Erratum *ibid.* **55**, 837.

Adler, S. L., Lieberman, J., and Ng, Y. J. (1977). Regularization of the stress-energy tensor for vector and scalar particles propagating in a general background metric, *Ann. Phys. (N. Y.)* **106**, 279–321.

Agullo, I., Navarro-Salas, J., Olmo, G. J., and Parker, L. (2008a). Two-point functions with an invariant Planck scale and thermal effects, *Phys. Rev. D* **77**, 124032 (10 pages); arXiv:0804.0513.

Agullo, I., Navarro-Salas, J., Olmo, G. J., and Parker, L. (2008b). Acceleration radiation at the Planck scale, *Phys. Rev. D* **77**, 104034 (8 pages); arXiv:0802.3920.

Albrecht, A. and Steinhardt, P. J. (1982). Cosmology for grand unified theories with radiatively induced symmetry breaking, *Phys. Rev. Lett.* **48**, 1220–1223.

Allen, B. (1985). Vacuum states in de Sitter space, *Phys. Rev. D* **32**, 3136–3149.

Allen, B. and Folacci, A. (1987). Massless minimally coupled scalar field in de Sitter space, *Phys. Rev. D* **35**, 3771–3778.

Anderson, P. R. and Eaker, W. (1999). Analytic approximation and an improved method for computing the stress-energy of quantized scalar fields in Robertson-Walker spacetimes, *Phys. Rev. D* **61**, 024003-1–0024003-5.

Anderson, P. R., Hiscock, W. A., and Samuel, D. A. (1995). Stress-energy tensor of quantized scalar fields in static spherically symmetric spacetimes, *Phys. Rev. D* **51**, 4337–4358.

Anderson, P. R. and Parker, L. (1987). Adiabatic regularization in closed Robertson–Walker universes, *Phys. Rev. D* **36**, 2963–2969.

Armendariz-Picon, C. and Lim, E. (2003). Vacuum choices and predictions of
   inflation, *J. Cosmol. Astropart. Phys.* **12**, 006; arXiv:hep-th/0303103.
Arnowitt, R. and Nath, P. (1976). Riemannian geometry in spaces with Grassmann
   coordinates, *Gen. Relativ. Gravitation* **7**, 89–103.
Ashmore, J. F. (1972). A method of gauge-invariant regularization, *Lett. Nuovo
   Cimento* **4**, 289–290.
Ashtekar, A. and Magnon, A. (1975). Quantum fields in curved spacetimes, *Proc. R.
   Soc. Lond. A* **346**, 375–394.
Avis, S. J. and Isham, C. J. (1979a). Lorentz gauge invariant vacuum functionals for
   quantized spinor fields in nonsimply connected space-times, *Nucl. Phys. B* **156**,
   441–455.
Avis, S. J. and Isham, C. J. (1979b). Quantum field theory and fibre bundles in a
   general spacetime, in *Recent Developments in Gravitation*, ed. M. Lévy and S.
   Deser (Plenum, New York), 347–401.
Avis, S. J. and Isham, C. J. (1980). Generalized spin structures on four-dimensional
   spacetimes, *Commun. Math. Phys.* **72**, 103–118.
Babelon, O. and Viallet, C. M. (1979). The geometrical derivation of the
   Faddeev–Popov determinant, *Phys. Lett. B* **85**, 246–248.
Babelon, O. and Viallet, C. M. (1981). The Riemannian geometry of the configuration
   space of gauge theories, *Commun. Math. Phys.* **81**, 515–525.
Balakrishnan, J. and Toms, D. J. (1991). Gauge-independent effective action for
   quantum electrodynamics in Euclidean de Sitter space, *Phys. Lett. B* **269**,
   339–344.
Balakrishnan, J. and Toms, D. J. (1992). Gauge-independent effective potential for
   minimally coupled quantum fields in curved space, *Phys. Rev. D* **46**, 4413–4420.
Bardeen, J. (1981). Black holes do evaporate thermally, *Phys. Rev. Lett.* **46**, 382–385.
Bardeen, J. M., Steinhardt, P. J., and Turner, M. S. (1983). Spontaneous creation of
   almost scale-free density perturbations in an inflationary universe, *Phys. Rev. D*
   28, 679–693.
Bargmann, V. (1932). Bemerkungen zur allgemein-relativistischen fassung der
   quantentheorie, *Sitzungsber. Preuss. Akad. Wiss., Phys.-math. Kl.* **28**, 346–354.
Barvinsky, A. O. and Vilkovisky, G. A. (1985). The generalized Schwinger–DeWitt
   technique in gauge theories and quantum gravity, *Phys. Rep. C* **119**, 1–74.
Bastero-Gil, M., Freese, K., and Mersini-Houghton, L. (2003). What can WMAP tell
   us about the very early Universe? New physics as an explanation of the
   suppressed large scale power and running spectral index, *Phys. Rev. D* **68**,
   123514.
Batalin, I. A. (1981). Quasigroup construction and first class constraints, *J. Math.
   Phys.* **22**, 1837–1850.
Batalin, I. A. and Vilkovisky, G. A. (1981). Gauge algebra and quantization, *Phys.
   Lett. B* **102**, 27–31.
Batalin, I. A. and Vilkovisky, G. A. (1984). Closure of the gauge algebra, generalized
   Lie equations and Feynman rules, *Nucl. Phys. B* **234**, 106–124.
Bekenstein, J. D. (1972). Black holes and the second law, *Lett. Nuovo Cimento* **4**,
   737–740.
Bekenstein, J. D. (1973). Black holes and entropy, *Phys. Rev. D* **7**, 2333–2346.
Bekenstein, J. D. (1974). Generalized second law of thermodynamics in black hole
   physics, *Phys. Rev. D* **9**, 3292–3300.
Bekenstein, J. D. and Parker, L. (1981). Path-integral evaluation of Feynman
   propagator in curved spacetime, *Phys. Rev. D* **23**, 2850–2869.

Belinfante, F. J. (1939). On the spin angular momentum of mesons, *Physica* **6**, 887–897.

Belinfante, F. J. (1940). On the current and the density of the electric charge, the energy, the linear momentum, and the angular momentum of arbitrary fields, *Physica* **7**, 449–474.

Berezin, F. A. (1966). *The Method of Second Quantization* (Academic Press, New York).

Berezin, F. A. (1987). *Introduction to Superanalysis* (D. Reidel, Dordrecht, Holland).

Berger, B. K. (1982). Singularity avoidance in the semi-classical Gowdy T^3 cosmological model, *Phys. Lett. B* **108**, 394–398.

Berger, B. K. (1984). Quantum effects in the Gowdy T^3 cosmology, *Ann. Phys. (N. Y.)* **156**, 155–193.

Bernstein, J. (1974). Spontaneous symmetry breaking, gauge theories, the Higgs mechanism and all that, *Rev. Mod. Phys.* **46**, 7–48.

Birrell, N. D. (1978). The application of adiabatic regularization to calculations of cosmological interest, *Proc. R. Soc. Lond. A* **361**, 513–526.

Birrell, N. D. and Davies, P. C. W. (1978). Massless thirring model in curved space: Thermal states and conformal anomaly, *Phys. Rev. D* **18**, 4408–4421.

Birrell, N. D. and Davies, P. C. W. (1982), *Quantum Fields in Curved Space* (Cambridge University Press, Cambridge).

Birrell, N. D. and Ford, L. H. (1979). Self-interacting quantized fields and particle creation in Robertson–Walker universes, *Ann. Phys. (N. Y.)* **122**, 1–25.

Birrell, N. D. and Taylor, J. G. (1980). Analysis of interacting quantum field theory in curved space-time, *J. Math. Phys.* **21**, 1740–1760.

Bjorken, J. D. and Drell, S. D. (1964). *Relativistic Quantum Mechanics* (McGraw-Hill, New York).

Bjorken, J. D. and Drell, S. D. (1965). *Relativistic Quantum Fields* (McGraw-Hill, New York).

Bollini, C. G. and Giambiagi, J. J. (1972). Dimensional renormalization: The number of dimensions as a regularizing parameter, *Nuovo Cimento B* **12**, 20–25.

Borchers, H. J. (1960). Über die mannigfaltigkeit der interpolierenden felder zu einer kausalen S-matrix, *Nuovo Cimento* **15**, 784–794.

Bozza, V., Giovannini, M., and Veneziano, G. (2003). Cosmological perturbations from a new-physics hypersurface, *J. Cosmol. Astropart. Phys.* **05**, 001–012; arXiv:hep-th/0302184.

Braaten, E., Curtright, T. L., and Zachos, C. K. (1985). Torsion and geometrostasis in nonlinear sigma models, *Nucl. Phys. B* **260**, 630–688.

Brandenberger, R. H. and Martin, J. (2002). On signatures of short distance physics in the cosmic microwave background, *Int. J. Mod. Phys. A* **17** 3663–3680.

Brauer, R. and Weyl, H. (1935). Spinors in n dimensions, *Am. J. Math.* **57**, 425–449.

Brout, R., Englert, F., Frere, J-M., Gunzig, E., Nardone, P., and Truffin, C. (1980). Cosmogenesis and the origin of the fundamental length scale, *Nucl. Phys. B* **170**, 228–264.

Brout, R., Englert, F., and Gunzig, E. (1978). Creation of the universe as a quantum phenomenon, *Ann. of Phys.* **115**, 78–106.

Brown, L. S. and Cassidy, J. P. (1977). Stress tensors and their trace anomalies in conformally flat spacetimes, *Phys. Rev. D* **16**, 1712–1716.

Brown, L. S. and Collins, J. C. (1980). Dimensional renormalization of scalar field theory in curved spacetime, *Ann. Phys. (N. Y.)* **130**, 215–248.

Brown, M. R., Ottewill, A. C., and Page, D. N. (1986). Conformally invariant quantum field theory in static Einstein space-times, *Phys. Rev. D* **33**, 2840–2850.

Brunetti, R., Fredenhagen, K., and Köhler, M. (1996). The microlocal spectrum condition and Wick polynomials of free fields on curved spacetimes, *Commun. Math. Phys.* **180**, 633–652.

Buchbinder, I. L. and Odintsov, S. D. (1985). Effective potential and phase transitions induced by curvature in gauge theories in curved spacetime, *Classical Quantum Gravity* **2**, 721–731.

Bunch, T. S. (1978). Calculation of the renormalized quantum stress tensor by adiabatic regularization in two-dimensional and four-dimensional Robertson–Walker space-times, *J. Phys. A* **11**, 603–607.

Bunch, T. S. (1980). Adiabatic regularization for scalar fields with arbitrary coupling to the curvature, *J. Phys. A* **13**, 1297–1310.

Bunch, T. S. (1981a). Local momentum space and two loop renormalizability of $\lambda\Phi^4$ field theory in curved space-time, *Gen. Relativ. Gravitation* **13**, 711–723.

Bunch, T. S. (1981b). BPHZ renormalization of $\lambda\phi^4$ field theory in curved spacetime, *Ann. Phys. (N. Y.)* **131**, 118–148.

Bunch, T. S. and Davies, P. C. W. (1979). Quantum field theory in de Sitter space: Renormalization by point-splitting, *Proc. R. Soc. Lond. A* **360**, 117–134.

Bunch, T. S. and Parker, L. (1979). Feynman propagator in curved spacetime: A momentum space approach, *Phys. Rev. D* **20**, 2499–2510.

Burgess, M. (2002). *Classical Covariant Fields* (Cambridge University Press, Cambridge).

Burgess, C. P. and Kunstatter, G. (1987). On the physical interpretation of the Vilkovisky–DeWitt effective action, *Mod. Phys. Lett. A* **2**, 875–886.

Burton, W. K. (1955). Equivalence of the Lagrangian formulations of quantum field theory due to Feynman and Schwinger, *Nuovo Cimento* **1**, 355–357.

Caldwell, R. R., Komp, W., Parker, L., and Vanzella, D. A. T. (2006). Sudden gravitational transition, *Phys. Rev. D* **73**, 023513-1–023513-8.

Calzetta, E. A. (1985). Ph.D. Thesis, Gravitational constants and the effective action in curved spaces, University of Wisconsin-Milwaukee.

Calzetta, E. A. (1986). The behavior of the effective gravitational constants for broken SU(5), *Ann. Phys. (N. Y.)* **166**, 214–233.

Calzetta, E., Habib, S., and Hu, B. L. (1988). Quantum kinetic field theory in curved space-time: Covariant Wigner function and Liouville-Vlasov equation, *Phys. Rev. D* **37**, 2901–2919.

Calzetta, E. and Hu, B. L. (1987). Closed-time-path functional formalism in curved spacetime: Application to cosmological back-reaction problems, *Phys. Rev. D* **35**, 495–509.

Calzetta, E. and Hu, B. L. (1988). Nonequilibrium quantum fields: Closed-time-path effective action, Wigner function, and Boltzmann equation, *Phys. Rev. D* **37**, 2878–2900.

Calzetta, E. and Hu, B. L. (1989). Dissipation of quantum fields from particle creation, *Phys. Rev. D* **40**, 656–659.

Calzetta, E. and Hu, B. L. (1994). Noise and fluctuations in semiclassical gravity, *Phys. Rev. D* **49**, 6636–6655.

Calzetta, E., Jack, I., and Parker, L. (1985). Curvature-induced asymptotic freedom, *Phys. Rev. Lett.* **55**, 1241–1243.

Candelas, P. (1980). Vacuum polarization in Schwarzschild spacetime, *Phys. Rev. D* **21**, 2185–2202.

Candelas, P. and Raine, D. J. (1975). General-relativistic quantum field theory: An exactly soluble model, *Phys. Rev. D* **12**, 965–974.

Carroll, S. (2004). *Spacetime and Geometry* (Addison Wesley, San Francisco, CA).

Carter, B. (1968). Global structure of the Kerr family of gravitational fields, *Phys. Rev. D* **174**, 1559–1571.

Case, K. M. (1955). Biquadratic spinor identities, *Phys. Rev.* **97**, 810–823.

Casimir, H. B. G. (1948). On the attraction between two perfectly conducting plates, *Proc. K. Ned. Acad. Wet.* **51**, 793–795.

Chakraborty, B. (1973). The mathematical problem of reflection solved by an extension of the WKB method, *J. Math. Phys.* **14**, 188–190.

Chandrasekhar, S. (1983). *The Mathematical Theory of Black Holes* (Cambridge University Press, Cambridge).

Cheng, T. P. and Li, L. F. (1984). *Gauge Theory of Elementary Particle Physics* (Clarendon Press, Oxford).

Cheng, T. P. and Li, L. F. (1988). Resource letter: GI-1 gauge invariance, *Am. J. Phys.* **56**, 586–600.

Chern, S. S. and Simons, J. (1974). Characteristic forms and geometric invariants, *Ann. Math.* **99**, 48–69.

Chernikov, N. A. and Tagirov, E. A. (1968). Quantum theory of scalar field in de Sitter space-time, *Ann. Inst. Henri Poincaré A* **9**, 109–141.

Chockalingham, A. and Isham, C. J. (1980). Twisted supermultiplets, *J. Phys. A* **13**, 2723–2733.

Chou, K., Su, Z., Hao, B., and Yu, L. (1985). Equilibrium and nonequilibrium formalisms made unified, *Phys. Rep. C* **118**, 1–131.

Christensen, S. M. (1978). Regularization, renormalization, and covariant geodesic point separation, *Phys. Rev. D* **17**, 946–963.

Christensen, S. M. and Fulling, S. A. (1977). Trace anomalies and the Hawking effect, *Phys. Rev. D* **15**, 2088–2104.

Chung, D. J., Kolb, E. W., and Riotto, A. (1999). Superheavy dark matter, *Phys. Rev. D* **59**, 023501.

Chung, D. J., Kolb, E. W., Riotto, A., and Tkachev, I. I. (2000). Probing Planckian physics: Resonant production of particles during inflation and features in the primordial power spectrum, *Phys. Rev. D* **62**, 043508.

Coleman, S. (1973). Dilations, in *Properties of the Fundamental Interactions*, ed. A. Zichichi (Editrice Compositori, Bologna), 358–399.

Coleman, S. and Weinberg, E. (1973). Radiative corrections as the origin of spontaneous symmetry breaking, *Phys. Rev. D* **7**, 1888–1910.

Collins, J. C. (1974). Scaling behavior of ϕ^4 theory and dimensional regularization, *Phys. Rev. D* **10**, 1213–1218.

Collins, J. C. and Macfarlane, A. J. (1974). New methods for the renormalization group, *Phys. Rev. D* **10**, 1201–1212.

Curtright, T. L. and Zachos, C. K. (1984). Geometry, topology, and supersymmetry in nonlinear models, *Phys. Rev. Lett.* **53**, 1799–1801.

Daniel, M. and Viallet, C. M. (1980). The geometrical setting of gauge theories of the Yang–Mills type, *Rev. Mod. Phys.* **52**, 175–192.

Danielsson, U. H. (2002). Note on inflation and trans-Planckian physics, *Phys. Rev. D* **66**, 023511.

Davies, P. C. W. (1975). Scalar particle production in Schwarzschild and Rindler metrics, *J. Phys. A* **8**, 609–616.

Davies, P. C. W., Fulling, S., and Unruh, W. G. (1976). Energy-momentum tensor near an evaporating black hole, *Phys. Rev. D* **13**, 2720–2723.

Deser, S., Jackiw, R., and Templeton, S. (1982a). Three-dimensional massive gauge theories, *Phys. Rev. Lett.* **48**, 975–978.

Deser, S., Jackiw, R., and Templeton, S. (1982b). Topologically massive gauge theories, *Ann. Phys. (NY)* **140**, 372–413.

Deser, S. and Levin, O. (1998). Equivalence of Hawking and Unruh temperatures and entropies through flat space embeddings, *Classical Quantum Gravity* **15**, L85–L87; arXiv:hep-th/9806223.

Deser, S. and Levin, O. (1999). Mapping Hawking into Unruh thermal properties, *Phys. Rev. D* **59**, 064004-1–064004-7; arXiv:hep-th/9809159.

Deser, S. and Woodard, R. P. (2008). Nonlocal cosmology, *Phys. Rev. Lett.* **99**, 111301; arXiv:0706.2151.

de Sitter, W. (1917). *Proc. K. Ned. Akad. Wet.* **19**, On the relativity of inertia: Remarks concerning Einstein's latest hypothesis, 1217–1225; *ibid.* **20**, On the curvature of space, 229–243; *ibid.*, Further remarks on the solutions of the field equations of Einsteins theory of gravitation, 1309–1312; *Mon. Not. R. Astron. Soc.* **78**, 3.

de Wit, B. and van Holten, J. W. (1978). Covariant quantization of gauge theories with open gauge algebra, *Phys. Lett. B* **79**, 389–393.

DeWitt, B. S. (1964a). Dynamical theory of groups and fields, in *Relativity, Groups and Topology*, ed. C. DeWitt and B. S. DeWitt (Gordon and Breach, New York), 585–820. (Reprinted as DeWitt (1965).)

DeWitt, B. S. (1964b). Theory of radiative corrections for non-Abelian gauge fields, *Phys. Rev. Lett.* **12**, 742–746.

DeWitt, B. S. (1965). *Dynamical Theory of Groups and Fields* (Gordon and Breach, New York).

DeWitt, B. S. (1967). Quantum theory of gravity. II. The manifestly covariant theory, *Phys. Rev.* **162**, 1195–1239.

DeWitt, B. S. (1972). Covariant quantum geodynamics, in *Magic Without Magic: John Archibald Wheeler, A Collection of Essays in Honor of His 60th Birthday*, ed. J. R. Klauder (Freeman, San Francisco), 409–440.

DeWitt, B. S. (1975). Quantum field theory in curved spacetime, *Phys. Rep. C* **19**, 295–357.

DeWitt, B. S. (1984a). *Supermanifolds* (Cambridge University Press, Cambridge).

DeWitt, B. S. (1984b). The spacetime approach to quantum field theory, in *Relativity, Groups and Topology II*, ed. B. S. DeWitt and R. Stora (North Holland, Amsterdam), 381–738.

DeWitt, B. S. (1987). The effective action, in *Quantum Field Theory and Quantum Statistics, Volume 1*, ed. I. A. Batalin, C. J. Isham, and G. A. Vilkovisky (Adam Hilger, Bristol), 191–222.

DeWitt, B. S., Hart, C. F., and Isham, C. J. (1979). Topology and quantum field theory, *Physica A* **96**, 197–211.

Dirac, P. A. M. (1933). The Lagrangian in quantum mechanics, *Phys. Z. Sowjetunion Band* **3**, *Heft* **1**, 64–72. (Reprinted in Schwinger (1958).)

Dirac, P. A. M. (1958). *Principles of Quantum Mechanics, Fourth Edition* (Oxford University Press, Oxford).

Dittrich, W. and Reuter, M. (1986). *Selected Topics in Gauge Theories, Lecture Notes in Physics*, **244** (Springer-Verlag, Berlin).

Dodelson, S. (2003). *Modern Cosmology* (Academic Press, New York).

Dolan, L. and Jackiw, R. (1974). Gauge-invariant signal for gauge-symmetry breaking, *Phys. Rev. D* **9**, 2904–2912.

Dowker, J. S. and Critchley, R. (1976). Effective Lagrangian and energy-momentum tensor in de Sitter space, *Phys. Rev. D* **13**, 3224–3232.

Drummond, I. T. (1975). Dimensional regularization of massless theories in spherical space-time, *Nucl. Phys. B* **94**, 115–144.

Drummond, I. T. and Shore, G. M. (1979a). Conformal anomalies for interacting scalar fields in curved spacetime, *Phys. Rev. D* **19**, 1134–1143.

Drummond, I. T. and Shore, G. M. (1979b). Dimensional regularization of massless quantum electrodynamics in spherical spacetime, *Ann. Phys. (N. Y.)* **117**, 89–120.

Duff, M. J. (1994). Twenty years of the Weyl anomaly, *Classical Quantum Gravity* **11**, 1387–1403.

Duistermaat, J. J. and Hörmander, L. (1972). Fourier integral operators. II, *Acta Math.* **128**, 183–269.

Easther, R., Greene, B. R., Kinney, W. H., and Shiu, G. (2002). Generic estimate of trans-Planckian modifications to the primordial power spectrum of inflation, *Phys. Rev. D* **66**, 023518-1–023518-9.

Eckart, C. (1930). The penetration of a potential barrier by electrons, *Phys. Rev.* **35**, 1303–1309.

Ecker, G. and Honerkamp, J. (1971). Application of invariant renormalization to the non-linear chiral invariant pion Lagrangian in the one-loop approximation, *Nucl. Phys. B* **35**, 481–492.

Eguchi, T., Gilkey, P. B., and Hanson, A. J. (1980). Gravitation, gauge theories and differential geometry, *Phys. Rep. C* **66**, 23–393.

Ellicott, P., Kunstatter, G., and Toms, D. J. (1989). Geometrical derivation of the Faddeev–Popov ansatz, *Mod. Phys. Lett. A* **4**, 2397–2407.

Ellicott, P., Kunstatter, G., and Toms, D. J. (1991). Geometrical interpretation of the functional measure for supersymmetric gauge theories and of the gauge invariant effective action, *Ann. Phys. (N. Y.)* **205**, 70–109.

Ellicott, P. and Toms, D. J. (1989). The new effective action in quantum field theory, *Nucl. Phys. B* **312**, 700–714.

Emch, G. (1972). *Algebraic Methods in Statistical Mechanics and Quantum Field Theory* (Wiley Interscience, New York).

Elster, T. (1983). Vacuum polarization near a black hole creating particles, *Phys. Lett. A* **94**, 205–209.

Endo, R. (1984). Gauge dependence of the gravitational conformal anomaly for the electromagnetic field, *Prog. Theor. Phys.* **71**, 1366–1384.

Epp, R. J., Kunstatter, G., and Toms, D. J. (1993). Path-integral quantization of scalar QED, *Phys. Rev. D* **47**, 2474–2482.

Epstein, P. J. (1930). Reflection of waves in an inhomogeneous absorbing medium, *Proc. Natl. Acad. Sci. (USA)* **16**, 627–637.

Fabbri, A. and Navarro-Salas, J. (2005). *Modeling Black Hole Evaporation* (Imperial College Press, London).

Faddeev, L. D. and Popov, V. N. (1967). Feynman diagrams for the Yang–Mills field, *Phys. Lett. B* **25**, 29–30.

Fawcett, M. S. and Whiting, B. F. (1982). Spontaneous symmetry breaking near a black hole, in *Quantum Structure of Space and Time*, ed. M. J. Duff and C. J. Isham (Cambridge University Press, Cambridge), 131–154.

Feynman, R. P. (1948). Space-time approach to non-relativistic quantum mechanics, *Rev. Mod. Phys.* **20**, 367–387. (Reprinted in Schwinger (1958).)

Feynman, R. P. (1963). Quantum theory of gravitation, *Acta Phys. Pol.* **24**, 697–722.

Feynman, R. P. and Hibbs, A. R. (1965). *Quantum Mechanics and Path Integrals* (McGraw-Hill, New York).

Fock, V. (1929). Geometisierung der Diracschen theorie des elektrons, *Z. Phys.* **57**, 261–277.

Ford, L. H. (1980). Vacuum polarization in a nonsimply connected spacetime, *Phys. Rev. D* **21**, 933–948.

Ford, L. H. (1985). Quantum instability of de Sitter spacetime, *Phys. Rev. D* **31**, 710–717.

Ford, L. H. and Parker, L. (1977a). Infrared divergences in a class of Robertson–Walker universes, *Phys. Rev. D* **16**, 245–250.

Ford, L. H. and Parker, L. (1977b). Quantized gravitational wave perturbations in Robertson–Walker universes, *Phys. Rev. D* **16**, 1601–1608.

Ford, L. H. and Roman, T. A. (1996). Quantum field theory constrains traversable wormhole geometries, *Phys. Rev. D* **53**, 5496–5507.

Fradkin, E. S. and Tseytlin, A. A. (1984). On the new definition of the off-shell effective action, *Nucl. Phys. B* **234**, 509–523.

Freeman, M. D. (1984). Renormalization of non-Abelian gauge theories in curved space-time, *Ann. Phys. (N. Y.)* **153**, 339–366.

Freese, K. and Kinney, W. H. (2004). On natural inflation, *Phys. Rev. D* **70**, 083512-1–083512-8.

Friedan, D. (1980). Nonlinear models in $2 + \epsilon$ dimensions, *Phys. Rev. Lett.* **45**, 1057–1060.

Friedman, J., Morris, M., Novikov, I., Echeverria, F., Klinkhammer, G., Thorne, K. S., and Yurtsever, U. (1990). Cauchy problem in spacetimes with closed timelike curves, *Phys. Rev. D* **42**, 1915–1930.

Friedman, J. L., Papastamatiou, N. J., and Simon, J. Z. (1992). Failure of unitarity for interacting fields on spacetimes with closed timelike curves, *Phys. Rev. D* **46**, 4456–4469.

Frolov, V. P. and Zel'nikov, A. I. (1987). Killing approximation for vacuum and thermal stress-energy tensor in static space-times, *Phys. Rev. D* **35**, 3031–3044.

Fujikawa, K. (1979). Path integral measure for gauge-invariant fermion theories, *Phys. Rev. Lett.* **42**, 1195–1198.

Fujikawa, K. (1980a). Comment on chiral and conformal anomalies, *Phys. Rev. Lett.* **44**, 1733–1736.

Fujikawa, K. (1980b). Path integral measure for gauge theories with fermions, *Phys. Rev. D* **21**, 2848–2858.

Fujikawa, K. (1981). Energy-momentum tensor in quantum field theory, *Phys. Rev. D* **23**, 2262–2275.

Fulling, S. A. (1972). Ph. D. Thesis, Princeton University (unpublished).

Fulling, S. A. (1973). Nonuniqueness of canonical field quantization in Riemannian space-time, *Phys. Rev. D* **7**, 2850–2862.

Fulling, S. A. (1989). *Aspects of Quantum Field Theory in Curved Space-Time*, (Cambridge University Press, Cambridge).

Fulling, S. A. and Parker, L. (1973). Unpublished.

Fulling, S. A. and Parker, L. (1974). Renormalization in the theory of a quantized scalar field interacting with a Robertson–Walker spacetime, *Ann. Phys. (N. Y.)* **87**, 176–203.

Fulling, S. A., Parker, L., and Hu, B. L. (1974). Conformal energy-momentum tensor in curved space-time: Adiabatic regularization and renormalization, *Phys. Rev. D* **10**, 3905–3924.

Geroch, R. (1968). Spinor structure of spacetimes in general relativity. I, *J. Math. Phys.* **9**, 1739–1744.

Gibbons, G. W. (1975). Vacuum polarization and the spontaneous loss of charge by black holes, *Commun. Math. Phys.* **44**, 245–264.

Gibbons, G. W. (1979). Quantum field theory in curved spacetime, in *General Relativity, An Einstein Centenary Survey*, ed. S. W. Hawking and W. Israel (Cambridge University Press, London), 639–679.

Gibbons, G. W. and Hawking, S. W. (1977). Cosmological event horizons, thermodynamics, and particle creation, *Phys. Rev. D* **15**, 2738–2751.

Gilkey, P. B. (1975). The spectral geometry of a Riemannian manifold, *J. Differ. Geom.* **10**, 601–618.

Gilkey, P. B. (1979). Recursion relations and the asymptotic behavior of the eigenvalues of the Laplacian, *Compos. Math.* **38**, 201–240.

Gliozzi, F., Scherk, J., and Olive, D. (1977). Supersymmetry, supergravity theories and the dual spinor model, *Nucl. Phys. B* **122**, 253–290.

Goldstone, J. (1957). Derivation of the Breuckner many-body theory, *Proc. R. Soc. Lond. A* **329**, 267–279.

Goodison, J. W. and Toms, D. J. (1993). No generalized statistics from dynamics in curved spacetime, *Phys. Rev. Lett.* **71**, 3240–3242.

Gradshteyn, I. S. and Ryzhik, I. M. (1965). *Table of Integrals, Series, and Products* (Academic Press, New York).

Green, M., Schwarz, J., and Witten, E. (1987). *Superstring Theory* (Cambridge University Press, London).

Gribov, V. N. (1978). Quantization of non-Abelian gauge theories, *Nucl. Phys. B* **139**, 1–19.

Grishchuk, L. P. (1974). Amplification of gravitational waves in an isotropic universe, *Zh. Eksp. Teor. Fiz.* **67**, 825; *Sov. Phys. JETP* **40**, 409 (1975).

Gross, D. and Wilczek, F. (1973). Ultraviolet behavior of non-Abelian gauge theories, *Phys. Rev. Lett.* **30**, 1343–1346.

Guth, A. H. (1981). Inflationary universe: A possible solution to the horizon and flatness problems, *Phys. Rev. D* **23**, 347–356.

Guth, A. H. and Pi, S.-Y. (1982). Fluctuations in the new inflationary universe, *Phys. Rev. Lett.* **49**, 1110–1113.

Guth, A. H. and Weinberg, E. J. (1983). Could the Universe have recovered from a slow first-order phase transition? *Nucl. Phys. B* **212**, 321–364.

Haag, R. (1992). *Local Quantum Physics: Fields, Particles, Algebras* (Springer-Verlag, Berlin).

Habib, S., Molina-Paris, C., and Mottola, E. (1999). Energy-momentum tensor of particles created in an expanding universe, *Phys. Rev. D* **61**, 024010-1–024010-22.

Hadamard, J. (1923). *Lectures on Cauchy's Problem in Linear Partial Differential Equations* (Yale University Press, New Haven, CT).

Hajicek, P. and Israel, W. (1980). What, no black hole evaporation? *Phys. Lett. A* **80**, 9–10.

Hartle, J. B. (2003). *Gravity* (Addison Wesley, San Francisco, California).

Hartle, J. B. and Hawking, S. W. (1976). Path integral derivation of black hole radiance, *Phys. Rev. D* **13**, 2188–2203.

Hathrell, S. J. (1982a). Trace anomalies and $\lambda\phi^4$ theory in curved space, *Ann. Phys. (N. Y.)* **139**, 136–197.

Hathrell, S. J. (1982b). Trace anomalies and QED in curved space, *Ann. Phys. (N. Y.)* **142**, 34–63.

Hawking, S. W. (1974). Black hole explosions? *Nature* **248**, 30–31.

Hawking, S. W. (1975). Particle creation by black holes, *Commun. Math. Phys.* **43**, 199–220.

Hawking, S. W. (1976). Breakdown of predictability in gravitational collapse, *Phys. Rev. D* **14**, 2460–2473.

Hawking, S. W. (1977). Zeta-function regularization of path integrals in curved spacetime, *Commun. Math. Phys.* **55**, 133–148.

Hawking, S. W. (1981). Interacting quantum fields around a black hole, *Commun. Math. Phys.* **80**, 421–442.

Hawking, S. W. (1982). The development of irregularities in a single bubble inflationary universe, *Phys. Lett. B* **115**, 295–297.

Hawking, S. W. (1992). Chronology protection conjecture, *Phys. Rev. D* **46**, 603–611.

Hawking, S. W. and Ellis, G. F. R. (1973). *The Large Scale Structure of Space-Time* (Cambridge University Press, Cambridge).

Hawking, S. W. , Hertog, T., and Reall, H. S. (2001). Trace anomaly driven inflation, *Phys. Rev. D* **63**, 083504-1–083504-23.

Hawking, S. W. and Moss, I. G. (1982). Supercooled phase transitions in the very early universe, *Phys. Lett. B* **110**, 35–38.

Heisenberg, W. and Euler, H. (1936). Folgerungen aus der Diracschen theorie des positrons, *Z. Phys.* **98**, 714–732.

Higuchi, A. (1987a). Ph. D. Thesis, Quantum fields of nonzero spin in de Sitter space-time, UMI87-29076-mc, (microfiche, Yale University).

Higuchi, A. (1987b). Forbidden mass range for spin-2 field theory in de Sitter spacetime, *Nucl. Phys. B* **282**, 397–436.

Higuchi, A., Matsas, G. E. A., and Peres, C. B. (1993). Uniformly accelerated finite-time detectors, *Phys. Rev. D* **48**, 3731–3734.

Higuchi, A., Parker, L., and Wang, Y. (1990). Consistency of Faddeev–Popov ghost statistics with gravitationally induced pair creation, *Phys. Rev. D* **42**, 4078–4081.

Hilbert, D. (1915). Die grundlagen der physik. I, *Nachr. Königlichen Ges. Wissensch. Göttingen, Math-phys Kl.* Heft 3, 395–407.

Hilbert, D. (1917). Die grundlagen der physik. II, *Nachr. Königlichen Ges. Wissensch. Göttingen, Mathe.-phys. Kl.* Heft 1, 53–76.

Hollands, S. and Wald, R. M. (2005). Conservation of the stress tensor in perturbative interacting quantum field theory in curved spacetimes, *Rev. Math. Phys.* **17**, 227–311.

Hollands, S. and Wald, R. M. (2008). Axiomatic quantum field theory in curved spacetime, arXiv:0803.2003.

Holman, R., Mersini-Houghton, L., and Takahashi, T. (2008). Cosmological avatars of the landscape. II. CMB and LSS signatures, *Phys. Rev. D* **77**, 063511; arXiv:hep-th/0612142v1(2006).

Honerkamp, J. (1972). Chiral multi-loops, *Nucl. Phys. B* **36**, 130–140.

Howard, K. W. (1984). Vacuum $\langle T_\mu{}^\nu \rangle$ in Schwarzschild spacetime, *Phys. Rev. D* **30**, 2532–2547.

Howe, P. S., Papadopoulos, G., and Stelle, K. S. (1988). The background field method and the non-linear σ-model, *Nucl. Phys. B* **296**, 26–48.

Hu, B. L. (1978). Calculation of the trace anomaly of the conformal energy momentum tensor in Kasner space-time by adiabatic regularization, *Phys. Rev. D* **18**, 4460–4470.

Hu, B. L. and O'Connor, D. J. (1984). Effective Lagrangian for $\lambda\phi^4$ theory in curved spacetime with varying background fields: Quasilocal approximation, *Phys. Rev. D* **30**, 743–755.

Hu, B. L. and Parker, L. (1977). Effect of gravitation creation in isotropically expanding universes, *Phys. Lett. A* **63**, 217–220.

Hu, B. L. and Parker, L. (1978). Anisotropy damping through quantum effects in the early universe, *Phys. Rev. D* **17**, 933–945.

Huggins, S. R., Kunstatter, G., Leivo, H. P., and Toms, D. J. (1987a). Unique effective action in five-dimensional Kaluza–Klein theory, *Phys. Rev. Lett.* **58**, 296–298.

Huggins, S. R., Kunstatter, G., Leivo, H. P., and Toms, D. J. (1987b). The Vilkovisky–DeWitt effective action for quantum gravity, *Nucl. Phys. B* **301**, 627–660.

Husemoller, D. (1966). *Fibre Bundles* (McGraw-Hill, New York).

Isham, C. J. (1978a). Twisted quantum fields in a curved spacetime, *Proc. R. Soc. Lond., A* **362** 383–404.

Isham, C. J. (1978b). Spinor fields in four-dimensional space-time, *Proc. R. Soc. Lond. A* **364**, 591–599.

Itzykson, C. and Zuber, J. B. (1980). *Quantum Field Theory* (McGraw-Hill, New York).

Jack, I. (1984). Background field calculations in curved spacetime II. Application to a pure gauge theory, *Nucl. Phys. B* **234**, 365–378.

Jack, I. (1985). Background field calculations in curved spacetime III. Application to a general gauge theory coupled to fermions and scalars, *Nucl. Phys. B* **253**, 323–352.

Jack, I. and Osborn, H. (1982). Two-loop background field calculations for arbitrary background fields, *Nucl. Phys. B* **207**, 474–504.

Jack, I. and Osborn, H. (1984). Background field calculations in curved spacetime (I). General formalism and application to scalar fields, *Nucl. Phys. B* **234**, 331–364.

Jack, I. and Osborn, H. (1985). General background field calculations with fermion fields, *Nucl. Phys. B* **249**, 472–506.

Jack, I. and Parker, L. (1985). Proof of summed form of proper time expansion for propagator in curved space-time, *Phys. Rev. D* **31**, 2439–2451.

Jack, I. and Parker, L. (1987). Response to "Two important invariant identities", *Phys. Rev. D* **35**, 771–772.

Jackiw, R. (1974). Functional evaluation of the effective potential, *Phys. Rev. D* **9**, 1686–1701.

Jacobson, T. (1991). Black-hole evaporation and ultrashort distances, *Phys. Rev. D* **44**, 1731–1739.

Jacobson, T. and Mattingly, D. (2001). Gravity with a dynamical preferred frame, *Phys. Rev. D* **64**, 024028-1–024028-9.

Jauch, J. M. and Rohrlich, F. (1980). *The Theory of Photons and Electrons, Second Edition* (Springer-Verlag, New York).

Jordan, R. D. (1986). Effective field equations for expectation values, *Phys. Rev. D* **33**, 444–454.

Junker, W. and Schrohe, E. (2002). Adiabatic vacuum states on general spacetime manifolds: Definition, construction, and physical properties, *Ann. Henri Poincaré* **3**, 1113–1181; arXiv:math-ph/0109010.

Kay, B. S. (1978). Linear spin-zero quantum fields in external gravitational and scalar fields. I. A one particle structure for the stationary case, *Commun. Math. Phys.* **62**, 55–70.

Kay, B. S. (1985). The double-wedge algebra for quantum fields on Schwarzschild and Minkowski spacetimes, *Commun. Math. Phys.* **100**, 57–81.

Keldysh, L. V. (1964). Diagram technique for nonequilibrium processes, *Zh. Eksp. Teor. Fiz.* **47**, 1515–1527; *Sov. Phys. JETP* **20**, 1018 (1965).

Kimura, T. (1969). Divergence of axial-vector current in the gravitational field, *Prog. Theor. Phys.* **42**, 1191–1205.

Kirsten, K. (2002). *Spectral Functions in Mathematics and Physics* (Chapman and Hall/CRC, Baton Roca, FL).

Kobes, R., Kunstatter, G., and Rebhan, A. (1990). QCD plasma parameters and the gauge-dependent gluon propagator, *Phys. Rev. Lett.* **64**, 2992–2995.

Kobes, R., Kunstatter, G., and Rebhan, A. (1991). Gauge dependence identities and their application at finite temperature, *Nucl. Phys. B* **355**, 1–37.

Kobes, R., Kunstatter, G., and Toms, D. J. (1988). The Vilkovisky-DeWitt effective action : Panacea or placebo? in *TeV Physics*, ed. G. Domokos and S. Kovesi-Domokos (World Scientific, Singapore), 73–109.

Kolb, E. W. and Turner, M. S. (1994). *The Early Universe* (Perseus Publishing, USA).

Korenman, V. (1966). Nonequilibrium quantum statistics – Application to the laser, *Ann. Phys. (N. Y.)* **39**, 72–126.

Kostelecky, V. A. (2004). Gravity, Lorentz violation, and the standard model, *Phys. Rev. D* **69**, 105009-1–105009-20.

Kulsrud, R. M. (1957). Adiabatic invariant of the harmonic oscillator, *Phys. Rev.* **106**, 205–207.

Kunstatter, G. (1987). Vilkovisky's unique effective action: An introduction and explicit calculation, in *Super Field Theories*, ed. H. C. Lee, V. Elias, G. Kunstatter, R. B. Mann, and K. S. Viswanathan (Plenum, New York), 503–517.

Lamoreaux, S. K. (1997). Demonstration of the Casimir force in the 0.6 to 6 micrometers range, *Phys. Rev. Lett.* **78**, 5–8.

Lamoreaux, S. K. (1999). Comment on "Precision measurement of the Casimir force from 0.1 to 0.9 micrometers", *Phys. Rev. Lett.* **83**, 3340.

Landau, L. D. and Lifshitz, E. M. (1958). *Statistical Physics* (Pergamon, Oxford).

Lee, C. (1982). Proper-time renormalization of multi-loop amplitudes in the background field method (I). ϕ^4-theory, *Nucl. Phys. B* **207**, 157–188.

Lee, C. and Rim, C. (1985). Background Fermi fields and Schwinger–DeWitt proper-time method, *Nucl. Phys. B* **255**, 439–464.

Leen, T. K. (1982). Ph.D. Thesis, Topics in gravitation and gauge fields, University of Wisconsin-Milwaukee.

Leen, T. K. (1983). Renormalization and scaling behavior of non-Abelian gauge fields in curved spacetime, *Ann. Phys. (N. Y.)* **147**, 417–444.

Liddle, A. R. and Lyth, D. H. (2000). *Cosmological Inflation and Large-Scale Structure* (Cambridge University Press, Cambridge).

Lifshitz, E. M. (1946). *Zh. Eksp. Teor. Fiz.* **16**, 587.

Linde, A. D. (1982a). A new inflationary universe scenario: A possible solution of the horizon, flatness, homogeneity, isotropy and primordial monopole problems, *Phys. Lett. B* **108**, 389–393.

Linde, A. D. (1982b). Coleman–Weinberg theory and the new inflationary universe scenario, *Phys. Lett. B* **114** 431–435.

López Nacir, D. and Mazzitelli, F. D. (2007). Backreaction in trans-Planckian cosmology: Renormalization, trace anomaly, and self-consistent solutions, *Phys. Rev. D* **76**, 024013-1–024013-14.

Lorenz, L. (1867). On the identity of the vibrations of light with electrical currents, *Phil. Mag.* **34**, 287–301.

Lüders, C. and Roberts, J. E. (1990). Local quasiequivalence and adiabatic vacuum states, *Commun. Math. Phys.* **134**, 29–63.

Lyth, D. H. (1985). Large-scale energy-density perturbations and inflation, *Phys. Rev. D* **31**, 1792–1798.

Magueijo, J. and Smolin, L. (2002). Lorentz invariance with an invariant energy scale, *Phys. Rev. Lett.* **88**, 190403; arXiv:hep-th/0112090.

Maldacena, J. M. (1997). Black holes and D-branes, *Nucl. Phys. B Proc. Suppl.* **61**, 111–123; (1998). *ibid.* **62**, 428–442.

Massar, S. (1995). Semiclassical back reaction to black hole evaporation, *Phys. Rev. D* **52**, 5857–5864.

Mattuck, R. D. (1976). *A Guide to Feynman Diagrams in the Many-body Problem, Second Edition* (McGraw-Hill, New York).

Maxwell, J. C. (1954). *A Treatise on Electricity and Magnetism, Third Edition* (Dover, New York).

Mersini, L., Bastero-Gil, M., and Kanti, P. (2001). Relic dark energy from trans-Planckian regime, *Phys. Rev. D* **64**, 043508-1–043508-9.

Messiah, A. (1961). *Quantum Mechanics* (North Holland, Amsterdam).

Milnor, J. W. and Stasheff, J. D. (1974). *Characteristic Classes* (Princeton University Press, Princeton, N. J.).

Minakshishundaram, S. and Pleijel, A. (1949). Some properties of the eigenfunctions of the Laplace operator on Riemannian manifolds, *Can. J. Math.* **1**, 242–256.

Misner, C. W., Thorne, K. S., and Wheeler, J. A. (1973). *Gravitation* (W. H. Freeman, San Francisco).

Mohideen, U. and Roy, A. (1998). Precision Measurement of the Casimir force from 0.1 to 0.9 micrometers, *Phys. Rev. Lett.* **81**, 4549–4552.

Mohideen, U. and Roy, A. (1999). Mohideen and Roy reply, *Phys. Rev. Lett.* **83**, 3341.

Morette, C. (1951). On the definition and approximation of Feynman's path integrals, *Phys. Rev.* **81**, 848–852.

Morris, M. and Thorne, K. S. (1988). Wormholes in spacetime and their use for interstellar travel: A tool for teaching general relativity, *Am. J. Phys.* **56**, 395–412.

Morris, M., Thorne, K. S., and Yursever, U. (1988). Wormholes, time machines, and the weak energy condition, *Phys. Rev. Lett.* **61**, 1446–1449.

Mukhanov, V. (2005). *Physical Foundations of Cosmology* (Cambridge University Press, Cambridge).

Mukhanov, V. F. and Chibisov, G. V. (1981). Quantum fluctuations and a nonsingular universe, *JETP Lett.* **33**, 532–535.

Mukhanov, V. F., Feldman, H. A., and Brandenberger, R. H. (1992). Theory of cosmological perturbations, *Phys. Rep. C* **215**, 203–333.

Mukhi, S. (1986). The geometrical background field method, renormalization and the Wess-Zumino term in nonlinear σ-models, *Nucl. Phys. B* **264**, 640–652.

Nelson, B. L. and Panangaden, P. (1982). Scaling behavior of interacting quantum fields in curved spacetime, *Phys. Rev. D* **25**, 1019–1027.

Nielsen, N. K. (1975). On the gauge dependence of spontaneous symmetry breaking in gauge theories, *Nucl. Phys. B* **101**, 173–188.

Nishijima, K. (1950). On the generalized transformation function and the integrability condition, *Prog. Theor. Phys.* **5**, 813–821.

Odintsov, S. D. (1990a). Vilkovisky effective action in quantum gravity with matter, *Theor. Math. Phys.* **82**, 45–51.

Odintsov, S. D. (1990b). The parametrization invariant and gauge invariant effective actions in quantum field theory, *Fortschr. Phys.* **38**, 371–391.

Omote, M. and Ichinose, S. (1983). One-loop renormalization in curved spacetimes using the background-field method, *Phys. Rev. D* **27**, 2341–2347.

O'Raifeartaigh, L. (1986). *Group Structure of Gauge Theories* (Cambridge University Press, Cambridge).

Osterwalder, K. and Schrader, R. (1973). Axioms for Euclidean Green's functions, *Commun. Math. Phys.* **31**, 83–112.

Osterwalder, K and Schrader, R. (1975). Axioms for Euclidean Green's functions II, *Commun. Math. Phys.* **42**, 281–305.

Page, D. N. (1976). Particle emission rates from a black hole: Massless particles from an uncharged, nonrotating hole, *Phys. Rev. D* **13**, 198–206.

Page, D. N. (1982). Thermal stress tensors in static Einstein spaces, *Phys. Rev. D* **25**, 1499–1509.

Pais, A. (1986). *Inward Bound* (Oxford University Press, Oxford).

Panangaden, P. (1980). Ph.D. Thesis, Propagators and renormalization of quantum field theory in curved spacetimes, University of Wisconsin-Milwaukee.

Panangaden, P. (1981). One-loop renormalization of quantum electrodynamics in curved spacetime, *Phys. Rev. D* **23**, 1735–1746.

Parentani, R. and Piran, T. (1994). Internal geometry of an evaporating black hole, *Phys. Rev. Lett.* **73**, 2805–2808.

Parker, L. (1965). Ph.D. Thesis, the creation of particles in an expanding universe, Harvard University (available from:University Microfilms Library Service, Xerox Corp., Ann Arbor, MI, USA).

Parker, L. (1968). Particle creation in expanding universes, *Phys. Rev. Lett.* **21**, 562–564.

Parker, L. (1969). Quantized fields and particle creation in expanding universes. I, *Phys. Rev.* **183**, 1057–1068.

Parker, L. (1971). Quantized fields and particle creation in expanding universes. II, *Phys. Rev. D* **3**, 346–356.

Parker, L. (1972). Backscattering caused by the expansion of the universe, *Phys. Rev. D* **5**, 2905–2908.

Parker, L. (1973). Conformal energy-momentum tensor in Riemannian space-time, *Phys. Rev. D* **7**, 976–983.

Parker, L. (1975a). Probability distribution of particles created by a black hole, *Phys. Rev. D* **12**, 1519–1525.

Parker, L. (1975b). Quantized fields and particle creation in curved spacetime (66 pages), in *Relativity, Fields, Strings and Gravity: The Second Latin American Symposium on Relativity and Gravitation (SILARG 2)*, ed. C. Aragone (Universidad Simon Bolivar, Caracas), 115–180.

Parker, L. (1976). Thermal radiation produced by the expansion of the Universe, *Nature* **261**, 20–23.

Parker, L. (1977). The production of elementary particles in strong gravitational fields, in *Asymptotic Structure of Space-Time*, ed. F. P. Esposito and L. Witten (Plenum Press, New York), 107–226.

Parker, L. (1979a). Path integrals for a particle in a curved space, *Phys. Rev. D* **19**, 438–441.

Parker, L. (1979b). Aspects of quantum field theory in curved spacetime: Effective action and energy-momentum tensor, in *Recent Developments in Gravitation, Cargèse 1978*, ed. M. Lévy and S. Deser (Plenum Press, New York), 219–273.

Parker, L. (1980). The one-electron atom as a probe of spacetime curvature, *Phys. Rev. D* **22**, 1922–1934.

Parker, L. (2007). Amplitude of perturbations from inflation, ArXiv:hep-th/0702216, (27 Feb 2007).

Parker, L. and Fulling, S. A. (1973). Quantized matter fields and the avoidance of singularities in general relativity, *Phys. Rev. D* **7**, 2357–2374.

Parker, L. and Fulling, S. A. (1974). Adiabatic regularization of the energy-momentum tensor of a quantized field in homogeneous spaces, *Phys. Rev. D* **9**, 341–354.

Parker, L., Komp, W., and Vanzella, D. A. T. (2003). Cosmological acceleration through transition to constant scalar curvature, *Astrophys. J.* **588**, 663–673.

Parker, L. and Raval, A. (1999a). Non-perturbative effects of vacuum energy on the recent expansion of the universe, *Phys. Rev. D* **60**, 063512-1–063512-21.

Parker, L. and Raval, A. (1999b). Vacuum effects of ultra-low mass particle account for recent acceleration of universe, *Phys. Rev. D* **60**, 123502-1–123502-8.

Parker, L. and Raval, A. (2000). New quantum aspects of a vacuum-dominated universe, *Phys. Rev. D* **62**, 083503-1–083503-9.

Parker, L. and Raval, A. (2001). A new look at the accelerating universe, *Phys. Rev. Lett.* **86**, 749–752.

Parker, L. and Tiomno, J. (1972). Pair-producing electric fields and pulsars, *Astrophys. J.* **178**, 809–817.

Parker, L. and Toms, D. J. (1984). Renormalization group analysis of grand unified theories in curved spacetime, *Phys. Rev. D* **29**, 1584–1608.

Parker, L. and Toms, D. J. (1985a). New form for the coincidence limit of the Feynman propagator, or heat kernel, in curved space-time, *Phys. Rev. D* **31**, 953–956.

Parker, L. and Toms, D. J. (1985b). Explicit curvature dependence of coupling constants, *Phys. Rev. D* **31**, 2424–2438.

Parker, L. and Toms, D. J. (1985c). Renormalization group and nonlocal terms in the curved space-time effective action: Weak field results, *Phys. Rev. D* **32**, 1409–1420.

Parker, L. and Vanzella, D. A. T. (2004). Acceleration of the universe, vacuum metamorphosis, and the large-time asymptotic form of the heat kernel, *Phys. Rev. D* **69**, 104009-1–10409-15.

Parker, L. and Wang, Yi. (1989). Statistics from dynamics in curved spacetime, *Phys. Rev. D* **39**, 3596–3605.

Pauli, W. (1940). The connection between spin and statistics, *Phys. Rev.* **58**, 716–722. (Reprinted in Schwinger (1958).)

Pauli, W. (1980). *General Principles of Quantum Mechanics* (Springer-Verlag, Berlin).

Pauli, W. and Heisenberg, W. (1929). Zur quantendynamik der wellenfelder, *Z. Phys.* **56**, 1–61.

Peiris, H. V. *et al.* (2003). First-year Wilkinson microwave anisotropy probe (WMAP) observations: Implications for inflation *Astrophys. J.* **148**, 213–231. See also, Spergel *et al.*, (2007).

Petrov, A. Z. (1969). *Einstein Spaces* (Pergamon, Oxford).

Pirk, K. (1993). Hadamard states and adiabatic vacua, *Phys. Rev. D* **48**, 3779–3783.

Politzer, D. (1973). Reliable perturbative results for strong interactions? *Phys. Rev. Lett.* **30**, 1346–1349.

Polkinghorne, J. C. (1955). On the Feynman principle, *Proc. R. Soc. Lond.* A **230**, 272–276.

Quigg, C. (1983). *Gauge Theories of the Strong, Weak and Electromagnetic Interactions* (Benjamin-Cummings, Reading, MA).

Radzikowski, M. J. (1996). Micro-local approach to the Hadamard condition in quantum field theory on curved space-time, *Commun. Math. Phys.* **179**, 529–553.

Ray, D. B. and Singer, I. M. (1973). Analytic torsion for complex manifolds, *Ann. Math.* **98**, 154–177.

Rebhan, A. (1987). The Vilkovisky–DeWitt effective action and its application to Yang–Mills theories, *Nucl. Phys.* B **288**, 832–857.

Rebhan, A. (1988). Feynman rules and S-matrix equivalence of the Vilkovisky–DeWitt effective action, *Nucl. Phys.* B **298**, 726–740.

Redlich, N. (1984). Parity violation and gauge noninvariance of the effective gauge field action in three dimensions, *Phys. Rev.* D **29**, 2366–2377.

Rindler, W. (1966). Kruskal space and the uniformly accelerated frame, *Am. J. Phys.* **34**, 1174–1178.

Rogers, A. (1980). A global theory of supermanifolds, *J. Math. Phys.* **21**, 1352–1365.

Roman, P. (1969). *Introduction to Quantum Field Theory* (John Wiley and Sons, New York).

Ruse, H. S. (1931). Taylor's theorem in the tensor calculus, *Proc. Lond. Math. Soc.* **32**, 87–92.

Russell, I. H. and Toms, D. J. (1989). Field-parametrization dependence of the effective action in scalar electrodynamics, *Phys. Rev.* D **39**, 1735–1742.

Sakai, T. (1971). On eigenvalues of Laplacian and curvature of Riemannian manifold, *Tôhoku Math. J.* **23**, 589–603.

Sakharov, A. D. (1967). Vacuum quantum fluctuations in curved space and the theory of gravitation, *Dokl. Akad. Nauk S. S. S. R.* **177**, 70–71; *Sov. Phys. Dokl.* **12**, 1040–1041 (1968).

Salam, A. and Strathdee, J. (1974). Super-gauge transformations, *Nucl. Phys.* B **76**, 477–482.

Sato, K. (1981). Cosmological baryon-number domain structure and the first order phase transition of a vacuum, *Phys. Lett.* B **99**, 66–70.

Schrödinger, E. (1932). Diracshes elektron im schwerefeld I, *Sitzsungsber. Preuss. Akad. Wiss., Phys.–Math. Kl.*, 105–128.

Schrödinger, E. (1956). *Expanding Universes* (Cambridge University Press, Cambridge).

Schweber, S. S. (1961). *An Introduction to Relativistic Quantum Field Theory* (Row, Peterson and Company, Evanston).

Schwinger, J. (1951a). On gauge invariance and vacuum polarization, *Phys. Rev.* **82**, 664–679. (Reprinted in Schwinger (1958).)

Schwinger, J. (1951b). The theory of quantized fields. I, *Phys. Rev.* **82**, 914–927. (Reprinted in Schwinger (1958).)

Schwinger, J. (1951c). On the Green's functions of quantized fields. I, *Proc. Natl. Acad. Sci. USA* **37**, 452–455. (Reprinted in Schwinger (1958).)

Schwinger, J. (1951d). On the Green's functions of quantized fields. II, *Proc. Natl. Acad. Sci. USA* **37**, 455–459. (Reprinted in Schwinger (1958).)

Schwinger, J. (1953). The theory of quantized fields. II, *Phys. Rev.* **91**, 713–728. (Reprinted in Schwinger (1958).)

Schwinger, J. ed. (1958). *Selected Papers on Quantum Electrodynamics* (Dover, New York).

Schwinger, J. S. (1961). Brownian motion of a quantum oscillator, *J. Math. Phys.* **2**, 407–432.

Schwinger, J. (1989). *Particles, Sources, and Fields, Volume II* (Addison-Wesley, USA).

Sewell, G. L. (1982). Quantum fields on manifolds: PCT and gravitationally induced thermal states, *Ann. Phys. (N. Y.)* **141**, 201–224.

Singer, I. M. (1978). Some remarks on the Gribov ambiguity, *Commun. Math. Phys.* **60**, 7–12.

Sparnaay, M. J. (1958). Measurements of attractive forces between flat plates, *Physica* **24**, 751–764.

Spergel, D. N. *et al.* (2007). First-year Wilkinson microwave anisotropy probe (WMAP) observations: Implications for inflation, *Astrophys. J.* **170**, 377–408.

Starobinsky, A. A. (1973). Amplification of electromagnetic and gravitational waves scattered by a black hole, *Zh. Eksp. Teor. Fiz.* **65**, 3–11; *Sov. Phys. JETP* **37**, 28 (1973).

Starobinsky, A. A. (1979). Relict gravitation radiation spectrum and initial state of the universe, *JETP Lett.* **30**, 719–723.

Starobinsky, A. A. (1980). A new type of isotropic cosmological models without singularity, *Phys. Lett. B* **91**, 99–102.

Starobinsky, A. A. (1982). Dynamics of phase transition in the new inflationary universe scenario and generation of perturbations, *Phys. Lett. B* **117**, 175–178.

Steenrod, N. (1951). *The Topology of Fibre Bundles* (Princeton University Press, Princeton, NJ).

Steinhardt, P. J. (1982). Natural inflation, in *The Very Early Universe*, ed. G. W. Gibbons, S. W. Hawking, and S. Siklos (Cambridge University Press, Cambridge), pp. 251–266.

Streater, R. F. and Wightman, A. S. (1964). *PCT, Spin & Statistics, and All That* (W. A. Benjamin, Inc., New York).

Suen, W. M. and Anderson, P. R. (1987). Reheating in the higher-derivative inflationary models, *Phys. Rev. D* **35**, 2940–2954.

Taylor, J. C. (1976). *Gauge Theory of Weak Interactions* (Cambridge University Press, Cambridge).

't Hooft, G. (1971a). Renormalization of massless Yang–Mills fields, *Nucl. Phys. B* **33**, 173–199.

't Hooft, G. (1971b). Renormalizable Lagrangians for massive Yang–Mills fields, *Nucl. Phys. B* **35**, 167–188.

't Hooft, G. (1973). Dimensional regularization and the renormalization group, *Nucl. Phys. B* **61**, 455–468.

't Hooft, G. and Veltman, M. (1972). Regularization and renormalization of gauge fields, *Nucl. Phys. B* **44**, 189–213.

Tkachev, I. (1992). Gravitational phase transition: An origin of the large-scale structure in the universe? *Phys. Rev. D* **45**, R4367–R4371.

Toms, D. J. (1982). Renormalization of interacting scalar field theories in curved spacetime, *Phys. Rev. D* **26**, 2713–2729.

Toms, D. J. (1983a). The effective action and the renormalization group equation in curved spacetime, *Phys. Lett. B* **126**, 37–40.

Toms, D. J. (1983b). Background-field method and the renormalization of nonabelian gauge theories in curved space-time, *Phys. Rev. D* **27**, 37–40.

Toms, D. J. (1987). Functional measure for quantum field theory in curved spacetime, *Phys. Rev. D* **35**, 3796–3803.

Toms, D. J. (1988). The effective action: A geometrical approach to quantum field theory, in *Proceedings of the 2nd Canadian Conference on General Relativity and Relativistic Astrophysics*, ed. A. Coley, C. Dyer, and T. Tupper (World Scientific, Singapore), 148–201.

Toms, D. J. (2007). *The Schwinger Action Principle and Effective Action* (Cambridge University Press, Cambridge).

Tsamis, N. C. and Woodard, R. P. (1998). Non-perturbative models for the quantum gravitational back-reaction on inflation, *Ann. Phys.* **267**, 145–192; arXiv:hep-ph/9712331.

Unruh, W. G. (1974). Second quantization in the Kerr metric, *Phys. Rev. D* **10**, 3194–3205.

Unruh, W. G. (1976). Notes on black-hole evaporation, *Phys. Rev. D* **14**, 870–892.

Unruh, W. G. (1995). Sonic analogue of black holes and the effects of high frequencies on black hole evaporation, *Phys. Rev. D* **51**, 2827–2838.

Utiyama, R. (1956). Invariant theoretical interpretation of interaction, *Phys. Rev.* **101**, 1597–1607.

Van Vleck, J. H. (1928). The correspondence principle in the statistical interpretation of quantum mechanics, *Proc. Nat. Acad. Sci. USA* **14**, 178–188.

Vanzella, D. A. T. and Matsas, G. E. A. (2001). Decay of accelerated protons and the existence of the Fulling–Davies–Unruh effect, *Phys. Rev. Lett.* **87**, 151301-1–151301-4.

Viallet, C. M. (1987). On the covariant quantization of anomalous gauge theories, in *Super Field Theories*, ed. H. C. Lee, V. Elias, G. Kunstatter, R. B. Mann, and K. S. Viswanathan (Plenum, New York), 399–405.

Vilenkin, A. and Ford, L. H. (1982). Gravitational effects upon cosmological phase transitions, *Phys. Rev. D* **26**, 1231–1241.

Vilkovisky, G. A. (1984). The Gospel according to DeWitt, in *Quantum Theory of Gravity*, ed. S. M. Christensen (Adam Hilger, Bristol), 169–209.

Wald, R. M. (1975). On particle creation by black holes, *Commun. Math. Phys.* **45**, 9–34.

Wald, R. M. (1977). The back reaction effect in particle creation in curved space-time, *Commun. Math. Phys.* **54**, 1–19.

Wald, R. M. (1978). Trace anomaly of a conformally invariant quantum field in curved space-time, *Phys. Rev. D* **17**, 1477–1484.

Wald, R. M. (1984). *General Relativity* (University of Chicago Press, Chicago).

Wald, R. M. (1994). *Quantum Field Theory in Curved Spacetime and Black Hole Thermodynamics* (University of Chicago Press, Chicago).

Wald, R. M. (2006). The history and present status of quantum field theory in curved spacetime, arXiv: gr-qc/0608018v1, (3 Aug 2006).

Weinberg, S. (1972). *Gravitation* (John Wiley & Sons, New York).

Weinberg, S. (1973). Perturbative calculations of symmetry breaking, *Phys. Rev. D* **7**, 2887–2910.

Weinberg, S. (1974). Recent progress in gauge theories of the weak, electromagnetic and strong interactions, *Rev. Mod. Phys.* **46**, 255–277.

Weinberg, S. (2005). Quantum contributions to cosmological correlations, *Phys. Rev. D* **72**, 043514-1–043514-19.

Weinberg, S. (2006). Quantum contributions to cosmological correlations. II. Can these corrections become large? *Phys. Rev. D* **74**, 023508-1–023508-5.

Wentzel, G. (1949). *Quantum Theory of Fields* (Interscience, New York).
Wess, J. and Zumino, B. (1971). Consequences of anomalous Ward identities, *Phys. Lett. B* **37**, 95–97.
Whittaker, E. T. (1951). *A History of the Theories of Aether and Electricity* (Thomas Nelson and Sons, London).
Wick, G. C. (1950). The evaluation of the collision matrix, *Phys. Rev.* **80**, 268–272.
Yang, C. N. and Mills, R. L. (1954). Conservation of isotopic spin and isotopic gauge invariance, *Phys. Rev.* **96**, 191–195.
York, J. W. (1983). Dynamical origin of black-hole radiance, *Phys. Rev. D* **28**, 2929–2945.
Zeldovich, Y. B. and Starobinsky, A. A. (1972). *Sov. Phys. JETP* **34**, 1159.
Zeldovich, Y. B. and Starobinsky, A. A. (1977). Rate of particle production in gravitational fields, *Pis'ma Zh. Eksp. Teor. Fiz.* **26**, 373; *JETP Lett.* **26**, 252.

Index

$Q^\alpha{}_\beta$
  scalar electrodynamics, 412
  Yang-Mills theory, 365
$Q_{\alpha\beta}$
  gravity, 368
R-summed action
  and acceleration of the universe, 132
R-summed form
  dimensionally invariant, 131
R-summed form of propagator, 130
R-summed heat kernel
  and effective action, 132
S_{int}, 297
$\hat{\Gamma}[\sigma[\varphi_\star;\overline{\varphi}];\varphi_\star]$, 290
Γ-function, integral representation, 208
$\stackrel{K}{=}$ definition, 377
$Q^\alpha{}_\beta$
  definition, 362
β-function
  definition, 319
  determined by simple pole, 320
  Yang-Mills theory, 407
δ-function, see Dirac δ-distribution
  periodic, 213
δ-function, defined, 270
ϵ in dimensional regularization, 196
$\gamma^{\alpha\beta}$
  definition, 360
  gravity, 367
  scalar electrodynamics, 410
  Yang-Mills theory, 364
γ_5 matrix, 146, 232
$\gamma_{\alpha\beta}$
  definition, 360
  gravity, 367
  interpretation, 361
  scalar electrodynamics, 410
  Yang-Mills theory, 364
$\hbar$ expansion, 296
$\langle\cdots\rangle$ defined, 281, 290
$\langle\cdots\rangle_G$ defined, 297
$\overline{\varphi}$, 281, 289
$\tilde{\delta}$-function, defined, 271
ζ-function, 197

ζ-function regularization, 196–199
  conformal anomaly, 219
  $\zeta(0)$ related to pole in dimensional regularization, 199
n-bein, 221
  covariant derivative, 223
s-wave approximation, 180
divp defined, 194

a-type numbers, 242
abelianization, 421
accelerated detector
  derivation of temperature observed, 92
  Fulling-Davies-Unruh effect, 126
  in Minkowski spacetime, 152
  interaction with quantized field, 92
accelerated detector in Minkowski spacetime, 91
  Rindler coordinates, 91
  Unruh detector, 91
accelerated proton
  decays by weak-interaction, 92
Acceleration radiation
  insensitive to Planck scale physics, 181
acceleration radiation
  from Page approximation in Minkowski spacetime, 125
action, 2, 37
action functional
  N scalar fields, 302
  $O(N)$ model, 322
  electromagnetism, 350
  gravitational, 302
  gravity, 352
  non-Abelian gauge theory, 352
  non-linear sigma model, 285
  scalar electrodynamics, 408
  scalar fields, 186
  under field redefinitions, 283
  Yang-Mills theory, quadratic order, 401
adiabatic bases
  any one can be used to "label" and evolve the set of physical expectation values, 97
adiabatic condition, 96
  and Hadamard condition, 89

adiabatic condition (cont.)
    determines unique adiabatic series that is generally not convergent, 96
adiabatic expansion, see adiabatic series
    of Green function in Riemann normal coordinate local momentum space, 139
adiabatic regularization, 94
    adiabatic order, 95
    definition of, 103
    energy-momentum tensor, 94
    of field squared, 104
    particle number in expanding universe, 94
    slowness parameter, 95
    trace anomaly, conformal anomaly, 106
    trace of energy-momentum tensor, 105
adiabatic series
    a second iterative derivation, 100
    explicit expressions to fourth order, 101
    of Feynman Green function in general spacetime, 111
    of field squared, 104
    of inhomogeneous Green function in general spacetime, 111
adiabatic series in closed form
    to any finite order, 99
adiabatic subtraction, see adiabatic regularization
adiabatic vacuum
    definition and non-uniqueness, 96
adiabatic vacuum state: not unique–examples from exact solution, 96
adiabatic vacuum states
    in general curved spacetime, 89
adjoint representation, 228
adjoint representation, Lie algebra generators, 404
adjoint spinor, 245
algebraic quantum field theory
    in curved spacetime, 90
algebraic quantum field theory
    interacting fields in curved spacetime, 91
angular momentum, 9
    total including spin, 32, 33
annihilation operator, 15
anticommutation relations, 28
anticommuting numbers, see Grassmann numbers
antiparticle, 29
antiperiodic boundary conditions, 211
area of event horizon
    Kerr black hole, 173
    relation with entropy, 174

asymptotic form
    and conserved Wronskian, 51
    early times, 49, 51
    late times, 51
asymptotic freedom, 407
asymptotically static cosmological spacetime, 47
auxiliary field, 398, 400
axial anomaly, 257–263
axial current, 258
    divergence related to heat kernel coefficient, 260
axial transformation, 257
axial vector or chiral current
    and chiral symmetry, 146
    conserved for massless unquantized Dirac field, 146

back reaction to black hole evaporation, 178
background field, 287
background-field gauge
    gravity, 368
    Yang-Mills theory, 365
Bardeen metric, 179
bare coupling constants, 302
    dimensions, 303
    in terms of renormalized ones, 318
    independent of renormalization length, 319
Berezin integration, 242–245
    change of variable, 244
biscalar, 270
bispinor Green function, 148
black hole luminosity
    massless scalars, 178
    neutrinos, photons, and gravitons, 178
Bogolubov transformation, 47, 55, 152
    finding coefficients in spherical gravitational collapse, 166
    in gravitational collapse spacetime, 164
    relating coefficients in gravitational collapse of rotating body, 172
    relating coefficients in Schwarzschild black hole, 167
Borchers' theorem, 284
boundary conditions
    relation to inequivalent spin structures, 240
Boyer-Lindquist coordinates, 154
Bunch-Davies vacuum
    from conformal vacuum, 72

c-type numbers, 242
canonical commutation relations, 2, 49, 50, 276
canonical commutators, 6
canonical momentum, Dirac field, 147

canonical momentum, field, 276
canonical quantization, 5, 6
   charged scalar field, 25
Casimir effect, 19, 21
   complex scalar field, 215
   confirmed experimentally, 24
Casimir invariant, 404
centre of algebra, 232
charge, 26
   conserved, 40
   electron, positron, 31
charge conjugate spinor, 237
Chern-Simons action, 266
Chern-Simons theory, 263–267
chiral anomaly, *see* axial anomaly
chiral current anomaly, 144–151
   of Dirac field in curved spacetime, 94
chiral current anomaly derived by adiabatic subtraction, 147–151
chiral spinors, 236
chiral symmetry
   and axial vector or chiral current, 146
chiral transformation, 257
chirality, 236
Christoffel connection
   gravity, 366
   non-linear sigma model, 332
   scalar electrodynamics, 409
   vanishes for Yang-Mills theory, 363
chronological product, 278
classical Schwarzschild and Kerr black holes, 153–162
Clifford algebra, 229
   non-trivial center in odd dimensions, 232
closed algebra, 354
closed timelike curves
   and quantum field theory, 182
   and traversable wormholes, 182
closed-time-path functional integral
   and in-in formalism, 183
   in curved spacetime, 182
CMB, 65
   temperature anisotropies, 75
Coleman-Weinberg effective potential, 325
complex conjugation of Grassmann numbers, 246
complex scalar field
   Cartesian form, 408
   complex form, 408
   polar form, 408
condensed notation, 186–188, 270–271
   gravity, 357
   Yang-Mills theory, 355

conformal vacuum
   reduces to Bunch-Davies vacuum in de Sitter, 71
conformal anomaly, *see* trace anomaly
   functional integral derivation, 216–221
   link to functional measure, 216
conformal invariance, 42
   action, 44
   and trace anomaly, 42
   and vanishing trace of energy-momentum tensor, 42
conformal invariance of Dirac equation
   requires masslessness, 251
conformal Killing vector field, 108
conformal time, 52
conformal transformation
   n spacetime dimensions, 216
   infinitesimal, 42
   spinor field, 250–251
conformal vacuum
   used to derive Bunch-Davies vacuum in de Sitter, 72
conformally-invariant massless scalar field
   proof of conformal invariance, 56
conjugate spinor and the ϵ-matrix in curved spacetime, 144
connection
   torsion-free condition, 223
   Vilkovisky-DeWitt, 378
connection one-form, 223
constant of the motion, 155
constants of the motion
   Kerr, 159
   Schwarzschild, 155
continuum limit, 19
conventions, 43
coordinate basis one-form, field space, 360
coordinate transformation
   infinitesimal, 352
   metric tensor under, 352
cosmic time, 47
Coulomb gauge, 349
Coulomb potential, 348
counterterms, 318
   expressed as sum of poles, 319
   interacting scalar field theory, 303
   loop expansion, 304
   scalar field, one-loop order, 305
   two-loop order, 317
covariant derivative, 26, 27, 37
covariant derivative of a spinor, 144, 228
covariant Taylor expansion
   derivation, 344–345
   for action functional, 287
   recursive relation for action functional, 332

covariantly constant gauge field, 403
creation and annihilation operators
　in spacetime of body collapsing to form a black hole, 162, 164
creation of particles, 47
　absence of, 47
　by black holes, 36
　by expanding universe, 36
　inflation, primordial fluctuations and CMB, 36
current, 13, 25, 40
　electron, positron, 31
curvature conventions, 43
curvature two-form, 224
curved spacetime γ-matrices
　and vierbein, 145

de Sitter spacetime, 48, 64–72
　10 parameter de Sitter group, 66
　Bunch-Davies vacuum, 71
　Bunch-Davies vacuum state, 65
　conformal vacuum and Bunch-Davies vacuum, 65
　conformal vacuum and Bunch-Davies vacuum agree, 72
　high frequency asymptotic form of scalar field, 69
　Hubble horizon, 68
　infrared divergence, 72
　isometries, 65
　scalar field, 67
DeWitt effective action
　gauge theories, 397
　Landau-DeWitt gauge, 399
DeWitt metric, 365
dimensional regularization, 112, 195–196, 303
Dirac δ-distribution, 186, 270
Dirac delta function, 40
Dirac equation, 28
　n-bein form, 227
　covariant derivative of a spinor, 144
　gamma matrices in curved spacetime, 144
　in curved spacetime, 144
　Lagrangian in curved spacetime, 145
　transformation, 34
　Weyl spinor, 236
Dirac field, 27
　covariant derivative, 228
Dirac field in curved spacetime, 94, 144–149
　affine connection and vierbein, 145
　vector current, 145
Dirac matrices
　curved spacetime, 144
　Majorana representation, 237
　Weyl representation, 234

dual electromagnetic field tensor, 207

effective action
　and acceleration of the universe, 132
　and derivatives of curvature tensors, 132
　definition, 281
　dependence on field parameterization in usual definition, 286
　DeWitt definition, 293, 294
　DeWitt definition, loop expansion, 302
　expansion to two-loop order, 301
　from R-summed heat kernel, 132
　functional integral representation, 282, 290
　gauge invariance, 390–393
　gauge theories, 392
　gauge theories, definition, 390
　general definition, 290
　imaginary part and particle creation, 133
　independence of gauge conditions, 393–395
　loop expansion, 298–302
　loop expansion, gauge theories, 395–399
　non-perturbative transition to constant Ricci scalar curvature, 132
　one-loop order, 283
　one-loop, gauge theory, 400
　two-loop expression, 301
　two-loop order, divergent part, 316
　two-loop order, interacting scalar field, 307
　zeroth order term, 296
effective Lagrangian
　related to effective action, 205
effective potential
　$O(N)$ model, 324
　curved spacetime, 328–331
　exact result, 327
　in terms of effective action, 322
　in terms of one-loop effective action, 191
　scalar electrodynamics, 417–419
Einstein equations
　needs in-in expectation values, 182
energy-momentum tensor
　by Hadamard regularization, 120
　by point splitting time-ordered product, 141
　canonical, 8, 38
　conserved result by modified Hadamard regularization, 121
　difference between vacuum states of Schwarzschild black hole spacetime, 176
　from trace anomaly in R-W universes for conformally invariant free fields, 108, 110
　of classical conformally invariant scalar field, 120
　of scalar field, 45

other approximations, 129
Page approximation to, 126
physical quantities quadratic in fields, 93
sign convention, 39
symmetric, 9, 39, 45
trace anomaly, 42, 93, 106, 142
trace anomaly by Hadamard
   regularization, 121
trace anomaly for higher spin, 107
trace anomaly for spin-0, 114
trace anomaly for spin-1/2 and spin-1, 114
trace anomaly in general curved spacetime,
   107
trace anomaly, covariant form of, 107, 113
energy-momentum tensor with Hawking flux,
   174–178
energy-momentum vector, 9
   Dirac field, 31
entropy
   Kerr black hole, 173
entropy and generalized second law, 172–174
entropy of black hole relation to area, 174
equation of motion
   for transition amplitude, 279, 287
equation of motion, for field operators, 275
equivalence class
   gauge fields, 359
Euclidean Green function, 122
Euler characteristic, 335
event horizon
   Kerr solution, 154
evolution operator, 4
exact solution
   black body radiation, 64
   particle creation, 63
exterior derivative, 223
external source, 277

Faddeev-Popov ansatz, 389
Faddeev-Popov ghosts, 396
Feynman diagrams, 301
Feynman Green function
   for Dirac field in curved spacetime, 147,
      150
   in terms of time ordered product, 186
   scalar fields, 186
   spinor field, 253
Feynman path integral
   for transition amplitude, 280
Feynman propagator, see Feynman Green
   function
field redefinitions, 285
field space, 285
field space connection, 286
   scalar electrodynamics, 410, 411

field space displacement, 360
field space line element, 361
   Kaluza-Klein form, 389
   scalar electrodynamics, 408
field space metric, 286, 360
   gravity, 365
   non-linear sigma model, 332
   scalar electrodynamics, 409
   Yang-Mills theory, 363
field strength tensor
   electromagnetism, 350
   in terms of electric and magnetic fields, 207
   non-Abelian gauge theory, 351
first law of thermodynamics
   Kerr black hole, 173
Fock representation, 16
Fourier series, 212
Fulling, Davies, Unruh effect
   from Page approximation, 125
functional derivative, 6
functional derivative, defined, 271
functional integral, see Feynman path
   integral
functional integral measure
   scalar electrodynamics, 412–413

gamma matrices, 27, 227, see Dirac matrices
gauge conditions, 348, 361
gauge coupling constant, 351
gauge coupling constant renormalization,
   406
gauge covariant derivative, 401
gauge covariant derivative operator, 364
gauge field
   constant electromagnetic field, 200
gauge field renormalization, 406
gauge group
   electromagnetism, 349
   non-Abelian, 349
gauge invariance
   classical action, 353
   electromagnetism, 348, 351
gauge transformation, 26
   electromagnetism, 350
   general form, 352
   generators, 352
   local, 352
   non-Abelian gauge theory, 351
   rigid, 352
   scalar electrodynamics, 408
gauge transformation generators
   scalar electrodynamics, 409
gauge-fixing conditions, 361
gauge-fixing term, 349
Gauss-Bonnet invariant, 114

Gaussian approximation, 93
　to Feynman propagator, 116
　and Page approximation, 117
　exact in Einstein static universe, good approximation in de Sitter universe, 116
　non-perturbative, 117
　relation to proper-time and heat kernel series, 117
general relativity, 36
generalized ζ-function, 197
generalized second law of thermodynamics, 174
generating functional
　$W[J]$ defined, 281
　$Z[J;\varphi_*]$ defined, 337
　$W[J;\varphi_*]$, 289
　$W[J;\varphi_*]$ contains only connected diagrams, 340–342
　relation of $W[J;\varphi_*]$ to $Z[J;\varphi_*]$, 337
generating functional $Z[J;\varphi_*]$
　sum of all vacuum diagrams, 338
generator, 10, 25, 33
　of infinitesimal isometry and field transformation, 39
　when conserved in curved spacetime, 40
generators
　Lie algebra, 351
generators of gauge transformations
　gravity, 357
　Yang-Mills theory, 356
geodetic interval, 286
　in Riemann normal coordinates, 331
Grassmann numbers, 242–245
gravity with dynamical preferred frame, 180
gray body spectrum
　of radiation from Kerr black hole, 172
　Schwarzschild black hole, 169
　wave packets, 171
Green function
　divergent parts of products, 309–315
　inverse of differential operator, 187
　manifestly covariant form in configuration space from Riemann normal coordinate momentum space, 143
　proper time series from normal coordinate momentum space expansion, 142
　proper-time adiabatic series coefficients, 111
　relation to heat kernel, 308
Green function bispinor
　momentum space expression in Riemann normal coordinates, 149

Green function in momentum space
　setting boundary conditions by displacing poles, 140
Gribov ambiguity, 363

Hadamard condition
　and adiabatic condition, 89
Hadamard form
　change with invariant Planck length, 180
Hadamard Green functions in different vacuum states, 176
Hadamard regularization, 118
　coincidence limit of point-split result, 120
　Hadamard Green function, 119
　Hadamard-regularized Green function, 120
　of energy-momentum tensor, 120
　point-splitting, 120
　trace anomaly, 121
Hadamard series, 119
Hamiltonian
　particle in constant electromagnetic field, 200
Hawking radiation
　from acceleration radiation in 6-dimension, 181
　insensitive to Planck scale physics, 181
　nonlinear realization of Lorentz group, 180
　trans-Planckian frequencies, 180
Hawking temperature
　Kerr black hole, 172
　Schwarzschild black hole, 169
heat kernel, 193–194, 198
　R-summed form, 130
　asymptotic expansion, 193, 199
　coefficients, 194
　constant electromagnetic field, 201–205
　constant gauge field, 211–213
　for $O(N)$ model, 322
　free scalar field, 204
　series coefficients, 111
Heisenberg picture, 4
homomorphisms and spin structure, 241
Hubble radius, 68
Hubble scale, 68

in-in formalism
　and higher-order correlations in cosmology, 183
in-out to in-in amplitudes
　by Bogolubov coefficients, 183
inequivalent spin structures, 239–242
inflation, 48, 73–88
　amplifies vacuum fluctuations by particle creation, 65

Einstein equations, 78
  history, 73
  inflaton field, 73
  new inflation, 74
  perturbations, 65, 71
  reheating, 75
  slow-roll parameters, 79
inflaton field, 65, 71
  dispersion and two-point function, 71
  energy density and pressure, 78
  inflation, 73
  inflaton potential, 75
  inhomogeneous quantized perturbations, 75
  spectrum of perturbations, 71
inflaton perturbations
  asymptotic condition on modes, 81
  field equation, 80
  momentum-space components, 83
  nearly scale-invariant, 75
  quantization, 80
  spectrum of, 82
inflaton potential
  conditions for slow-roll
     inflation, 79
integration of Gaussian
  Berezin case, 244
  finite dimensional case, 189
  infinite dimensional case, 189
interacting fields
  in curved spacetime, 90
  local momentum-space representation, 90
  operator product expansions in curved spacetime, 91
interaction part of the action, 297
interaction picture, 5
invariants for electromagnetic field, 207
involution, 246
isometry, 39
  and Killing vector, 41

Jacobi identity, 354
Jacobi relation, 351
Jacobian
  conformal transformation, 218–220
  general gauge theory, 389
  in polar coordinates, 412

Kerr solution, 154
Killing vector
  and conserved generalized momentum, 41
  and isometry, 41
Killing's equation, 371
  DeWitt metric, 375
  Yang-Mills theory, 373

Lagrangian
  for Dirac equation in curved spacetime, 145, 247
Landau gauge, 412
Landau-DeWitt gauge
  gravity, 368
  importance, 384
  Yang-Mills theory, 365, 401
Legendre transformation, for effective action, 281, 290
Levi-Civita tensor, 207, 334
Lie algebra
  non-Abelian gauge group, 351
Lie bracket, 354
Lie derivative
  and infinitesimal coordinate
     transformation, 38
  metric tensor, 108
linked cluster theorem, 340, 342
local Lorentz transformation
  defined, 222
  for fields, 224
  infinitesimal form, 224
local momentum space method, 134, 308
local momentum-space representation
  interacting fields, 90
local orthonormal frame, 221
local quantum field theory, 271
loop expansion
  effective action, 298–302
Lorentz group Lie algebra, 225
Lorentz invariance violation
  and thermal spectrum, 180
  and trans-Planckian physics, 180
Lorentz transformation, 32
Lorenz gauge, 348

Majorana representation, 237
Majorana spinor, 237
  restrictions on spacetime dimension, 237
Maxwell field, 26
metric, 37
metric of evaporating black
     hole, 179
minimal coupling, 26, 37
minimal operator, 367
Minkowski space
  accelerated detector, 125
Minkowski spacetime, 1
  initial, 49, 51
  Rindler coordinates, 125
models of black hole evaporation, 181

Noether's theorem, 10
non-Abelian Lie groups, 26

non-linear sigma model
  invariance properties of classical theory, 334
  not renormalizable in more than two spacetime dimensions, 335
  renormalization, 336
non-local divergences, 315
non-perturbative effects
  R-summed propagator, 93
normal coordinate local momentum-space
  expansion, 94
  interacting fields, 94
normal coordinate momentum space
  adiabatiic expansion of Green function, 139
  derivation of configuration space DeWitt-Schwinger or heat kernel series, 142
  derivation of trace anomaly, 141
  dimensional regularization in, 141
  Green function boundary conditions by displacing poles, 140
  Green function in, 138
normal coordinates, see Riemann normal coordinates
normal neighbourhood, 134
null geodesics
  affine parameter, 158
  Carter constant and principal null congruence in Kerr, 159
  from past to future null infinity in Kerr, 162
  from past to future null infinity in Schwarzschild, 158
  in Kerr, 159
  in Schwarzschild, 156
  in Schwarzschild and Kerr, 154
  principal null congruence in Kerr, 159

one-loop effective action
  as determinant of differential operator, 188
  as functional integral, 188–190
  comparison of cut-off method and dimensional regularization, 196
  comparison of dimensional and ζ-function regularization, 199
  Dirac spinor, 248
  divergent part, 194
  divergent part, spinors, 249
  divergent part, Yang-Mills theory, 405
  functional integral in Riemannian space, 197
  general definition, 299
  in terms of heat kernel, 193
  in terms of in-out transition amplitude, 185

  pole part, scalar electrodynamics, 417
  scalar electrodynamics, 413, 416
one-loop effective Lagrangian
  constant electric field, 209–210
  constant electric field, imaginary part, 209, 256
  constant electromagnetic field, 205–210
  constant electromagnetic field, spinors, 255
one-loop effective potential
  scalar electrodynamics, 419
  scalar electrodynamics, massless limit, 419
one-particle irreducible, 301
open algebra, 354
orbit space, 359

Page approximation, 94, 118
  and Fulling, Davies, Unruh effect, 125
  and Gaussian approximation, 117
  and Hadamard regularization, 121
  de Sitter and Schwarzschild metrics, 127
  derivation, 122
  derivation of temperature observed by accelerated observer in Minkowski spacetime, 125
  energy-momentum tensor and field fluctuations exact for de Sitter invariant thermal state, 128
  energy-momentum tensor and field fluctuations in thermal state of Schwarzschild black hole, 128
  exact in Minkowski spacetime, 125
  Hadamard regularization, 124
  relation to Gaussian approximation, 123
  ultrastatic metric, optical metric, 122
parallel propagation, 332
parallelizable
  defined, 239
  global definition of spinors, 239
partially summed form of amplitude or heat kernel, see R-summed form
particle creation
  black body radiation, 64
  coherent superposition of pairs, 64
  exact solution, 61, 63
  from imaginary part of effective action, 133
  gravitons, 58
  in anisotropically changing universes, 58
  none for conformally-invariant free fields in Robertson-Walker universes, 57
  none for free massless neutrinos and photons in Robertson-Walker universes, 58
  of conformal scalar field with quartic self-interaction, 110

probability distribution, 60
stimulated creation of bosons, 60
particle creation by black holes
   Hawking flux pressure, energy density and current, 177
   Hawking thermal radiation, 153
   history, 152
   probability distribution, 153
   wave packets, 169
particle creation by Schwarzschild and Kerr black holes, 165–172
particle number
   early times, 53
   expectation value, 52
   measurement, 53
particle number operator, 16, 25
particle properties, 17
path integral, see Feynman path integral
Pauli matrices, 234
Penrose conformal diagram, 157
periodic δ-function, 213
periodic boundary conditions, 211
periodically identified spacetime, 210
physical quantities quadratic in fields
   energy-momentum tensor, 93
point splitting regularization
   using time-ordered product, 141
point-splitting
   coincidence limit, 120
   Hadamard regularization, 120
Poisson bracket relations for classical mechanics, 268
positive frequency
   ambiguity, 53
   early times, 51
   late times, 52
principle of stationary action, for operators, 275
probability distribution
   of created particles, 60
projection operator $P^i{}_j$, 359
propagator, see Green function
   R-summed form, 130
proper time, 52, 200
proper time series
   from normal coordinate momentum space expansion of Green function, 142

quantized field in black hole spacetimes, 162–165
quantized inflaton field
   discrete and continuous representations, 423
quantized inflaton perturbations
   and ensemble average over random phases, 423
   and regularization, 84
   power spectrum in discrete and continuous representations, 423
   properties in discrete and continuous representations, 422

reduction of gauge theory to non-gauge theory, 410
regularization, 18
   ζ-function, 196–199
   cut-off, 195
   dimensional, 195–196
   dimensional, zeta-function, point-splitting, Hadamard, adiabatic, 93
renormalization, 18
   $\lambda\phi^4$ in curved spacetime, 90
   interacting field in curved spacetime, 302–317
   non-Abelian gauge theories in curved spacetime, 90
   of interacting fields in curved spacetime, 90
   quantum electrodynamics in curved spacetime, 90
   scalar field in constant electromagnetic field, 208–209
   Yang-Mills theory in curved spacetime, 399–407
renormalization conditions, 324
renormalization counterterms, 318
renormalization factors, related in Yang-Mills theory, 406
renormalization group
   in curved spacetime, 91
   related to rescaling of background metric, 320
renormalization group equation
   effective action, $O(N)$ model, 325
   effective potential, $O(N)$ model, 326
   effective potential, curved spacetime, 328
   for coupling constants, 321
   for effective action, 321
   Yang-Mills theory, 407
renormalization length, 188, 303, 319
   dimensional regularization, 206
renormalization of scalar field
   one-loop order, 305
   two-loop order, 306–317
renormalized gauge field, 209
Ricci scalar, see scalar curvature
Riemann curvature
   from curvature two-form, 224

Riemann normal coordinates, 134
　coordinate transformation, 137
　geodesics, 134
　Green function in, 137
　metric, 135, 136

scalar curvature, 43
scalar electrodynamics
　curved spacetime, 407–417
　effective potential, 417–419
scalar field, 13, 24, 43
　arbitrary mass and coupling
　　constant, 54
　de Sitter spacetime, 67
scalar product, 14, 28, 30, 46
Schrödinger picture, 3
Schrödinger representation, 3, 7, 19
Schwarzschild solution, 153
Schwinger action principle, 1, 10
　for general fields, 271–276
　for in-out transition amplitude, 185
　in curved spacetime, 39
　variation of external source, 277
Schwinger effective Lagrangian
　scalars, 209
　spinors, 256
Schwinger-Keldysh closed-time-path
　formalism
　in curved spacetime, 183
second law of thermodynamics
　Kerr black hole, 173
sonic black hole
　and thermal spectrum, 180
spectrum of quantized inflaton perturbations
　nearly scale-invariant, 77
spin, 29, 33
spin connection, 144
　definition, 225
　transformation, 225
spin-statistics, 30, 36, 47
　Bose-Einstein statistics from dynamics, 55
　Fermi-Dirac statistics from dynamics, 55
　from dynamics in curved spacetime, 54
　generalized statistics, 55
　ghost fields, 55
　higher spin fields, 55
spinorial affine connections
　and vierbein, 145
Steiffel-Whitney class, 239
string theory black hole
　Hawking radiation and gray-body factor, 181
structure constants, 27
　Lie algebra, 351

structure functions
　definition, 354
　gravity, 359
　Yang-Mills theory, 357
super-radiant scattering
　by Kerr black hole, 172
surface gravity
　Kerr black hole, 161, 171
　Schwarzschild black hole, 161
surface of last scattering, 65
　recombination, 75

thermal state
　Green function, 121
time reversal, 53
　correlations, 53
time-ordered product, 278
topologically massive gauge theory, 267
trace anomaly, 106, 142
　conformally-coupled scalar field, 106
　derived in normal coordinate momentum
　　space, 141
　energy-momentum tensor, 93
　energy-momentum tensor from, 108, 110
　for higher spin, 107
　for spin-0, 114
　for spin-1/2 and spin-1, 114
　functional integral derivation, 220
　manifestly covariant expression, 107, 113
　manifestly covariant expression in general
　　curved spacetime, 107
　same in any state satisfying adiabatic
　　condition, 141
　spinor field, 253
trace of an operator, 188
trans-Planckian physics
　and CMB fluctuation spectrum, 181
transformation function, 272
transition amplitude, 272
　expressed as functional integral, 287–289

ultrastatic metric, 122
unitary gauge, 411
units, 2

vacuum diagrams, 301
vacuum energy, 18, 21, 24
vacuum state, 1, 16, 19
　adiabatic vacuum state, 96
　ambiguity, 1, 47, 53
　Bunch-Davies vacuum, 82
　Bunch-Davies vacuum, 71
　conformal, 53
　conformal invariance, 47
　Hartle-Hawking thermal equilibrium state, 174

late times, 58
of eternal black hole with Hawking flux; Unruh vacuum, 175
Poincaré invariance, 1
unique for conformally coupled massless scalar, Dirac, and two-component neutrino fields, 53
unique for conformally-invariant free fields in Robertson-Walker universes, 57
with no incoming particles from past null infinity, 175
with no particles incoming from past null infinity in gravitational collapse, 164
Van Vleck-Morette determinant, 111, 288
vector current, 263
　of Dirac field in curved spacetime, 145
vector potential, 348
vierbein field, 145

volume element
　gauge theories, 385

wedge product, 223
Wess-Zumino term, 334
Weyl representation, 234
Weyl spinor
　left-handed, 236
　not possible in odd spacetime dimensions, 236
　right-handed, 235
Weyl tensor squared, 113
Weyl-Majorana spinor, 238
　restriction on spacetime dimension, 238
Wick's theorem, 298

Yang-Mills field, 27
Yang-Mills theory, 349